ACAROLOGY

The old and the new

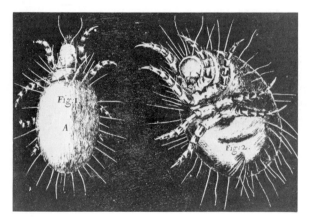

Acarus geniculatus from Robert Hooke's *Micrographia*.

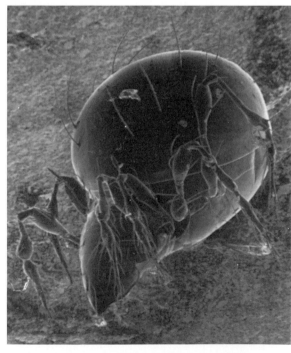

Living *Oppia coloradensis* (Woolley, SEM).

ACAROLOGY

MITES AND HUMAN WELFARE

Tyler A. Woolley

Colorado State University

A Wiley-Interscience Publication

JOHN WILEY & SONS

New York / **Chichester** / **Brisbane** / **Toronto** / **Singapore**

Library of Congress Cataloging-in-Publication Data:

Woolley, Tyler A.
  Acarology: mites and human welfare.

  "A Wiley-Interscience publication."
  Includes bibliographies and index.
  1. Mites. I. Title.
QL458.W66 1987   595.4′2   87-13701
ISBN 0-471-04168-8

Printed in the United States of America

10 9 8 7 6 5 4 3 2 1

To Lucile and Margaret, beloved eternal companions;
and to Ed Baker and George Wharton,
my principal benefactors in acarology

Preface

The science of acarology has reached a level of global accomplishment that merits distinction apart from the broader aspects of arachnology and entomology. Acarological problems have increased to the point that the science makes important contributions to human welfare. There is a need for a general textbook to provide students and professionals with basic information on the morphology, classification, biology, and ecology of mites.

George Wharton told the First International Congress of Acarology in 1963 that acarology was a somewhat new branch of biology, but the science has now definitely come of age. The convening of five such congresses since the first—in England, Czechoslovakia, Austria, the United States, and Scotland—plus one held in India in 1986—emphasizes the distinctiveness of the specialty.

As in the case in every modern science, the burgeoning literature defies complete review; but every attempt has been made to provide readers with a foundation of reference materials that will enable them to learn the fundamentals about mites and to find sources in the literature for further study.

This book is designed to identify, define, and place the subclass Acari, or mites, within the systematics of the arthropods and arachnids of both past and present forms. It shows mites in the hierarchy of arachnids (spiders, scorpions, etc.), but demonstrates their nearly unique and distinctive status. Insects now are the undisputed tenants of first place in the animal kingdom for numbers of species and individuals. Entomologists have been sufficiently numerous and active over the centuries to assure this position. Vying for second place are mites and nematodes. A general textbook on the acarines should emphasize their importance in animal biology; someone else will have to champion nematology.

Part I is an introduction to acarology. It begins with an overview (Chapter 1) that provides some general information about the Acari, their ecological locations, and the major ways in which they affect people. Behavior, parasitism, and distributional mechanisms of these tiny creatures are described briefly, with the intent to elicit further reading. Chapter 2 is a characterization of the morphological features that make the subclass Acari distinctive. The major tagmata are compared with those of other arachnids. Internal systems and functions are summarized briefly.

Relationships of the mites and their arachnid cousins are delineated in Chapter 3. The orders are described briefly in a historical way to enable the reader to place them within the contexts of modern classification and zoological history. Evolution and phylogeny are discussed.

Form and function are the focus of Part II. Much of the anatomy of mites, although basically arachnid in plan, must be described and interpreted in the context of the unique distinctions of the subclass and its separate orders and subordinate groups.

Size, integument and tagmatosis, gnathosoma (chelicerae and pedipalps), legs, and musculature are discussed in the section on external morphology. Tagmatization and chaetotaxy, which are important to identification and classification, are descriptively employed. The section on internal morphology involves the various organ systems and their known functions. These are described with emphasis on new information that clarifies or embellishes current understanding. Systematic comparisons are included, together with selected references.

The reproductive functions of acarines show a diversity of interesting phenomena, including heritable differences, a variety of transport stages (e.g., hypopi), and an equally interesting array of host relationships among parasitic forms. Life cycles and genetics, the latter only recently elaborated upon in detail, are nearly as diverse as the types of mites that have been described.

The classification of mites is presented in Part III. Each of the two cohorts (Parasitiformes, Acariformes) and seven orders (Opilioacarida, Holothyrida, Gamasida, Ixodida, Actinedida, Astigmata, Oribatida) are identified and characterized. The general hierarchical classification of the orders is listed. Many families are described briefly, distinguished, and exemplified. This part of the text is meant to be a companion account and supplement to Krantz' *A Manual of Acarology* and McDaniel's *Guide to the Families of Mites and Ticks*. It is intended to complement and support, not to compete with these outstanding works. The taxonomic keys and classification in these two books are not repeated. Classification is summarized telegraphically, although some details of taxonomy and phylogeny are included. Structural features important to the understanding of classification are illustrated by line drawings and some scanning electron micrographs. The illustrations have two purposes: first, to augment the descriptive narrative of the text; and second, to provide illustrative means for identification of acarines in the laboratory and the field. Except at the ordinal level, little use is made of keys because of the rapidly changing systems and taxonomy. The reader is referred to Krantz, McDaniel, and other literature for such assistance. Recent taxonomic keys are found in the references.

In Part IV, the history of acarology (Chapter 22) and the ecology of mites (Chapter 23) are discussed briefly. Extensive references are provided to direct readers to other sources for further information.

The Appendix contains references and annotations on techniques of collection, preservation, and preparation for studies using light and electron microscopy, and on the control of mites.

A glossary is not included because Hammen, in his *Glossary of Acarological Terminology* has completed a volume on general terminology and one on the Opilioacarida. Seven other volumes are in preparation and will form a more complete glossary than could be attempted here. Many terms are defined or explained in the text. References are placed topically at the ends of the chapters for convenience, although this may duplicate some citations. An index completes the book.

No general textbook on the Acari is still in print, nor has a book been published recently that replaces the comprehensive work of Baker and Wharton's *An Introduction to Acarology. Mites or the Acari* by Hughes and *Terrestrial Acari of the British Isles* by Evans et al. are important contributions to the science, but are limited in scope. Krantz's *A Manual of Acarology* beautifully summarizes many important facts about mites, their biology, and particularly their classification. It has excellent keys, illustrations, and synoptic taxonomy as well as an extensive bibliography. McDaniel's 1979 *How to Know the Mites and Ticks* is a welcome help for identification. Important bibliographies and indexes can be found in Hammen's multi-volume compendium *Oevres Acarologiques Completes* (1972–1976), in Grandjean's writings, and in A. Berlese, *Complete Acarological Works: Published in Redia 1903–1923* (1977) and *Acari Myriapoda et Scorpiones 1882–1903* (1980). Balogh and Mahunka's *Soil Mites of the World* is a welcome volume and the

new journal, *Experimental and Applied Acarology,* adds another dimension to the journal literature of the science. Most recent and enlightening publications are Griffiths and Bowman (eds.), *Acarology VI*, Vols. 1, 2 (1984); Helle and Sabelis (eds.), *Spider Mites: Their Biology, Natural Enemies and Control*, Vols. 1A, 1B (1985); and Sauer and Hair (eds.), *Morphology, Physiology, and Behavioral Biology of Ticks* (1986). Every year will bring new and exciting additions to the literature of acarology that will broaden and embellish the science. The prospects of the future are indeed bright.

Acarology is an important, though relatively undeveloped science. Mites affect the agriculture, food, health, and physical comfort of all members of our society in significant ways. It is a challenging yet exciting science, and everyone should learn something about acarology, which is the purpose of this book.

Tyler A. Woolley

*Fort Collins, Colorado
October 1987*

Acknowledgments

There is always a possibility that with advancing age one might inadvertently omit or temporarily forget to acknowledge the services or help of a person from whom one has received assistance. I sincerely hope that there has been a minimum of such neglect in this writing.

My wife, Margaret, has been extremely understanding and patient through the several years of this labor. Her contributions through sacrifices of time and peripheral activities are substantial in and of themselves.

I am indebted to many people for their encouragement and assistance in this writing. I look first to those who read specific chapters and parts of the early manuscript and made valuable comments and suggestions, namely Edward W. Baker, Richard J. Elzinga, G. Owen Evans, Gerald W. Krantz, Herbert W. Levi, Rob Longair, William C. Marquardt, Roy A. Norton, Kevin O'Neill, and William Rubink.

Special thanks are due to the late Harold G. Higgins, who read most of the manuscript in its formative and draft stages and made valuable corrections; to William B. Nutting who incisively edited the entire manuscript and suggested valuable and constructive alterations; to Barry OConnor for his helpful and instructive review of the Astigmata; to John Kethley for timely suggestions on the classification and the Actinedida (Prostigmata) as well as collections and ecology; and to Daniel Sonenshine for vital suggestions on pheromones. Rodger Mitchell provided lecture materials extremely valuable for descriptions of size and muscles.

Gratitude is expressed to Mary M. Conway, Editor, for many comments and assistance in the development of the manuscript and the details attendant to the final production. Thanks are also due to the production personnel at John Wiley & Sons for their very able assistance.

Thanks are extended to Sonja Evans and Peter Eades for help with some illustrations; to Ruth O'Neill for her sterling efforts in placing the rough draft in the word processor, making corrections, and executing the printing of the manuscript.

I am grateful for permission to use scanning electron micrographs and other illustrations from Cambridge Instruments, Dr. Richard Axtell, Dr. L. B. Coons and Dr. M. A. Roshdy, Dr. R. J. Elzinga, Dr. Robert Husband, Dr. Gerald W. Krantz, Dr. F. S. Lukoschus, Dr. P. Robaux, Dr. R. V. Southcott, Dr. V. A. Tsvileneva, Dr. W. A. Willis III and Dr. H. L. Cromroy, and Dr. William T. Wilson.

Contents

ACAROLOGY

PART I

INTRODUCTION

Acarology—An Overview

The most incomprehensible fact of nature is the fact that nature is comprehensible.

EINSTEIN

Acarology is the study of mites—the Acari—and is probably one of the fastest growing disciplines in zoology. A general and commonly used definition of acarology is "the study of mites and ticks." A more correct and inclusive definition is used here, however, for in the strict acarological sense all *ticks* are mites but not all *mites* are ticks. Ticks (Ixodida) as a group are no more remarkably different than are some of the other assemblages of mites. They are comparatively larger in size, generally more visible to the naked eye, and have been studied longer than some other mites groups because of their medical importance. They have been closely associated with humans and domestic animals so that over time they perhaps have become better and more generally known than others of their fellows. The volume of literature written about them through the centuries is also extensive. Fur, feather, and cheese mites, chiggers, and oribatids comprise assemblages that could be separated by description; yet, like ticks,

they technically still belong under the significant canopy of "mites."

We are becoming more knowledgeable about the impact of mites in natural environments, in soil, in forestry, in human and veterinary medicine, in agriculture, in biological control, in physiology, and in other basics of zoology. As the study and discipline grows, acarology will find its most important place among the specialties of zoological science.

Mites affect five major areas of concern to humans: health, agriculture, stored products, biological control, and esthetics. Human health is adversely influenced by transmission of disease by mites and by the irritation of their bites. Mites are most important vectors of some of the most devastating and lethal disease agents known, namely, bacteria, rickettsiae, viruses, and pathogenic protozoa. It has been reported that more soldiers died from scrub typhus in the South Pacific during World War II than were killed in battle (Micherdzinski, 1966). Scrub typhus is transmitted by chiggers (Trombiculidae). Some mites cause allergenic reactions by their bites, by their exuviae, and by their presence in various stages, for example, house dust mites.

Some mites may cause a psychological reaction in human behavior, as illustrated by the following incident. Some time ago a letter was received from a woman intensely anxious about what she could do to control some extremely small "pests" that were causing her discomfort, itching, and concern. She enclosed in the letter two sheets of white paper on which she had pasted (under cellophane tape) examples of these "pests." Examination of the materials under a microscope disclosed pieces of dirt, hair, dander, and other extraneous matter, but no mites or insects of any kind. It appeared that this was an instance of imagined pests, an "entomophobia," which might rightly be called an "acarophobia." Recently "symbiophobia" was proposed as an umbrella term to cover these and other such conditions (Nutting and Beerman, 1983).

One of the larger orders of mites is free-living in the soil. These ground inhabitants are called beetle mites because of their heavily armored bodies. Although these oribatid mites live in the soil, climb trees (Aoki, 1973), and are not usually seen, they may cause uncomfortable feelings when they are overly abundant indoors. Large numbers creeping on tatami (mats), pillars, crossbars, and ceilings of a Buddhist temple in Sakata, Japan, cause concern (Aoki, 1963).

Efforts and economics in agriculture are constantly affected by acarines that infest crops and farm animals. Damage by mites of various types can occur in field and fruit crops, ornamental shrubs, forest trees, and houseplants. Domestic animals are afflicted with varieties of ecto- and endoparasitic mites. Sheep, cattle, swine, poultry, dogs, and cats each have populations of acarines that may adversely affect their health and well-being. A mite-infested animal may suffer from loss of vigor, mangy fur or wool that may have a matted, unsightly appearance, or seeping sores. Laboratory animals are not immune to attacks of mites either. Colonies of laboratory mice are frequently infested with *Myobia*, a specialized mite that clings to hairs with its modified legs (Fig. 1-1A). Dorsal hairs of guinea pigs and white laboratory mice may be heavily infested with *Campylochirus caviae* (=*Chirodiscoides*), a hair mite in the family Listrophoridae.

Stored products and food materials, whether in the warehouse, the elevator, or the home, may harbor populations of mites as serious economic pests. Expensive depradations occur in grains and cereals, copra, cheese, potatoes, and tuberous vegetables.

Relatively little is known about the beneficial roles played by mites in the environment, but more and more disclosures are made that give us insight into such useful behavior. Oribatid mites are among the most important secondary decomposers in the soil. Their contributions to the formation of humus are vital to soil tilth and plant growth. An extensive variety of predatory mites assists in the control of other mite pests and insects (Beer, 1963). The phystoseid predators of the cyclamen mite on strawberries offer an excellent example. Pyemotid mites may be good control agents of red imported fire ants (Bruce and LeCato, 1980). As we learn more of the biology of acarines, we will know how extensively they operate in the biological control balance within nature.

In terms of artistic beauty, arthropods are spectacular animals; one never ceases to marvel at the intricacies of their structure, functions, behavior, and ecology. Colorful displays of variety and complexity are most easily seen in insects and larger crustaceans, yet may be as striking, though less readily observable, in many arachnids. There are beautiful spiders, and we have discovered many colorful mites. Esthetically, mites contribute, if we are observant and sensitive enough, to human understanding and culture by giving us an appreciation of minute spectra of beauty and outstanding structure. The scarlet – red, velvety skin of trombidiid mites has real artistic appeal. The

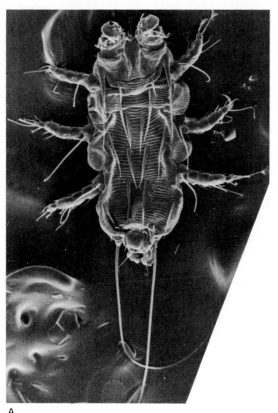

A

B

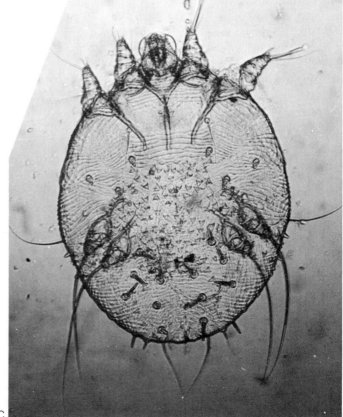

Figure 1-1. (A) *Myobia musculi,* a fur mite (with permission of Dr. F.S. Lukoschus); (B) *Dermatophagoides farinae,* house dust mite (with permission of Dr. G.W. Wharton, *Science* 167:1382–1383); (C) *Sarcoptes scabei,* human itch mite (with permission of Dr. R.V. Southcott).

C

irridescent integument of many scutate ticks reflects light as colorfully as any metallic moiré. Intricate, lacy integumental patterns on predaceous mesostigs or the delicately folded "fingerprinting" of the skin of fur and feather mites demonstrate other appealing superficial characteristics. Corresponding assortments of interesting form apply to general body shapes, kinds of hairs, legs, sensory structures, and body shields.

Behavorial and ecological differences of mites are much less known, but portend for intricate activities, interrelationships, peculiar actions, and an almost unending series of habitats related to their actions.

The term *ubiquitous* was used by Jacot to describe the locations of mites in the natural environment. The implication is that mites are universal in distribution, that they are present everywhere in the macro- and microhabitats of the world. Thus, they rival insects and nematodes in their occupation of terrestrial and aquatic habitats.

If success of animals is measured by the variety of their habitats and niches, then the arthropods have achieved a successful status. If the same yardstick is applied to groups within the phylum Arthropoda, then the Acari—mites—are among the most successful arachnids and also live up to their ubiquitous reputation. Banks (1915) thought that mites were most numerous in temperate regions, but both poles of the globe have been shown to have substantial acarine faunas; the tropics have a variety of other forms indigenous to that zone. Savory (1977) says of mites, "They appear not to be as rich in species as the spiders, but this inequality is annually disappearing as mites attract more and more attention; and like all specialists their successes are limited by the fact of their specialization. Yet within these limits it must be admitted that they reign supreme."[1]

The natural locations of mites include an al-most endless list of habitats. Like their cockroach relatives, mites are present everywhere—in nearly every conceivable and in some inconceivable locations. Their domiciles range from earth to water to air, from sea to mountain peak. They are considered the commonest mesofauna of the earth, yet remain relatively inconspicuous in the soil. Their terrestrial locations range from lichen-studded rocks to sparse amounts of soil on mountain tops; they are found in powder-dry duff and arable soils, and in moss clumps, fungal masses, and mushroom gills. Some make their homes wherever organic debris exists. They can be extracted from solidly frozen soil or moss in what appears to be a miraculous release from an icy prison. In forest soils rich with humus, certain mites may comprise up to 7% of the invertebrate fauna by weight. Other mites may be arboreal, remaining in the leafy treetop or wandering from tree to tree and back to the soil again.

Mites may be in marine water, tidal flats, and estuaries and on sandy beaches. In fresh water they may be in cold lakes or streams, warmer paludal areas or potholes, bogs, and ponds; some assumedly terrestrial forms (*Hydrozetes*) withstand complete immersion in lakes. A few even live in thermal springs with temperatures of 50.8°C. Others, water mites, live in highly alkaline waters with a pH in excess of 9.6.

Many mites live in close association with plants. In fact, mites are the only plant-feeding arachnids. They are found on grass, shrubs, deciduous trees, and conifers. The leaves, stems, bark, and roots of many types of plants, wheat, strawberries, and several ornamentals have specific inhabitants. Most of us recognize (though perhaps not by name) the spider mites that range over African violets and other indoor plants. Some mites live in homes of their own making, in galls on leaves, in ornate "witches' brooms" on trees, and in burrows in pine needles. Others dwell in crevices in bark, in rotting punk, or tree holes, or peacefully coexist in the carved, tenuous galleries of wood-

[1]Published with permission from *Arachnida* by Theodore Savory, 1977. Copyright by Academic Press, London, Ltd.

boring beetles. Pitch-pits, special treeholes, and forks of larger tree branches all display peculiar occupants of the mite world. *Larvacarus transitans* lives in a "prison" from which it can exit only when a microlepidopteran bores a hole in the gall (E.W. Baker, personal communication, 1983).

Mites may live in the germ of stored wheat or in other cereal grains. Specific forms are limited to dried fruits, stored potatoes, cheese, and other food products. One may find still others in the blossom end of a fresh apple at the grocery store, on the surface of an orange, or in berry fruit.

Household furniture is not immune to habitation. Mites may literally be our bedfellows, residing in the dust and lint of the mattress and pillows, an upholstered couch, or a favorite easy chair while they subsist on the dander of human skin. The principal allergen of human sniffles and sneezes is the house dust mite (*Dermatophagoides farinae* or *D. pteronyssus*, Fig. 1-1B).

Examples of mites associated with animals extend from conditions of phoresy (temporary, passive transport) and commensalism (passive temporary association) to active and sometimes peculiar parasitism (living at the expense of one's host). Mites are found in mammalian hair follicles in spaces so small one wonders how they survive. There are flat and fat external parasites on the feathers of birds and the skin of mammals, slender-bodied mites inside the quill cavities of feathers, and grasping forms hanging tightly to mammal fur.

G.L. Van Eyndhoven (personal communication) found an oribatid mite (*Punctoribates punctum*) within a cyst in a shrew. When he opened the cyst, he was unable to determine how the mite became encased. Perhaps the shrew, a ground-dwelling animal like the mite, was wounded and the mite became lodged in the wound, which later formed the cyst.

Residents of the midwestern and eastern United States will have experienced the bites of "red bugs" or chiggers, a larval stage of trombiculid mites. People may also be infested with *Sarcoptes scabei* (Fig. 1-1C) a skin-itch mite that burrows in the dermis of human skin. The nasal and respiratory cavities of birds, snakes, dogs, seals, and monkeys have special acarine inhabitants. Uropeltid shield-tail snakes of southwestern India have specialized, flattened mites beneath their scales. Some chigger mites (Trombiculidae) infest the trachea of amphibious sea snakes (Nadchatram and Radovsky, 1971).

William Nutting's excellent works on Demodicidae, the hair follicle mites, have disclosed fascinating information about the location and biology of these tiny cryptic residents of human skin. Most people are unaware that they harbor and carry populations of *Demodex* (Fig. 1-2) in their facial follicles, particularly at the base of the nose, in the eyelids, and in the ear canals. Such infestations are reflected in the terms *pityriasis folliculorum* and *blephariasis acaria* (Parish et al., 1983). The variety of adaptations of the diverse acarine parasites is exceeded only by that of protozoans, nematodes, and flatworms (Nutting, 1968).

Ticks have plagued people and animals for centuries. They are ectoparasitic acarines and are known the world over as the most important disease-transmitting group of arthropods other than mosquitoes. Some carry the most virulent of pathogens among which are the viral agent of Colorado tick fever, *Rickettsiae*, the causative organism of Rocky Mountain spotted fever, bacterial and spirochaetal infections (relapsing fever), and protozoan-caused Texas cattle fever (*Babesia bigemina*).

Two species of tarsonemids (*Stenotarsonemus fulgens* and *S. nitidus*) provide examples of social parasitism. These mites commandeer the abodes of eriophyids (*Phyllocoptes didelphis*) on poplar trees and feed upon the galligenous tissues caused by the gall mites. Invasion by the tarsonemids causes abandonment by the original tenants, leaving both house and pantry available to the new owners. A similar relationship exists between *S. nitidus* and a

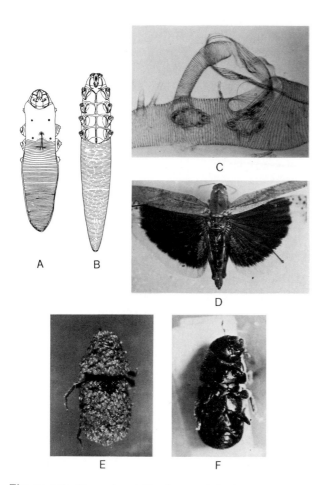

Figure 1-2. *Demodex folliculorum,* follicle mite: (A) dorsum of male; (B) venter of female. (C) *Acarapis woodi,* tracheal mite of honeybees (with permission of W.T. Wilson). (D) Bark beetle (*Dendroctonus ponderosae*) with hypopi on elytra (courtesy USDA, Rocky Mountain Forest & Range Experiment Station, Ft. Collins); (E) grasshopper (*Dissosteira carolina*) with mites on hind wing; (F) bark beetle (*Dendroctonus ponderosae*) with tarsonemid mites in gular space.

gall-forming species, *Eriophyes laevis,* on birch trees.

One very tiny mite (150 μm, *Acarapis woodi,* Tarsonemidae) lives in the tracheae of honeybees (Fig. 1-2C). Another tiny form (Podapolipidae) lives in the tracheal stigmata of curculi-

onid beetles (Feldman-Muhsam and Havivi, 1977), or under the elytra of palm weevils (Husband and Flechtmann, 1972). Yet another lives in the ears of certain moths, where selective destruction of one ear by the developing stages of the mites proceeds in an orderly fashion, but allows the moth to hear on the other side where the eardrum is not pierced and broken (Treat, 1975). Elytra of some beetles are covered by multitudes of phoretic nymphs or shelter similar populations beneath them (Fig. 1-2D). The unfolded hind wing of a grasshopper may harbor a large population of small red mites that become visible only when the tegmina is lifted and the underwing is stretched out (Fig. 1-2E). Coprophagous beetles carry large numbers of shiny, brown, tortoiselike mites on their backs. Bark beetles may have a collar of tiny tarsonemid mites beneath the head in the gular space (Fig. 1-2F). Nearly all of the members of one subfamily of tarsonemids (Coreitarsoneminae) live in the odiferous glands of bugs (Hemiptera: Coreidae) (Fain, 1979).

An extremely small mite is venereally transmitted by the males of some wasps when the mites are "galvanized" into action during the mating process (Cooper, 1955). A similar circumstance occurs in the family of mites (Cloacaridae) that inhabits the cloaca of turtles: mites are transferred from one host to another during the copulatory process.

Nests of birds and mammals have a varied entourage of small arachnids. In 1975 Johnston postulated that some nidicolous (nest-loving) mites are precursory evolutionary stem forms for external parasitic mites on several vertebrates. Lagutenko (1967) suggests that blood-feeding gamasines developed from nonparasitic ancestors with corrolary adaptive modifications of mouthparts.

Distribution of some spider mites is accomplished by "ballooning" on long silky threads and sailing on the air currents, similarly to spiders (Huffaker et al., 1969).

The literature (Fain, 1971; Treat, 1975) discloses an astonishing assortment of "hitchhiking" forms. Three main groups of these hypopi (phoretic or passive transport stages) are (1) entomophilic forms that attach by a suctorial plate to the smooth surface of insect integument (Fig. 1-2E), (2) terophilic or pilicolous stages that grip the hairs of mammals with special claspers, and (3) subcutaneous stages that live temporarily in the skin of hosts from pigeons to reptiles (Fain, 1971). Some of these hypopi invade the cocoons of earthworms (Oliver, 1962).

In the future of our rocket age when the recollections of Mariner's exploration of Mars have waned, when the penetrations of Voyager and Pioneer into space are dimmed in memory, and when Columbia and Challenger have shuttled into space innumerable times, another missile launched into space will likely contain populations of organisms not yet carried in any of our stellar craft. Insects, amebae, bacteria, spores, plant seedlings, and flowers have been orbited in space to give us information on weightlessness and how it affects growth and reproduction. Mice and other vertebrates have been lofted for similar studies. None of the space craft has carried populations of mites, however—unless they were cryptically harbored in or on some of the astronauts or other organisms. Mites hold keys to the future of spaceship earth; they have had impact in the past and vitally affect the present. They possibly will affect the future in ways not yet understood. These tiny denizens of forest, field, and home, pests of plants and animals, can bite, pierce, cut, claw, pinch, bruise, burrow, scratch, and scrape in such a variety of ways that people may not fully understand the mechanisms, but surely experience the effects or observe them in domestic animals, forests, field crops, and ornamentals.

These minute, cryptic creatures are overlooked by or unknown to most travelers on the planet. In fact, Banks (1915) remarked that "To the ordinary observer of nature mites do not exist." Yet, one such "wee beastie" prompted the late Robert Frost (1942–1970) to describe what he assumed was an intelligent life form in one of his poems.

A CONSIDERABLE SPECK
(Microscopic)

A speck that would have been beneath my
  sight
On any but a paper sheet so white
Set off across what I had written there.
And I had idly poised my pen in air
To stop it with a period of ink
When something strange about it made me
  think.
This was no dust speck by my breathing
  blown,
But unmistakably a living mite
With inclinations it could call its own.
It paused as with suspicion of my pen,
And then came racing wildly on again
To where my manuscript was not yet dry;
Then paused again and either drank or
  smelt—
With loathing, for again it turned to fly.
Plainly with an intelligence I dealt.
It seemed too tiny to have room for feet,
Yet must have had a set of them complete
To express how much it didn't want to die.
It ran with terror and with cuning crept.
It faltered: I could see it hestitate;
Then in the middle of the open sheet
Cower down in desperation to accept
Whatever I accorded it of fate.
I have none of the tenderer-than-thou
Collectivistic regimenting love
With which the world is being swept,
But this poor microscopic item now!
Since it was nothing I knew evil of
I let it lie there till I hope it slept.
I have a mind myself and recognize
Mind when I meet with it in any guise.
No one can know how glad I am to find
On any sheet the least display of mind.[2]

In the study of zoology three groups of invertebrates appear to vie for supremacy in numbers of species and occupation of different habitats. The Insecta are already established

---

as the main contenders for top place with close to a million species described from a near-equal multitude of ecological niches. Though no less established in nature than insects, but relatively little known by comparison, nematodes are being found, observed, and studied more extensively than ever before. With the extension of information about numbers and habitats, these worms are expected to compete for a possible second place. The Acari—mites—constitute a third group of animals in this cryptic "battle" and by current expectations and estimates bid fair to equal nematodes and insects in numbers and species as well as in diversity of habitats. When the science of acarology has reached the status of development now enjoyed by its sister science, entomology, it is estimated that nearly a million species of mites may be known.

This "contest" may never be settled, for each of these scientific specialties is an ongoing procession and profession. Each discipline will have its spokesmen, articles, and books. It is sufficient to note, however, that interest in and study of acarology presage the future of the science, its growth, its understanding, and its potential.

This parade of multiple acarine activities provides an introduction to the subject and to the detailed information that follows. Acarology lends itself to history, to science, to philosphy, to poetry, to human relations, to evolution and phylogeny, to behavior, and to ecology. In fact, study of this discipline will open new vistas of learning and excitement to the reader that should prove informative, useful, and appealing.

REFERENCES

Aoki, J. 1963. An observation on the prevalence of *Liodes* sp. (Oribatei) in a main hall of a Buddhist temple in northern Japan. *Jpn. J. Sanit. Zool.* 14(3):183–185.

Aoki, J. 1973. Soil mites (Oribatids) climbing trees. *Proc. Int. Congr. Acarol., 3rd, 1971,* pp. 59–65.

Banks, N. 1915. *The Acarina or Mites,* Rep. No. 108. U.S. Dep. Agric., Washington, D.C.

Beer, R.E. 1963. Social parasitism in the Tarsonemidae, with a description of a new species of tarsonemid mite involved. *Ann. Entomol. Soc. Am.* 56(1):153–160.

Bruce, W.A. and G. LeCato. 1980. *Pyemotes tritici*: A potential new agent for biological control of the red imported fire ant, *Solenopsis invicta* (Acari: Pyemotidae). *Int. J. Acarol.* 6(4):271–274.

Cooper, K.W. 1955. Veneral transmission of mites by wasps and some evolutionary problems arising from the remarkable association of *Ensliniella trisetosa* with the wasp *Ancistrocerus antilope*. Biology of eumenine wasps. II. *Trans. Am. Entomol. Soc.* 80:119–174.

Fain, A. 1971. Evolution de certains groupes d'hypopes en fonction du parasitisme (Acarina: Sarcoptiformes). *Acarologia* 13(1):171–175.

Fain, A. 1979. Les Coreitarsoneminae Parasites de la gland odiférante d'Hémiptères de la famille Coreidae. *Acarologia* 21(2):279–290.

Feldman-Muhsam, B. and Y. Havivi. 1977. Two new species of *Podapolipus*, redescription of *P. sharouii* Hust. 1921, and some new notes on the genus. *Acarologia* 14(4):657–674.

Frost, R. 1942–1970. *The Poetry of Robert Frost.* Holt, New York.

Huffaker, G.B., M. van de Vrie and J.A. McMurtry. 1969. The ecology of tetranychid mites. *Annu. Rev. Entomol.* 14:125–173.

Husband, R.W. and C.H. Flechtmann. 1972. A new genus of mites, *Rhynchopolipus*, associated with the palm weevil in Central and South America (Acarina: Podapolipidae). *Rev. Bras. Biol.* 32:519–522.

Johnston, D.E. 1975. The evolution of mite-host relationships. *Misc. Publ. Entomol. Soc. Am.* 9(5):227–254.

Lagutenko, Y.P. 1967. Structural peculiarities of the mouthparts in certain gamasid mites (Parasitiformes, Gamasoidea) in relation to their transition to hematophagy. *Entomol. Rev.* 46(4):461–466 (translated from the Russian).

Micherdzinski, W. 1966. Historia Akarologii. *Zesz. Probl. Postpepow Nauk Wydz. Nauk Roln. Lesn.* 65:7–21.

Nadchatram, M. and F. Radovsky. 1971. A second

species of *Vatacarus* (Prostigmata, Trombiculidae) infesting the trachea of amphibious sea snakes. *J. Med. Entomol.* 8(1):37–40.

Nutting, W.B. 1968. Host specificity in parasitic acarines. *Acarologia* 10(2):165–180.

Nutting, W.B. and H. Beerman. 1983. Demodicosis and symbiophobia: Status, terminology, and treatments. *Int. J. Dermatol.* 22(1):13–17.

Oliver, J.H., Jr. 1962. A mite parasitic in the cocoons of earthworms. *J. Parasitol.* 48:120–123.

Parish, C.L., W.B. Nutting and R.W. Schwartzman. 1983. *Cutaneous Infestations of Man and Animal.* Praeger, New York.

Savory, T. 1977. *Arachnida,* 2nd ed. Academic Press, London.

Treat, A. 1975. *Mites of Moths and Butterflies.* Cornell Univ. Press (Comstock), Ithaca, New York.

2

Subclass Acari—A Characterization

The least of Reptiles I have hitherto met with is a Mite.
ROBERT HOOKE, *Micrographia* (1665), 1961

Frost's poem in Chapter 1 elicits recollections of several occasions at my basement desk at home, when I watched as a bright red whirligig mite scurried over the page I was studying. Crablike in appearance, yet faster in gait, this mite darted about, avoided friendly touches with pen or pencil (for I wished not to molest), and eventually scurried back whence it came. At other times, usually in the spring and fall when temperatures were moderating, a spider mite, less colorful and bright, yet just as interesting, entered the scene, sporadically tapping its front pair of legs on the page, and jerkily transported itself over similar surfaces.

Uncovered in my digging in a leaf mat or in litter, there have been large, rake-legged mites (Caeculidae) and numerous oribatids that cowered down in a death-feint (letisimulation) and "played possum" until I stopped the disturbance of their earthly home; then after a short pause they moved gingerly on again. Predaceous orange-colored cheyletids (Cheyletidae) and reddish snout mites (Cunaxidae) with large

scissored chelicerae and enormous raptorial palps remind one of the tiger beetles of the insect world. Any of these mites enlarged to the size of beetles would be formidable foes for any prey forms. Fortunately, their size is limited by their hydrostatics and exoskeletons.

The diversity of mites makes a simple characterization difficult. One mite cannot be exemplary of all forms and increased knowledge expands our understanding of the characterization. Even though the lead line in this chapter implies an erroneous placement of mites in the animal kingdom, it is certain that Hooke did not intend to compare them with snakes and turtles. In retrospect, however, two short excerpts from *Micrographia* (Fig. 2-1) and a figure (Fig. 2-2) serve to introduce the characterization that follows.

One general description of an absolutely typical mite is impossible. There are too many modifications and variations to allow short delineation without some qualifications of differences from a general form. The variations and

Obſerv. L. *Of the wandring* Mite.

IN *September* and *October*, 1661. I obſerv'd in *Oxford* ſeveral of theſe little pretty Creatures to wander to and fro, and often to travel over the plains of my Window. And in *September* and *October*. 1663. I obſerv'd likewiſe ſeveral of theſe very ſame Creatures traverſing a window at *London*, and looking without the window upon the ſubjacent wall, I found whole flocks of the ſame kind running to and fro among the ſmall groves and thickets of green moſs, and upon the curiouſly ſpreading vegetable blew or yellow moſs, which is a kind of a Muſhrome or Jewsear.

Theſe Creatures to the naked eye ſeemed to be a kind of black Mite, but much nimbler and ſtronger then the ordinary Cheeſe-Mites; but examining them in a *Microſcope*, I found them to be a very fine cruſted or ſhell'd Inſect, much like that repreſented in the firſt Figure of the three and thirtieth *Scheme*, with a protuberant oval ſhell A, indented or pitted with an abundance of ſmall pits, all covered over with little white briſtles, whoſe points all directed backwards.

It had eight legs, each of them provided with a very ſharp tallon, or claw at the end, which this little Animal, in its going, faſtned into the pores of the body over which it went. Each of theſe legs were beſtuck in every joynt of them with multitudes of ſmall hairs, or (if we reſpect the proportion they bore to the bigneſs of the leg) turnpikes, all pointing towards the claws.

Obſerv. L V. *Of* Mites.

THe leaſt of *Reptiles* I have hitherto met with, is a Mite, a Creature whereof there are ſome ſo very ſmall, that the ſharpeſt ſight, unaſſiſted with Glaſſes, is not able to diſcern them, though, being white of themſelves, they move on a black and ſmooth ſurface; and the Eggs, out of which theſe Creatures ſeem to be hatch'd, are yet ſmaller, thoſe being uſually not above a four or five hundredth part of a well grown Mite, and thoſe well grown Mites not much above one hundredth of an inch in thickneſs; ſo that according to this reckoning there may be no leſs then a million of well grown Mites contain'd in a cubick inch, and five hundred times as many Eggs.

Figure 2-1. Excerpts from Hooke's *Micrographia* (1961).

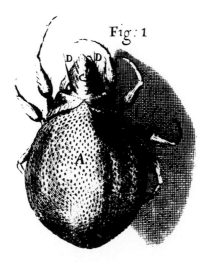

Figure 2-2. An Oribatid mite from Hooke's *Micrographia* (1961).

diversities of structure in the Acari can be correlated with specializations of function, behavior, and ecology (Jeppson et al., 1975). What follows is, perhaps, a somewhat simplified characterization.

The Acari, which are potentially the largest group of chelicerate arthropods, do not appear to be a natural group, but rather a heterogeneous assemblage of different forms derived along varying, yet distinctly phylogenetic lines. From present information they rival insects in adaptability to terrestrial and aquatic habitats. As chelicerates, mites are distant relatives of insects, but closer cousins of spiders, which have a range of comparable variability.

Embryogenic information is relatively scant, and this restricts an understanding of those processes that could give us clues to origins and to basic developmental differences between mite groups. Biology, ecology, and behavior are also poorly understood. External morphology shows some general conditions that contribute to a characterization of distinguishing features. We must deal with this type of preliminary generalization first; then more specific details and involved descriptions can follow.

Acarines are notably chelicerate arthropods. They are usually octopod as adults (exceptions would be the four-legged gall mites, Eriophyidae, some of the six-legged false spider mites, Tenuipalpidae, or two-legged examples of the Podapolipidae). Distinctively, the larvae of mites are hexapod, a characteristic shared with the Ricinuleida and some myriapods. Life cycles of most major orders of mites have similar stages from larva through nymphal stadia to adult. Special phoretic (transport) stages may be present.

EXTERNAL MORPHOLOGY

As a most distinctive morphological feature, the body tagma of the Acari comprise two regions: the gnathosoma (jaw-body) and the idiosoma (distinct-body). Mites are the only arachnids in which the chelicerae are located in a body region—the gnathosoma—separate from the one with the legs attached—idiosoma.

Gnathosoma

The anterior division of the body, the gnathosoma (Fig. 2-3A), is called the *capitulum* (false head) by some, especially when describing ticks (Ixodida). It is not a true head because the brain is located posterior to the gnathosoma as part of a larger nerve mass (synganglion) situated between the legs.

The gnathosoma is distinctly separated from the idiosoma by a suture and is usually movable at this intersegmental connection. The region is never fused with the idiosoma as in some arachnids. It may be shielded by a hood and be retractable. The anterior dorsal prominence is referred to as an *epistome* (*tectum capituli* by earlier authors). If the dorsal shield is formed into a cavity into which the mouthparts can be retracted, the space is called a *camerostome* (vaulted chamber) (Figs. 2-4 and 2-6C) and is found in groups such as Uropodina, Speleognathida, Camerobiidae, and Oribatida. The gnathosoma also may take the form of a trun-

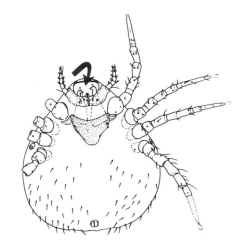

Figure 2-4. Camerostome (Speleorhynchidae).

cated, tunnellike tube (Cryptognathidae, Fig. 2-3B), a cylindrical stalk (Smaridiidae, Fig. 2-3C), or a narrowed snout (Cunaxidae, Bdellidae Fig. 2-3D).

Segmentally the gnathosoma is thought to consist of the ventral portions of three principal segments (and their appendages): (1) the precheliceral somite without an appendage, (2) a cheliceral somite with the chelicerae attached,

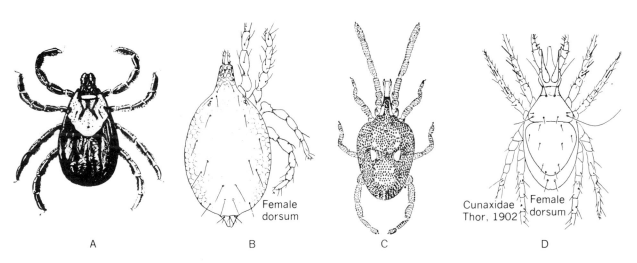

A   B   C   D

Figure 2-3. Gnathosoma: (A) Tick showing capitulum; (B) truncated form of Cryptognathidae; (C) cylindrical stalk of Smaridiidae; (D) snout form of Cunaxidae.

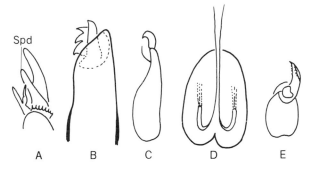

Spd

A B C D E

Figure 2-5. Cheliceral types: (A) chelate type with spermadactyl (*Androlaelaps*, after Till (1963)); (B) tick chelicera (after Nuttall et al., 1908); (C) sickle-shaped (Anystidae); (D) needlelike, stylettiform (Tetranychidae, after Pritchard and Baker (1955)); (E) hooklike (Trombiculidae).

and (3) the pedipalpal somite where the palps arise. Chelicerae of acarines are variable, (Fig. 2-5) but they exhibit the basic arachnid form of three segments and usually terminate in pincers or chelae. Chelate chelicerae are found in all orders of mites; each chela is composed of a fixed digit and a movable digit.

In the Holothyrida, where the chelicerae are two-jointed, the movable digit is ventral or lateral for grasping and piercing (Petrunkevitch, 1955). In some other mites the distal piece of the chela also may be modified in the males (Gamasida) for sperm transfer (spermadactyl) (Fig. 2-5A), may become needle-like (Tetranychidae, spider mites) (Fig 2-5D), may be strongly hooked and sickle-shaped (Anystidae) (Fig. 2-5C), or may be variably formed into cutting tips (Ixodida, Fig. 2-5B) or sickle-shaped, serrated piercing organs (Trombiculidae, chiggers) (Fig. 2-5E).

Pedipalps are usually five-segmented and leglike, but variations demonstrate remarkable adaptations for several different functions. The proximal segments of the palps form the base of the gnathosoma. The palps may be heavily sclerotized with terminal claws in primitive forms (Opilioacarida) or tined setae, homologs of claws (Gamasida), or they may be reduced to

vestigal size (Scutacaridae, Demodicidae). Some parasitic mites have modified or reduced pedipalps (Ixodida). Many palps show remarkable adaptations for different functions, such as predation (Cheyletidae). Buccal organs and apparatus associated with the palps vary in relationship to body form, food habits, and ecology. Like their silk-secreting relatives (Araneida from abdominal spinnerets and Pseudoscorpionida from cheliceral glands) some mites (Tetranychidae) secrete silk from the palps.

Idiosoma

The major body region (tagma) posterior to the gnathosoma is the *idiosoma* to which the legs are attached. It is not possible to demonstrate primary segmentation in the acarines and they are said to have retrograde metamerism. The segmentation is obscure, and the regionation of the body (tagmatization) is not comparable to that of other arachnids where the 6-segmented prosoma and the 12-segmented opisthosoma usually pertain. It is assumed that the acarines have 13 somites (Andre and Lamy, 1937; Baker and Wharton, 1952); a basic set of 14 somites has been proposed by Grandjean (1954), and 19 by Hammen (1968). There seem to be vestiges of primitive segmentation, but with the paucity of embryogenic evidence, verified details are lacking. The primary segmentation and its sclerites (plates and shields) are usually replaced by secondary plates of diverse origin. The idiosoma may be covered by a single sclerotized shield, a divided shield, or many shields. The entire body of the beetle mite (Oribatida) is usually sclerotized.

The *idiosoma* may be subdivided into regions determined principally by the insertion or location of the legs in relationship to each other. In some instances the identification of mites depends upon the type of leg attachments, their positions on the idiosoma, the types of transverse sutures present, and other distinguishing boundaries between the somal tagmata (regions). The general body shape of the idiosoma is ovoid, but may be modified or

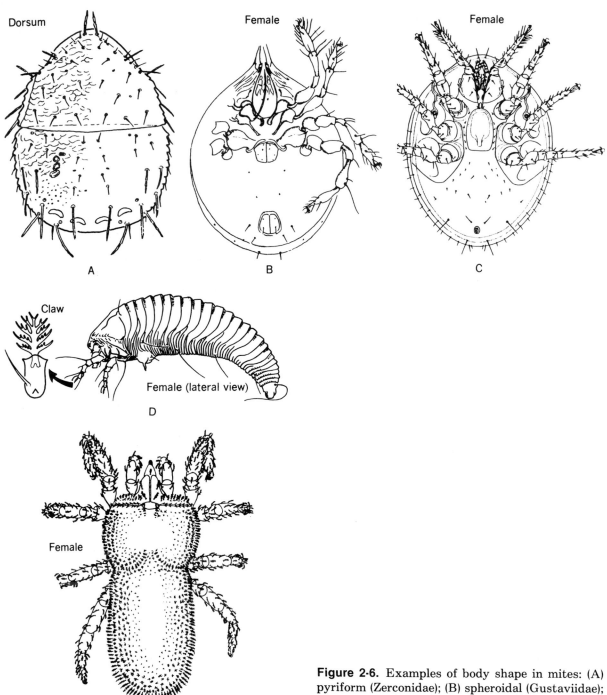

Figure 2-6. Examples of body shape in mites: (A) pyriform (Zerconidae); (B) spheroidal (Gustaviidae); (C) discoidal (Uropodina); (D) vermiform (Eriophyidae); (E) hourglass-shaped (Trombiculidae).

specialized, depending on the order: It may be pyriform (Zercondiae, Fig. 2-6A, Speleorhynchidae—Gamasida), spheroidal or globular (Gustaviidae, Damaeidae—Oribatida) (Fig. 2-6B), rectangular (Pyemotidae—Actinedida), discoidal and dorsally convex (Uropodidae—Gamasida) (Fig. 2-6C), vermiform (Eriophyidae), vermiform (Eriophyidae, Fig. 2-6D, Demodicidae, Fig. 1-2, Syringophilidae—Actinedida), or hourglass shaped (Trombiculidae—Actinedida) (Fig. 2-6E).

Different regions of the idiosoma are recognizable and have been delineated (Vitzthum, 1940). These are detailed in Chapter 4, but in the most elaborate separation the regions are designated as *propodosoma* (with legs I and II), *metapodosoma* (with legs III and IV) and *opisthosoma*. The terms *prosoma* and *opisthosoma*, which are used to describe ordinary arachnids, are not applicable to the acarines because there is no cephalothorax or abdomen, although Hammen (1968) disputes this. The gnathosoma of mites is distinctively separate from the idiosoma and is never fused with the posterior tagmata as in other arachnids. *Speleorchestes* is an example of a mite whose body is divided into the tagma of gnathosoma, propodosoma, metapodosoma, and opisthosoma (Figs. 5-8, 5-9).

Legs are attached to the idiosoma in different locations. Leg segments vary from two to eight; they may consist of four, six, and rarely eight segments and reflect the modifications of the mites involved (e.g., short and telescoped legs of four segments in the follicle mites—Demodicidae). The acarines have no extensor muscles. As in most arachnids, the locomotor muscles are limited to long-tendoned flexors in the legs. Because of their small size, mites efficiently use hydrostatic pressure (turgor) to extend limb joints. The tarsi are sufficiently modified to be diagnostic for the order or other taxonomic subdivisions and contribute, along with leg chaetotaxy, to the identification of family, genera, and species. Claws, caruncles, apoteles, and ambulacra are important features differentiating the acarine orders, as we will see

in Chapter 6. Major internal body muscles are directed dorsoventrally and the larger muscle masses are associated in the leg region (podosoma) of the idiosoma.

INTERNAL MORPHOLOGY

The internal systems of mites are almost as varied as the external features. Variations of the respiratory apparatus, for example, have been correlated with some systems of classification (Chapter 14, Classification). Among the arachnids, Ricinuleida and Solpugida exhibit prosomal stigmata, which compare somewhat to the gnathosomal respiratory openings of many prostigmatid Actinedida. These are mainly tracheate arachnids, but many mites also exhibit idiosomal stigmata at different locations. They have been classified in part by schemes that reflect the different positions of the stigmata and the attendant tracheal locations. Some mites (Acariformes—Astigmata) are without stigmata and tracheae. Others (Acariformes—Oribatida) have hidden stigmata and tracheae. Circulatory, digestive, muscular, nervous, and reproductive systems have their variations and modifications in these different orders.

Most arachnids are carnivorous as are some mites. The Acari, however, is the only subclass of Arachnida in which plant-feeding and plant-parasitic species are found. Most mites are fluid feeders and have little musculature associated with the alimentary canal except for the pharynx. Where particulate food is consumed (Oribatida) some gut musculature is present. Digestive systems (Chapter 7) in mites have three typical divisions of the gut with modifications related to the food consumed. The foregut includes the pharynx and esophagus. Salivary glands associated with this region are highly developed in ticks and a few other species, and may be modified to secrete silk.

The foregut and hindgut have a chitinous intima that precludes absorption. The midgut (ventriculus) is most diverse in morphology, as can be observed in its diverticulate, lateral caeca. The midgut is the area of absorption. The hindgut conforms in most instances to a simple, similarly structured rectal sac preceding the anus. The anus is ventral and terminal in most mites, but it may be dorsal or absent in a few. Where the anus is absent, the hindgut may be excretory in function and expel accumulated feces and metabolic waste by means of direct rupture through the exoskeleton to the outside (schizeckenosy).

Circulatory systems (Chapter 8) are reduced or absent in many mites. When present, the features are similar and consist of a pulsating dorsal heart with ostia and limited blood vessels. The open (lacunar) hemocoel may exhibit hemocytes in the hemolymph.

Respiration in mites usually involves tracheae, although some mites respire integumentally. Variations of this system are described in Chapter 9.

The excretory system (Chapter 10), as presently understood, comprises paired Malpighian tubules at the junction of the midgut and hindgut. In some species, one to four pairs of coxal glands associated with the legs may also function in excretion.

The nervous system extends from the synganglion (a fused major ganglion) and innervates the gnathosoma and idiosoma from the leg region (podosoma) (Chapter 11). If eyes are present, they are simple and located on the idiosoma (a condition that differs from that of other arachnids, where the eyes are usually prosomal in location; e.g., spiders, solpugids). Sensory structures, such as trichobothria, that are common to other arachnids are found in mites in the anterior region of the idiosoma rather than on the legs. Hollow sensory hairs (solenidia) on the legs of mites have different structures than those of common arachnid orders.

Reproductive systems (Chapter 12) have some general similarities in all the orders, but modifications of reproductive structures are associated with the life cycle and behavior that characterize each particular group of mites. The sexes are separate, and dimorphic in most

instances. An intromittent organ (aedeagus or penis) may be present in males; an ovipositor is distinctive in many females. The gonopore is usually ventral and associated with the second sternite between coxae II and IV as in most arachnids, but in some mites it is secondarily displaced. In a few instances the male gonopore is mid-dorsal above the legs (Demodicidae). In the Podapolipidae the male organs are disproportionately large (Fig. 12-22, SEM). Sperm transfer by spermatophores is common among the mites. This usually involves a stalked sperm packet deposited by the male and picked up by the female. Where an intromittent organ is present, copulation occurs.

Development

Again, it is difficult to generalize mite life cycles and stages, for variations occur even within an order. The stages may include an egg, a hexapod larva, usually two nymphs (in some orders one may be phoretic for passive transport), and an adult. Notable modifications of this arrangement are found in ticks and velvet mites, where nymphal distinctions occur. For example, in some velvet mites dormant stages may be found in both the nymph (nymphochrysalis) and a resting stage (imagochrysalis) prior to adulthood. All of the nymphal stages of *Pyemotes ventricosus* occur within the egg so that after hatching, active young may be found inside the swollen body of the gravid female—an example of ovoviviparous hatching.

Although various designations are made for the regions of the idiosoma and the segmentation is very obscure, the acarines are generally assumed to be epimorphic in their development (the number of segments established or fixed at hatching, i.e., no segments added). Life cycles differ and in the oribatids (beetle mites) the development is anamorphic (segments are added between the idiosoma and the pygidial segment after hatching); thus both epimorphic and anamorphic development occurs. It is postulated that the Anactinotrichida (mites without birefringent setae) are epimorphic and that the Actinotrichida (mites with birefringent setae) are anamorphic (Zakhvatkin, 1959; Evans et al., 1961). Again, embryonic evidence is needed to clarify the matter.

REFERENCES

Andre, M. and E. Lamy. 1937. Les idées actuelles sur la phylogénie des Acarines. Published by Authors, Paris.

Baker, E.W. and G.W. Wharton. 1952. *An Introduction to Acarology.* Macmillan, New York.

Evans, G.O., J.G. Sheals and D. MacFarlane. 1961. *Terrestrial Acari of the British Isles,* Vol. 1 Adland & Son, Bartholomew Press, Durking, England.

Grandjean, F. 1954. Etude sur les Palaeacaroides. *Mem. Mus. Natl. Hist. Nat., Ser. A* (Paris) 7:179–272.

Hammen, L. van der. 1968. Introduction générale à la classification, la terminologie morphologique, l'ontogénèse et l'évolution des Acariens. *Acarologia* 10(3):401–412.

Hooke, R. (1665) 1961. *Micrographia.* Reprint by Cramer-Weinheim, U.K., Wheldon & Wesley, London, and Hafner, New York.

Jeppson, L.R., H.H. Keifer and E.W. Baker. 1975. *Mite Pests of Economic Plants.* Univ. of Calif. Press, Berkeley.

Nuttall, G.H.F., C. Warburton, W.F. Cooper and L.E. Robinson. 1908. *Ticks: A Monograph of the Ixodoidea.* Part I. Argasidae. pp. 1–104. Cambridge University Press, Cambridge.

Petrunkevitch, A. 1955. Part P. Arthropoda 2. Arachnida. *In*: Moore, R.C., ed. *Treatise on Invertebrate Paleontology.* Geol. Soc. Am., Boulder, Colorado, and Univ. of Kansas Press, Lawrence.

Pritchard, A.E. and E.W. Baker. 1955. A revision of the spider mite family Tetranychidae. Memoirs series, Vol. 2. Pacific Coast Entomol. Soc., San Francisco.

Till, W.M. 1963. Ethiopian mites of the genus *Androlaelaps* Berl. *s. lat.* (Acari: Mesostigmata). *Bull. Brit. Mus. (nat. hist.) Zoology* 10(1):1–104.

Vitzthum, H. 1940. Acarina. *Bronns Klassen* 5:1–1011.

Zakhvatkin, A.A. 1959. *Fauna of USSR. Arachnoidea.* Vol. 6, No. 1, Tyroglyphoidea (ACARI). Translation AIBS, Washington, D.C.

3

Arachnid Relatives Past and Present

All that has gone before involves the proposition that evolution is orientation.

—G. G. Simpson

CHELICERATA

Arthropods are arranged into Chelicerata, Crustacea, Uniramia (Myriapoda and Insecta), and Pentastomida. The Chelicerata comprise many relatives of the acarines that merit consideration in this brief summary of arachnology.

Most chelicerates are terrestrial, although some are marine and freshwater animals. In most chelicerates the body consists of two regions (tagmata): a prosoma (=cephalothorax) of six metameres and the acron, and an opisthosoma (=abdomen) of varying numbers of metameres. The prosoma usually has six pairs of appendages: the chelicera (preoral appendages), which vary widely in structure; paired pedipalps, which may be ambulatory, prehensile, mastigatory, or sensory in function; and four pairs of legs. The gonopore is found on the venter of the eighth somite (second opisthosomal segment) (Fig. 3-1A, B). Relatively few append-ages occur on the opisthosoma of chelicerates. Obscure segmentation, the presence of two distinctive body regions (gnathosoma, idiosoma), and differences in locations of the gonopore make the Acari variants among the Chelicerata and homologies are obscure, as can be seen in the comparisons that follow.

The Chelicerata probably arose in the Precambrian, presumably from the Cambrian trilobite family Olenellidae (Raw, 1957; Sharov, 1966). Evidence of these connections lie in the morphological similarities of this family and the most ancient Chelicerata, the Aglaspida. Aglaspida and early Merostomata have structural similarities.

Regarding the phylogeny of the chelicerates, Evans et al. (1961) state: "Their origins are problematical in the absence of a complete fossil record, and many aspects of their relationships are obscured by convergence. The recent theories on the evolution of Arthropoda differ according to whether the phylum is considered

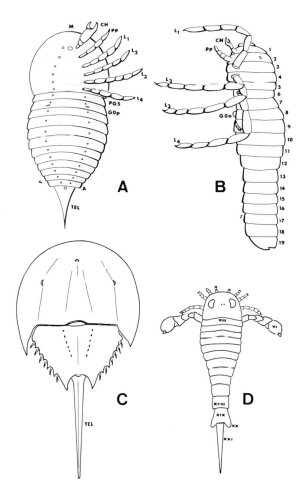

Figure 3-1. (A) Generalized chelicerate; (B) generalized arachnid (A, anus; CH, chelicerae; GEN OP, genital operculum; L1,2,3,4, legs; M, mouth; P, pectines; PG, pregenital somite; PP, pedipalps; TEL, telson) (after Petrunkevitch, 1955); (C) Merostomata – Xiphosura, general features of horseshoe crab from dorsal aspect; (D) Merostomata – Eurypterida, dorsal view of water scorpion (after Raw, 1957).

ons. An aquatic origin is also postulated for the tracheate arachnids and implies an independent origin of the tracheal system arachnids and the insects – myriapods (Kacstner, 1968).

Chelicerates show two major lines of evolution: (1) an assemblage of Eurypterida – Merostomata – Scorpionida, and (2) a grouping of Uropygi – Amblypigi – Araneae (Fig. 3-2). Palpigradi are separated by themselves; Pseudoscorpionida and Solifugae are grouped together, as are the Opiliones, Ricinulei, and Acari (Weygoldt and Paulus, 1979a).

Two types of tagmatization occurred among the chelicerates of these groups, one arachnoid, the other acaroid (Zakhvatkin, 1952), with the Acariform mites considered closest to the trilobites. A slightly different arrangement of these relationships represented by paleontological and stratigraphical illustration is shown in Figure 3-3 (Petrunkevitch, 1955). Most of the extinct groups are omitted, but Eurypterida, Architarbida and Kustarachnida are retained for reference. [See Sharov (1966), Yokishima (1975), Savory (1977a), and Hammen (1977a) for additional refinements of the relationships.]

The Chelicerata are divided into three classes: the Merostomata, the Arachnida, and the Pycnogonida (Stormer, 1955; Barnes, 1980). The Pycnogonida are aberrant marine forms—sea spiders—and will not be considered here. The class Merostomata consists of two subclasses, the Xiphosura and the Eurypterida. The class Arachnida and its orders follow with special reference to those most closely related to the subclass Acari.

Class Merostomata—Order Xiphosura

Xiphosura are known as horseshoe crabs and are the only living merostomes. Fossils of this order are more widely known than any other chelicerate and are found in parts of Europe, Africa, North and South America, and Australia. Xiphosurida are represented from the Lower Cambrian to Recent and the fossil suborder Aglaspida includes forms from the upper Cambrian.

to be monopyletic (Snodgrass, Waterlot, Heegaard) or polyphyletic (Ivanov, Stormer, Vandel and Tiegs and Manton)."

A principal postulated phylogenetic derivation involves homologizing the book gills of the class Merostoma with the book lungs of scorpi-

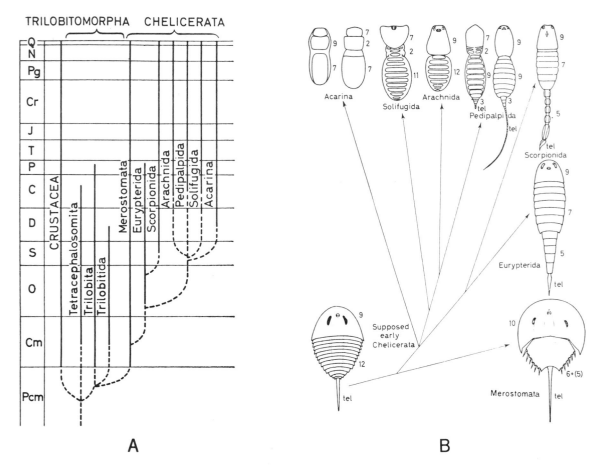

Figure 3-2. (A) A phylogeny of Trilobitomorpha and Chelicerata; (B) morphological variations of chelicerates in evolution (after Sharov, 1966).

The largest number of fossil xiphosurans occurs in Palaeozoic rocks and a relatively smaller group has persisted to recent times. Morphological changes are so relatively slight that the merostomes provide excellent information on the phylogeny of an extant arthropod group over a long span of geologic time. Living forms are not much different than the fossils from the Cambrian to Recent. Living species are generally littoral bottom dwellers found along the eastern seacoasts of North America and of Asia.

Limulus polyphemus, a living species of this order, has a prosoma and opisthosoma with a posterior spine (incorrectly referred to as a telson) (Fig. 3-1C). Compound eyes and ocelli are present on the dorsal shield. Six pairs of ventral, prosomal appendages are present. The most anterior (preoral) pair is the chelicerae, three-jointed and chelate. The other prosomal appendages are ambulatory in function and bear characteristic coxal gnathobases that act as jaws in grinding food. The dorsum of the prosoma is covered by a distinctive shieldlike carapace. Six metameres of the opisthosoma are usually ankylosed (fused) into a continuous shield with lateral spines reflective of the metamerism. Book gills are found on the venter of

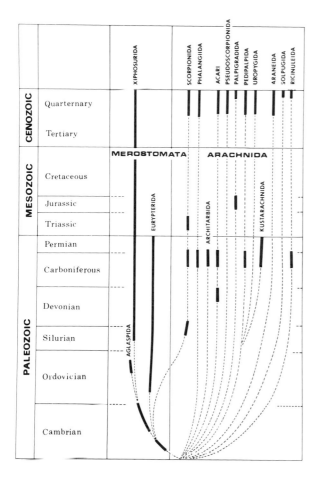

Figure 3-3. Postulated phylogenetic and stratigraphic distribution of main groups of chelicerates. The abrupt appearance of orders in the Carboniferous indicates conditions that allowed preservation rather than origin in the late Paleozoic era (after Petrunkevitch, 1955).

the opisthosoma. The genital operculum covers the gills anteriorly and carries the gonopores on the dorsal surface of the 8th somite.

The ontogenetic development of horseshoe crabs reflects the phylogeny of Merostomata and the postulated connections with the Trilobita (Fig. 3-2A). The four primary somites of the prosoma resemble those of the trilobite cephalon, and the larva of Xiphosura, called

"trilobite larvae," resemble the "protaspis" in trilobite development.

Class Merostomata— Subclass Eurypterida

Eurypterida, the giant water scorpions, are the largest of the fossils and were greater in size than any living Arthropoda. They existed from mid-Ordovician to the Permian times. (They were called Gigantostraca by Haeckel because of their size and were originally assumed to be crustaceans.) Eurypterida are characterized by a widened prosoma bearing dorsal compound eyes and ventral appendages (Fig. 3-1D). The opisthosoma is commonly 12-segmented with a 7-segmented preabdomen and a 5-segmented postabdomen, which resembles the opisthosomal arrangement found in scorpions. A styliform-to-spatulate telson is also found. Water scorpions ranged from the usual length of 10–20 cm to larger forms of Devonian specimens that reached 140 and 180 cm in length. The mouth was central behind the typical chelicerae and was bordered posteriorly by an endostoma and a metastoma. These ancient water scorpions were probably estuarine. Because of their covered gills it is postulated that they may have been able to reside for brief periods on land. Eurypterigid water scorpions are considered to be the ancestral stock of scorpions in the class Arachnida (Fig. 3-2) (Petrunkevitch, 1955; Sharov, 1966).

ARACHNID RELATIVES PAST AND PRESENT

The following key of the subclasses and orders of arachnid chelicerates identifies the distinctive features. Morphological descriptions that follow relate to both extinct and extant forms. [More specific information concerning the orders may be found in Barnes (1980), Barth and Broshears (1982), Kaestner (1968), Lindquist (1984), Meglitsch (1972), Savory (1974, 1977a, b), Weygoldt and Paulus (1979a, b), and Parker (1982).]

KEY TO THE SUBCLASSES AND ORDERS OF ARACHNIDS

1. Prosoma and opisthosoma broadly joined; first opisthosomal segment as wide as prosoma . . . . Subclass Latigastra, 2

1a. Prosoma and opisthosoma usually joined by a narrow waist or pedicel; first opisthosomal segment usually narrower than prosoma . . . . Subclass Caulogastra, 5

2. Opisthosoma with two regions, a broad mesosoma and a narrow metasoma with a terminal sting; with ventral pectines on mesosomal segment II; pedipalps chelate . . . . Superorder Pectinifera, Order Scorpiones, the scorpions (Fig. 3-4A)

2a. Opisthosoma not divided into broad mesosoma and narrow metasoma; without pectines . . . . Superorder Epectinata, 3

3. Body with two regions, gnathosoma and idiosoma, idiosoma indistinctly segmented or unsegmented . . . . Subclass Acari, the mites (Fig. 3-4J)

3a. Opisthosoma distinctly segmented 4

4. Opisthosoma of 12 metameres, rounded posteriorly, without a sting; chelicerae with comblike serrulae and distal spinneret; pedipalps large, chelate; legs not extremely long . . . . Order Pseudoscorpiones, the pseudoscorpions (Fig. 3-4B)

4a. Prosoma with unsegmented carapace; opisthosoma of 9 or 10 metameres; chelicerae with three articles; pedipalps leglike and may have terminal claws; eyes on a dorsal, prosomal tubercle; legs variable in length, tarsi multiarticulate . . . . . Order Opiliones, the harvestmen, daddy longlegs (Fig. 3-4I)

5. Opisthosoma terminating in a narrowed, segmented flagellum 6

5a. Opisthosoma otherwise 7

6. Body with bipartite prosoma and pedicel;

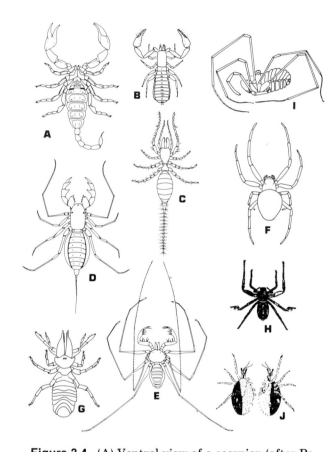

Figure 3-4. (A) Ventral view of a scorpion (after Petrunkevitch, 1955); (B) dorsal view of pseudoscorpion (after Grassé, 1949); (C) dorsal view of a palpigrade, microwhipscorpion (after Hansen from Kaestner; drawing by Sonya Evans); (D) dorsal view of whip scorpion (Uropygi) (after Pocock from Kaestner; drawing by Sonya Evans); (E) dorsal view of amblypygid (Amblypygi) (after Millot; drawing by Sonya Evans); (F) dorsal view of a spider (Araneae = Araneida), *Araneus* (after Bristoe, 1958); (G) dorsal view of *Rhagodes*, a solpugid (Solifugae) (after Grassé, 1949); (H) dorsal view of male *Cryptocellus pelaezi*, a ricinuleid (Ricinulei) (after Pittard and Mitchell, 1972); (I) lateral view of a phalangiid (*Opiliones*) (after Grassé, 1949); (J) dorsal, ventral view of a mite (Dermanyssidae).

opisthosoma with terminal flagellum of 14–15 rings; without eyes; chelicerae chelate with three articles; pedipalps leglike; body not over 3 mm long . . . .

. . . . Order Palpigradi, palpigrades or microwhipscorpions (Fig. 3-4C)

6a. Prosoma undivided dorsally; opisthosoma with terminal, segmented flagellum; chelicerae with three articles, unchelate; pedipalps robustly chelate, coxae fused medially to form a camerostome . . . . Order Uropygi, the whip scorpions (Fig. 3-4D)

7. Prosoma with a movable cucullus or hood (carapace); second abdominal metamere forms a pedicel; pedipalps chelate; eyes absent . . . . Order Ricinulei, the ricinuleids (Fig. 3-4H)

7a. Without a cucullus or hood covering the mouthparts; pedipalps leglike 8

8. Carapace (propeltidium) covers only first two metameres of prosoma, with two or more eyes; chelicerae disproportionately large, chelate; pedipalps leglike, adhesive, coxae large with endites (gnathobases); legs with raquette organs on trochanters I . . . . Order Solifugae, the wind spiders, sun spiders, solpugids (Fig. 3-4G)

8a. Carapace covers all prosomal segments; pedipalpal coxae contiguous, but free . . . . Superorder Labellata, 9

9. Pedipalps spinous, tarsi pointed; chelicerae stout, unchelate, without poison fangs; legs I extremely long, whiplike, multiarticulate; opisthosoma with 12 metameres, without ventral spinnerets . . . . Order Amblypygi, the amblypygids (Fig. 3-4E)

9a. Pedipalps leglike; chelicerae with movable fangs; opisthosoma usually unsegmented, with ventral spinnerets . . . . Order Araneae, the spiders (Fig. 3-4F).

CLASS ARACHNIDA

A principal, postulated phylogenetic derivation of the arachnids requires homologizing the

book gills of the class Merostoma with those of scorpions. An aquatic origin is also postulated for the tracheate arachnids and implies an independent origin of the tracheal system of arachnids and the insects – myriapods (Kaestner, 1968).

Fossils of the class Arachnida are relatively rare, which limits knowledge of their past history and vertical distribution. They were more prevalent in the past than at the present. Sixteen orders are known from the fossil record (Fig. 3-2). Scorpions, the only arachnids in the Silurian, and pseudoscorpions are well represented (Petrunkevitch, 1955).

Locations of arachnids in time and strata probably were influenced by specific physical factors, means of reproduction, and distribution of immatures and adults. The variations of external forms of arachnid chelicerates evidences the large number of orders that have undergone extensive adaptive radiation; secondary flows of adaptation occurred within the larger groups (Fig. 3-3). Some are more widely distributed geographically; others are mainly denizens of the upper soil layers in warmer temperate or tropical areas. Water loss and relative humidity play important roles in determining location, especially in smaller forms that are more abundant in humid areas. Larger arachnids may be more common in warm deserts.

Differences of the prosomal appendages exist in different orders of Arachnida (Fig. 3-4). Various names have been used for the articles of these appendages and the homology of the articles is not fully agreed upon. The chelicerae are three-jointed in Scorpionida, Opiliones, Palpigradi, (Architarbida).[1] All other orders, Pseudoscorpionida, Uropygi, Amblypygi, Araneae (Araneida), Solifugae, Ricinulei, and the subclass Acari have chelicerae with two arti-

[1]Petrunkevitch (1955) indicates that the name Phalangiida is preferable to Opiliones because it has priority over the latter and contains the type genus of the Order, *Phalangium* Linne, but Shear (1982) uses Opiliones and H.W. Levi (personal communication, 1984) indicates that this is the preferred name.

cles. Chelate (retrovert) chelicerae are most common. From embryological information and phylogenetic assessment the movable digit should be dorsal to the fixed digit. Embryonic torsion, however, causes different positions in different orders, namely, lateral or ventrolateral (Scorpiones, Pseudoscorpiones, Palpigradi), medial (Haptopodida), and ventral (Solifugae, Acari). Other modifications of structure occur in the chelicerae, especially among mites.

The pedipalps usually have six articles, but in the Acari the number of pedipalpal segments varies from seven in the fossil *Protacarus crani* Hirst (Fig. 3-6), to two in the stored products mites (Astigmata). The pedipalps may be chelate and raptorial (Scorpiones, Pseudoscorpiones, Acari – Gamasida), subchelate (Uropygi), pediform (Palpigradi; Acari – Actinedida), styliform (Acari – Actinedida), or bear a fang (Araneae).

The number of leg segments usually show coxa, trochanter, femur, patella, tibia, metatarsus, and tarsus; divided femora or multiple tarsal segments may occur. The acarine leg has these basic segments but *genu* replaces the term patella.

Order Scorpiones. The two most striking characteristics of scorpions are the chelate pedipalps and segmented opisthosoma with a narrowed metasoma and sting (Fig. 3-4A). The anterior prosoma joins the mesosoma, which bears the spiracles and the internal book lungs. A pair of ventral pectines is found on the third abdominal metamere. The prosoma has a carapace with two median and a variable number of lateral ocelli on the dorsum. The chelicerae are three-segmented. The powerful, chelate pedipalps are six-segmented. Each chela has a movable digit modified for lateral movements. The genital operculum (usually trapezoidal in shape) is anterior to the pectines in somite 11 in both sexes (Francke, 1982).

Paired ostia of the heart are restricted to the mesosoma. Coxal glands open at the bases of legs III. Males have paraxial organs that produce the spermatophore. Females are vivipa-

rous and many have placentalike structures for the developing eggs; the newborn young climb onto the mother's back and remain there until their first molt.

A scorpion was the first fossil Paleozoic arachnid discovered and was described in 1835 by Corda from Czechoslovakia. Later fossil scorpions were uncovered in Illinois (Mazon Creek) (Petrunkevitch, 1955). About 2000 extant species are found in the warmer parts of the world except for New Zealand, Patagonia, and most tropical oceanic islands. They range in size from 31 to 200 mm.

Order Pseudoscorpiones. The opisthosoma is broadly joined to the prosoma and has 12 metameres (11 in fossils); the twelfth is rudimentary and is easily overlooked. The absence of a "tail" (metasoma) and sting, as well as small size distinguish pseudoscorpions from scorpions (Fig. 3-4B). Chelicerae of pseudoscorpions have a comblike serrula and a silk gland spinneret (galea) at the tip. Pedipalps are six-jointed, chelate, and powerful. The finger of the pincer is ventral in position; the palpal coxa has a maxillary lobe used in feeding. Poison glands may also open in the movable digit. The dorsal prosomal carapace is shieldlike, undivided, and may bear one or two pairs of simple anterolateral ocelli; some pseudoscorpions are blind. A

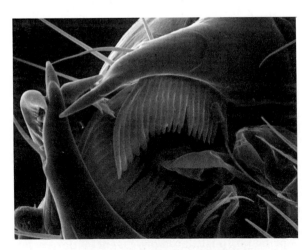

Serrula of a pseudoscorpion SEM.

sternum is absent. Leg tarsi bear claws and a midclaw, padlike areolium (=empodium). Two pairs of tracheae occur; book lungs are absent; Malpighian tubules are present (Kaestner, 1968; Weygoldt, 1969; Muchmore, 1982).

About 2000 species of pseudoscorpions are known. Fossils have been found only in the Oligocene and Miocene ages. Representatives have been found in Baltic and Mexican amber. Extant examples are found worldwide except for frigid regions of the Arctic and Antarctic.

Order Palpigradi. The Palpigradi are small, soft-bodied arachnids distinguished by widely separated pedipalpal coxae and devoid of endites (or coxae). The palpigrades are micro-whip scorpions. Behind the propeltidium are four sclerites, the smaller sclerites (mesopeltidium and metapeltidium) are posterolateral and mesal, respectively. Eleven or more metameres comprise the opisthosoma which bears a terminal flagellum of 14 or 15 segments (Savory, 1977b; Levi, 1982) (Fig. 3-4C).

A stomach is present; a heart, Malpighian tubules, book lungs, and tracheae are absent; respiration is assumed to be integumental (Kaestner, 1968; Meglitsch, 1972).

Four genera and less than 50 species are found mainly in southern Europe and from the southern United States to Paraguay and Chile. Asia and Africa have one and two species, respectively. Rocky shale of Germany was the collection site of a single species in 1890; another was found in Jurassic shale by Oppenheim in 1887.

Order Uropygi. The chelicerae are chelate (retrovert) in whip scorpions and schizomiids. Pedipalps are subchelate, six-jointed, and robustly raptorial. The terminal joint is pointed and used for piercing prey by flexion of the joint against the next palpal segment. Pedipalpal coxae are fused medially and form the base of a camerostome in which the mouth opens. The prosoma is covered by a rectangular carapace over its entire dorsal surface (Fig. 3-4D). Eyes may be present in clusters or absent. Of 12

opisthosomal metameres, the first is very short and pedicellike; the last three somites are fused into a postabdomen (metasoma) that ends in a dorsal, multiarticulate whip (flagellum) of 30–40 segments. Legs I are tactile, clawless, and without a patella. Tarsal segments vary (Kaestner, 1968).

Book lungs are in segments II and III; coxal glands open at the bases of coxae I. Repugnatorial glands open in opisthosomal segment XII near the anus and secrete a repellent fluid composed of acetic and caprylic acids. The acrid odor and irritating effect has resulted in the name "vinegaroons" for these animals.

About 150 species are found in America and Asia (Kaestner, 1968; H.W. Levi, personal communication, 1984).

Order Amblypygi. The tailless whip scorpions, or amblypygids and whip spiders, have raptorial pedipalps and highly modified legs I. The body is rounded posteriorly without a terminal flagellum (cf. Palpigradi, Uropygi). The prehensile and raptorial pedipalps are pointed distally and armed with many spines on the mesial surfaces (Fig. 3-4E). Pedipalpal coxae are movable, without coxal lobes (endites) or poison glands. Legs I are narrow, thin, and whiplike with widely separated coxae. The tarsi of legs I appear antennalike with multiple joints and are thought to be tactile organs (Kaestner, 1968).

About 80–100 living and 4 extinct species are known (Levi, 1982). The name *Tarantula*, commonly applied to the hairy mygalomorph spiders, is one of the generic names of amblypygids.

Order Araneae—(Araneida). The Araneae have a narrow pedicel between the prosoma and the opisthosoma. The opisthosoma is usually unsegmented (except in one group of primitive spiders) and has distinctive spinnerets. Four pairs of simple eyes on the dorsal prosomal carapace are very common among spiders (Fig. 3-4F). The movable digit of the chelicere is a fang usually with an opening for the poison gland. Pedipalps are usually six-jointed, with or without claws in females; in males the pedipalp is formed into a complicated structure used in copulation. Legs have seven articles; each leg may have two or three claws that handle silk, or claw tufts that act as adhesive organs. The distinctive spinnerets are varied among the spider groups, with four pairs in segmented spiders, three pairs in the majority of spider families, and sometimes fewer. A perforated plate, the cribellum, is found anterior to three pairs of spinnerets in cribellate spiders and is used for the passage of special silk. Such spiders have row of spines (calimistrium) on tarsi IV to brush the silk out of the cribellum.

Malpighian tubules are present; coxal glands open at legs I and II. One or two pairs of book lungs may be present with varied combinations of tracheae or completely absent. The complex circulatory system has the tubular heart with two to five ostia in the opisthosoma; blood vessels extend through the pedicel to the brain and posteriorly.

Eyes, tactile structures, and lyriform organs comprise the sense organs of the nervous system. Eyes may be direct, with the retinal apparatus facing the source of light, or indirect, with a tapetum and sensory cells facing away from the lens of the eye. Trichobothria are typical of this order and consist of hairs extending from vaselike pits. They are sensitive to air currents and gentle touch. Stridulating, sound-producing structures are common. The lyriform sensilla (slit-organs) are thought to be auditory as well as stress indicators for the exoskeleton. Open, hollow setae, found at the tips of the tarsi are olfactory, as also are the disc-shaped tarsal organs (Foelix and Muller-Vorbolt, 1983; Foelix and Schabronath, 1983; H.W. Levi, personal communication, 1984).

The araneids are mostly terrestrial; only one species out of 35 000 spiders is aquatic (Kaestner, 1968).

Fossil spiders from sedimentary deposits are difficult to classify; specimens in amber are somewhat easier. Nearly 2800 genera of living spiders are known and 84 genera of fossils. The first fossil spider was describe in 1866 by Romer from Caroniferous deposits in upper Silesia. The first fossil spider in America came from Illinois; others are known from Florissant, Colorado, and Chiapas (Mexico) amber (Petrunkevitch, 1955, 1963, 1971).

Order Solifugae. Sun spiders, wind spiders, and camel spiders or solfugids (Fig. 3-4G) have massive, powerful chelicerae. Pedipalps are leg-like with an adhesive organ at the distal tip of each. The prosomal carapace (of three thoracic tergites) bears one or two pairs of median eyes and rudimentary eyes on the propeltidium (the remains of the carapace; the last two somites of the prosoma are free). The opisthosoma has 11 metameres. Sexes can be distinguished mainly by cheliceral characteristics. The first pair of legs is narrowed thinly in contrast to the pedipalps; the remaining legs are more robust and have distal claws. The trochanters of legs II, III, and IV are divided. The ventral surfaces of legs IV have distinctive raquette organs (malleoli), not found in other arachnids. Solfugids lack book lungs and have tracheae opening into seven spiracles. Coxal glands open at the base of the pedipalps. The dorsal heart is regionated with characteristic ostia (Kaestner, 1968; Muma, 1982).

Solifugae range in size from 9 to 70 mm. About 1000 species are known. Fossil forms have been collected in Illinois (Petrunkevitch, 1955).

Order Ricinulei. Ricinuleids are tropical and subtropical arachnids of relatively small size (4–10 mm). They exhibit the largest number of specialized characteristics among the orders of the Arachnida (Savory, 1977a). They are without eyes, yet respond to light. The chelicerae are two-jointed and chelate, located beneath a hood or cucullus. In some the movable digit is longer than the fixed digit. The chelate pedipalps have fused coxae at the midline to form a part of a camerostome (as in Uropygi and Amblypygi) into which the mouth opens and acts as filter. Tarsi have sunken sensilla. A pumping stomach is present as well as diverticulate midgut and hindgut (Petrunkevitch, 1955). Tracheae extend inward from the spiracles above the coxae of legs III, but book lungs are absent.

The pedicel is obscured and the prosomal – opisthosomal coupling gives the appearance of a broad joint. Dorsal tergites are divided into median and lateral plates (not present in one fossil family). Extant species have three telescoped, cylindrical segments at the posterior end of the opisthosoma; in fossils these segments are fully exposed. Legs III of males are modified for sperm transfer (Fig. 3-4H).

Ricinulei and Acari are considered sister groups. In Ricinulei, like the Acari, larvae are hexapod, followed by proto-, deuto-, and tritonymphal stages (Kaestner, 1968; Pittard and Mitchell, 1972; Levi, 1982). The other synapomorphic features (shared character states) include the movable gnathosoma separated by a circumcapitular suture, denticulate labrum above the mouth, and division of femora III and IV (Lindquist, 1984).

Ricinuleids are found in forest litter of the Amazon basin, in western Africa, and in some caves of Texas and Mexico (Levi, 1982). About 30 living species in two genera are known; two fossil species are described from Carboniferous times (Petrunkevitch, 1955; H.W. Levi, personal communication, 1984).

Order Opiliones. The opilionids are the daddy longlegs, harvestmen, or shepherd spiders and are considered by some to be most closely related to primitive acarines. The apomorphic (derived character states) include the compact, unsegmented prosoma that is broadly joined and similar ovipositors. The compact, six-segmented prosoma is broadly joined to a segmented opisthosoma (some metameres may be fused or missing). The prosoma is completely covered by a carapace that has a distinctive dorsal tubercle with an eye on each side (except in some Cyphophthalmi). Chelicerae are three-jointed; each of the six-jointed pedipalps has a terminal claw. Distinctive odiferous, repugnatorial glands open at the anterolateral margins

of the prosoma (Kaestner, 1968; Shear, 1982) (Fig. 3-4I).

Opilionids are generally predators of snails, worms, other arachnids, and insects; some are scavengers—the only nonacarian arachnids that are (Shear, 1982; H.W. Levi, personal communication, 1984). A pair of spiracles opens in sternite I of the opisthosoma; accessory spiracles occur on the tibia.

Four pairs of long legs have multiarticulate tarsi, which may act in some as prehensile organs. Males have larger chelicerae and an extrusible penis (similar to some mites, e.g., Oribatida, but unlike other arachnids) and engage in direct copulation. A tubular ovipositor of females is used to place eggs in crevices in the soil. The young resemble the adults (Shear, 1982).

The harvestmen of the Cyphophthalmi and Eupnoi resemble mites in the Opilioacarida (=Notostigmata) (Savory, 1977b). Indeed, distinguishing features of these harvestmen include the broadly joined prosoma and opisthosoma, the anterior genital opening, the diverticulate caeca of the midgut, and the penis of males.

Frederickson (personal communication) postulates that the opilionids represent the closest arachnid relatives of the Acari because of body similarities and the presence of pedipalpal claws. Such claws are found in all harvestmen, but in only one family of primitive mites (Opilioacaridae), probably closest to the family Sironidae (in the order Cyphophthalmi of Savory, 1977b) (Fig. 3-5).

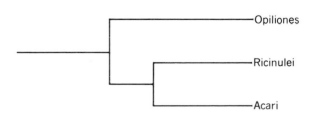

Figure 3-5. Postulated relationships of Opiliones, Ricinulei, and Acari (after Weygoldt and Paulus, 1979a).

The following comparison demonstrates some of the features that support this assumed relationship of Acari with Opiliones.

COMPARISON OF OPILIONES AND ACARI

Opiliones	Acari
Prosoma, opistho-soma broadly joined; segmented in some; carapace	Gnathosoma, idiosoma broadly joined; or obscure segmentation
Chelicerae chelate, 3-jointed	Chelicerae chelate or variously modified; 2-jointed
Pedipalps leglike; with large, heavy gnathobases	Pedipalps leglike in primitive forms; variously modified; without gnathobases
Tracheae present	Tracheae or cutaneous respiration
Eversible ovipositor	Eversible ovipositor in Opilioacarida, Acariformes
Protrusible penis in males	Aedeagus in some males
Ingestion of solid food	Diverse ingestion, solid food in Acariformes

SUBCLASS ACARI[2]

The third principal branch is the subclass Acari, which constitutes a main branch of arachnid relationship.

[2]Order Acari Nitsch, 1818. *(nom. correct.* Engelmann, 1860; (*ex* Acari Sundevall, 1833, *ex* Acarina Nitsch, 1818 = Monomerostomata Leach, 1815; Acarides van Heyden, 1826; Acarenses Duges, 1834; Acari *auctt.* Type: *Acarus* Linne, 1758). Nitsch was the first to use the name Acarina as an order. Brunnich used the name Acari in 1772 as the plural of the genus *Acarus*, the only genus recognized by him. Latreille used the name Acaridiae in 1806 as a family name and included only five genera. Leach substituted the name Acari for his Order Monomerostomata in 1819, but raised it to the rank of class. Although the laws of priority do not apply to categories of this level, the name Acari has been used here because of its nearly universal appeal.

The Acari (Fig. 3-4J) belong to the assemblage of terrestrial chelicerates in Arachnida, whose representatives have secondarily adapted to an aquatic habitat. Estimates assume 2800 species of aquatic mites out of approximately 20 000 known species compared to a single aquatic species among 30 000 species of spiders (Kaestner, 1968). The varied modes of life have influenced diversified morphology and functions. Mites are the only phytophagous and parasitic arachnids.

Evolution of the Acari

Much has been written about the evolution and phylogeny of the Acari. The subject is extensive and, if reviewed from the cladistic standpoint of phylogenetic systematics, would require more space than can be allotted. The reader is advised to seek additional information from the literature cited.

The lack of identifiable mite fossils in sedimentary formation is probably due to size and the substrate of fossilization. The oldest fossil mite recorded is *Protacarus crani* Hirst, 1923 (Fig. 3-6) a species of the superfamily Pachygnathoidea (Evans et al., 1961; Southcott, 1961; Gupta, 1979; Krantz and Lindquist, 1979) from red sandstone deposits of Aberdeenshire, Scotland, and

represents a genus not differing materially from recent mites of the family Eupodidae; it is known from the Devonian. A fair number of more recent species have been found in Oligocene amber and many of them, especially

Figure 3-6. *Protacarus crani* a primitive mite from Devonian sandstone (after Lawrence, 1953).

among the Oribatoidea, are indistinguishable from living genera. (Lawrence, 1953).

This mite is diminutive and although the feeding habits of modern pachygnathids are not known the styletlike mouthparts of this ancient acarine could have pierced either plant or animal tissue. Possibly by the Devonian period, algae, fungi, and bacteria preceded vascular plants in the terrestrial habitats and provided a substantial source of microfloral food (Kevan et al., 1975). An extant pachygnathoid genus, *Nanorchestes,* feeds on green algae (Schuster and Schuster, 1977).

P. crani is the only example from the Palaeozoic, but surviving examples exist of 60 genera and 107 species in the Tertiary, and 1389 genera and 6000 species in the Quarternary; 10 of the genera from the Tertiary are extinct (Petrunkevitch, 1955, 1971; Krivolutsky, 1979).

Current acarine families represented in the fossil record include the Holothyridae (Holothyrida); Phytoseiidae (Gamasida); Ixodidae (Ixodida); Eupodidae, Bdellidae, Anystidae, Cheyletidae, Tetranychidae, Eriophyidae, and Trombidiidae (Actinedida). The Tetrapodili or Eriophyoidea (blister, rust, and gall mites) appear to have been long-time successful residents of the earth. Fossil rust mites have been found in the old North Maslin Sands in Australia. The geologic age of the Sands is 37 million years, so it is estimated that the Eriophyoids are over 50 million years old (Jeppson et al., 1975).

Fossils in amber prove less difficult to identify (Woolley, 1971). Fossil mites were first described from amber in 1854 (Berendt and Koch). In such a matrix the oldest described Erythraeoid mite is a larval form from the Cretaceous amber of Canada (Ewing, 1937, cited by Southcott, 1961). The species is unnamed and was thought by Ewing to be partially engorged at the time of death, yet in proximity to a small dipteran which he discounted as a possible host. The inference is that the small fly could have been the host, since these mites frequently parasitize insects smaller than themselves, especially when the mites become moderately engorged (Southcott, 1961).

> The evolution of the ticks proceeded via reduction in the size of the body, loss of segmentation, reduction of the number of trunk segments and in the organ of vision to the point of complete disappearance of both the lateral eyes—vestiges of compound eyes—and the dorsal eyes—vestiges of the aboral complex. The fullest set of organs of vision are found in the lower representatives of Acariformes, where it consists of one or two pairs of lateral eyes and an unpaired dorsal eye. (Sharov, 1966, p. 491)

The ancestral progenitors of the obligatory phytophagous mites were edaphic and predatory. The morphological specialization related to plant feeding are found in four major groups (superfamilies) where it is an obligate condition. Intermediate steps to phytophagy include mycophagy, found in Gamasida, Actinedida and Oribatida, even though this practice arose independently, repeatedly, and in different ways in each order. Gamasids probably became phytophagous through necrophagy and phycophagy; plant-feeding in the Acaridida and Oribatida assumedly came through saprophagy. Where edaphic mites fed by saprophagy and mycophagy rather than predation, limited phytophagy developed. This also accounts for the morphological homogeneity of the saprophagous - mycophagous and phytophagous forms (Krantz and Lindquist, 1979).

Speculating on the ancestry of the Astigmata and the Oribatida, these same authors assume progenitors came from the actinedid - pachygnathine line, but suggest that phytophagy is incidental at the present level of evolutionary development. The acarine groups that left the soil - litter environment for more phytophagous habits (Gamasida, Actinedida) show differences from their edaphic cousins in mouthparts, structure of the digestive tract, life histories, and behavior. The "aerial gamasids" retain the predatory k-type strategies typical of their edaphic ancestors. This implies they have low fecundity, spend most of their energies on food gathering, and have retained basically unaltered life cycles.

> The chelate – dentate chelicerae of aerial gamasids are comparable in structure to the mandibulate chelicerae of insect herbivores, but herbivory by aerial gamasids is rarely observed and may be limited too by mechanical constraints associated with diminutive size. Similarly, feeding by aerial gamasids on products of living higher plants, such as pollen, nectar and sap, is little more than incidental to predation in most cases, possibly due to lack of appropriate mouthpart modification. Thus, a major adaptative radiation to feeding on the aerial parts of higher plants has occurred only in the Actinedida. The remarkable success of many Actinedida as phytophages may relate to the development of styliform chelicerae in their predaceous and mycophagous progenitors, chelicerae that were ideally suited for the transition to phytophagy. Furthermore, it is significant that only in the Actinedida have phytophagous mites acquired modified life cycles, specializing overwintering morphs, specialized methods of dispersal, and modified genetic mechanisms and sex ratios, all of which may be regarded as factors in the evolution of *r*-type strategies for survival. (Krantz and Lindquist, 1979, p. 15)

Soft-bodied acarid mites (Astigmata) are represented by few fossils. Oribatid mites have a much heavier exoskeleton in most instances, which would leave fossil remains more easily than their soft-bodied cousins. The Astigmata are more evolved than the Oribatida (W. Knulle, personal communication), although there are homologies in the setation of both. The structures of the mouthparts, the arrangement of the body setae, and the segmentation are similar. The acaridids are considered to be related to the Actinedida in the simple maxillary setae as also found in the Speleorchestidae (OConnor, 1984).

The evolution of feather mites parallels the evolution of their hosts. Feather mites are obligatory ectoparasites of birds. The mites are transmsitted from parents to offspring of the same species of host and each order of birds has a distinct acarofauna (Gaud and Atyeo, 1979). It is likely that fur and itch mites have a similar relationship with their hosts.

Fossil oribatid mites from the Upper Jurassic period of Europe represent recent families, that is, Cymbaeremaeidae (*Jureremus*), Trhypochthniidae (*Palaeochthonius*), Oribatulidae (*Cultroribula*), and Achipteriidae (*Achipteria*). Numerous oribatids are described from Baltic amber (Sellnick, 1918, 1927). A species of *Dendrolaelaps* (Gamasina, Digamasellidae) is identified from amber in the same location (Hirschmann, 1957–1976). Seven genera and eight species of oribatids are described from Chiapas amber (Mexico) (Woolley, 1971). At least 18 families of Oribatida are represented among the fossil acarines, some from terrestrial deposits (Krivolutsky, 1973, 1979, 1986).

Mites of Acariformes were numerous in the Mesozoic era and their evolution was a slow process, "one of the slowest in the animal kingdom especially among terrestrial animals" (Krivolutsky, 1979).

Phylogeny of the Acari

Both the Chelicerata and the Arachnida are considered monophyletic taxa. Changes of early aquatic chelicerates evidently led to efficient predation and then emergence of terrestrial lines. In the Arachnida body size may be correlated with the differentiation of taxa and adaptation to various ecological niches (Weygoldt and Paulus, 1979a). [Additional information of arachnid evolution and phylogeny may be found in Kaestner (1968) and Savory (1977a).]

It is difficult to construct a general definition and phylogenetic summary of the Acari as a natural group based on morphological and biological bases that does not require extensive clarification (Zakhvatkin, 1952). It is stated:

> The Acari do not constitute a natural ensemble, but comprise several heterogeneous groups derived from phylogenetically distinct

lines. Their metamery is strongly degraded; and in rare cases where vestiges of segmentation are preserved, these with some exceptions do not correspond at all to that of other arachnids. (Andre, 1949)

The somal characteristics of mites strongly suggest that the primitive segmentation of Acari has been extensively suppressed, although remnants of it persist in the Opilioacarida, and in the Ixodida, where pleural festoons of hard ticks suggest six segments, but segmentation of the Argasidae is not as definite (Raw, 1957).

Because of the great age and early divergence of the major arachnid orders, striking and profound differences exist in the morphology, development, and life-styles of extant forms. Adaptive radiation was followed

by an arrestment that has persisted since the late Palaeozoic, such that representatives of most arachnid orders look essentially the same as they did 250 million years ago. Even some mites, fossils from the Devonian, show family-level similarities with extant representatives of relatively "primitive" (early derivative) endeostigmatic and oribatid mites. (Lindquist, 1984)

It appears, however, that in mites secondary radiation occurred without the arrestment typical among their arachnid relatives. Subgroups of mites have separately and repeatedly evolved from predatory, scavenging, and fungivorous habits to become plant-feeders, parasites, and commensals associated with both invertebrate and vertebrate animals. Some families (e.g., Laelapidae, Tydeidae, Tarsonemidae) exhibit continuous diversification; others, such as the phytoparasitic Eriophyoidea and the parasitic Parasitengona, were set in a life pattern long ago (Lindquist, 1984).

The subclass Acari is considered by some to be at least diphyletic in origin (Boudreaux, 1979; Dubinin, 1956, 1958, 1959a,b; Bekker, 1959) and possibly polyphyletic (Sharov, 1966; Grandjean, 1935; Andre and Lamy, 1937;

Zakhvatkin, 1952; Woolley, 1961; Hammen, 1969). A differing opinion is that the Acari (Acaromorpha) originated as a monophyletic group from "primitive forms of the subclass Pedipalpides . . . in particular from organisms of the type that exist in Palpigradi and Schizopeltids" (Dubinin, 1957). The monophyletic origin is defended on the basis of internal structures (Bekker, 1959) and on cladistic analysis (OConnor, 1984; Lindquist, 1984). Assertion of monophyletic ancestry of the subclass, with closer relationships to other arachnids, is not convincing (Weygoldt and Paulus, 1979a; Hammen, 1979).

For over 30 years Grandjean (1935) hypothesized that two major acarine lineages existed, with remote but common ancestry. No other group of the Arachnida is as closely related to either the Actinotrichida and Anactinotrichida as these are to each other. The separation of those major lineages is profound because (1) apomorphies indicated for one do not readily compare with alternative conditions in the other; (2) the fossil record of the Devonian shows recognizable and diverse forms of early endeostigmatic and oribatid mites (cf. Hirst, 1923; Dubinin, 1962; Rolfe et al., 1982; cited by Lindquist, 1984). This implies the establishment of acarine stock in the later Silurian (400 million years ago) and separation into sister lineages (an opinion held by Dubinin, 1962, that the Acari arose as the earliest group of terrestrial arachnids).

It is considered that

two large groups within a large and diverse taxon such as the Acari, may have numerous and consistent internal and external differences . . . can be attributed equally well to divergence and convergence. The greater number of differences, even if certain characters are not mutually exclusive, between Anactinotrichida and Actinotrichida indicates either an earlier divergence or earlier convergence. A diphyletic origin (convergence) of the Acari simply places an earlier date on the common ancestor than does a monophyletic origin, because one or both lines must have arisen from arachnid stock. We believe that the Anactinotrichida represent separate evolutionary lines of Acari, and the only means of deciding convergent or divergent lines is an extensive comparative study of all arachnids. Based on comparative external studies of the Class Arachnida, Zakhvatkin (1952) concluded that the Acari are diphyletic. A like comparative study of arachnid internal morphology could be very helpful to confirm or refute the diphyletic theory. (Woodring and Galbraith, 1976)

The ancestral Acari may be among the earliest forms of terrestrial arachnids and "sister- or out-group" relationships with orders of Arachnida may be precluded from consideration with mites if any doubt exists about acarine lineages extending back into the Silurian times (Lindquist, 1984).

The question of relationships is succinctly and quite completely reviewed by Lindquist (1984, p. 44):

> The question of whether oplioacarid-like Anactinotrichida or the endeostigmatic-like Actinotrichida is more "primitive" or closer to the ancestral acarine stock, is a red herring, a distraction. Theoretically, as sister-groups, neither is more early derivative than the other. Not surprisingly, "primitive" members of both groups retain a considerable variety of plesiomorphic (primitive) characteristics. Comparison of the *number* of plesiomorphies retained in early-derivative members of each group is *not* necessarily a reliable index of degree of primitiveness according to cladistic methodology.

The implications on classification of the Acari, as a monophyletic group . . . are straightforward. First, we can mercifully continue to use "Acari" and "mites" as meaningful names denoting the entire subclass as a natural group. Second, we can recognize the two major lineages by continuing to use Actinotrichida (or Acariformes) for the one and Anactintrichida for the other; the latter includes the Opilioacarida along with the Holothyrida, Ixodida and Gamasida. Third, we can apply the name

Parasitiformes to the major grouping *within* the Anactinotrichida that *excludes* the Opilioacarida. Note that this concept of Parasitiformes *includes* the Holothyrida along with the Ixodida and Gamasida, in contrast to a more traditional use of this name for just the latter two suborders as found in general works of the 1950s (e.g., Andre, 1949; Hughes, 1959). In fact, this suggested usage of Parasitiformes goes back to Reuter's (1909) original concept of the group, which included the Holothryida but excluded Opilioacarida.

The implications of a diphyletic classification of mites would be more confusing and unsettling. "Acari" could not be used readily alone for either assemblage, and we would see further use of "true mites" for Acariformes, and perhaps "other mites" for Anactinotrichida. "Acari" would then refer to a polyphyletic or paraphyletic group in much the same way as we continue to conceptualize "reptiles."

The hypothesis of monophyly of the Acari will be subject to further testing as addditional critical data become available, especially embryonic, early postembryonic (e.g., prelarval and larval), and fine structural data from key groups such as the Palpigradi, Ricinulei, Opilioacarida, and Holothryrida. More careful recognition, description, and homologization of structures, such as lyrifissures, solenidia, actinopilous setae, leg segments, ventral opisthosomal structures, ovipositor, male reproductive organs, and lateral (Claparede) organs, done in a comparative way among all arachnid orders, is indispensable.

Another cladistic study (OConnor, 1984, p. 24) treats the phylogeny of the Acariformes in a slightly different way:

> A phylogenetic analysis of relationships among the higher taxa in the order Acariformes results in a hypothesis of relationships quite different from prior conceptions. The order is divided into two suborders, for which the names Trombidiformes (including the Prostigmata and Sphaerolichida) and Sarcoptiformes (including the Astigmata, oribatid groups and remaining endeostigmatid groups)

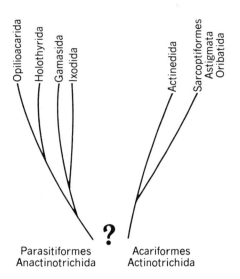

Figure 3-7. Postulated phylogeny of the Acari (modified after Hammen, 1972, and OConnor, 1984).

are retained. The Astigmata are proposed to have had a most recent common ancestor with oribatid groups possessing the lateral opisthosomal gland, elongate tibial solendidia, and reduced setation on opisthosomal segment F. Apparently, ancestral character states in the Astigmata all relate to neotenous trends already present in the related oribatid groups. In a truly phylogenetic classification, the Oribatei is seen as a paraphyletic group (excluding some descendants of a common ancestor, the Astigmata) and must be rejected as a formal group. The suborder Sarcoptiformes, as presently defined, is a natural group and should serve to replace Oribatei, particularly as several early derivative lineages formerly considered in the Trombidiformes are now placed with their nearest sarcoptiform relatives.

Kethley (1979) reviews higher categories of Trigynaspida cladistically.

Phylogenetic dendrograms at best are suppositional and illusory, but may show natural relationships, which, with future investigations will become changed and more instructive (Fig. 3-7).

REFERENCES

Akimov, I.A. 1979. Morphological and functional characteristics of the mouthparts of the Acaridae mites (Acaridae Ewing and Nesbitt, 1942). *Proc. Int. Congr. Acarol., 4th, 1974,* pp. 569–574.

Andre, M. 1949. Acari. *In* Grassé, P.-P., ed. *Traité de Zoologie 1949.* Masson, Paris, Vol. 7, pp. 794–892.

Andre, M. and E. Lamy. 1937. *Les idées actuelles sur la phylogénie des Acariens.* Published by Authors, Paris.

Barnes, R. 1980. *Invertebrate Zoology,* 4th ed. Holt, New York.

Barth, R.H. and R.E. Broshears. 1982. *The Invertebrate World.* Saunders, Philadelphia, Pennsylvania.

Bekker, E.G. 1959. Concerning the Acarina as a normal grouping. *Trud. Inst. Morf. Zhiv.* 27:151–162.

Boudreaux, H.B. 1979. *Arthropod Phylogeny with Special Reference to Insects.* Wiley, New York.

Dubinin, V.B. 1956. Acarina and their position in the Chelicerate system. *Trud. 2. Sci. Conf. Parasit. U.S.S.R.,* pp. 46–51.

Dubinin, W.B. 1957. New system of the superclass Chelicerata. *Bull. Soc. Mosc. Erp. Nat.* 62:25–53.

Dubinin, V.B. 1958. On new classification of mites Class Acaromorpha W. Dubinin and its position in the system of chelicerate animals, Chelicerophora. W. Dubinin. *9th Conf. Parasit. Probl.* 1957, pp. 82–85 (in Russian).

Dubinin, V.B. 1962. *In*: Rodendorf, B.B., ed. *Osnovy paleontologii* (Fundamentals of Paleontology). Moscow, Academy Sciences U.S.S.R., pp. 375–530 [in Russian].

Evans, G.O., J.G. Sheals and D. MacFarlane. 1961. *The Terrestrial Acari of the British Isles.* Adlard & Son, Bartholomew Press, Dorking, England.

Foelix, R.F. and G. Muller-Vorholt. 1983. The fine structure of scorpion sense organs. 2. Pecten sensilla. *Bull. Br. Arachnol. Soc.* 6(2):68–74.

Foelix, R.F. and J. Schabronath. 1983. The fine structure of scorpion sense organs. 1. Tarsal sensilla. *Bull. Br. Arachnol. Soc.* 6(2):53–67.

Francke, O.F. 1982. Scorpiones. *In*: Parker, S., ed. *Synopsis and Classification of Living Organisms.* McGraw-Hill, New York, pp. 73–75.

Gaud, J. and W.T. Atyeo. 1979. Coévolution des

Acariens Sarcoptiformes plumicoles et de leur hô-
tes. *Acarologia* 21(3):291–306.

Grandjean, F. 1935. Observations sur les Acariens.
Bull. Mus. Natl. Hist. Nat. (2)7:201–208.

Grassé, P. 1949. *Traité de Zoologie*, Vol. 6. Chelicerates. Masson, Paris.

Gupta, A.P., ed. 1979. *Arthropod Phylogeny*. Van
Nostrand-Reinhold, New York.

Hammen, L. van der. 1969. Notes on the mouthparts
of *Euoekenia mirabilis* (Grasse) (Arachnida +
Palpigradida). *Zool. Meded.* 44:41–45.

Hammen, L. van der. 1971. La phylogenèse des Opi-
lioacaride, et leurs affinitiés avec l'autre Acariens.
Acarologia 12(3):465–473.

Hammen, L. van der. 1972. A revised classification
of the mites (Arachnidea, Acarida) with diagnoses, a key, and notes on phylogeny. *Zool. Meded.
Leiden*, 47:273–392.

Hammen, L. van der. 1977a. A new classification of
Chelicerata. *Zool. Meded.* 51:307–319.

Hammen, L. van der. 1977b. The evolution of the
coxa in mites and other groups of the Chelicerata.
Acarologia 19:12–19.

Hammen, L. van der. 1979. Evolution in mites and
the pattern of evolution in Arachnidea. *Proc. Intl.
Congr. Acarol., 4th, 1974*, pp. 425–430.

Hedgpeth, J.W. 1955. Part P. Arthropoda 2. Pycnogonida. In: Moore, R. C., ed. *Treatise on Invertebrate Paleontology*. Geol. Soc. Am., Boulder, Colorado, and Univ. of Kansas Press, Lawrence.

Hirschmann, W. 1957-1976. *Gangsystematik der
Parasitiformes*, Ser. 1-22, Parts 1-232. Acarologie. Schriftenreihe fur vergleichende Milben-
kunde. Hirschmann-Verlag Inh., Furth/Bayern.

Hirst, S. 1923. On some arachnid remains from the
Old Red Sandstone (Rheynie Chert Bed, Aberdeenshire). *Ann. Mag. Nat. Hist.* 12(9):455–474.

Hughes, T.E. 1959. *Mites or the Acari*. Athlone,
London, pp. 1–225.

International Code of Zoological Nomenclature.
15th Int. Trust Zool. Nomenclature, London.

Jeppson, L.R., H.H. Keifer and E.W. Baker. 1975.
Mite Pests of Economic Plants. Univ. of California
Press, Berkeley.

Kaestner, A. 1968. *Invertebrate Zoology*. Vol. 2. Wiley (Interscience), New York.

Kethly, J. 1979. A cladistic analysis of the Trigynaspida (Acari: Parasitiformes) with a review of the
higher categories and nominate taxa. *Proc. Intl.
Congr. Acarol., 4th, 1974*, pp. 459–465.

Kevan, F.G., W.G. Chaloner and D.B.O. Savile.
1975. Interrelationships of early terrestrial arthropods and plants. *Paleontology* 18:391–417.

Krantz, G.W. and E.E. Lindquist. 1979. Evolution
of phytophagous mites (Acari). *Annu. Rev. Entomol.* 24:121–158.

Krivolutsky, D.A. 1973. The evolutionary ecology,
trends and tempo of evolution in palearctic Oribatei. *Proc. Intl. Congr. Acarol., 3rd, 1971*, pp. 91–94.

Krivolutsky, D.A. 1979. Some Mesozoic Acarina
from the U.S.S.R. *Proc. Intl. Congr. Acarol., 4th,
1974*, pp. 471–475.

Krivolutsky, D. A. and V. A. Druk. 1986. Fossil Oribatids. *Annu. Rev. Entomol.* 31:533—545.

Lawrence, R.F. 1953. *The Biology of the Cryptic
Fauna of Forests*. A. A. Balkema, Amsterdam.

Levi, H.W. 1982. Uropygi, Amblypygi, Palpigradi,
Araneae, Ricinulei. *In*: Parker, S., ed. *Synopsis
and Classification of Living Organisms*. McGraw-
Hill, New York, pp. 75–96.

Lindquist, E.E. 1984. Current theories on the evolution of major groups of Acari and their relationships with other groups of Arachnida, with consequent implications for their classification. *In*:
Griffiths, D.A. and C.E. Bowman, eds. *Acarology
VI*, Vol. 1, pp. 28–46.

Manton, S.M. 1973. Arthropod phylogeny—a modern synthesis. *J. Zool.* 171:111–130.

Manton, S.M. 1977. *The Arthropods. Habits, Functional Morphology, and Evolution*. Oxford Univ.
Press (Clarendon), London and New York.

Martens, J. 1969. Webercknechte. *Zool. Jahrb.*
96:184–185

Meglitsch, P.A. 1972. *Invertebrate Zoology*, 2nd ed.
Oxford Univ. Press, London and New York.

Muchmore, W.B. 1982. Pseudoscorpionida. *In*:
Parker, S., ed. *Synopsis and Classification of Living Organisms*. McGraw-Hill, New York, pp. 96–
102.

Muma, M. 1982. Solpugida. *In*: Parker, S., ed. *Synopsis and Classification of Living Organisms*. Mc-
Graw-Hill, New York, pp. 102–104.

OConnor, B.M. 1984. Phylogenetic relationships

among higher taxa in the Acariformes, with particular reference to the Astigmata. *Acarol. [Proc. Int. Congr. Acarol.], 6th, 1982,* Vol. 1, pp. 603–608.

Parker, S.P. ed. 1982. *Synopsis and classification of Living Organisms.* Mc-Graw Hill, New York.

Petrunkevitch, A. 1955. Part P. Arthropoda 2. Arachnida. *In:* Moore, R.C., ed. *Treatise on Invertebrate Paleontology.* Geol. Soc. Am., Boulder, Colorado, and Univ. of Kansas Press, Lawrence, pp. 42–162.

Petrunkevitch, A. 1963. Chiapas amber spiders. *Univ. Calif. Publ. Entomol.* 31:1–40.

Petrunkevitch, A. 1971. Chiapas amber spiders. II. *Univ. Calif. Publ. Entomol.* 63:1–44.

Pittard, K. and R.W. Mitchell. 1972. Comparative morphology of the life stages of *Cryptocellus paleazi* (Arachnida: Ricinulei). *Grad. Stud. Tex. Tech. Univ.* 1:1–77.

Raw, F. 1957. Origin of Chelicerates. *J. Paleontol.* 31(1):139–192.

Rolfe, W.D.I., J.K. Ingham, E.D. Currie, S. Neville, J.S. Brannan and E. Campbell. 1981. Type specimens of fossils from the Hunterian Museum and Glasgow Art Gallery and Museum. Hunterian Museum, *Univ. of Glasgow,* Vol. 1, pp. 1–8.

Savory, T. 1974. On the arachnid order Palpigradi. *J. Arachnol.* 2:43–45.

Savory, T. 1977a. *Arachnida,* 2nd ed. Academic Press, London.

Savory, T. 1977b. Cypophthalmi: The case for promotion. *Biol. J. Linn. Soc.* 9:299–304.

Schuster, R. and I.J. Schuster. 1977. Ernahrungs- und forplanzlungs-biologisches studien an der Milbenfamilie Nanorchestidae. *Zool. Anz.* 199: 89–94.

Sellnick, M. 1918. Die Oribatiden der Bernstein- sammlung der Universitat Konigsberg i Pr. *Schr. Phys.-Okon. Ges. Konigsb.* 59:21–42.

Sellnick, M. 1927. Rezente und Fossile Oribatiden. *Schr. Phys.-Okon. Ges. Konigsb.* 65:114–116.

Sharov, A.G. 1966. *Basic Arthropodan Stock.* Pergamon, Oxford.

Shear, W.A. 1982. Opiliones. *In:* Parker, S., ed. *Synopsis and Classification of Living Organisms.* Mc- Graw-Hill, New York, pp. 104–110.

Snodgrass, R.E. 1962. *Anatomy of Arthropoda.* Mc- Graw-Hill, New York.

Southcott, R.V. 1961. Studies on the systematics and biology of the Erythraeoidea (Acarina), with a critical revision of the genera and subfamilies. *Aust. J. Zool.* 9(3):367–610.

Stormer, L. 1955. Part P. Arthropoda 2. Chelicerata. *In:* Moore, R.C., ed. *Treatise on Invertebrate Paleontology.* Geol. Soc. Am., Boulder, Colorado, and Univ. of Kansas Press, Lawrence.

Tiegs, O.W. and S.M. Manton. 1958. The evolution of the Arthropoda. *Biol. Rev. Cambridge Philos. Soc.* 33:255–337.

Tuxen, S.L. 1974. The African genus *Ricinoides* (Arachnida: Ricinulei). *J. Arachnol.* 1:85–106.

Weygoldt, P. 1969. *The Biology of Pseudoscorpions.* Harvard Univ. Press, Cambridge, Massachusetts.

Weygoldt, P. and H.F. Paulus. 1979a. Untersu- chungen zur Morphologie, Taxonomie und Phy- logenie der Chelicerata. I. Morphologische Unter- suchungen. *Z. Zool. Syst. Evolutionsforsch.* 17(2):85–116.

Weygoldt, P. and H.F. Paulus. 1979b. Untersu- chungen zur Morhologie, Taxonomie und Phyloge- nie der Chelicerata. II. Cladogramme und die Ent- faltung der Chelicerata. *Z. Zool. Syst. Evolutionsforsch.* 17(3):177–200.

Wiley, E.O. 1981. *Phylogenetics: The Theory and Practice of Phylogenetic Systmatics.* Wiley, New York.

Woodring, J.P. and C.A. Galbraith. 1976. The anat- omy of the adult uropodid *Fuscouropoda agitans* (Arachnida: Acari) with comparative observations of other Acari. *J. Morphol.* 150(1):19–58.

Woolley, T.A. 1961. A review of the phylogeny of mites. *Annu. Rev. Entomol.* 6:263–284.

Woolley, T.A. 1971. Fossil oribatid mites in amber from Chiapas, Mexico. Part II. *Univ. Calif. Publ. Entomol.* 63:91–99.

Yoshikura, M. 1975. Comparative embryology and phylogeny of Arachnida. *Kumamoto J. Sci. Biol.* 12:71–142.

Zakhvatkin, A.A. 1952. The division of the Acarina into orders and their position in the system of the Chelicerata. *Parazitol. Sb.* 14:5–46.

FORM AND FUNCTION— EXTERNAL MORPHOLOGY

What we are after is the nature of living processes
—SIR HANS KREBS

In mites, as in other animals, morphological features determine to a great degree the specifics of the functions that occur. In this part of the text an attempt is made to describe the major anatomical features by systems and to discuss the known functions as related to morphology. The reader is referred to references for extensions of the general information. For some functions there is much more in the literature than can be reproduced here, even in summary fashion. In other instances there is a paucity of available data.

A comparative systematic approach is used for discussion of functions as they relate to orders. In this way the reader may compare functions and morphology in a correlative way.

4

Size, Integument, and Tagmatosis

Natura nusquam magis quam in minimis tota est.
—Pliny

SIZE

Mites have an overall range of 100 μm to 3 cm. Relatively few mites reach beyond a millimeter in length. The vast majority range between 250 and 750 μm; the average size varies with the order involved.

The word "mite" implies diminutive. Some mites are enormous compared to certain protozoans; tardigrades and some parasitic and phytophagous mites may be very small (Rosicky, 1973). One of the smallest mites is *Acarapis woodi* (Fig. 1-5) a tiny tarsonemid parasite (120 μm) that lives in the tracheae of honeybees and grasshoppers. Transport nymphs (hypopi) are dwarfed in size by the large velvet mite *Dinothrombium* or when contrasted to the head of a termite (Fig. 4-1A SEM). Hair follicle mites are tiny when measured against the diameter of the hair on which they live (Fig. 4-1B). (See also Fig. 4-2.)

The following poem (attributed to the late Dr. H. B. Hungerford) has humorous and scientific implications relating to size and animal interactions:

> The thing called a chigger
> Is really no bigger
> Than the smaller end of a pin.
> But the bump that it raises
> Just itches like blazes
> And that's where the rub sets in.

The chigger is of importance to humans and other animals as an irritating pest and a crucial vector of disease. Its medical impact is related to its size.

Among the consequences of being small are the following:

1. A wider range of resources, life-styles, and locations is available to mites in myriads of habitats and niches. Habitats and life-styles affect population size; for example, from 600–1000 scab mites/cm^2 of skin to 1000–3000 mites/cm^2 of cheese rind or tens of thousands of

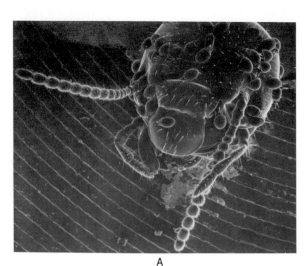

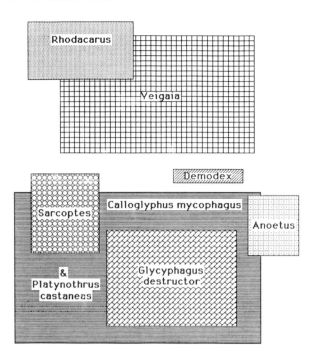

Figure 4-2. Schematic comparison of size in mites (computer illustration by John S. Olsen).

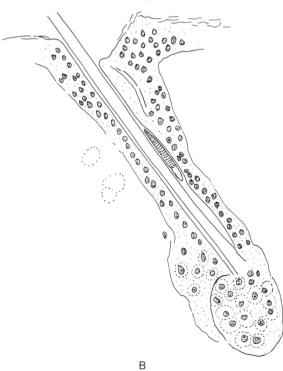

Figure 4-1. (A) Mites on the head of a termite (*Reticulotermes tibialis*), SEM; (B) diagrammatic reconstruction of *Demodex aurati* in hair follicle (drawn from photomicrograph by William B. Nutting).

mites per kilogram of stored products (Rosicky, 1973).

2. Dispersal mechanisms are extensive within a habitat, together with the behavioral techniques employed; for example, phoresy of hypopi on insects and mammals, external and integumental transfer of mites on birds, and parasitism of various types within a broad range of behavioral adaptations.

3. Surface-to-volume ratio affects function, for example, respiration. Respiration by diffusion is related to size. The smaller the size, the greater the ratio and the increased surface area in which cutaneous respiration can occur. Terrestrial arthropods of about 1 mm or less in length (e.g., tardigrades, Pauropoda, and many mites) lack respiratory and circulatory organs (Krogh, 1968). The distance from the absorptive surface to the point of oxygen utilization is vital. Gases diffuse short distances with ease, but influence survival when translated into molecular movement (Kennedy, 1927). A minute

cuticle may lose water more rapidly and water balance is essential to survival.

4. Structural complexity may be diminished with decreased size; and reduction or elimination of body parts may be correlated with extremely small forms of relative complexity.

Animals obviously differ in size and little attention has been paid to these differences in an analytical sense. "For every type of animal there is a most convenient size, and a larger change in size inevitably carries with it a change in form." The complex comparative anatomy of vertebrates is largely the odyssey of the interaction of surface area proportionate to volume, for "The higher animals are not larger because they are more complicated. They are more complicated because they are larger" (Haldane, 1925–1926, 1956).

Exoskeleton imposes limitations on smaller arthropods. The surface to volume ratio relates to respiration, to the type and size of the nervous system, to the extent of the circulatory system, to temperature relations involved in the blood supply, and to the senses of touch, hearing, and sight (Kennedy, 1927). Size relates to the mass of muscle, body tissue, and organs, which cannot be larger than what the exoskeleton will support or what the available energy for movement and other functions will allow (Savory, 1968).

Water balance and dessication relate to size and the ratio of surface to volume. Ratio of surface to volume influences the absorption of food internally, where the absorption is proportional to the surface area occupied by the mass of cells through which the dissolved nutrients must pass before distribution in the hemolymph. The mass of these cells and the structural configuration of their aggregate surfaces greatly influence absorption in both quantity and rate.

Size reduction in mites is correlated with the nervous system, where the consolidation of nervous components into the synganglion reflects this relationship and economy of innervation from this body "brain." The cells of the nervous system are relatively the same size throughout the animal kingdom, but the synganglion does not become more complex by developing smaller cells or by reducing the size of the ganglion or increasing the number of cells (Kennedy, 1927).

During growth the surface area increases with the square of the linear dimensions, while weight increases as the cube. The formula below identifies the size relationship for mites where S is the surface, V is the volume (weight), and k is a constant that relates to units and shape. The surface to volume ratio becomes progressively greater as size diminishes, which is of great significance to terrestrial mites for water relationships and heat exchange (Edney, 1977).

$$S = kv^{2/3}$$

$$\text{volume} = 1\,[\times]\,1\,[\times]\,1 = 1$$
$$\text{surface} = 1\,[\times]\,1\,[\times]\,6 = 6$$
$$S/V = \qquad 6/1 = 6$$

The reasons why mites are the size they are are elusive, but size relates to morphological specializations of various groups and correlates with ecological phenomena. Size in mites directly involves their ability to cope with or to "control" their microhabitats and hence affects their dispersal. As organisms decrease in size to that of smaller insects and mites, their ability to control their environment diminishes correspondingly, but the heterogeneity of habitats in which they may survive increases. Because they are able to manipulate circumstances in the microhabitats to their advantage, they remain small and survive. Their food resources, though reduced in amount, are varied enough to provide a miniscule but heterogeneous smorgasbord. Organisms the size of mites and smaller have a greater capacity to exploit heterogeneity of microhabitats and to manipulate circumstances to their advantage; size is also useful in dispersal (Janzen, 1977).

The general arachnid feature of preoral digestion is of prime importance among the factors that influence and determine the small size of mites. Most mites are fluid feeders; the food is selectively digested exteriorly to form a liquid that may be sucked in. Thus, the mass of ingested food is reduced and the size of the digestive tract is correspondingly smaller. Such features are also correlated with the gnathosoma of mites, which is a relatively simple structure. Reduced, modified cheliceral musculature, the presence of a suctorial pharyngeal pump, and a reduction of the "cranial" armature are necessary for the muscular movement of the chelicerae and their retraction without an excessively large "head" structure and a correspondingly complex skeleton.

The ingestion of liquid food corrolates with changes in the structure of the digestive tract other than the absorptive cells. A muscular layer around the tract for movement of its lumenary contents is absent (except in oribatids). Hemocoelic hydrostatic pressure probably influences the movements of the gut contents. Also, the volumetric capacity of the gut is determined by the mass of cells that function in digestion and absorption. Combined with the liquid food ingested, the surface area of the gut determines the digestive capacity of the tract, not the volume of the tissues involved. Where liquification of food materials is efficient, there is less need for extensive digestive cells, and for storage capacities in the midgut and hindgut, which allows for reduction of the size of these organs. Where the capacity of the digestive tract increases, the ratio of surface area to volume also increases. When that occurs there is less need for a circulatory system for distributional purposes and, except for the hemocoel, the normal organs of the circulatory system may be absent.

The limitations that result from the general ratio of body surface area to body volume may be exemplified by a sphere 6 mm in diameter which has a S/V ratio of 1.0 compared to a 0.2 mm sphere with a S/V ratio of 30. This means that the area available for absorption or release of gases (a desirable feature) is increased. At the same time such a surface has potential hazard through a much greater loss of water (an undesirable feature). Where the body S/V ratio is 1.0 reduction of the internal respiratory mechanisms is possible. In a body with S/V ratio of 30, the organisms would have to compensate structurally, physiologically, and behaviorally to affect water retention (R. Mitchell, personal communication, 1977).

Mite size is also affected by integument. As the body becomes more diminutive the integument (exoskeleton) need not be as sturdy. The tensile strength of the internal tissues and the hydrostatic pressure of the hemocoel are sufficient support for the internal organs as well as the simpler, reduced exoskeleton. In smaller, terrestrial forms the tensile strength of the tissues and the hydrostatic pressure of the hemocoel provide the means to retain the general shape when the exoskeleton is soft, and even during ecdysis (R. Mitchell, personal communication, 1977; Kennedy, 1927).

Muscles help to generate and control hydrostatic pressure, which adds to the overall efficiency of their functions while maintaining optimal mass. Mites are known to have only flexor muscles in their legs. Extension of the leg segments is accomplished by hydrostatic pressure. Absence of extensor muscles reduces by nearly one-half the mass of muscle necessary for functions associated with muscles (R. Mitchell, personal communication, 1977) (Figs. 6-1, 6-3).

With diminutive size, energy demands are greatly reduced, to the point that in some instances a particular organ system may be eliminated. In some mites the midgut serves in both digestion and circulation (gastrovascularly). Where systems are absent and the remaining organs acquire more than one function, a concomitant reduction of the complexity of internal organs or a combination of their functions occurs (R. Mitchell, personal communication, 1977).

A striking example is the use of the blind hindgut in some chigger mites for defecation where no anus is present. A pouch in the hindgut accumulates waste materials until replete.

This waste pouch is eliminated by rupture of the integument and expulsion of these pouches over time (as shown by integumental scars) and serves a dual function of elimination and excretion. The loss of wastes in this manner is called "schizeckenosy" (Mitchell and Nadchatram, 1969).

Because of "preadaptations" of the arachnids for size reduction, few limitations are imposed upon mites. Because cellulose cell walls of plants pose distinct barriers to preoral digestion, the only herbivorous mites are those that are small enough, that have the buccal mechanisms necessary to suck plant juices from individual plant cells, or that have styletlike chelicerae (R. Mitchell, personal communication, 1977).

Size and locomotion are relatively limiting where movement is necessary from one mass of food to another. Water loss is a greater hazard to mites than to insects because the S/V ratio of mites is nearly "two orders of magnitude greater than most insects." Size thus restricts mites to niches with higher relative humidity and allows only relatively narrow limits of movement for foraging (R. Mitchell, personal communication, 1977).

Mites outside the soil habitats may have remained small because of other factors. If mites use other arthropods as transport hosts (phoresy), the mites themselves must be relatively small to effect such transport. (Such is the case in the hypopi of the stored products mites where the phoretic nymphs are numerous on the transport host; Fig. 4-1 SEM.) What applies to phoretic species would also be true of parasitic species, especially ectoparasites, where size is of the utmost importance in avoiding defensive actions by the host. With internal parasites such as *Acarapis*, tracheal parasites of bees, sarcoptid scab mites in the skin, and hair follicle mites, size would be vital to habitat location. The principle also applies to nasal mites, quill mites, lung mites, and others (R. Mitchell, personal communication, 1977).

Size, food reserves, and length of life are related. In mites probably the length of life is in inverse ratio to activity (Loeb, 1919). Mite size is correlated with relatively short life spans, although few studies have been directed specifically to determine those facts. In comparison with larger forms, the smaller mites have fewer food reserves and the tissues are probably worked actively for the relatively short life.

Size, an exoskeleton, and food reserves affect the temperature relations of acarines. Most small animals are poikilothermous, and so are mites. They apparently lack a control mechanism for temperature regulation and have vital thermal limits related to size and environment. As with dessication and water balance, this temperature relation involves the S/V ratio, for the small size limits the production of heat within the body.

INTEGUMENT

> If I was asked to single out a system which dominates Arthropod ways of life, it would have to be the cuticle.
> —NEVILLE, 1975, p. 4.

Mites have an integument similar to that of other arthropods, a remarkable phenomenon considering the diversity and probably polyphyletic nature of these animals. Such an integument acts both as a skin and as external support for the internal muscles and organs. The multilayered exoskeleton is secreted by a single-layered, cellular hypodermis (Fig. 4-3), thus consisting of living (cellular) and nonliving (inert cuticle) components. The composition of the laminated procuticle above the hypodermis is a complex of chitin and protein (possibly mucopolysaccharides). Chitin is a nitrogenous polysaccharide identical to materials in the walls of fungi (fungine) and is believed to have the form of long chains of acetylated glucosamine residues (Edney, 1977).

The integumental exoskeleton is the most important feature of the body; it is anything but inert material and serves a variety of functions. In mites with hard exoskeletons it affords protection against abrasion. In both hard- and soft-bodied mites the integument is a means of conserving the internal moisture to prevent dessication. In those with soft bodies the exoskeleton functions as a limiting boundary to hydrostatic pressure; that is, the fluid of the hemocoel is the main force by which the integument is distended and the shape of the body is maintained.

It is the molecular cross-linkage in the integument of arthropods (including mites) that gives the exoskeleton its rigidity and strength. Chemical and physical interlinkage in the polymer networks contribute to the strength of the integument. The linkage involved may be between proteins, for example, quinones and free amino acids, between phenolic rings, or between proteins and chitin, as well as between proteins and lipids. It has been shown that the structural proteins secreted by the hypodermis have unusual amino acid content, although the primary sequences have not been worked out. A cross-link with sulfur and quinone groups is also recognized (Neville, 1975); sulfur occurs in the cuticle of *Acarus siro* (Hughes, 1959).

The greater proportion of the integument is considered within the metabolic pool. Only the outer, chemically inert layers are shed, while the inner layers are reorganized and resorbed in the metabolic changes associated with ecdysis (Locke, 1964, 1974; Nathanson, 1967, 1970). The cells of tick hypodermis become hypertrophied, with enlarged nuclei, distinct nucleoli, and evidence of abundant RNA. New cuticle is secreted during feeding in *Ixodes ricinus*, and separated into two discrete phases. The first is a period of 7 days, in which the synthesis and deposition of new cuticle occurs, and is concurrent with the continuous ingestion of blood by the ticks. The second phase, a short period of 24 h, follows, during which rapid engorgement is accompanied by the stretching of the integument (Lees, 1952). Similar conditions are noted for *Ixodes ricinus* and *Hyalomma asiaticum* (Balashov, 1972).

Layers

Classically the integument of mites consists of five layers from the exterior to the interior (Legendre, 1968). (Corollary terms are indicated parenthetically in the scheme below.)

1. Epicuticle
   (a) Tectostracum: external colorless layer, perhaps consisting of a waxy covering or pellicle)
   (b) Epiostracum: chitinous, colored layer)
2. Procuticle
   (a) Exocuticle
      (1—Ectostracum: a lamellated chitinous layer with acidophilic characteristics)
   (b) Endocuticle
      (2—Hypostracum: a lamellated chitinous layer with basophilic characteristics)
3. Hypodermis or Epidermis
   The matrix of living cells that secretes or forms the outer layers. (In Opilioacarida and some of the Actinedida this cellular layer may confer the bright coloration that is present.)

The laminated cuticle (procuticle) varies in thickness, in the distinctive laminar regions present, in the presence and number of pore canals, in coloration, and in a wide variety of surface structures (Fig. 4-3). While relatively little is known about the mite integument, compared to information available on insect cuticle, there are definite similarities (and possible homologies) of the mite cuticle to features observed in arachnids as well as in insects. In the discussion that follows the words exoskeleton, integument, and cuticle are used interchangeably to avoid the boring repetition of a single term.

Relatively few articles have been written on the fine structure of mite cuticle, but the reader is directed to Anwarullah (1963), Gibbs and Morrison (1959), and Henneberry et al. (1965) on spider mite integument; Daniel and Ludvic

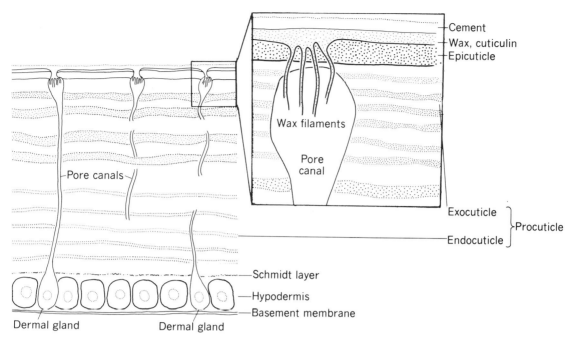

Figure 4-3. Diagrammatic representation of acarine cuticle with enlarged window for detail (drawing by Sonya Evans; after Wharton et al., 1968).

(1954) and Nathanson (1967, 1970) regarding tick integument; and Wharton et al. (1968) on the integument of the spiny rat mite (Fig. 4-4).

All mite integument presumably has a cellular hypodermis that interfaces internally with the hemocoel and secretes the laminated regions comprising the exoskeleton. These cells are flattened in the adult spiny rat mite with a thickness of approximately 1 μm. A well-developed endoplasmic reticulum occurs next to the Schmidt layer. Numerous ribosomes and mitochondria are indications of the secretory activity of these cells (Wharton et al., 1968).

Not all layers of the integument can be detected in all mites and various combinations of laminary components are present, but certain layers may be missing from the structural complex (Andre, 1949). In the velvet mite, *Allothrombium fuliginosum*, a distinct endocuticle is absent and in the tropical rat mite, *Ornithonyssus bacoti*, folded cuticle accompa-

nies a tanned or sclerotized cuticle (Hughes, 1959).

Immediately adjacent to the hypodermis in some mites is a poorly defined layer of pigment granules called the Schmidt layer (Wharton et al., 1968; Krantz, 1970, 1978) (Fig. 4-3). According to the former author the Schmidt layer is 0.025–0.050 μm thick in the spiny rat mite, *Echinolaelaps echidninus*, but the extent of this layer and its function is poorly understood.

External to the hypodermis and the pigmented Schmidt layer (if present) is a large homogeneous matrix that contains chitin. The layer is called the endocuticle (formerly the hypostracum) and stains with basic dyes and is said to be the location of the birefringent *actinochitin=actinopiline* that the late Grandjean (1935, 1947) described as the principal component in the cuticle and setae of the majority of mites (the Actinotrichida or Actinochaeta). [The nonlaminated endocuticle of the

 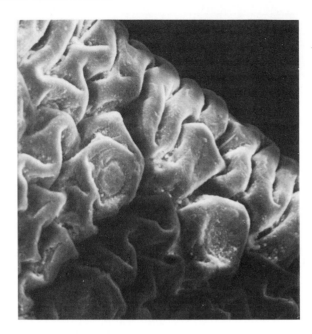

Figure 4-4. Enlarged integument of (A) Labidostommidae; (B) fossettes of soft tick (*Otobius megnini*), SEM.

spiny rat mite is about 0.4 μm thick (Wharton et al., 1968).] Sinuous pore canals 0.5–1 μm in diameter course outward from the hypodermis through the endocuticle and other layers of the integument to the surface (Hughes, 1959). The chemicals carried for deposit in the matrix or on the exterior flow through these canals to the sites of deposition in the laminae or to the surface; this is especially true of waxy secretions that cover the exterior of the exocuticle (Brody, 1969).

The exocuticle (formerly called ectostracum) is external to the endocuticle, but not always identifiably separate from it. In the spiny rat mite it is 10 μm thick and consists of about 100 laminae each about 0.1 μm thick and separated from each other by electron-dense material 0.01–0.02 μm in thickness. This layer is replete with pore canals and their branches (Fig. 4-3) and is frequently more laminated than the endocutle. Several types of pores occur in the cuticle and are thought to secrete the cement layer above the epicuticle (Wharton et al., 1968).

Recently the endocuticle and the exocuticle have been considered parts of the same general layer, the procuticle. The more external laminae of this procuticle usually stain with acidic dyes.

Tanning

The exocuticle may become sclerotized through orthoquinone tanning of the layers (Brody, 1969; Hughes, 1959). This hardening of the cuticle by percolation and permeation of these specific proteins into the exocuticle causes a more leathery or more horny consistency for the layers involved. Relative degrees of hardness are associated with the tanning proteins present, not specifically with the proportionate amount of chitin in the layer. "Chitinization" implies that chitin is present in the integument. This is not the same condition as "sclerotization," which results from the tanning process apart from the presence of chitin.

Frequently the tanning process is accompanied by a deepening in color, but pigmentation associated with tanning may be restricted to large shields or plates (Fig. 4-5), to smaller surface plaques, to genital structures, or to setal bases, and may not affect the general integument.

Tanning may also be related to the stage in the life cycle. The cuticle of the hypopial stage of *Histiostoma polypoii* is hardened and brown compared to the softer integument of other stages in the development of this species (Wallace, 1960). The hypopial stages of *Caloglyphus micophagus* have considerably thicker cuticle than other stages in the life cycle and show many more pore canals in the integument.

In the egg cases of spider mites sulfur-

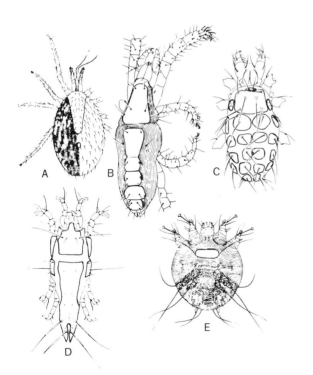

Figure 4-5. Dorsal integumental sclerites of (A) female Dermanyssidae (Gamasida); (B) Pomerantziidae (Actinedida); (C) larval Johnstonnianidae (Actinedida); (D) female Proctophyllodidae (Astigmata); (E) female Sarcoptidae (Astigmata) (after Baker et al., 1958).

tanned proteins resemble similar substances in the coverings of certain insect eggs (Beament, 1951). Such integument is white and flexible, contrasted to the darkened coverings that result from orthoquinone tanning. The whitish exoskeletons of tyroglyphid mites contain sulfur (as determined by chromatographic analysis), but in amounts of less than 5%. This tanning of mite cuticle varies in relation to the specific chemical reactions and components (e.g., quinones or sulfydril chemicals) within the the cuticular layers (Hughes, 1959).

Paleness or whiteness of the exoskeleton is not necessarily an indication of thinness, for in some of the white-bodied forms the exocuticle is thicker than the same layer in darker colored mites. In parasitic mites that feed by engorging on the blood of their hosts (e.g., gamasid mites and ticks), the integument is usually capable of stretching, which permits engorgment. Quinone scleroproteins are thought to facilitate this elasticity and distension.

The exocuticle also demonstrates the external beauty of mites, for it may be folded, thickened, or erupted into a variety of sculptural effects on the surface. Pits, puncta, bumps, tubercles, bosses, reticulations, fingerprinting, ridges, and other ornamentations may be present. The integument of a large ground mite (Labidostommidae) is intricately and geometrically sculptured (Fig. 4-4A SEM); in soft ticks rosettes may be present (Fig. 4-4B SEM). Many oribatids have an extensive cerotegument—an elaborate, secreted covering above the epicuticle that acts as a camouflage and a protection. The cerotegument appears to be composed mainly of epicuticular wax blooms, some of very intricate design.

The epicuticle (epiostracum and tectostracum of some authors) is the outermost layer of the integument of mites. It is usually a rather thin layer of achromatic chitin, additionally composed of cuticulin and/or waxes and polyphenols. A thin cement layer may exist on the outer surface, but is usually visible only in TEM sections. Three distinct layers of substances occur above the epicuticle in the spiny rat mite, namely, cuticulin, waxy layer, and cement (a total thickness of 0.15 μm) (Wharton et al., 1968). Where other layers may accept stains, the epicuticle appears impervious to the usual dyes. It is the waterproofing layer of the integument as well as the abrasion-resistant layer. Water loss from the body is regulated principally by this layer. The exterior cement and wax layers are included in the designation of epicuticle of arachnids, so the above description is consistent for mites (Savory, 1977).

Types of Integument

A variety of integumental conditions exists within the Acari (Andre, 1949; Baker and Wharton, 1952; Hughes, 1959; Evans et al., 1961). The general conditions are indicated below.

Anactinotrichida

Opilioacarida: thin, granulated cuticle, transversely wrinkled.

Parasitiformes

Holothyrida: thickened, armored cuticle over entire body surface.

Gamasida: sclerotized, sometimes hardened dorsal plates, scuta, and genital sclerites; armored in some (Uropodina).

Ixodida: secondary shields or plates; scuta in hard ticks; integumental fossettes in soft ticks.

Actinotrichida

Acariformes

Actinedida: generally thin, flexible, and transparent.

Tarsonemina: thin cuticle; sclerites seemingly related to segmentation; thin cuticle elastic, relatively transparent; some with sclerotization.

Tetrapodili: thin, wrinkled, annulated cuticle.

Astigmata

Acarida: thin, flexible, transparent cuticle; shields and setal bases sclerotized; genital sclerotization in some.

Oribatida: heavily sclerotized cuticle in most; Palaeacaroidea with thin integument, sclerotization not limited to setal bases; with or without cerotegument.

Sclerites of many arachnids are related directly to the segmentation of the body. It is characteristic of mites, however, where there is an absence of primary segmentation, for the integumental sclerites or plates to be independent of metamerism (Fig. 4-5). The sclerites thus appear to have relatively little phylogenetic significance. The Tarsonemoidea seem to show a retention of remnants of primary segmentation in the positions of their sclerites, but are one of the very few groups of mites with such a condition. Some of the different sclerites found in mites are shown in Figures 4-17, 4-18, and 4-19, but generally, the sclerites or plates are associated with muscle attachments, are secondarily developed, or are secondary sex characteristcs. Classificatory value varies with the orders.

Shields or plates in the spiny rat mite are divided into polygonal areas on the surface. It is assumed that the marginal outlines and ridges of these areas represent the outlines of the secretory hypodermal cells that underly the cuticle (Wigglesworth, 1965; Wharton et al., 1968; Neville, 1975). Small dotlike pits or puncta are within the surface of the polygons and appear to be pore canals that extend from the hypodermis up to but not through the cuticulin layer (Fig. 4-3). These pore canals branch about 2 μm from the surface. Electron-dense wax filaments and wax blooms extend from the distal tips of the canals on the epicuticle. Electron-dense materials are also found in the canals themselves (Nadchatram, 1970). Distinct laminations of alternating light and dark bands occur in both the exocuticle and endocuticle of Parasitus sp. (Brody, 1969). The epicuticle is a single, nonlaminated layer in this mite. Branching ducts of pore canals occur in *Echinolaelaps echidninus* (Wharton et al., 1968).

The scutum of hard ticks (Ixodidae) is composed of a definite endocuticle with extensive pore canals that are bunched together in both the endocuticle and exocuticle layers. A wrinkled epicuticle covers these other layers. Numerous pore canals and ducts of dermal glands also lace these layers (Hughes, 1959). Argasid ticks (Argasidae) exhibit a softer cuticle than the ixodid ticks and appear to lack an endocuticular layer. Small sclerotized plates, called fossettes, are imbedded in the cuticle of soft ticks and are used for muscle attachment (Fig. 4-4B). Small, seta-bearing plaques, called scutella, are also found on the surface of soft ticks. Scanning electron microscopy of larval development in *Ornithodoros moubata* shows a number of changes in the layers of the integument, including numerous furrows, ridges, and integumental embellishments in these soft ticks (Vogel, 1975).

Two structurally and functionally distinct types of integument are found in *Haemaphysalis leporispalustris*. The "hard" cuticle is sclerotized and provides skeletal surface for muscle attachment. It remains constant in size and unaltered during engorgement. An unsclerotized "soft" cuticle has the capability of extensive stretching during the final stages of engorgment. During the final 24 h of feeding the flattened hypodermal cells become cuboidal or columnar in shape. The cells quadruple in size to a height of 20 μm, and show an increase in the amount of endoplasmic reticulum present. Free ribosomes are found at the cell–cuticle interface concurrent with an increase of mitochondrial and cytoplasmic granules of electron-dense materials. Instead of a plasmalemma interface between the procuticle and the hypodermal cells, a series of interdigitating, branching, anastomosing double membranes are present (Nathanson, 1970).

Both sexes exhibit pores, pits and hairs of both sensory and other functions as well as hypodermal or integumentary glands. In section these glands comprise a basal, modified hypodermal cell from which a protoplasmic process extends into the pore for about half its length and connects with a chitinous tubule. The tubule is produced from the side walls of the pore

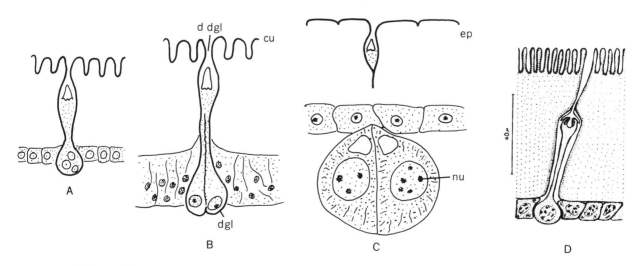

Figure 4-6. Dermal glands of female *Hyalomma asiaticum*: (A) unfed; (B) middle of feeding; (C) engorged (cu, cuticle; d dgl, duct of dermal gland; dgl, dermal gland; ep, epidermis; nu, nucleoli of gland cell) (after Balashov, 1972).

and exits on the surface of the integument (Fig. 4-6) (Douglas, 1943; Balashov, 1972).

In *Ixodes ricinus*, cuticular growth may occur after nymphal stadia are reached and some adults may show such growth. These changes occur during feeding. Hypodermal cells are greatly enlarged and show histological signs of secretion in the first phase. As the second phase of feeding occurs (usually about a day later) the gut becomes greatly enlarged and the resultant stretching of the cuticle reduces its thickness to about half (Lees, 1952).

Many authors agree that changes in the cuticle occur during feeding in most ixodid ticks, a change that enables the scutate ticks to ingest one massive blood meal in each instar. Argasid ticks show this type of cuticular change only in the slow-feeding larval stage (Lees, 1952).

Four main types of integument are observed among the Actinedida. Among the water mites the genus *Lebertia* exhibits weak to leathery integument and a dorsum without a shield. The relatively soft cuticle is differentiated into a thickened exocuticle above a rather thin layer of endocuticle. The pore canals are spiral and twisted; ducts of the dermal glands also course sinuously through these layers.

A second type is described for *Leptus* (Erythraeidae), parasitic mites found on spiders and lizards (Baker and Wharton, 1925; also cited as the first record of an erythraeid mite on a vertebrate). An endocuticle is absent; the exocuticle is laminated and rests directly on the hypodermis. A similar condition exists in *Linopodes*, a phytophagous pest of mushrooms (Eupodidae).

A third type of integument where the epicuticle is thicker and folded into ridges is exemplified by two prostigmatid mites in widely different taxonomic groups: *Erythraeus* (Erythraeidae), a parasitengonid mite similar to *Leptus*, and *Tydeus* (Tydeidae). The Cheyletidae also exhibit this feature. In *Tydeus* sp. the integument has three distinct layers, but without laminations in each layer. The endocuticle is slightly sclerotized, relatively unordered, but with a fibrous condition in some cases. The exocuticle is unusually thick, a possible protective

mechanism for the softer layers below. Glandular material secreted by the hypodermis is contained in pore canals, which branch just beneath the epicuticle, similar to conditions in insects (Brody, 1969).

The fourth arrangement is found in those mites in which the integument exhibits external specializations of the exocuticle. In one of the velvet mites, *Trombidium*, the surface is divided into polygonal areas thought to correspond to the arrangement of the hypodermal cells beneath. A fibrillostracum layer (with plentiful twisted fibrils) occurs in the endocuticle of *Calyptostoma*. The endocuticle of these mites includes setal bases joined by interlocking fibrils as well (Thor, 1903).

Gall mites (Eriophyidae) and hair follicle mites (Demodicidae), as well as a few other prostigmatid mites, have a thin, flexible cuticle. The majority of the stored products mites (Astigmata) are similar, but others of this order show sclerotization in a scutum or plate on the prodorsum, especially over the first pair of legs (propodosoma). Such a scutum may be found even in soft-skinned forms. In some this scutum is small and restricted (crista metopica of the Trombiculidae) (Fig. 4-9). In others it may be divided into smaller plaques (Sejidae, Fig. 4-17A).

Sclerotization of the genital area in many Actinedida and Astigmata appears to be independent of other body sclerites. Genital plates seem to have a similar independence, but are sexually distinguishing (e.g., fur and feather mites).

The integument of a typical tetranychid mite (*Tetranychus (T.) urticae* is essentially a thin layer with a pattern of external ridges and troughs. The inner surface of the cuticle exhibits ridges opposing the external troughs. The outer layer lacks exocuticle, but shows a dark-staining lipoid covering. Beneath this layer is a dark-staining chitinless layer of epicuticle. The unstained procuticle consists of two main layers, both of which contain chitin (Gibbs and Morrison, 1959). Other spider mites have a thick-layered epicuticle, a procuticle, and an underlying hypodermis with a basement membrane (Henneberry et al., 1965).

The outer, waterproofing layer of wax is subject to melting at specific temperatures and such melting results in water loss. If the loss occurs abruptly at a particular transition temperature, it may adversely affect the mites. These effects are correlated with life stages and the species of mite involved. As one would expect, some species are more resistant to the effects of temperature than are others.

Most beetle mites (Oribatida) possess a dense, heavily sclerotized integument. The cuticle has an opaque, dense appearance and a somewhat leathery-to-stiff composition. Narrow or broadened plates (lamellae) are found on the prodorsum; others are expanded into winglike shields (pteromorphs) over the legs for protection. The integument in *Oppia coloradensis* shows it is laminated (5–7 μm in thickness) in both the endocuticle and exocuticle, but the layers in the exocuticle are more compacted and less well-defined. The epicuticle is nonlaminated. Pore canals traverse the layers both vertically and horizontally. Infolded ventral projections of the cuticle (apodemes) are used for muscle attachment and form distinct rugosities (Brody, 1969).

Internal apodemes of sclerotized integument are also characteristic of the oribatid skeleton. When observed by electron microscopy, these infolded projections for muscle attachment and structural support are observed to form distinct rugosities in mites (Brody, 1969).

Coloration

Coloration of mites is of two types: extrinsic and intrinsic. When mites have a relatively translucent, unpigmented cuticle, the ingested food of the gut or other internal pigments may be the principal source of color rather than pigment incorporated in the integument itself. Such vital coloration is called *extrinsic (exogenous)*. Chlorophyll in the food of phytophagous mites shows through the integument of several groups and lends a greenish tint to the body (e.g., Tydeidae,Tetranychidae, Tenuipalpidae). Coloration in the tetranychids may vary with the seasons, diverse conditions of the habitat, sexual behavior, and reproduction. In bloodsucking mites the red color intensifies as the food is ingested (Dermanyssidae, Pterygosomidae), then changes to darker shades of red or brown as the blood coagulates, or becomes gray in engorged female ticks.

A second type of pigmentation occurs within the tissues of the body, particularly the integument, but may be associated with other internal organs as well. This condition is called *intrinsic (endogenous)* because it is natural to the tissue where it occurs, even if the original source may be either exogenous, having been obtained from outside the body and incorporated into the tissues, or endogenous within the tissues and produced by means other than alimentation. Endogenous pigments commonly occur in the hypodermis (e.g., *Opilioacarus*, reds of velvet mites, reds, blue-greens, and yellows of water mites). Bright reddish-orange coloration (aposematic) is common to the large velvet mites (Trombidiidae). These pigments resemble crustacena zooerythrin and may be receptive to light in modifying hypodermal colors. Reddish pigments are common to chiggers or "red bugs" (larval Trombiculidae). In *Trombicula autumnalis* the colors are either changed or broken down chemically during feeding. Loss of pigments during the larval stages is unexplained (Jones, 1950).

Coloration differs among the species of *Tyroglyphus* where one observes a pinkish hue in the integument of the rostrum, the legs, and the genital plates in both sexes and in all stages of development. The rest of the body is usually colorless or may be yellowish-white. The translucence of the integument enables one to see the food contents as they move through the alimentary canal. A predator of scale insect eggs, *Hemisarcoptes malus*, is an ivory color, but where insect eggs are purple, a lavender hue is incorporated in the body and legs of this predatory mite.

Coloration may be spotty lavender, purple,

or yellow in primitive mites. The Opilioacarids have colors that range from gold to blue or violet (Cloudsley-Thompson, 1958). The purple coloration of *Opilioacarus* is located in the hypodermis as well as in the cuticle (Hammen, 1969). Bluish or violet granules, usually widely separated, occur around the eyes and are clustered so closely that they appear black. *Adenacarus abicus* has a bluish pattern on the dorsum with dark violet color surrounding the eye patches. Vague, bluish, transverse stripes appear segmentally on the body and the legs; the gnathosoma is bluish, the ventral surface pale.

Hard ticks (Ixodidae) are more colorful than soft ticks (Argasidae); for example, *Amblyomma* has beautiful brown-to-yellow coloration with some red and green. *Dermacentor* species are more reddish-brown. The surface coloration in many of the hard ticks exhibits a moiré effect in bright light. Tropical species of ticks are known to produce irridescent or bright color in the midgut.

Endeostigmatid mites, for example, Nanorchestidae, exhibit what appears to be an intrinsic organ pigment within or beneath the somewhat silvery integument. Cunaxidae and Bdellidae, the predaceous snout mites, are bright red, orange, or bordering on the pinkish or lavender hues.

Dermal Glands

Glands of various kinds are associated with the integument and their pores may be seen in various positions on the body and legs. The pore openings may be round, elliptical, or lyreshaped. The venter and the legs exhibit peculiarly shaped lyrfissures (Fig. 4-6). Dermal glands of ticks are particularly significant during feeding (Balashov, 1972).

Setae

All mites have setae associated with the integument. Setal components, setal patterns, and spatial relationships of the gnathosoma, idosoma, and legs vary, but are usually diagnostic and important taxonomically. Chaetotaxy, the study of location and arrangement of setae, is an essential element of identification and classification.

The most primitive type of seta is assumed to have been attenuated and pilose—a simple hairlike form. It is thought that the setal arrangement on the body corresponded to the primary segmentation, each segment possessing a row of two to six setae (Fig. 4-17G). While this primitive arrangement appears to have been maintained without much modification in some groups, there is no certainty as to the original arrangement. Evidently reduction or multiplication of setae has occurred so many times in the course of the evolution of mites that the basic primitive setal arrangement is obscure (Baker and Wharton, 1952).

Two fundamental types of setae are present: tactile (setae proper) and sensory forms. The majority of sensory setae are tactile, but chemoreceptors are present as well. Both tactile and chemoreceptive setae may have an internal layer or solid core of optically active material called *actinopiline* (=actinochitin of Grandjean, 1935) that is birefringent under polarized light. Those mites with such optically active setae (Actinedida, Astigmata, Oribatida) are classified as Actinotrichida. Mites lacking actinopiline (Opilioacarida, Holothyrida, Gamasida, Ixodida) are classified as Anactinotrichida (Hammen, 1973; Krantz, 1978).

Most of the body setae in both of these groups of mites are hollow, tactile hairs. Some setae respond to pressure; others are protective in a manner similar to the quills of a porcupine (Baker and Wharton, 1952). Body setae range from simple filiform hairs to a variety of examples: pilose, barbed, pectinate, plumose, capitate, clavate, spatulate, lanceolate, cordate, palmate, pilidiform, furcate, pinnate, sickleshaped, dentate, serrated (Fig. 4-7). Mixtures of some forms of setae occur on the palps, legs, and idosoma. Gamasids exhibit tined setae on the palp tarsi. Spider mites (Tetranychidae) have duplex setae on the legs and T-shaped tenent hairs on the tarsi. Trombidiid mites

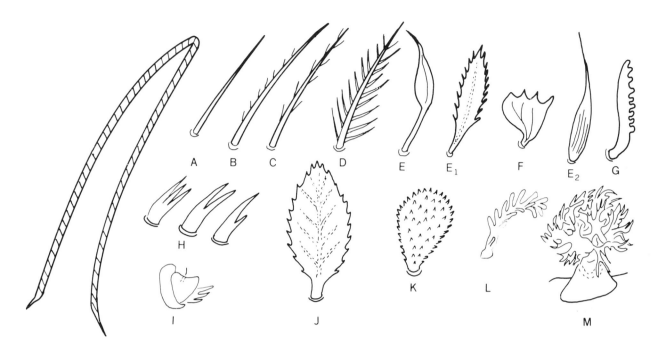

Figure 4-7. Tactile setae, hollow, without core of protoplasm or actinopiline: (A) (B) (C) pilose setae; (D) pectinate; (E) (E₁) lanceolate, (E₂) (Myobiidae); (F) palmate (Rosensteiniidae); (G) serrate (Cheyletidae); (H) palpal apoteles, Gamasida; (I) palpal apotele of Veigaiidae; (J) pectinate (Ctenoglyphidae); (K) Tuckerellidae; (L) Camisiidae; (M) *Dendrothrombidium* (Trombidiidae).

have ornately plumose or antlered setae. Astigmata exhibit a wide range of setal types.

Trichobothria. Trichobothria are special tactile hairs within a peculiar vase-shaped bases (Fig. 4-8). The bothridium is a pit-shaped or bowl-shaped capsule from which a tactile hair, the sensillum, (sensilla, pl.) arises. Structurally, each sensillum has a solid core of actinopiline. Trichobothria are found on various parts of the body and on the legs of spiders. They were first named "Horhaare" (Dahl, 1911) and described as sonotactic or auditory organs. In mites the capsules were termed pseudostigmata and the sensilla were called pseudostigmatic organs on the assumption that they were associated with respiration. These organs exhibit a variety of forms and probably as many complex functions

(Beklemishev, 1969; Haupt and Coineau, 1975; Krantz, 1978).

Trichobothria are found on the idiosoma and legs of mites. Those on the body may be paired, simple, or more elaborate sensilla associated with special types of sclerotized plates. In families of the Actinedida simple paired sensilla may arise from the propodosoma (Tydeidae). Two widely separated pairs of sensilla, one propodosomal, the other hysterosomal, may occur (Ereynetidae). A single pair of clavate or capitate sensilla is characteristic of Tarsonemidae and Pyemotidae.

The sensilla of some velvet mites are associated with distinctly sclerotized plates (Fig. 4-9A–C). Chiggers (Trombiculidae) have a pair of capitate or clavate sensilla in the scutum of the larva that is distinctive for the species. When

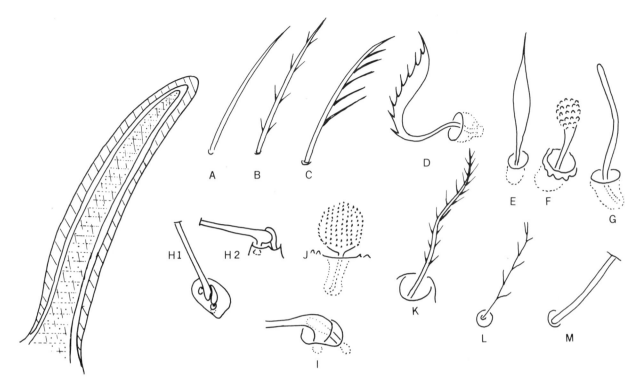

Figure 4-8. Trichobothria, sensilla with a core of actinopiline: (A) Calyptostommidae; (B) Epilohmanniidae; (C) Perlohmanniidae; (D) Eremobelbidae; (E) *Gustavia, Liacarus*; (F) *Camisia*; (G) *Hermannia*; (H$_1$, H$_2$) *Nanorchestes*; (I) *Smaris*; (J) *Bimichaelia*; (K) *Petralychus*; (L) *Lordalychus*; (M) *Caeculus* (on leg).

the sensilla are found in the expanded ends of narrowed plates, the complex is called a *crista metopica* (*Balaustium sp.*, Trombidiidae). Broadened plates with sensilla (*crista scutella*) and elongated "bone-shaped" arrangements (*crista ossiforma* both are found in the Smaridiidae (Fig. 4-9A–C). Some of the stored products mites have sensory setae associated with a crista metopica but not as a typical trichobothrium (Hughes, 1961) The trichobothria of the legs are usually simple and very long and are assumed to be tactile in function, but may be sensitive to vibrations (vibroreceptors) and to air currents (anemoreceptors) (Oribatida: Belbidae; Pauly, 1948; Krantz, 1978).

A single pair of trichobothria on the dorsolateral aspects of the prodorsum distinguishes the Oribatida from other orders of mites. The variety of forms are correlated with ecological habits as illustrated by Aoki (1973). Arboreal species of oribatids exhibit a fairly uniform type of clavate or capitate sensillum. Wandering species and soil species show a greater variety (Fig. 4-8B–G, N). Function is poorly understood.

Solenidia. Solenidia, hollow, chemosensory sensory setae (solenidions, Grandjean; solendien, Vitzthum), are found on the distal segments of appendages (e.g., tarsi). A protoplasmic extension of the trichogen cell is in the center of these setae, but they lack the core or sheath of actinopiline (Fig. 4-10) and are connected directly to the central nervous system (Baker and Wharton, 1952). Solenidia usually have striated walls, although this is not always

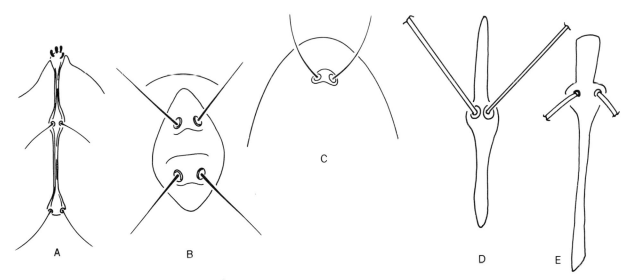

Figure 4-9. Types of crista in Parasitengona; (A) crista ossiforma; (B) crista scutellata, (C) Calyptostomidae; in Acaridida: (D) *Glycyphagus ornatus*; (E) *G. privatus* (after Hughes, 1961).

a demonstrable feature. They range in form from short, peglike bacilliform types (Tarsonemidae, Pyemotidae) to longer, erect bacilliform examples (Anystidae, Labidostommidae, Acaridae). Some are simple, adpressed, and procumbent setae in simple depressions (Eupodidae, Pachygnathidae), while others are L-shaped and procumbent (Lordalychidae). Rhagidial organs (=T-shaped solenidia) are characteristic of the family Rhagidiidae. Examples of long, filiform types of solenidia occur in a number of families of Astigmata and the Oribatida. Some of these whiplike solenidia extend from the insertion of the tibia far beyond the distal tip of the tarsus (Fig. 4-10G).

Microsolenidia. Microsolenidia, or famuli, are usually confined to the tarsus of leg I (Fig. 4-11). Characterized by a sheath or core of actinopiline, they thus resemble eupathidia, but are very much smaller. Their form varies from dendritic, filiform types (Labidostommidae) to sickle-shaped or hooked forms and some with expanded, ornate, distal tips (Sphaerolichidae). They also may be partially recessed (Ereyneti-

dae) or found in pits below the integumental surface (Caeculidae). Their function is not understood (Grandjean, 1941).

Eupathidia. Eupathidia (=acanthoides, pseudoacanthoides of Grandjean, 1946) are spinelike setae with a sheath of actinopoline around the protoplasmic core. This type of seta is found on the palpi and legs. While most are on the surface of the integument, a few eupathidia are found in pits, sometimes in association with other eupathidia or with recessed famuli (Caeculidae) (Fig. 4-12A–D).

TAGMATOSIS

Tagmatosis (tagmosis) implies the coalescence or grouping of segments into body regions, and is of importance in classification. Metamerism of primitive arthropods shows each segment with a dorsal tergite, a ventral sternite, and lateral pleura. The tagmata (body regions) vary between classes and are distinct. The subdivision of the arachnid body into two principal re-

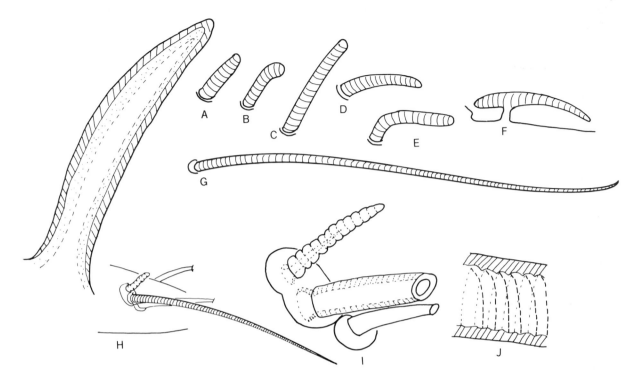

Figure 4-10. Solenidia, sensory setae with a core of cellular protoplasm, spiral in appearance: (A) Tarsonemidae, Pyemotidae; (B) Acaridae, Saproglyphidae; (C) Anystidae, Labidostommidae; (D) Eupodidae; (E) Lordalychidae; (F) Rhagidiidae; (G) Oribatida; (H) enlarged view of solenidial base and famulus; (I) enlarged view; (J) detail of spiral interior (after Grandjean, 1935).

gions, the prosoma and the opisthosomal, is based on inferences regarding the primary segmentation (Hammen, 1968). Mites do not exhibit typical arthropodan or arachnid metamerism and this difference is reflected in their tagmata (Hughes, 1959). The gnathosoma and idiosoma are diagnostic for acarines and separate them from the other arachnids.

The segmental designations of the prosomatal sterna of mites are shown below, but give only suppositional relationships and are not fully established from the embryogeny (Borner, 1903; Hughes, 1959):

Protosternum = cheliceral segment
Deutosternum = pedipalpal segment

Tritosternum = leg segment I
Tetrasternum = leg segment II
Pentasternum = leg segment III
Metasternum = leg segment IV

The first segment of the acarine body is the precheliceral segment (I), followed by the cheliceral somite (II) bearing the preoral chelicerae and the pedipalpal somite (III). The precheliceral, cheliceral, and pedipalpal segments thus comprise the anterior tagma—the gnathosoma, which is usually tubular in form with different accessory parts characteristic of each order (Hammen, 1968; Krantz, 1978). This region differs from the head of insects and from the prosoma of most arachnids because the coalesced

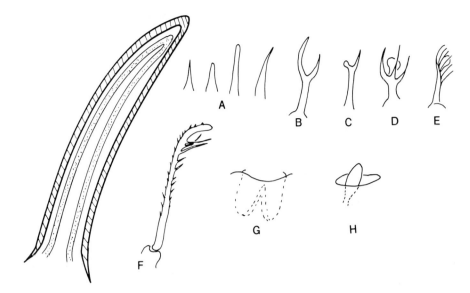

Figure 4-11. Microsolenidia or famuli, with a core-sheath of actinopiline (Grandjean, 1941): (A) common types; (B) *Heterochthonius*; (C) *Eniochthonius*; (D) *Parhypochthonius*; (E) Labidostommidae; (F) *Acaronychus*; (G) Caeculidae; (H) Ereynetidae.

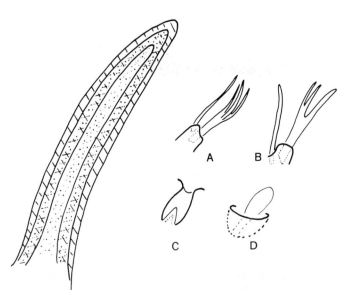

Figure 4-12. Eupathidia (=acanthoides, pseudoacanthoides of Grandjean (1946)) with a core of actinopiline around a protoplasmic core: (A) *Homocaligus*; (B) *Hypochthonius*; (C) *Caeculus*; (D) *Caligonella* (supracoxal).

central nervous system of mites, the synganglion, is podosomal in location between the legs. The origin of this tagma (gnathosoma) is obscure and difficult to derive from arachnid parts even though the region is distinctive for mites.

Inasmuch as no other arachnid has a head structure comparable to the capitulum (gnathosoma) of the Acarina, it becomes somewhat of a problem to understand how the acarine capitulum has been evolved from ordinary arachnid parts. (Snodgrass, 1965)

Segmentation

Arachnids vary in the number of total body segments, with a maximum of 13 opisthosomal segments in most; 17 in Solpugida, Phalangiida, Palpigradi (Savory, 1977); 13 metameres in mites (3 gnathosomal and 10 idiosomal) (Andre and Lamy, 1937); 14 segments in the primitive *Opilioacarus texanus* (Fig. 4-13A); and 16 somites in the endeostigmatan *Alycus roseum* (Hammen, 1970a).

Another opinion is that mites have no use for more than six somites. The primitive segmentation has been extensively suppressed, and posteriorly the segments failed to develop. This

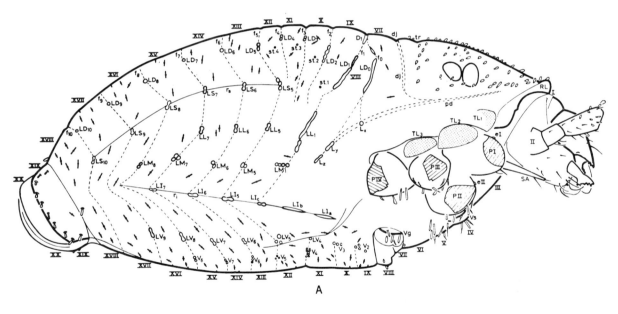

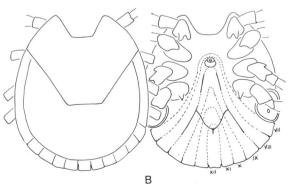

Figure 4-13. Segmentation of (A) *Opilioacarus texanus*, lateral view (after Hammen, 1966, 1970a); (B) segmentation of *Amblyomma paulopunctatum*, ventral and dorsal aspects (after Raw, 1957).

related to size and to the evolution of dwarfs to fill special niches in the environment. "Indeed, Nature's call for dwarfs will explain minuteness effected in other ways than by merocyclism, e.g., Opilioacaridae" (Raw, 1957).

Suggestions of segmentation survive in Opilioacaridae and Ixodidae, the hard ticks, with a subdivision of the posterior margin into 11 subequal parts—the festoons. If the median division is fused or stands alone, there is the suggestion of six segments by extending arched lines anteriorly (Fig. 4-13B). Such segmental remnants, unfortunately, are more definite in the Ixodidae than in the Aragaside, although caeca of the gut reflect about the same number in the latter (Raw, 1957).

It is impossible to establish with certainty the segmentation of the original acarine forms because the organization of the body has been so profoundly modified (Hammen, 1970a; Millot, 1949); metamerism of mites is strongly degraded (Andre, 1949; Evans et al., 1961). Their form has been shortened concurrent with a loss of segments resulting in conical, pyriform, or cylindrical body shapes. The immmatures may conserve some semblances of metamerism, but such is usually lost in the adults. The original body form of mites has adapted to many lifestyles. Mites are considered to be specialized arachnids that have lost many typical arachnoid characters. Future thorough study of morphology, embryogeny, organogenesis, and histological derivation of segmentation will help to establish homologies with the other arachnids.

Variations of Body Regions

Variations in the mite body are identified below (Fig. 4-14A):

    I. Gnathosoma
       Idiosoma
          Gamasida (Fig. 4-14B).

    II. Gnathosoma
       Prosoma*

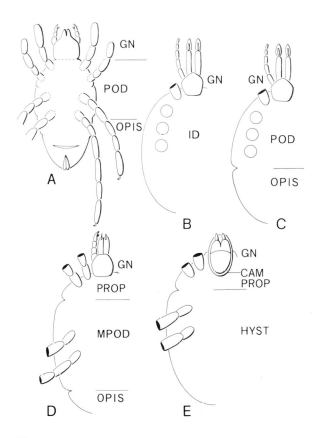

Figure 4-14. Diagrammatic tagmata: (A) acarine body regions; (B) gnathosoma, idiosoma (Gamasida, Ixodida); (C) gnathosoma, podosoma, opisthosoma (some Actinedida); (D) gnathosoma, propodosoma, metapodosoma, opisthosoma (some Actinedida); (E) proterosoma (camerostome), hysterosoma (Oribatida).

       Opisthosoma
         *Comparable to cephalothorax of arachnids, but somites are not fused into distinct regions (Fig. 4-14C).

    III. Gnathosoma
       Podosoma*
       Opisthosoma

         Halarachnidae; Nematalycidae; Demodicidae. *Dorsal parts of segments I–VII each with a pair of appendages (Fig. 4-14C; 4-15B).

IV. Gnathosoma
   Propodosoma*
   Metapodosoma**
   Opisthosoma

   Tenuipalpidae, Tydeidae, Pseudocheylidae, Syringophilidae, Heterocheylidae, Pomerantziidae, Stigmaeidae; Proctophyllodidae *Legs I, II; **Legs III, IV (Fig. 4-14D).

V. Proterosoma*
   Hysterosoma**

   *Gnathosoma in a camerostome, propodosoma). (**Metapodosoma and opisthosoma fused; in the Oribatida, dorsal aspect of propodosoma called prodorsum or aspis; dorsum of united metapodosoma and opisthosoma called notogaster or notaspis, if divided with lateral pleuraspis areas (Fig. 4-14E, F).

The gnathosoma and idiosoma are the principal regions in the Anactinotrichida. It is rare to find any division into other somal regions. In contrast, most of the Actinotrichida possess a well-defined sejugal suture (furrow) that divides the body into a propodosoma (or proterosoma) and hysterosoma. In a few instances there is a distinct (or slight) postpedal furrow that divides the body into a narrowed opisthosoma behind the last pair of legs (e.g., Demodicidae, *Demodex*; Halarachnidae, *Orthohalarachne attenuata*; Nematalycidae) (Fig. 4-15).

Dorsal Sclerites

Inasmuch as the primary segmentation is not discernable in most of the acarines, the sclerites and plates of the tergal surface vary in placement and size. Each order differs; some are diagnostic, others are not as significant. Tergal sclerites of free-living mites may be single, entire or divided; in parasitic mites the sclerites may be only slightly differentiated and used for muscle attachments (Evans et al., 1961; Hammen, 1963).

The Opilioacarida show relatively little sclerotization and a lack of plates or shields. Holothyrida have a heavily sclerotized exoskeleton (Hammen, 1961). Both immatures and adults of Gamasida have idiosomal shields that are useful in identifying stages and families. The larvae and protonymph frequently have an anterior podonotal shield and a posterior pygidial shield (Fig. 4-16B, C). Pairs of mesonotal scutellae occur on the idiosoma of some. Fusion of the pygidial and the mesonotal shields in the deutonymphs and adults results in an opisthonotal shield (Fig. 4-16C), which may remain separate and distinct, or may fuse partially or completely with the propodosomal shield to form the entire dorsal shield (Fig. 4-16D). Variations occur in some families (Fig. 4-17A–F).

Generally, ticks have two types of dorsal idiosomal sclerotization. Hard ticks (Ixodidae) exhibit a small, anterior, partial shield or scutum in the larvae, nymphs, and adult females; the scutum covers the entire dorsum in males. In soft ticks (Argasidae) larvae exhibit an irregular, middorsal shield, but nymphs and adults have wrinkled, mammillated, or folded integument devoid of major shields. *Nothoaspis reddeli* (Argasidae) is sclerotized more like ixodids; Nuttalliellidae are intermediate in sclerotization as well.

The dorsal sclerites are correlated with the flexibility and independent movement at the dorsosejugal suture between the propodosoma and hysterosoma (Actinotrichida). A single propodosomal shield may be accompanied by one or more medial hysterosomal shields or smaller hysterosomal sclerites (Fig. 4-17B, C, L). The extent of the shield formation and sclerotization usually increases during the developmental stages. Most Actinedida are weakly sclerotized with a finely "fingerprinted" surface of parallel ridges coursing over the cuticle in a uniform pattern. Relatively few families (e.g., Halacaridae, Caeculidae, Labidostommidae) exhibit any extensive shields. Water mites have a broad range of sclerotization and shield formation (Fig. 4-17N, O). The Tarsonemida show the

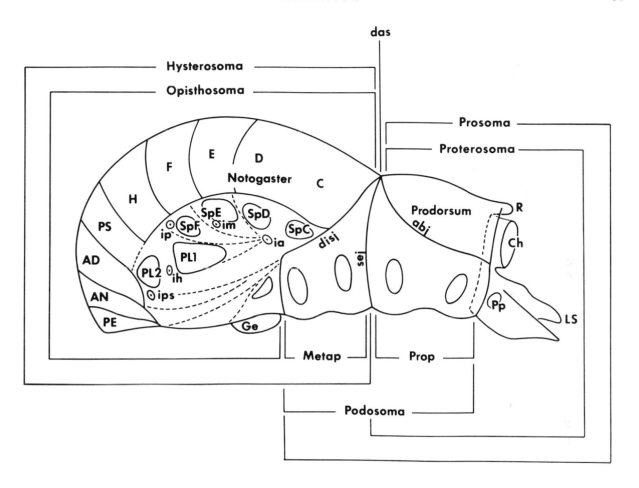

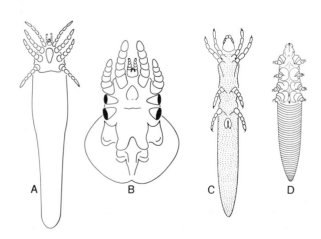

Figure 4-15. Diagrammatic tagmata of selected acarines: (A) Halarachnidae (Gamasida); (B) *Periglischrus* (bat ectoparasite; Gamasida); (C) Nematalycidae (Actinedida); (D) Demodicidae, follicle mite (Actinedida); (E) Oribatida, lateral view, gnathosoma (camerostome), proterosoma (gnathosoma, propodosoma), hysterosoma (metapodosoma, opisthosoma) (CAM, camerostome; GN, gnathosoma; HYST, hysterosoma; ID, idiosoma; MPD, metapodosoma; OPIS, opisthosoma; PROP, propodosoma; PROT, proterosoma); (F) diagrammatic lateral view of Oribatida showing prodorsum, notogaster, postulated segments (C-PE); sutures (abj, abjugal; das, dorsal abjugal; sej, sejugal; disj, disjugal; fissures *ia, ih, im, ip, ips*; PLl, PL2, pleural shields; R, rostrum; SpC, SpD, SpE, SpF, suprapleural shields; CH, chelicera; Pp, palp; LS, labial lobe) (drawing by Sonya Evans; after Moritz, 1976).

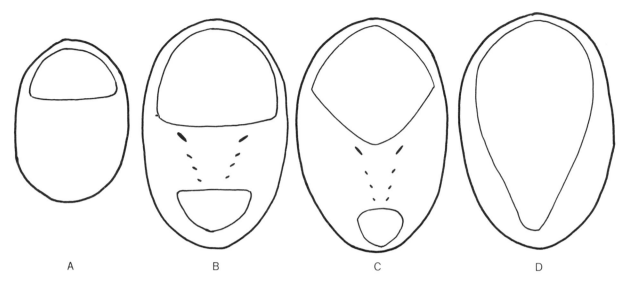

Figure 4-16. Diagrams of dorsal shields in Dermanyssidae (Gamasida): (A) larva; (B) protonymph; (C) deutonymph and adult (after Evans and Till, 1966).

idiosoma with overlapping shields postulated to indicate primary segmentation (Fig. 4-17G). Other groups (e.g., Cheyletidae) have variously sclerotized shields (Fig. 4-17K), yet also show the "fingerprint" of the integument.

In both free and parasitic mites of the Astigmata some sclerotization exists in the form of a prodorsal shield. Podonotal and opisthonotal shields are found in phoretic deutonymphs (Chaetodactyllidae) as well as in some adults (Proctophyllodidae, Dermoglyphidae) (Fig. 4-18).

The sclerotized beetle mites (Oribatida) have few separate, discernable sclerites. The dorsal covering of the propodosoma is called the prodorsum; the dorsal covering of the hysterosoma is called the notogaster and the latter may be divided by transverse or longitudinal sutures into median notaspis shields and lateral pleuraspis shields (Fig. 4-15). Some of the more primitive oribatids (Palaeacaroidea and immatures of higher oribatids) show sclerites restricted to a dorsal pygidial shield, to scutellae, or to both (Evans et al., 1961).

Ventral Sclerites

What has been described for dorsal (tergal) sclerites is also somewhat true for the ventral (sternal) sclerites, but with little evidence of segmentation in their placement or size. The legs are the most conspicuous evidences of segmentation and in many instances, the ventral sclerotization is found in association with the placement of the coxae and their apodematal attachments. Mites in the Anactinotrichida show movable coxae; coxae of the Actinotrichida are usually fused with a coxisternum, which leaves the trochanter as the first movable joint of the legs. The sclerites associated with reproductive openings of adult mites, particularly females, are pronounced in some orders, less conspicuous in others. The anal opening may exhibit a sclerite or shield (e.g., Gamasida, Fig. 4-19C–G).

Ventral sclerotization in Gamasida is taxonomically diagnostic for families of both free-living and parasitic mites. The sternites of the podosomal area include a tritosternum in the

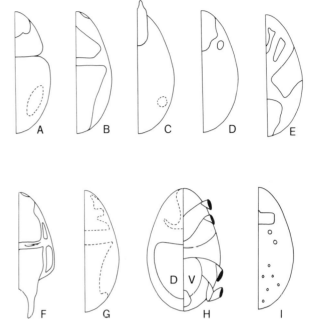

Figure 4-18. Dorsal sclerites of the Astigmata: (A) *Tyrophagus*; (B) *Chaetodactylus*; (C) *Dermatophagoides*; (D) *Caparinia*; (E) *Analges*; (F) Trouessartidae; (G) *Pterolichus*; (H) dorsum and venter of *Pneumocoptes*; (I) Sarcoptidae (drawings by Sonya Evans).

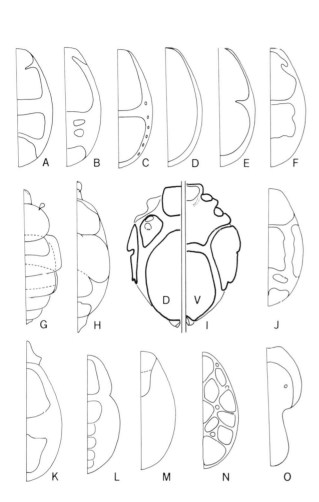

Figure 4-17. Variations in dorsal sclerites in Gamasida: (A) *Sejus* spp.; (B) Uropodellidae; (C) Microgyniidae; (D) Gamasellidae; (E) Otopheidomenidae; (F) Phytoseiidae (drawings by Sonya Evans; after Krantz, 1978); dorsal sclerites in selected Actinedida; (G) Tarsonemidae; (H) Pediculochelidae; (I) Halacaridae (dorsal and ventral); (J) Caeculidae; (K) Cheyletidae; (L) *Pomerantzia*; (M) *Leptus*; (N) *Sperchon*; (O) *Arrenurus* (drawings by Sonya Evans from various sources; water mites after Cook, 1974).

sternite between the first pair of legs and a prominent sternal shield. The sternal shield, its setation, and the female epigynial shield help to differentiate between families of these mites. The late Dr. Joseph Camin (personal communication) postulated that the ventral sclerotization in Gamasida did not follow segmentation, but that the forward movement of the sternal pores and sternal setae correlated with the fusion of the sternal shields and other skeletal changes (Fig. 4–19A, B). Variations in the types of genital sclerites of female gamasids are distinctive (Krantz, 1978) (Fig. 4-20). Males have no special genital sclerites, but have the genital orifice in the anterior or middle of the sternoventral shield.

The Ixodidae vary in the sclerotization of the

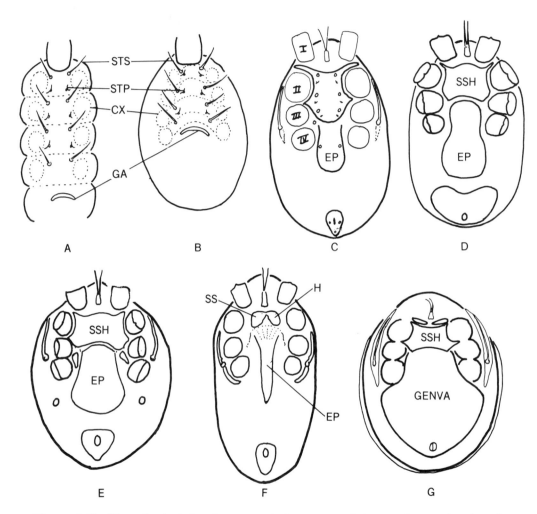

Figure 4-19. Hypothetical development of the venter of Gamasida (A) showing the sternal pores (STP) and sternal setae (STS) in relation to the genital aperture (GA) and coxae (CX) (after Camin); (B) diagrammatic representation of ventral sclerites in Gamasida. (C) *Hypoaspis*; (D) *Laelaps agilis*; (E) *Myonyssus gigas*; (F) *Ophionyssus natricis*; (G) *Ololaepas placentula* (SSH, sternal shield; EP, epigynial shield; GENVA, genitiventrianal shield) (after Evans and Till, 1966).

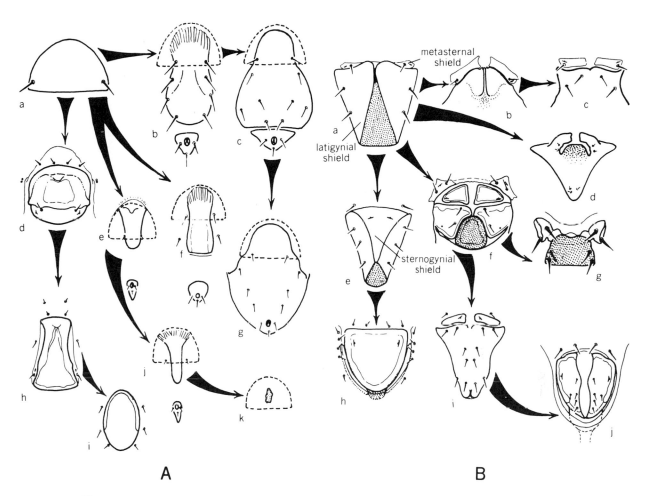

Figure 4-20. (A) Schematic of epigynial and genitiventral shield types in monogynaspids. Arrows indicate possible evolutionary trends in expansion or reduction of shield size from a simple truncate type (a); b, c, e, g, LAELAPIDAE; d, h, POLYASPIDAE; f, ASCIDAE; i, DIARTHROPHALLIDAE; j, MACRONYSSIDAE; k, RHINONYSSIDAE. (B) Schematiic of genital shield types and associated structures in the trigynaspids. Arrows indicate possible evolutionary trends in expansion, fusion, or loss of various components, with the diplogyniid genital apparatus (a) considered as the basic type; b, EUZERCONIDAE; c, CELAENOPSIDAE; d, NEOTENOGYNIIDAE; e, CERCOMEGISTIDAE; f, PARAMEGISTIDAE; g, PARANTENNULIDAE; h, FEDRIZZIIDAE; i, HOPLOMEGISTIDAE; j, MEGISTHANIDAE (after Krantz, 1978, with permission).

venter, which may be used for specific identification (Fig. 4-21). These sclerites are not related to segmentation, but represent different patterns in the ventral armor of these ticks, for example, festoons of some hard ticks (Fig. 4-13B) (Raw, 1957). Soft ticks (Argasidae) do not show ventral sclerites of any significance, for the venter resembles the dorsum in surface texture.

Relatively little evidence of ventral sclerotization exists in Actinedida, although some Tarsonemida are said to show remnants of primary segmentation in the coxisternum (Fig. 4-21E–G). Those families with dorsal sclerites (e.g.,

Halacaridae, Caeculidae, Labidostommidae) also show heavy ventral sclerotization and some shields, principally with the coxae and coxisternum. Water mites have shown heavily sclerotized coxal apodemes, coxisternal regions, and acetabulae (Fig. 4-21G).

Ventral sclerotization varies among the Astigmata, but varied coxal apodemes occur. The Analgidae, a family of feather mites, have large

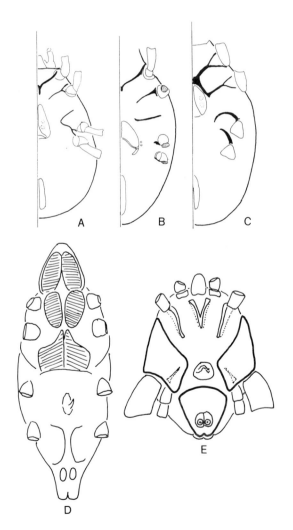

Figure 4-22. Schematics of venters of Astigmata: (A) Glycyphagidae; (B) Chortoglyphidae; (C) Carpoglyphidae; (D) Listrophoridae (after McDaniel, 1979); (E) Analgidae (after Krantz, 1978).

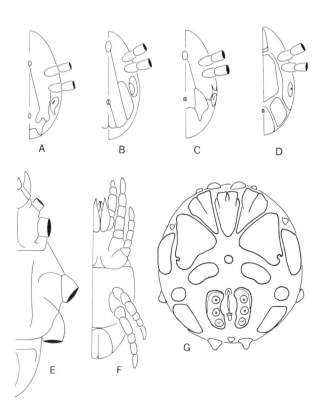

Figure 4-21. Ventral plates of Ixodida: (A) *Hyalomma*; (B) *Boophilus*; (C) *Rhipicephalus*; (D) *Ixodes* (drawing by Sonya Evans; after Hoogstraal). Representative venters in Actinedida: (E) *Steneotarsonemus* (after Beer, 1954); (F) Stigmaeidae; (G) *Hygrobatella papillata* (after Cook, 1974).

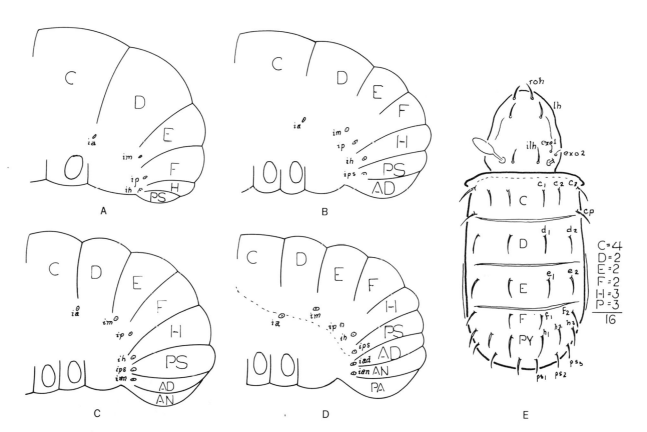

Figure 4-23. Diagrammatic representation of the postulated addition of segments in Oribatida: (A) larva; (B) protonymph; (C) deutonymph; (D) tritonymph and adult (after Grandjean, 1939b) (see Fig. 4-15). (E) Chaetotaxic arrangement of setae in Oribatid reflecting the postulated segmentation (*Haplochthonius* after Grandjean) (roh, rostral hair; lh, lamellar hair; ilh, interlamellar hair; exo1, exo2, exobothridial hairs).

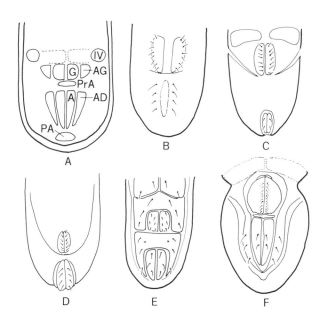

Figure 4-24. Ventral sclerotization of the Oribatida: (A) hypothetical arrangement of the sclerites (G, genital, AG, aggenital; PrA, preanal; A, anal; AD, adanal; PA, postanal); (B) agastry, Palaeacaroidea; (C) diagastry, Nanhermanniidae; (D) pseudo-diagastry, Eulohmanniidae; (E) schizogastry, Epilohmanniidae; (F) eugastry, Nothridae (after Grandjean, 1957).

coxal apodemes as well as distinct lateral and anal plates (Fig. 4-22E). Fur mites (e.g., Listrophoridae) have sclerotized coxal fields that are modified for grasping hairs (Fig. 4-22D).

Ventral sclerotization of the Oribatida is varied, yet diagnostic for the the principal groups within the order. The segmentation of the hysterosoma results in the addition of somites posteriorly (anamorphic) (Fig. 4-23). The conditions of ventral sclerotization extend from the primitive condition of *agastry* in the Palaeacaroidea, to *diagastry* in the Nanhermanniidae, *pseudodiagastry* in Eulohmanniidae, *schizogastry* in the Epilohmanniidae, and varied conditions of *eugastry* in the higher oribatids (Fig. 4-24). Most of the higher oribatids have a completely fused ventral plate

and variations in the sclerotization of the genital and anal apertures.

REFERENCES

Andre, M. 1949. Ordre des Acariens. *In*: Grassé, P.-P., ed. *Traité de Zoologie*. Masson, Paris, Vol. 6, pp. 794–892.

Andre, M. and E. Lamy. 1937. *Les idées actuelles sur la phylogénie des Acariens*. Published by Authors, Paris.

Anwarullah, M. 1963. Beitrage zur Morphologie und Anatomie einiger Tetranychiden. *Z. Angew. Zool.* 50(4):385–426.

Aoki, J. 1973. Oribatid mites from Iriomote-jima, the Southernmost island of Japan. *Mem. Nat. Sci. Mus.* 6:85–101.

Atyeo, T. and N.L. Braasch. 1966. The feather mite genus *Proctophyllodes*. *Bull. Univ. Nebr. State Mus.* 5:1–353.

Audy, J.R., F.J. Radovsky and P.H. Vercammen-Grandjean. 1972. Neosomy: Radical intrastadial metamorphosis associated with Arthropod symbioses. *J. Med. Entomol.* 9(6):487–494.

Balashov, Y.S. 1972. Bloodsucking ticks (Ixodoidea)—Vectors of Diseases of Man and Animals. [Transl. 500 (T500)]. Med. Zool. Dept. USNMRU No. 3. Cairo, Egypt, U.A.R.

Baker, E.W. and G.W. Wharton. 1952. *An Introduction to Acarology*. Macmillan, New York.

Baker, E.W., J.H. Camin, F. Cunliffe, T.A. Woolley and C.E. Yunker. 1958. Guide to the families of mites. Contrib. No. 3, Inst. Acarology, Zoology, Univ. Maryland. College Park, MD, pp. 1–230.

Beament, J. 1951. The structure and formation of the eggs of the fruit tree red spider mite *Metatetranychus ulmi* Koch. *Ann. Appl. Biol.* 38:1–24.

Beament, J. 1959. The water-proofing mechanism of Arthropods. *J. Exp. Biol.* 36:391–422.

Beament, J. 1968. The insect cuticle and membrane structure. *Br. Med. Bull.* 24:130–134.

Beer, R.E. 1954. A revision of the Tarsonemidae of the Western Hemisphere (Order; Acarina). Univ. Kansas Sci. Bull. 36, Pt. II, No. 16, pp. 1091–1387.

Beklemishev, W.N. 1969. *Principles of Comparative Anatomy of Invertebrates*, Vol. 2. Univ. of Chicago Press, Chicago, Illinois.

Borner, C. 1903. Die Beingliederung der Arthropoden. *Sitzungsber. Ges. Naturforsch. Freunde Berlin, Jahr. 1903.*

Brody, A.R. 1969. Comparative fine structure of Acarine Integument. *J. N. Y. Entomol. Soc.* 77(2):105–116.

Clarke, K.U. 1973. *The Biology of the Arthropoda.* Arnold, London.

Cloudsley-Thompson, J.L. 1958. *Spiders, Scorpions, Centipedes and Mites.* Pergamon, Oxford.

Condoulis, W. and M. Locke. 1966. The deposition of endocuticle in an insect, *Calpodes ethlius* (Lepidoptera, Hesperiidae). *J. Insect Physiol.* 12:311–323.

Cook, D. 1974. Water mite genera and subgenera. *Mem. Amer. Entomol. Inst.* No. 21, 1–860.

Dahl, F. 1911. Die Hörhaare (Trichobothriden) und das system der Spinnentiere. *Zool. Am.* 37:522–532.

Daniel, M. and J. Ludvic. 1954. Oberflachenstrukturen des scutums von *Ixodes ricinus. Z. Parasitenkd.* 16(3):241–252.

Douglas, J.R. 1943. Internal anatomy of *D. andersoni* Stiles. *Univ. Calif. Publ. Entomol.* 7(10):207–272.

Edney, E.B. 1977. *Water Balance in Land Arthropods.* Springer-Verlag, Berlin and New York.

Evans, G.O. and W.M. Till. 1966. Studies on the British Dermanyssidae (Acari: Mesostigmata). Part II. Classification. *Bull. Br. Mus. (Nat. Hist.) Zool.* 14(5):109–370.

Evans, G.O., J.G. Sheals and D. MacFarlane. 1961. *Terrestrial Acari of the British Isles,* Vol. 1. Adland & Son, Bartholomew Press, Dorking, England.

Gibbs, K.E. and F.O. Morrison. 1959. The cuticle of the two-spotted mite, *Tetranychus telarius* (Linn.) (Acarina Tetranychidae). *Can. J. Zool.* 37:633–637.

Gibbs, K.E. and F.O. Morrison. 1959. The cuticle of the two-spotted spider mite, *Tetranychus telarius* (Linnaeus). *Can. J. Zool.* 31:633–637.

Grandjean, F. 1935. Les poils et les organes sensitifs portés par les pattes et le palpe chez les Oribates. *Bull. Soc. Zool. Fr.* 60:6–39.

Grandjean, F. 1939a. La chaetotaxie des pattes chez les Acaridiae. *Bull. Soc. Zool. Fr.* 64:50–60.

Grandjean, F. 1939b. Les segments postlarvaire de l'hysterosoma. *Bull. Soc. Zool. Fr.* 64:2173–284.

Grandjean, F. 1941. Observation sur les Acariens (6ᵉ ser). *Bull. Mus. Nat. Hist. natur.* (1), 13:532–539.

Grandjean, F. 1946. Au sujet de l'organe de Claparede, des eupathidies multiples et des taenidies mandibulaires chez les Acariens actinochitineux. *Arch. Sci. Phys. Nat.* 28:63–87.

Grandjean, F. 1947. Observations sur les Acariens. *Bull. Mus. Natl. Hist. Nat.* (2) 18:337–344.

Grandjean, F. 1954. Etude sur les Palaeacaroides. *Mem. Mus. Natl. Hist. Nat., Ser. A* (Paris) 7:179–272.

Grandjean, F. 1957. L'Infracapitulum et la manducation chez Oribates et d'autres Acariens. *Ann. Sci. Nat. Zool.* 11:233–281.

Grandjean, F. 1968. The gnathosoma of Hermannia convexa and comparative remarks on its morphology in other mites. *Zool. Verh.* 94:1–45.

Grandjean, F. 1970a. La segmentation primitif des Acariens. *Acarologia* 5(3):443–454.

Grandjean, F. 1970b. Remarques générale sur la structure fondamentale du Gnathosoma. *Acarologia* 12(1):16–22.

Halaskova, V. 1969. Zerconidae of Czechoslovakia. *Acta Univ. Carol., Biol.,* pp. 175–352.

Haldane, J.B.S. 1925–1926. On being the right size. *Harper's Mag.* 152:424–427.

Haldane, J.B.S. 1956. On being the right size. *In:* Newman, J.R., ed. *The World of Mathematics.* Simon & Schuster, New York, Vol. 2, pp. 952–957.

Hammen, L. van der. 1961. Description of *Holothyrus grandjeani* nov. spec., and notes on the classification of mites. *Nova Guinea, Zool.* 9:173–194.

Hammen, L. van der. 1963. The addition of segments during postembryonic ontogenesis of the Actinotrichida (Acarida) and its importance for the recognition of the primary subdivision of the body and original segmentation. *Acarologia* 5(3):443–454.

Hammen, L. van der. 1966. Studies on Opilioacarida (Arachnida) I. Description of *Opilioacarus texanus* (Chamberlin & Mulaik) and revised classification of the genera. *Zool. Verh.,* pp. 1–88.

Hammen, L. van der. 1968. Introduction generale a la classification, la terminologie morphologique, l'ontogenes et l'evolution des Acariens. *Acarologia* 10(3):401–412.

Hammen, L. van der. 1969. Studies on Opilioacarida (Arachnida) III. *Opilioacarus platensis* Silvestri and *Adenacarus arabicus* (With). *Zool. Med.* 44(8):113–131.

Hammen, L. van der. 1970a. La segmentation primitive des Acariens. *Acarologia* 12(1):3–10.

Hammen, L. van der. 1970b. La segmentation des appendices chez les Acariens, remarques generales sur la structure fondamentale du gnathosoma. *Acarologia* 12(1):11–22.

Hammen, L. van der. 1973. Classification and phylogeny of mites. Proc. 3rd Int'l. Congress of Acarology, Prague. pp. 275–282.

Haupt, J. and Y. Coineau. 1975. Trichobathien und Tastborsten der milbe *Microcaeculus* (Acari: Prostigmata). *Z. Morphol. Tiere* 81:305–322.

Henneberry, T.J., J.R. Adams and G.E. Cantwell. 1965. Fine structure of the integument of the two-spotted spider mite, *Tetranychus telarius*. *Ann. Entomol. Soc. Am.* 58(4):532–535.

Hughes, A.M. 1961a. The mites of stored food. Ministry Ag., Fish and Food. Tech. Bull 9(NSI), pp. 5–198.

Hughes, A.M. 1961b. Terrestrial Acarina III. *In*: Fridriksson, A. and S.L. Tuxen. *The Zoology of Iceland*, Vol. 3, Part 57c, pp. 1–12.

Hughes, R.D. and C.G. Jackson. 1958. A review of the Anoetidae (Acari). *Virginia J. Sci.* 9(1):5–198.

Hughes, T.E. 1959. *Mites or the Acari*. Athlone, London.

Janzen, D.H. 1977. Why are there so many species of insects? *Proc. Int. Congr. Entomol., 15th, 1976*, pp. 84–94.

Jones, M.B. 1950. Penetration of host tissues by harvest mites. *Parasitology* 40:247–260.

Kaestner, A.M. 1968. *Invertebrate Zoology*, Vol. 2. Wiley (Interscience), New York.

Kennedy, C.H. 1927. The exoskeleton as a factor in limiting and directing the evolution of insects. *J. Morphol. Physiol.* 44(2):267–312.

Kethley, J.B. and R.E. Johnston. 1975. Resource teaching patterns in bird and mammal ectoparasites. *Misc. Publ. Entomol. Soc. Am.* 9:231–236.

Knulle, W. 1959. Morphologische und Entwicklungsg-geschichtliche untersuchungen zum phylogenetischen System der Acari: Acariformes. *Mitt. Zool. Mus. Berlin* 35(2):347–417.

Krantz, G.W. 1970. *A Manual of Acarology*. Oregon State Univ. Book Stores, Corvallis.

Krantz, G.W. 1978. *A Manual of Acarology*. Oregon State Univ. Book Stores, Corvallis.

Lees, A.D. 1952. The role of cuticle growth in the feeding process of ticks. *Proc. Zool. Soc. London* 121:759–772.

Legendre, R. 1968. La nomenclature anatomique chez les Acariens. *Acarologia* 10(3):413–417.

Locke, M. 1964. The structure and formation of the integument of insects. *In:* Rockstein, M., ed. *Physiology of Insecta.* Academic Press, New York, Vol. 3, pp. 379–470.

Locke, M. 1974. The structure and formation of the integument of insects. *In*: Rockstein, M., ed. *The Physiology of Insecta.* Academic Press, New York, Vol. 6, pp. 124–213.

Loeb, J. 1919. Natural death and duration of life. *Sci. Mon.* 9:578–585.

Marples, M.J. 1969. Life on the human skin. *Sci. Am.* 220(1):108–115.

McDaniel, B. 1979. *How to Know the Mites and Ticks.* Wm. C. Brown, Dubuque, IO, pp. 1–355.

Millot, J. 1949. Classe de Arachnides (Arachnida). I. *In:* Grassé, P.-P., ed. *Traité de Zoologie.* Masson, Paris, Vol. 6, pp. 263–319.

Mitchell, R. and M. Nadchatram. 1969. Schizeckenosy: The substitute for defecation in chigger mites. *J. Nat. Hist.* 3:121–124.

Moritz, M. 1976. Revision der europäischen Gattungen und Arten der Familie Brachichthoniidae. *Mitt zool. Mus. Berlin* 52(3):227–319.

Nadchatram, M. 1970. Correlation of habitat, environment and color of chiggers, and their potential significance in epidemiology of scrub typhus in Malaya (Prostigmata: Trombiculidae). *J. Med. Ent.* 7(2):131–144.

Nathanson, M.E. 1967. Comparative fine structure of sclerotized and unsclerotized integument of the rabbit tick *Haemaphysalis leporispalustris*. *Ann. Entomol. Soc. Am.* 60(6):1125–1135.

Nathanson, M.E. 1970. Changes in the fine structure of the integument of the rabbit tick *Haemaphysalis leporispalustris* which occur during feeding (Acari: Ixodides, Ixodidae). *Ann. Entomol. Soc. Am.* 63:1768–1774.

Neville, A.C. 1975. *Biology of the Arthropod Cuticle. Zoophysiology and Ecology*, Vol. 4/5. Springer-Verlag, Berlin and New York.

Nutting, W.B. 1976. Hair follicle mites (Acari) of man. *Int. J. Dermatol.* 15(2):79–98.

Pauly, F. 1948. Zur Biologie einiger Belbiden und zur Funktion uber pseudostigmatischen Organe.

Zool. Jahrb. 84:328 (cited by Tarman, 1961; Krantz, 1978).

Raw, F. 1957. Origin of the Chelicerates. *J. Paleontol.* 31(1):139–192.

Reuter, E.R. 1909. Zur Morphologie und Ontongenie der Acariden. Mit Besonderer Berucksigchtigung von *Pediculopsis graminum*. *Acta Soc. Sci. Fenn.* 36(4):1–288.

Richards, A.G. 1951. *The Integument of Arthropods*. Minnesota Univ. Press, Minneapolis.

Robaux, P. 1974. Recherches sur le développement et la biologie des Acariens "Thrombidiidae." *Mem. Mus. Natl. Hist. Nat. Ser. A (Paris)* 85:1–186.

Rosicky, B. 1973. Acarology and its practical importance. *Proc. Int. Congr. Acarol., 3rd, 1971,* pp. 21–32.

Savory, T. 1968. Hidden lives. *Sci. Am.* 219(1):108–115.

Savory, T. 1977. *Arachnida,* 2nd ed. Academic Press, London.

Sharov, A.G. 1966. *Basic Arthropodan Stock,* Int. Ser. Monogr. Pure Appl. Biol. Pergamon, Oxford.

Snodgrass, R.E. 1965. *A Textbook of Arthropod Anatomy*. Hafner, New York.

Tarman, K. 1961. Uber Trichobothrien und Augen bei Oribatei. *Zool. Anz.* 167:51–58.

Thompson, D.W. 1942. *Growth and Form*. Macmillan, New York.

Thor, S. 1903. Recherches sur l'anatomie comparée des Acariens prostigmatiques. *Ann. Sci. Nat. Zool.* 19:146.

Vitzthum, H.G. 1931. Acari-Milben. Metzger & Wittig, Leipzig.

Vogel, B.E. 1975. Morphologische Untersuchungen uber Integument und Hautung sowie Entwicklung des cierten Beinpaares bei Ornithodoros moubata (Murray) 1877 (Acarina, Ixodoidea, Argasidae). Inaugural dissertation, Organization Kolb, Basel.

Wallace, D.R.J. 1960. Observations on hypopus development in the Acarina. *J. Insect Physiol.* 5:216–229.

Wharton, G.W., W. Parrish and D.E. Johnston. 1968. Observations on the fine structure of the cuticle of the spiny rat mite, *Laelaps echidnina*. *Acarologia* 10(2):206–214.

Wigglesworth, V.B. 1965. *The Principles of Insect Physiology*. Methuen, London.

Zakhvatkin, A.A. 1952. The division of the Acarina into orders and their position in the system of the Chelicerata. *Parazitol. (Sb.)* 14:5–46.

5

Gnathosoma

The gnathosoma is the most distinctive and special feature of mites because no other arachnid has a comparable tagma. It is difficult to understand how the gnathosoma evolved from ordinary arachnid parts (Snodgrass, 1965). It may be likened partially to the head end (prosoma) of some chelicerates, because it does have the mouthparts attached to it, but its other features preclude homologous comparisons. Also, because the synganglion ("brain") of mites is located posteriorly in the idiosoma, and because ocelli, if present, are found dorsolaterally on the idiosoma, the gnathosoma cannot be called a head, even though it is headlike (e.g., Ixodida, Erythraeidae, water mites). It is termed a capitulum in ticks (e.g., Snodgrass, 1965; Balashov, 1972).

The attachment of the chelicerae and pedipalps to the gnathosoma and their structural contributions to its tubular form make this tagma principally a region of the body where the food is handled and passes into the digestive tract. Technically speaking, it is thus an external part of the overall digestive system as well as a distinctive body region (Figs. 5-1, 5-2).

Identification of segmentation in the gnathosoma is impossible at present because little embryological evidence for homologies of the various parts exists. Some believe it to be composed of three fused segments: a precheliceral segment (I) devoid of appendages; a cheliceral segment (II) bearing the paired chelicerae; and a pedipalpal segment (III) with the pair of pedipalps. The mouth is located between segments II and III (Evans et al., 1961). In another view the somite of the cheliceral frame (I) and the somite of the pedipalpal frame (II) are part of the prosoma rather than in a distinct tagma (Hammen, 1961). I prefer to consider the gnathosoma as a distinctive body region without a distinguishable homolog in other arachnids.

The sterna of the gnathosoma have different parts (Borner, 1903), namely, the prosternum with the chelicerae and the deutosternum with the pedipalps. The deutosternum may also be

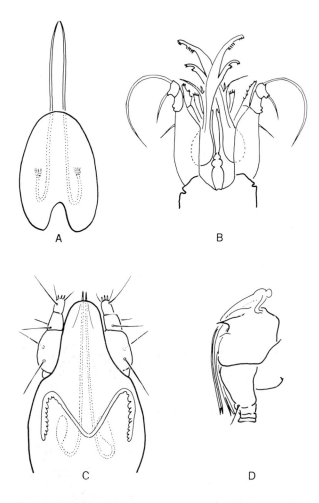

Figure 5-1. (A) Stylophore and styletiform chelicerae Tetranychidae; (B) gnathosoma of Histiostomatidae (Anoetidae); (C) gnathosoma of Syringophilidae with stylet chelicerae and surface peritreme; (D) lateral view of gnathosoma of Eriophyiidae.

extended forward to become part of the hypostome. Tergites are indistinguishable, but the transverse rows of setae suggest metamerism (Reuter, 1909), but unreliable because of variations of setal positions (Hughes, 1959).

The gnathosomal features of *Opilioacarus segmentatus* are used as a basis for describing the primitive and generalized features of the gnathosoma (Snodgrass, 1965; Grandjean, 1936a, 1957a,b). Compared with other arach-

nids the gnathosoma is specialized even in primitive mites (Snodgrass, 1965; Grandjean, 1936; Hammen, 1970c). Positional relationships of the mouthparts are used to differentiate primitive and advanced Anactinotrichida and Actinotrichida. Theoretically, two main arrangements of the mouth, mouthparts, and gnathosoma occur. The acarine gnathosoma has conserved some primitive arachnoid characters, but primitive mites (*Opilioacarus, Alycus*) in two different groups have retained different arachnid features (Hammen, 1970c).

The principal basal part (basis capituli) of the gnathosoma in the Anactinotrichida results from the fusion of the pedipalpal coxae and possibly the sternites and tergites of other gnathosomal segments. This tubular fusion takes the form of a sclerotized ring enclosing the chelicerae between the palps (Figs. 5-2A). Dorsally, the anterior extensions of the roof become the epistome (=tectum capituli). Ventrally, the extensions of the pedipalpal coxae meet in the midline to form the hypostome and deuterostome of gamasids and others. The endites of the palpal coxae become the hypostome and anteriorly may be divided into internal malae and corniculi (=external malae) (Fig. 5-2B). The corniculi may be stalked or sessile (Evans, 1957) (Fig. 5-2C, D). In gamasid mites the ventral surface of the hypostome and basis capituli (=gnathosomal base) bears four pairs of setae (Evans and Till, 1965). Three pairs of hypostomal setae are anterior; one pair of capitular setae is located on the hypostomal base (Fig. 5-2B).

Hypostomal development is specialized in the ixodids in the form of a tonguelike anterior projection with retrorse (recurved) teeth on the ventral surface.

The ventral aspects of the gnathosoma are different from the dorsal. Authors call this ventral region by different names, that is, infracapitulum, subcapitulum, and most recently hypognathosoma (Brody et al., 1972, for *Dermatophagoides farinae*).

The musculature of the gnathosoma is highly variable among the orders of mites and

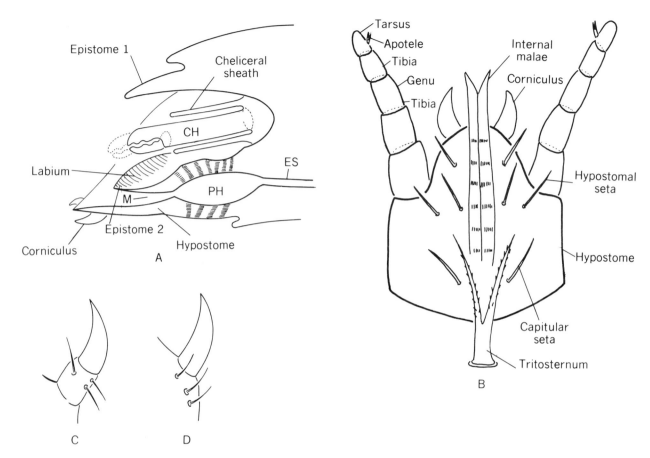

Figure 5-2. Schematics of general mite gnathosoma: (A) saggital view showing chelicerae, cheliceral sheaths, hypostome, pharynx, pharyngeal muscles epistome l, epistome
2, corniculi (after Snodgrass); (B) ventral view of gamasid gnathosoma showing hyposotome, hypostomal setae, capitular setae, deutosternum; (C) stalked corniculus; (D) sessile corniculus.

is of little use in determining metamerism. The
gnathosomal musculature of a generalized
arachnid (Fig. 5-3A) differs from the muscles in
a generalized gamasine mite (Fig. 5-3B), in a
specialized form (*Sauronyssus saurarum*) (Fig.
5-3C), and in an ixodid (*Ornithodoros papillipes*) (Fig. 5-3D). Flexor muscles of the gnathosoma in the hard ticks originate on the posterior edge of the scutum. Levator muscles
originate on the hind part of the scutum as well.
The gnathosomal musculature of the bee tra-

cheal parasite, *Acarapus woodi* (Tarsonemidae), shows a specialized and peculiar arrangement. There is an intricate lever system for the
extension of the stylets in this mite (Fig. 5-3E).
Gnathosomal musculature of the Oribatida is
different from that of the Astigmata although
basic similarities exist.

Because the homologies of the various parts
of the gnathosoma have not been established
between orders of mites, some confusion persists regarding the structures and terminology.

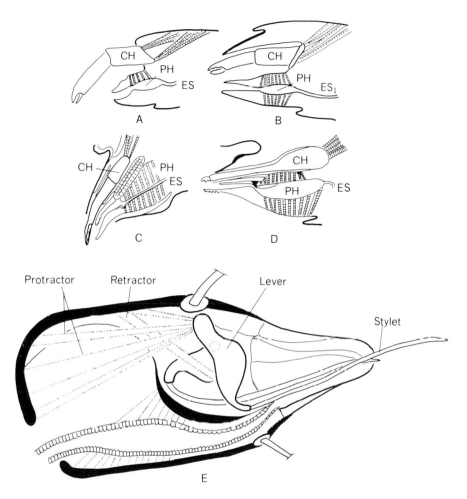

Figure 5-3. Musculature of the gnathosoma: (A) generalized arachnid; (B) generalized gamasid; (C) *Sauronyssus sauratum*; (D) *Ornithodoros papillipes* (after Snodgrass and Balashov) (CH, chelicerae; PH, pharynx; ES, esophagus); (E) schematic drawing of the gnathosoma of *Acarapis woodi*, bee trachea mite, showing the lever system, protractor and retractor muscles, stylets, and pharynx (after Hirschfelder and Sachs, 1952).

SYSTEMATIC REVIEW

What follows are some general descriptions of the characteristic gnathosomas in the different orders of mites. Many variations of this tagma, types of chelicerae, and pedipalps occur; however, an overview of the varied arrangements in the orders will clarify how diversified this single characteristic of mites has become and why it is so important in classification.

Opilioacarida. The gnathosoma of primitive mites in the Opilioacarida and Holothyrida is figured by numerous authors (Grandjean, 1936a; Evans et al., 1961; Hammen, 1970a; Krantz, 1978). Combinations of figures from these sources (and others for advanced orders) are used in the descriptions and illustrations that follow.

The gnathosoma of *Opilioacarus texanus* is cone-shaped, broadly based at the idiosoma, inclined ventrally, and probably extrusible (Hammen, 1966) and very mobile (Grandjean, 1936a). It is formed of the cheliceral frame and the infracapitulum (including the palpal bases). Dorsally the large chelicerae and cheliceral sheaths obscure the greater part of the structures beneath. The cheliceral sheaths pass proximally into the short, flexible epistome (=tectum capituli) (Hammen, 1966) (Fig. 5-4A).

Ventrally (Fig. 5-4B) the labrum is the anterior – mesal part of the gnathosoma bordered posterolaterally by rounded lateral lips. On the medial surface of the infracapitulum a distinct subcapitular groove lies between the lateral lips (Hammen, 1966; Grandjean, 1936b). This groove passes between the extensions of the tritosternum (sternapophysis of Hammen, 1970a) (Fig. 5-4B). Fused, paired plates on either side make up the floor of the infracapitulum, the anterior gena (G), and the posterior mentum (H). Absence of a genolabial suture makes the infracapitulum anarthric (Grandjean, 1957a). The anterolateral face of each gena has an atelebasic rutellum (RU) in association with With's organ (OW), thought to be a secondary rutellum (Fig. 5-4A). Both are considered to be food handlers. (This type of rutellum also occurs in some primitive oribatid mites, such as *Parhypochthonius, Camisia, Hermannia.*) The exact shape of the Organ of With is not always determinable because, as Hammen (1966) explains, it expands enormously when placed in lactic acid.

The infracapitular (paralabial and circumbuccal) setae are specifically placed (Fig. 5-4). Paralabial setae appear to be possible homologs of the corniculi (external malae) of gamasid mites. Sinuous salivary glands empty on the dorsum of the infracapitulum (=cervix of Hammen, 1966).

Holothyrida. One of the other primitive groups of mites, the Holothyrida, has a gnathosoma that resembles the same tagma of *Opilioacarus*, but shows a reduced cheliceral frame and the cheliceral sheaths attached to the infracapitulum (Hammen, 1961, 1965, 1970c). The gnathosoma of Holothyrida is in a camerostome (cavelike recess) into which the mouthparts and adjacent structures may withdraw. The dorsal vault above the chelicerae forms a small tectum (*coniculus* of Hammen, 1970c) (Fig. 5-4B). The cheliceral tectum is the most primitive complete tectum known and is considered the precursor of the gamasid tectum (Hammen, 1965). The infracapitulum is homologous with the ventral part of segment II (Hammen, 1961). Because there is no labiogenal articulation, the term mentum (H) applies to the whole posteroventral face of the gnathosoma (Fig. 5-5B). Hammen (1970c) introduced another term, *cervix*, for the dorsal surface of the infracapitulum because of the variations and confusion of terms applied to this surface.)

Gamasida. The gnathosoma of gamasid Parasitiformes is tubular; the cheliceral frames form the dorsal vault of segment I and the subcapitulum (segment II) (Hammen, 1964). The coxae of the pedipalps are enlarged to form the primary part of the basis capitulum (Hughes, 1959) (Fig. 5-2B). The external coxal walls become the lateral aspects of the tubular gnathosome and fuse dorsally to form the epistome (=tectum capituli) (rostrum or coniculus in Holothyrida) over the chelicerae. Its shape varies sufficiently in different families to be of taxonomic value among gamasid mites (Fig. 5-6).

Ventrally, the palpal coxae do not meet medially. The area of separation probably represents the deutosternum, or capitular groove, with transverse denticles on the floor of the groove. At the level of the mouth, the mesial walls of the coxae fuse to form the subcheliceral plate or shelf (Hughes, 1959), which Snodgrass

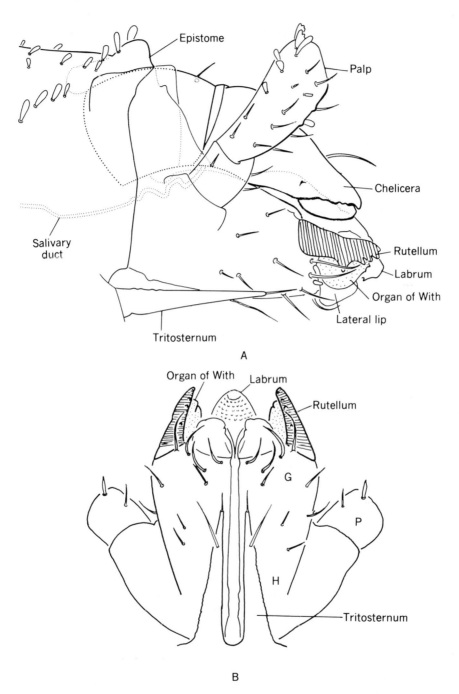

Figure 5-4. Gnathosoma of *Opilioacarus texanus*: (A) lateral view showing palps, chelicerae, rutellum (shaded), Organ of With (stippled) and salivary duct; (B) ventral view of same showing tritosternum (after Hammen, 1969).

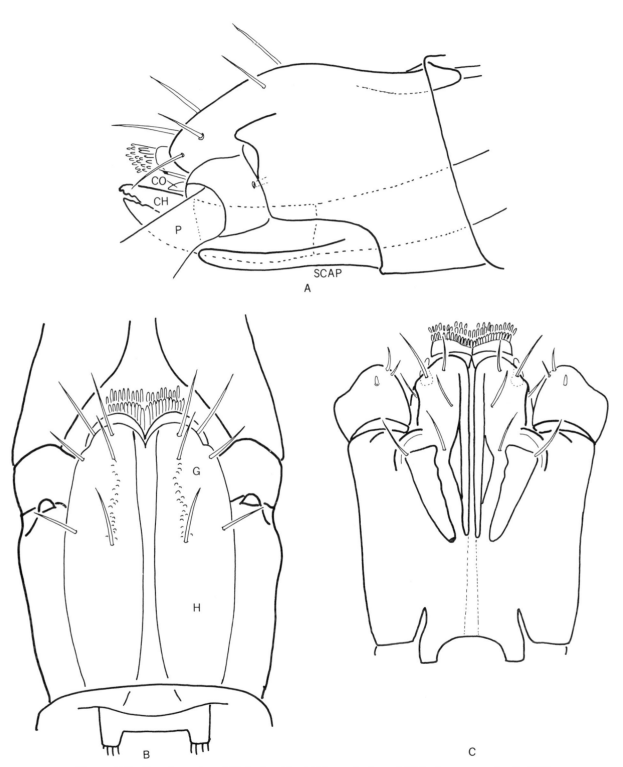

Figure 5-5. Gnathosoma of Holothyrida: (A) lateral view of *Holothyrus coccinella* (CH, chelicerae; P, palp; CO, corniculus; L, labrum; SCAP, subcapitular apodeme); (B) ventral of same; (C) ventral view of male *H. grandjeani* (after Hammen, 1961).

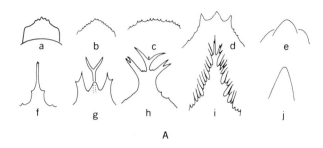

A

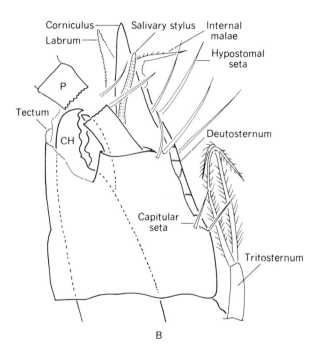

B

Figure 5-6. (A) Tectum capituli = epistome in Gamasida: a, *Parasitus* sp.; b, *Hypoaspis* sp.; c, *Hypoaspis praesternalis*; d, *Pseudolaelaps*; e, *Holothyrus coccinella*; f, *Geholaspis longisupinosi*; g, *Veigaia transisalae*; h, *Macrocheles tridentinus*; i, *Haemogamasus ambulans*; j, *Dermanyssus gallinae* (after Evans and Till; Hammen); (B) gnathosoma of *Glyptholaspis confusa*, lateral view showing corniculi, chelicerae, palps, salivary styli, and tritosternum.

(1965) calls the real epistome. This plate has some of the dorsal dilator muscles of the pharynx attached. Dorsal to the subcheliceral plate the walls form a thin cheliceral sheath around the base of each chelicera. The sheaths are thin and flexible, and are most pronounced in mites with protrusible chelicerae.

On the venter of the gnathosoma a median projection extends forward divided by a hypostomal groove. Two main parts comprise this area: the basis capitulum, with a single pair of capitular setae, and the anterior hypostome with three pairs of hypostomal setae anterior to the capitular groove. Larval mites exhibit only two pairs; the third pair is added in the protonymphal stage (Evans and Till, 1965). Two anterior extensions of the hypostome occur: the pair of mesial projections are the internal malae; the pair of lateral projections are the corniculi (=external malae, or corniculus maxillarus). In obligatory parasites the internal malae and corniculi may be elongated to form a preoral trough. While the corniculi are thought to be analagous to the rutella of oribatid mites, specializations do occur, as discussed by Evans and Till (1965) (Fig. 5-2C, D).

Salivary styli (siphunculi) are found lateral and ventral to the chelicerae. They carry the ducts of the salivary glands and tend to be rather large, sclerotized, and extended in parasitic forms (Fig. 5-6B).

Many other examples of gamasid gnathosomes are illustrated in the literature (e.g., Hammen, 1964; Evans, 1957; Furman, 1959; Evans and Till, 1965; Treat, 1975).

Ixodida. Soft ticks, such as *Argas boueti* have a folded, membranous collar at the base of the capitulum that may be extended for feeding (Roshdy, 1962) (Fig. 5-7A).

The mouthparts of the hard ticks represent some of the most specialized features among parasitic mites and the most perfectly adapted for bloodfeeding. Even with this specialization, the basic structure of arachnid mouthparts can be identified. Much has been written about the capitulum and these mouthparts (see Gregson, 1967).

The sclerotization of the dorsal parts of the capitulum forms the short rooflike epistome or tectum capituli, which fuses laterally with the pedipalpal bases and forms the ring of the basis capitulum. Characteristic of ticks, the hypostome is specialized and tonguelike, with many ventral retrorse (recurved) teeth, and forms the vental floor of the buccal cavity. The labrum lies over the mouth opening and attaches posteriorly to the subcheliceral plate. The subcheliceral plate may also be called epistome and is homologous with the epistome of other arachnids. From it is formed the intracoxal bridge to which the dilator muscles of the phayrnx attach (Balashov, 1972) (Figs. 5-3; 5-7C).

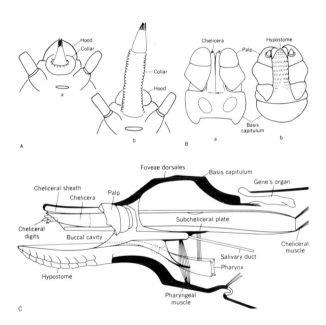

Figure 5-7. Gnathosoma of (A) soft ticks: a, collar of *Ornithodoros boueti* retracted; b, same extended (after Roshdy, 1962); (B) capitulum in Ixodidae: a, dorsal view of *Dermacentor variabilis* showing palps, chelicerae, areae porosae; b, ventral view showing hypostome with retrorse teeth and palps (after Arthur and Snow, 1967); (C) cutaway diagram of gnathosoma of *D. andersoni* showing the cheliceral digits, chelicerae, cheliceral sheath, subcheliceral plate, buccal cavity, toothed hypostome, salivary duct, pharynx, and pharyngeal muscles (after Gregson, 1967; drawing by Sonya Evans).

Cheliceral sheaths of *D. variabilis* are tubular extensions of the gnathosomal cuticle. The dorsal walls are weakly sclerotized; the ventral walls are membranous. A subcheliceral plate (=epistome) may be present in argasid ticks and generalized acarines (Snodgrass, 1965).

Actinedida. The greatest number of variations of the gnathosoma occur in this assemblage of mites. In the heterostigmatid family Pyemotidae the gnathosoma is almost cuboidal in shape with stubby palps and styletiform chelicerae (Fig. 5-8A). Tarsonemids have tiny stylets and adpressed palps against a square gnathosome. Snout mites (Bdellidae and Cunaxidae) and lizard mites (Pterygosomidae) have an elongated gnathosome with palps and chelicerae to match (Fig. 5-8B, C). Syringophilidae exhibit an elongated gnathosome with styletlike chelicerae (Fig. 5-1C). The gnathosomes of Psorergatidae and Harpyrhynchidae are square in outline (Fig. 5-8E).

The tiny gnathosoma of gall, rust, or blister mites (Eriophyidae) is directed ventrally. The tiny stylets are correlated with the other minute features of the tagma (Jeppson et al., 1975) (Fig. 5-1D).

The differing features among the families of actinedid mites include the palpal endites, hypostome, and chelicerae, which may be specialized for different types of feeding. In the spider mites (Tetranychidae) the epistome is small. Beneath the epistome is a fused structure called the stylophore, comprising the cheliceral bases which bear the styletlike chelicerae. The chelicerae lie in the grooved labrum and by muscular contraction may be protracted to pierce the plant tissue for release of the liquid food upon which these mites subsist (Kaestner, 1968; Blauvelt, 1945) (Fig. 5-1A; 5-8H).

The gnathosoma of *Demodex folliculorum* is extremely reduced in size (Fig. 5-8E). It appears short and stubby, with the palps pressed at the anterolateral corners on either side of the coxal endites and the stylophore. Each palp is three-segmented with a large coxa and two smaller, distal, free segments. The coxal en-

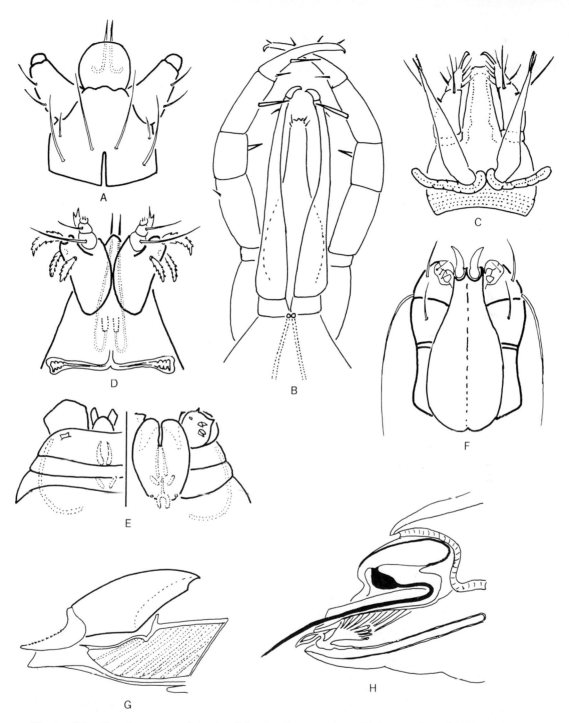

Figure 5-8. Gnathosomas of Actinedida: (A) Pyemotidae; (B) Cunaxidae; (C) Pterygo-somidae; (D) Harypyrhynchidae; (E) *Demodex folliculorum*, dorsal and ventral aspects; (F) larval *Piona* (Hyrdrachnida) (after Prasad and Cook, 1972); (G) chelicerae, pharyngeal muscles of common chigger (after Jones); (H) schematic diagram of tetranychid mouth parts showing stylets (blackened), pharyngeal muscles, pharynx, and trachea (after Blauvelt, 1945; Kaestner, 1968).

dites project anteriorly between the distal segments of the palps. The stylophore is cone-shaped and extends anterodorsally between the palps, somewhat ankylosed with the palpal coxae. The hypostome (the ventral surface of the coxal endites) is flattened and the preoral opening is slitlike. The tapered, curved stylets project downward into the food canal, ending in fine points that may become fused. Apparently the chelicerae lack fixed digits (Desch and Nutting, 1977).

Fifteen different characteristics distinguish 28 genera of water mites (Prasad and Cook, 1972). There is little difference, however, between the gnathosomas of the water mites and the terrestrial Trombiculidae and Trombidiidae. Both primitive and specialized conditions occur (Mitchell, 1962).

Because of the confusion in terminology related to the gnathosoma, the capitulum and the buccal capsule of chigger mites (Trombiculidae), Mitchell (1962) uses the word capitulum for the discrete functional entity that has the chelicerae and palps and a sigmoid piece (an S-shaped sclerite). This compound sclerite supplies both the skeletal support for the appendages and their muscles and for the pharynx and pharyngeal muscles as well. The sclerotized capsule is conical in shape with a troughlike floor. Independent movements of the chelicerae and palps are regulated by muscles that originate in the capsule. Elevator and depressor muscles occur in both the palps and the chelicerae (Mitchell, 1962). The roof of the capitulum encloses the chelicerae. An *atrium*, a middle *subcheliceral space,* and *oral cavity* are identified (Brown, 1952), but little use is made of these distinctions (Mitchell, 1962).

The mouth leads into the pharynx from which dorsal dilator muscles extend to the roof to the epistomal apodeme and eight flexor muscles are interspersed laterally between the dilator muscles (Mitchell, 1962).

Astigmata. Stored products mites (Acaroidea) show a simple gnathosoma with larger, exposed chelicerae and short, adpressed four-seg-

mented palps. Ectoparasitic and endoparasitic mites (e.g., Analgidae, Proctophyllodidae, Sarcoptidae), have mouthparts adapted to the varied food situations.

The gnathosoma of many free-living cheese mites (Acaridia) is only partially retractible into the body (Hughes, 1961a). Living mites hold this tagma at an angle to the idiosoma, which places the chelicerae and palps close to the food. Cheliceral muscles enable independent movement of each chelicera. The hypostomal part of the infracapitulum is broadly triangular, formed from the pedipalpal coxae and extended forward at the terminal external malae between the palps. A rutellum is absent. Hypostomal hairs are reduced in number. The labrum (=true epistome) is hollow and forms the roof over the mouth. The posterior extensions of the epistome are continuous with the subcheliceral plate and the lateral walls of the gnathosoma (Fig. 5-9).

In the feather mite genus *Proctophyllodes* (generally ectoparasitic on Passeriformes) the gnathosoma comprises the subcapitulum, the chelicerae, and the palps. The cheliceral fixed digit is short with a subterminal, bifurcate tooth. The single-toothed movable digit is strongly curved and operates in a vertical plane by levator and depressor muscles. Adpression of these digits forms a small pincer. A paraxial cheliceral hood extends anteriorly to the level of the cheliceral teeth. Paraxial setae (chx) and paraxial spurs (spur) insert posterior to the base of the fixed digit. An apophysis (apo) is also present (Fig. 5-9E).

The subcapitulum in this genus is rather short and broad. It has transverse ridges on the ventral surface. A distinctive feature is the prominent pseudorutellar process (psrp) and the pseudorutella (psr). The hypostome is a simple, triangular lobe (hyp) and the labrum (lb) is smooth. The palp has two segments or podomeres. Each palp has three setae and a single solenidion (Fig. 5-9).

Specialization of the gnathosoma is exhibited in certain families. The Cytoditidae are found in the respiratory passages of birds and

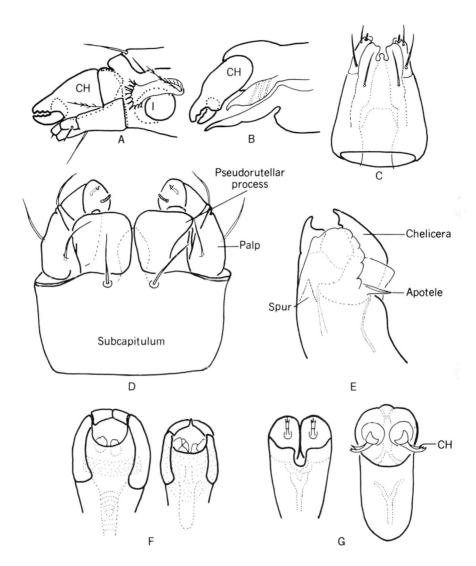

Figure 5-9. Gnathosomas of Astigmata: (A) lateral view of *Acarus siro* (after Evans et al., 1961); (B) schematic of acarid gnathosoma showing chelicerae, pharynx, and pharyngeal muscles (after Hughes, 1959); (C) ventral view of *Acarus siro* (after Evans et al., 1961); (D) ventral view of *Proctophyllodes quadrisetosus* showing palps, pseudorutella, pseudorutellar process; (E) chelicerae of same showing cheliceral hood, apophysis, and antaxial spur; (F) ventral view of *Cytodites* sucking gnathosoma without chelicera; (G) *Cytonyssus* ventral view of gnathosoma with chelicerae.

the endoparasitic habits are reflected in the gnathosoma present, as well as in other features. The gnathosoma is of two general types: (1) relatively large and transformed into a sucking mouth without chelicerae (*Cytodites*; Fig. 5-9F); and (2) very small with small tridentate, strongly curved chelicerae directed ventrally (*Cytonyssus*; Fig. 5-9G) (Fain and Bafort, 1964). It can be conjectured that the types of food and the feeding mechanisms for each of these genera are different, but details are not known.

Oribatida. A pair of rutella is distinctive for mouthparts of oribatids. These assumedly resulted from the hypertrophication of a pair of seta on the infracapitulum (Grandjean, 1957a) and facilitate feeding. They vary in form but are mainly cylindrical with narrow bases in the "lower" oribatids. In the "higher" oribatids the rutella are more massive and help to form a more distinctive preoral tube.

> It is probable that the ability to consume solid food particles is restricted to mites having rutella and that those without such structures must either feed on liquids or practise external digestion. (Evans et al., 1961)

The structure of the infracapitulum of these mites represents the palp-bearing epimere and shows the peculiar specializations that differentiate the oribatids from most other mites (Woolley, 1969). It is assumed that the infracapitulum is derived from the coxisternum of the primitive pedipalpal segment and that the coxisternal tergite is incorporated with the cheliceral segment into the proterosoma of the primitive mites (Grandjean, 1957a).

Grandjean also developed a partially hypothetical system for distinguishing three basic structural situations in the infracapitulum (Fig. 5-10). The nonspecialized arrangment is without labiogenal articulation and is called *anarthry*. Where the mentum is triangular and the labiogenal articulation is well posterior of the base of the palp, the condition is known as

stenarthry. When the mentum is square and the labiogenal articulation is directed laterally toward the base of the palps, the situation is called *diarthry*. Transitions from anarthry to diarthry illustrate possible relationships of these mites based on the infracapitular arrangement (Woolley, 1969) (Fig. 5-10G).

CHELICERAE

The chelicerae and pedipalps are the typical appendages of the gnathosoma. With the diversification of feeding habits found in the acarines it is not unusual that the mouthparts correlate with feeding functions and demonstrate a wide variety of structural and functional differentiation. Arachnid (and mite) chelicerae "by their position in the front of the body . . . are well placed to meet a variety of needs and perform a variety of functions, and indeed the versatility of the arachnid chelicerae is equalled by few arthropod appendages" (Savory, 1977).

The chelicerae are the first and only preoral pair of appendages that lie dorsal to the mouth. They are usually three-segmented in both the Anactinotrichida and Actinotrichida, but homologies between segments in the chelicerae of these two groups are hard to define (Evans et al., 1961). The primitive state of the chelicerae is chelate – dentate in the majority of mites. With such a pincerlike distal end, the appendage is capable of grasping, pinching, shearing, and sometimes even piercing. In more specialized chelicerae, there is a reduction to two segments that results in a piercing organ without the grasping capability.

Primitively, the cheliceral limb comprised seven segments, including an apotele (the equivalent of the ambulacrum of a walking appendage) (Grandjean, 1947). In this hypothesis, the apotele is homologized with the movable digit and the fixed digit is derived from four birefringement setae on the distal segments (Fig. 5-11, after Evans et al., 1961). The cylindrical fixed digit (palm of the chela) is a fusion of the femur, genu, and tibia, which is joined

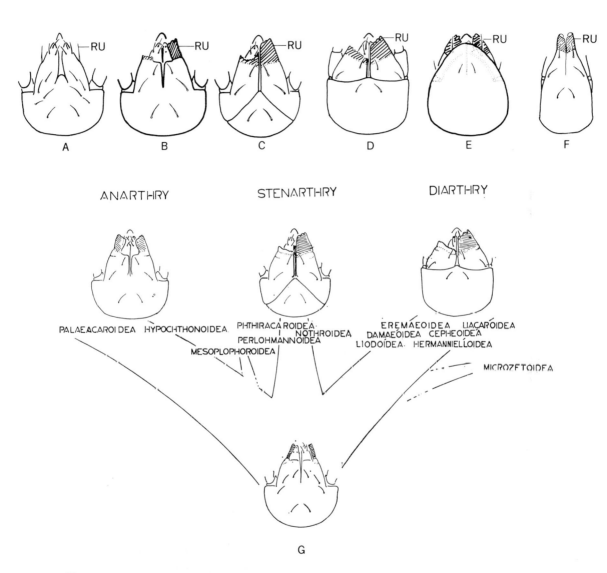

Figure 5-10. Subcapitulum of Oribatida: (A) hypothetical; (B) anarthry (without genomental suture); (C) stenarthry, with genomental suture; (D) diarthry with transverse genomental suture; (E) diarthry with subcapitulum enclosed in a camerostome; (F) diarthry of suctorial type (*Pelops*) (after Grandjean) (RU, rutellum); (G) dendrogram to show possible relationships of superfamilies of inferior oribatids based on the infracapitulum (after Woolley, 1967).

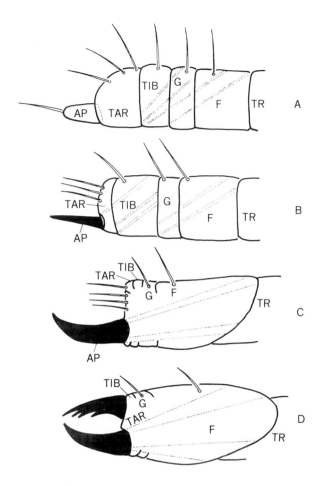

Figure 5-11. Postulated origin of chelate chelicerae in Acari (AP, apotele; F, femur; G, genu; TIB, tibia; TAR, tarsus; TR, trochanter) (after Evans et al., 1961).

distally by the opposing tarsus or movable digit (thumb of the chela) (Hammen, 1968).

Feeding Habits and Chelicerae

The following types of chelicerae are related to the feeding habits of mites:

1. Fungivorous. massive chelicerae with closely set teeth.

2. Predaceous. massive chelicerae with large teeth.

3. Parasitic. edentate chelicerae (Rhynonyssidae), some reduced chelicerae or reduced fixed digit.

4. Grasping and clamping. (Eviphididae) for clinging to erect hairs.

5. Elongated, dentate. (Veigaiidae); collembole feeders.

6. Styletiform. plant feeders (Tetranychidae; Tuckerellidae).

7. Sicklelike. predaceous and ectoparasitic (Anystidae; Trombiculidae).

The chelate chelicerae of *O. texanus* are three-segmented, relatively higher dorsoventrally than long. The movable digit (apotele) is shorter than half the length of the fixed digit. Two tendons attached to muscles move the apotelic bit. If the first segment of the chelicera is the trochanter; the fixed digit must be a fusion of the femur, genu, tibia, and tarsus. The movable digit is essentially the apotele of the tarsus (Hammen, 1966) (Fig. 5-12A).

The palpal coxae are fused into the structure of the gnathosoma. The five remaining segments are the trochanter, femur, genu, tibia, and tarsus (Hammen, 1966). Two claws terminate the palp tarsus. Similar to *Holothyrus*, the tibia and tarsus appear to be one large segment. Varieties of setae are found, depending on the segment of the palp.

The three-segmented chelicerae (*H. coccinella, H. grandjeani*) are long, tubelike, pale, and weakly sclerotized (Hammen, 1965). The fixed and movable digits have some malocclusion and it is assumed that the movable digit has lateral as well as vertical movement. The movable digit is considered to be an apotele. Below the chelicerae and between the bases of the corniculi is the radular organ of Thon (=labrum), which in cross section appears to be the dorsal prolongation of the pharynx. This structure is beset with chitinous teeth. The lateral lips also have many prolongations at the distal end (Hammen, 1961) (Fig. 5-12B).

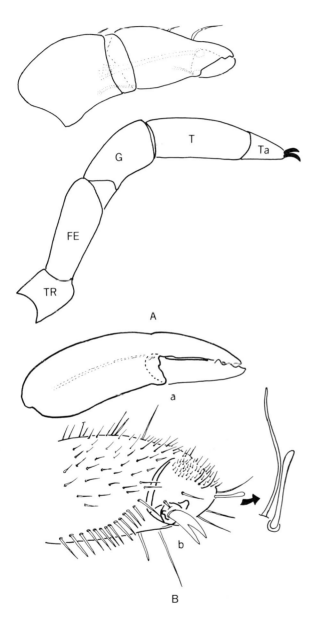

Figure 5-12. Chelicera and palp of (A) *Opilioacarus texanus*; (B) chelicera (antaxial) and palp of *Holothyrus grandjeani* showing distal specialized setae.

The palps consist of six joints; the tibia and tarsus appear as a single large segment. Lyrfissures are absent. In *Holothyrus grandjeani* there is a paraxial area with abundant, hollow hairs, two peculiar terminal seta, and a two-tined apotele. Behind the apotele is a comblike row of serrated, straight setae (Hammen, 1966) (Fig. 5-12B). Tarsus I of *H. coccinella* exhibits distal Haller's organs.

A correlation between cheliceral structure and food consumed is described for gamasid mites (Karg, 1971). Those with long and slender cheliceral digits and small, recurved teeth feed on other mites and small insects such as Collembola (Evans et al., 1961). Gamasids with short cheliceral digits and long teeth prey upon nematodes. In some parasitic gamasines like *Dermanyssus,* the chelae are very small and edentate at the ends of long shafts (Fig. 5-13A, D). These tiny chelae may be used to pinch tissue and form a pool of blood or possibly to pierce the skin of the host (Evans et al., 1961).

Most of the Gamasida, Astigmata, and Oribatida exhibit the chelate – dentate type of chelicerae (Evans et al., 1961). Where mites are blood-feeders or phytophagous, some modifications occur. It usually means the loss of teeth on the cheliceral digits (*Ornythonyssus*, Fig. 5-13A, C), reduction in the size of the movable digits (*Blattisocius*, Fig. 5-13A, B), or elongation of the cheliceral shafts (*Neomolgus, Dermanyssus,* Fig. 5-13A, D, E). Trigynaspid mites have elaborate excrescences on the chelicerae (Fig. 5-13A, F, G).

In many instances, the chelicerae of male gamasids possess a spermatophoral process (=spermadactyl) used in copulation and the transfer of sperm to the female. This process (together with crassate processes on leg II for clasping the female) are distinguishing dimorphic sexual characteristics (Figs. 5-13B, 5-14D–F). Also attached to the fixed digit of the chelicera is a *pilus dentilus,* a short, stout bristle in *Hypoaspis* (Fig. 5-14E) or a long and spatulate one in *Laelaps hilaris* (Fig. 5-14F). Other noticeable features are the lyrifissures and the arthrodial brush.

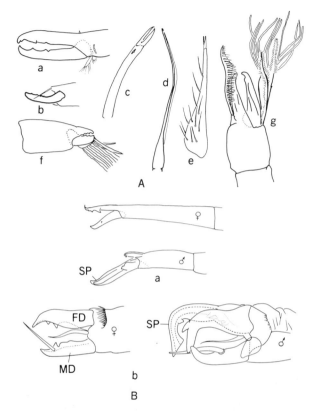

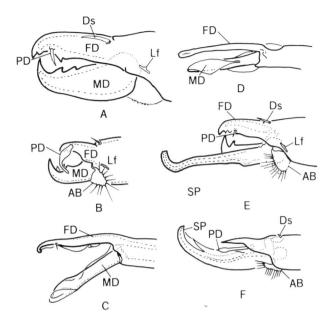

Figure 5-13. (A) Gamasida chelicerae: a, *Macrocheles*; b, *Blattisocius*; c, *Ornithonyssus*; d, *Dermanyssus*; e, *Neomolgus* (after Evans et al., 1961); trigynaspid chelicerae with excrescences: f, *Klinckowstroemius* (after Krantz, 1978); g, *Micromegistus bakeri*, protonymph (after Nickel and Elzinga). (B) Dimorphic characters of chelicerae: a, *Macronyssus ellipticus* female, male (after Rudovsky); b, *Haemogamasus ambulans* female and male (SP, spermadactyl) (after Furman, 1959).

Figure 5-14. Chelicerae of (A) *Eulaelaps stabularis* female; (B) *Androlaelaps fahrenholzi* female; (C) *Haemolaelaps hirsutus* female; (D) *Ornithonyssus bacoti* female; (E) *Hypoaspis krameri* male; (F) *Laelaps hilaris* male (AB, arthrodial brush; DS, dorsal seta; FD, fixed digit; LF, lyrifissure; MD, movable digit; PD, pilus dentilis; SP, spermadactyl (=spermatophoral process)).

The chelicerae in both hard and soft ticks are similar in structure, but typically are two-segmented and lack the chelate form typical of the gamasids and other mites. They lie dorsal to the hypostome and beneath the palps. The proximal shaft of each chelicera is swollen and somewhat membranous. Internally, this bulbous part contains the antagonistic muscles that move the digits by means of long tendons.

The anterior end of the shaft is tapered and ends in articulation with the modified, but slightly compound digit. As the chelicerae are observed from the dorsal aspect, the movable digits are directed laterally, so that movement results in horizontal incisions in the host tissue. The digit consists of two main articles, the internal article, with a dorsal process, and an external article (Fig. 5-15). Each of these triangular, sclerotized segments carry cutting denticles along their lateral (outer) margins. Flexor muscle tendons and extensor tendons are attached to the bases of these articles and in turn connect to the intracheliceral muscles that move these parts of the digit. A membranous hood or mantle is expanded dorsally and laterally, but leaves the cutting edges of the

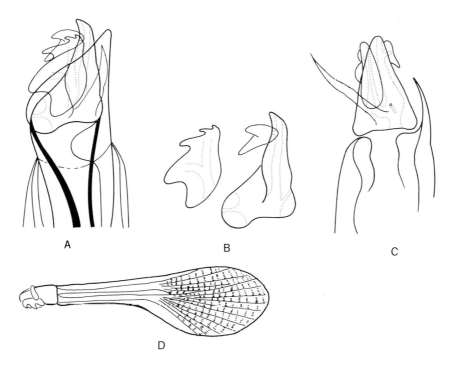

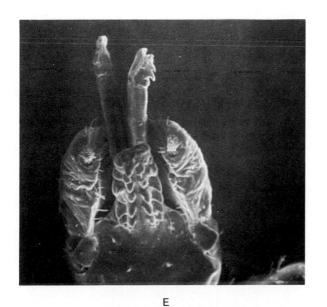

Figure 5-15. Chelicerae of *Haemaphysalis punctata* female: (A) right chelicera; (B) distal article isolated; (C) right chelicera from median aspects (after Nuttall et al., 1908; drawing by Sonya Evans); (D) chelicera and musculature of *Ornithodoros papillipes* (after Balashov, 1972); (E, F) SEMs of gnathosoma, chelicerae and cheliceral digits of *Ixodes ricinus;* (E) 100[×]; (F) 400[×].

denticles exposed (Balashov, 1972; Arthur, 1960; Nuttall et al., 1908; Nuttall and Warburton, 1911).

Exemplary chelate chelicerae of endeostigmatid mites in the Actinedida are shown in the accompanying figures from Baker et al. (1958) and Krantz and Lindquist (1979). The Pachygnathidae show finely toothed, chelate forms of chelicerae (Fig. 5-16A). In Lordalychidae the narrowed, truncate, fixed digit is reduced and the movable digit extends beyond the tip (Fig. 5-16B). Sphaerolichidae and Alicorhagiidae have other modifications of the chelae (Fig. 5-16C). Rhagidiidae exhibit long,

robust chelicerae, swollen at the base and tapered somewhat to relatively small chela at the end (Fig. 5-16G). The short and strong chelicerae of the Labidostommidae appear nearly sphaerical in outline with a curved, movable digit articulating with the narrowed tip of the fixed digit (Fig. 5-16E). Small styletlike, type B chelicerae are evident in Eupodidae and Ereynetidae (Fig. 5-16J, K).

There is a range of form in the chelicerae of bdellid mites (Atyeo, 1960). Several species have chelicerae with a truncate base, an attenuated and elongated fixed digit, and a small movable digit at the distal end. In others, the

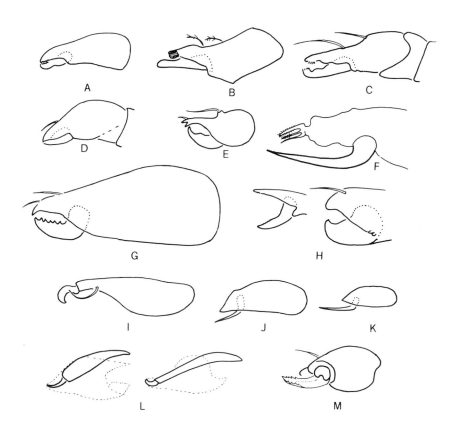

Figure 5-16. Chelicerae of Actinedida: (A) Pachygnathidae; (B) Lordalychidae; (C) Sphaerolichidae; (D) Alicorhagiidae; (E) Nicoletiellidae; (F) Penthalodidae; (G) Rhagidiidae; (H) Bdellidae; (I) Anystidae, Pterygosomidae; (J) Eupodidae; (K) Ereynetidae; (L) Hydrachnida (*Torrenticola* sp.); (M) Trombiculidae.

fixed-digit base is short and swollen with a stout, hooklike, movable digit at the end (Fig. 5-16H).

Chelicerae of whirligig and lizard mites are similar in form to those of ticks. The bases are enlarged and extend forward in a narrowed shaft. At the end of the shaft the movable digit is hooked or dentate and at least partially covered by a hood. Movable digits extend laterally as is the case with ticks (Fig. 5-16I). Water mites have similar chelicerae (Fig. 5-16L). Chigger mites have small, serrated, hooklike chelicerae (Fig. 5-16M).

Chelicerae in the tyroglyphid mites are chelate, swollen basally, and attenuated anteriorly. Each chelicera is slightly compressed laterally and exhibits anteriorly a conical spur and a mandibular spine (Fig. 5-17A). The movable digit and the fixed digit are frequently serrated or toothed, and work in a vertical plane. The chelicerae move independently of each other in the gnathosoma, which is directed at angle anteroventrally. Modifications of the chelate condition are noted in Rosentsteiniidae, Guanolichidae, and Faculiferidae (Fig. 5-17C–E).

Possibly the chelicerae of oribatid mites exhibit the widest variety of structure commensurate with their diverse feeding habits. They range in structure from the primitive oribatids like *Cosmochthonius* and *Amnemochthonius* to the standard chelate forms of *Oribatula* and *Ceratozetes*. The specialized chelicerae of *Pelops* and *Gustavia* are extreme (Figs. 5-18A, B).

PEDIPALPS

Generally speaking, the palps in mites represent the endites of the pedipalpal segments. The palps are frequently simple, but exhibit as many variations as in the chelicerae, again as related to the types of food and other habits. Some palps may be quite bizzare.

The number of segments in palps varies among the orders of mites from two to six. Most mites, except for the Opilioacaridae, lack a claw on the palp tarsus. Gamasids have six movable segments (trochanter, femur, genu, tibia, tarsus and apotele) including a distinct apotele in the form of a tined seta. Two- and three-tined apoteles are present, some with specializations (Fig. 5-19A). In Macrochelidae, the palps manipulate the food and move it closer to the mouth. Palpal size and chaetotaxy vary from larva through adult (e.g., *Ornithonyssus bacoti*). In some families (Rhynonyssidae, Halarachnidae) the palps are much reduced and closer to the mouth. Evans and Till (1965) illustrate the differences in the pedipalp of several gamasids and show the differences in the palpal chaetotaxy from larva through adult for *Ornithonyssus bacoti*. In some of the highly differentiated forms, like Rhinonyssidae or Halarachnidae, the palps are much reduced.

In addition to the sensory and grasping (raptorial) functions, the pedipalps may be used also for cleaning the chelicerae. On the antaxial surface of the pedipalps of some gamasines there are ctenelike setae used in this way, in a manner reminiscent of the serrulae of pseudoscorpions (Evans et al., 1961).

In Ixodida, the usual four palpal segments

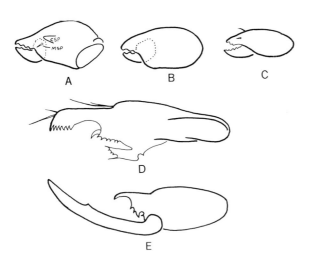

Figure 5-17. Chelicerae of Astigmata: (A) Acaridae; (B) Chortoglyphidae; (C) Rosentsteiniidae; (D) Guanolichidae; (E) Faculiferidae (after Hughes, 1959, Krantz, 1978).

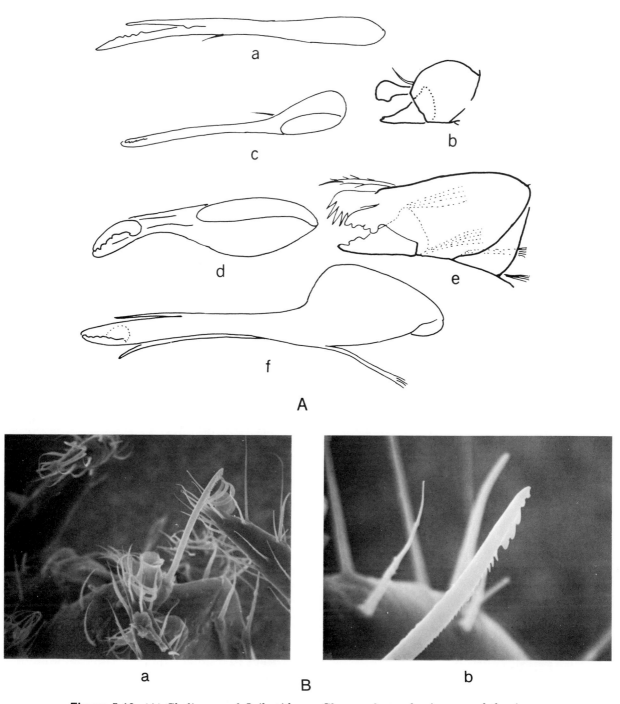

Figure 5-18. (A) Chelicerae of Oribatida: a, *Charassobates*; b, *Amnemochthonius*; c, *Rhychoribates*; d, *Plasmobates*; e, *Cosmochthonius*; f, *Pelops* (after Grandjean). (B) Chelicerae of Gustaviidae SEMs: a, mouth tube, chelicerae (410[×]); b, enlarged tip of chelicera (1000[×]).

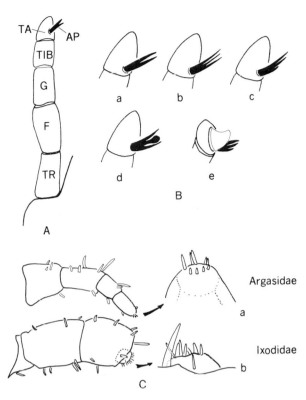

Figure 5-19. Gamasid palp and tarsi showing tined apoteles. (A) Palp segments (TR, trochanter; F, femur; G, genu; TIB, tibia, TA, tarsus, AP, tined apotele). (B) Variations of apoteles: a, two-tined apotele, Phytoseidae, Ascidae; b, three-tined apotele, Parasitidae, Epicriidae; c, Oligomasidae; d, Eviphididae; e, Veigaiidae with membranous projection; (C) Tick palps: a, Argasidae (*Ornithodoros savignyi*); b, Ixodidae (*Hyalomma dromedarii* (after Hammen, 1968).

chaetotaxy of the palps are obscure (Hammen, 1964).

The greatest variation of palpal form occurs in the Actinedida (Trombidiformes). Pyemotidae, Tarsonemidae, Eriophyidae (vermiform gall mites), and Demodicidae (vermiform follicle mites) all have very small palpi. The palp of Pyemotidae consists of a single segment, although suturelike lines occur in some species (Cross, 1965) (Fig. 5-20).

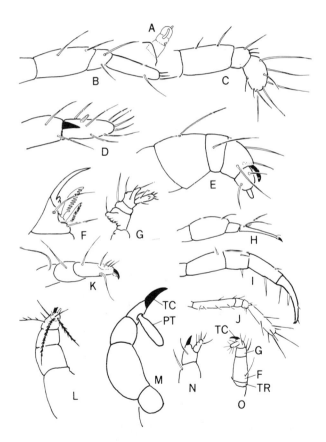

Figure 5-20. Palps of Actinedida: (A) Pyemotidae; (B) Tydeidae; (C) Rhagidiidae; (D) Rhaphignathidae; (E) Tetranychidae; (F) Cheyletidae; (G) Caeculidae; raptorial, (H) Halacaridae, (I) Cunaxidae, (J) Bdellidae, (K) Pterygosomidae; thumb-claw, (L) Trombiculidae; (M) Trombidiidae; (N) Caligonellidae; (O) Stigmaeidae (TC, tibial claw; PT, palpal tarsus; G, genu; F, femur; TR, trochanter).

exist, but slight differences are noted between the palps of soft ticks and those of hard ticks. In soft ticks the articles of the palp are about subequal and normally unmodified, somewhat comparable to those in gamasid and holothyrid mites (Fig. 5-19C, a). In hard ticks the palpal segments are of different lengths, are modified in shape, enclosing the hypostome, and the tarsal segment (IV) is sunk into the penultimate segment (III) (Fig. 5-19C, b). Homology and

Simple palps occur in the Tydeidae, Rhagidiidae, and Bdellidae. The relatively swollen palp tarsus in Rhagidiidae is distinctive. The long, rectangular palp tarsus of the bdellid mites, peculiarly angled from the relatively short tibia, is helpful in identifying these snout mites (Fig. 5-20). The palps of bdellids are probably tactile and chemosensory in function (Fig. 5-21). Raptorial palps are distinctive for predatory mites (e.g., Halacaridae, Cunaxidae, Cheyletidae, and Pterygosomidae) (Fig. 5-20F, H, I, K).

Spider mites (Tetranychidae), rake-legged mites (Caeculidae), many of the Parasitengona, and numerous other trombidiform mites have palps with a distinctive thumb – claw complex. In this condition the tibia of the palp exhibits a terminal spur or stout claw adjacent to or in apposition to the tarsal segment, which makes the end of the palp appear almost chelate (Fig. 5-20E, L–O). The chaetotaxy of the palpal segments associated with this thumb – claw complex varies. Cheyletidae have characteristic large, serrate setae on the tarsus and tibia (Fig. 5-20F). Some velvet mites, *Trombidium*, are devoid of setae on the palp (Fig. 5-20M).

The palps of *Histiostoma* (Anoetidae) are flattened and terminate in long flagella (T.E. Hughes, 1953; R.D. Hughes, 1958; and Jackson, 1958). Rapid movement of these flexible structures pulls microorganisms toward the mouth as the mites stand in the liquid of decaying organic material to feed (Fig. 5-1B). Palps of the tyroglyphid mites are two-segmented (Zakhvatkin, 1959; A.M. Hughes, 1961a). Adpressed to the subcapitulum, they are simple in form with few modifications (Fig. 5-9A).

One of the most pronounced specializations of the pedipalps occurs in some of the fur mites. Hughes (1954b) shows that in *Listrophorus* (Listrophoridae) the pedipalpal coxae are spread laterally into two enlarged, lobelike processes that curve mesad. The medial surface of these endites is corrugated (Fig. 5-22C) so that the hair of the host is grasped firmly.

Palpi of the Oribatida are usually five-segmented and simple with acanthions on the distal tips. The palps are associated with the rutellum and chelicerae in a specific cavelike camerostome that houses the infracapitulum of these mites (Fig. 5-22D, E).

Figure 5-21. Distal tip of bdellid palp (SEM) 1500[×].

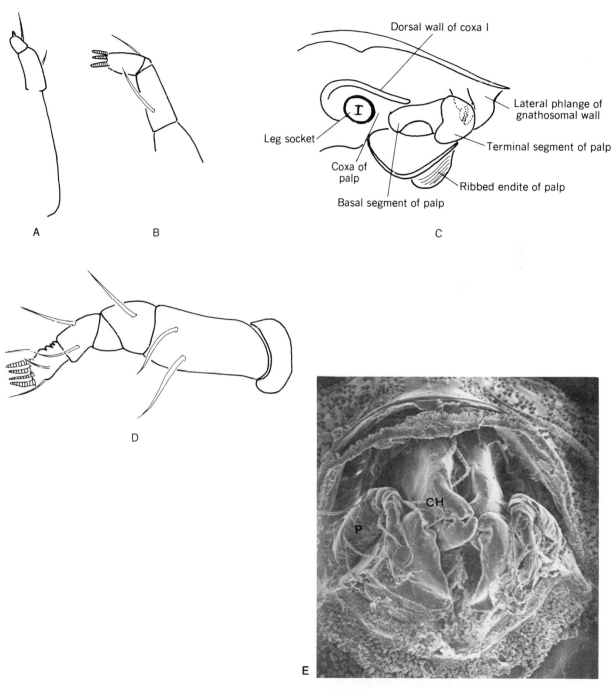

Figure 5-22. Palps of Astigmata and Oribatida: (A) *Acarus siro* and (B) *Glycyphagus destructor*; (C) schematic of palp of *Listrophorus leuckarti* showing modification of ribbed palpal endite for grasping hairs (after A.M. Hughes, 1961a; T.E. Hughes, 1954b); (D) palp of oribatid mite showing distal acanthions (after Grandjean); (E) camerostome of *Gymnodamaeus chalazionus* (CH, chelicera; P, palp) (SEM).

REFERENCES

Arthur, D.R. 1946. The feeding mechanism of *Ixodes ricinus* L. *Parasitology* 37:154–162.

Arthur, D.R. 1951. The capitulum and feeding mechanism in *Ixodes hexagonus* Leach. *Parasitology* 41:66–81.

Arthur, D.R. 1960. *Ticks,* Part V. Cambridge Univ. Press, London and New York.

Arthur, D.R. and K.R. Snow. 1967. The implications of size as shown in *Hyalomma anatolicum anatolicum* (Koch, 1844). (Ixodoidea: Ixodidae) *Wiad. Parazyt.* 13:497–309.

Atyeo, W.T. 1960. A revision of the mite family Bdellidae in North and Central America (Acarina, Prostigmata). *Univ. Kans. Sci. Bull.* 40(8):345–499.

Baker, E.W., T.M. Evans, D.J. Gould, W.B. Hull and H.L. Keagan. 1956. A manual of parasitic mites of medical and economic importance. Nat'l. Pest Control Assoc. Tech. Publ., 170 pp.

Baker, E.W., J.H. Camin, F. Cunliffe, T.A. Woolley and C.E. Yunker. 1958. *Guide to the Families of Mites.* Inst. Acarol., Dep. Zool., University of Maryland, College Park.

Balashov, Y.S. 1972. *Bloodsucking Ticks (Ixodoidea)—Vectors of Diseases of Man and Animals,* Transl. 500 (T 500). Med. Zool. Dep., USNAMRU No. 3, Cairo, Egypt, U.A.R.; (and *Misc. Publ., Ent. Soc. Amer.* 8(5):159–376.)

Bertram, D.S. 1939. The structure of the capitulum of Ornithodoros: a contribution to the study of the feeding mechanism in ticks. *Ann. Trop. Med. Parasitol.* 33:229–258 (cited by Evans *et al.,* 1961).

Blauvelt, W.E. 1945. The internal morphology of the common red spider mite *(Tetranychus telarius). Mem.-N.Y. Agric. Exp. Stn. (Ithaca)* 270:1–35.

Borner, C. 1903. Die Beingliderung der Arthropoden. *Sitzungsber. Ges. Naturforsch. Freunde Berlin, Jahr. 1903.*

Brody, A.R., J.C. McGrath and G.W. Wharton. 1972. *Dermatophagoides farinae:* the digestive system. *J. N. Y. Entomol. Soc.* 80(3):152–177.

Brown, J.R.C. 1952. The feeding organs of the adult of the "common chigger." *J. Morphol.* 91:15–51.

Brucker, E.A. 1901. Monographie de *Pediculoides ventricosus* Newport et theorie des pieces buccales des acariens. *Bull. Sci. Fr. Belg.* (6) 35(4):365–451 (cited by Cross, 1965).

Cross, E.A. 1965. The generic relationships of the family Pyemotidae (Acarina, Trombidiformes). *Univ. Kansas Sci. Bull.* 45(2):29–275.

Desch, C.E. and W.B. Nutting. 1977. Morphology and functional anatomy of *Demodex folliculorum* (Simon) of man. *Acarologia* 19(3):422–462.

Evans, G.E. 1957. An introduction to the British Mesostigmata (Acarina) with keys to families and genera. *J. Linn. Soc. Zool.* 43(291):203–259.

Evans, G.O. and W.M. Till. 1965. Studies on the British Dermanyssidae (Acari: Mesostigmata). Part I. External morphology. *Bull. Br. Mus. (Nat. Hist.) Zool.* 13(8):249–294.

Evans, G.O., J.G. Sheals and D. MacFarlane. 1961. *The Terrestrial Acari of the British Isles.* Adlard & Son, Bartholomew Press, Dorking, England.

Fain, A. and J. Bafort. 1964. Les Acariens de la Famille Cytoditidae (Sarcoptiformes). Descriptions de sept espèces nouvelles. *Acarologia* 6(3):504–528.

Furman, D.P. 1959. Observations on the biology and morphology of *Haemogamasus ambulans* (Thorell) (Acarina: Haemogamasidae). *J. Parasitol.* 45(3)274–280.

Grandjean, F. 1936a. Un acarien synthétique: *Opilioacarus segmentatus* With. *Bull. Soc. Hist. Nat. Afr. Nord.* 27:413–444.

Grandjean, F. 1936b. Observations sur les Acariens (3 série). *Bull. Mus. Natl. Hist. Nat.* (2), 8:84–91.

Grandjean, F. 1938. Description d'une nouvelle prelarva et remarques sur la bouche des Acariens. *Bull. Soc. Zool. Fr.* 68:58–68.

Grandjean, F. 1939. Quelques genres d'Acariens appartenant au groupe des Endeostigmata. *Ann. Sci. Nat., Zool. Biol. Anim.* (11) 2:1–122.

Grandjean, F. 1946. Observations sur les Acariens (9e série). *Bull. Mus. Natl. Hist. Nat.* (2) 18:337–344.

Grandjean, F. 1947. Au sujet des Erythroides. *Bull. Mus. Natl. Hist. Nat.* (2) 19:327–334.

Grandjean, F. 1957a. L'infracapitulum et la manducation chez les Oribates et d'autres Acariens. *Ann. Sci. Nat., Zool. Biol. Anim.* (11) 19:233–281.

Grandjean, F. 1957b. Observations sur les Oribates (37e série). *Bull. Mus. Natl. Hist. Nat.* (2) 29:88–95.

Gregson, J. D. 1967. Observations on the movement

of fluids in the vicinity of the mouthparts of naturally feeding *Dermacentor andersoni* Stiles. *Parasitology* 57:1–8.

Hammen, L. van der. 1961. Descriptions of *Holothyrus grandjeani*, nov. spec., and notes on the classification of the mites. *Nova Guinea, Zool.* 9:173–194.

Hammen, L. van der. 1964. The morphology of *Glyptholaspis confusa*. *Zool. Verh.* 71:1–56.

Hammen, L. van der. 1965. Further notes on Holothyrina (Acarida). I. Supplementary description of *Holothyrus coccilonella* Gervais. *Zool. Meded.* 40:253–276.

Hammen, L. van der. 1966. Studies on Opilioacarida. I. Description of *Opilioacarus texanus* (Chamberlain and Mulaik) and revised classification of the genera. *Zool. Verh.* 86:1–80.

Hammen, L. van der. 1968. The gnathosoma of *Hermannia convexa* (C.L. Koch) and comparative remarks on its morphology in other mites. *Zool. Verh.* 94:1–45.

Hammen, L. van der. 1969. Studies on Opilioacarida. III. *Opilioacarus patensis* Silvestri, and *Adenacarus arabicus* (With). *Zool. Meded.* 44(8):113–131.

Hammen, L. van der. 1970a. La segmentation primitive des Acariens. *Acarologia* 12(1):3–10.

Hammen, L. van der. 1970b. La segmentation des appendices chez les acariens. *Acarologia* 12(1):11–15.

Hammen, L. van der. 1970c. Remarques générale sur la structure fondamentale du gnathosoma. *Acarologia* 12(1):16–22.

Hirschfelder, H. and H. Sachs. 1952. Recent research on the acarine mite. *Bee World* 33(12):201–209.

Hughes, A.M. 1961a. The mites of stored food. Ministry Agr. Fish and Food. Tech. Bull. 9(N.S. 1), pp. 5–198.

Hughes, A.M. 1961b. Terrestrial Acarina III. Acaridiae III. *In*: Fridriksson, A and S.L. Tuxen. *The Zoology of Iceland.* Vol. 3, Part 57C, pp. 1–12.

Hughes, R.D. and C.G. Jackson. 1958. A review of the Anoetidae (Acari). *Virginia J. Sci.* 9(1):5–198.

Hughes, T.E. 1949. The functional morphology of the mouthparts of *Liponyssus bacoti*. *Ann. Trop. Med. Parasitol.* 43(3/4):349–360.

Hughes, T.E. 1950. The physiology of the alimentary canal of *Tyroglyphus farinae*. *Q. J. Microsc. Sci.* 91(1):45–61.

Hughes, T.E. 1953. The functional morphology of the mouthparts of the mite *Anoetus sapromyzarum* Dufour, 1839, compared with those of the more typical Sarcoptiformes. *Proc. Acad. Sci. Amst.* 56C:278–287 (cited by Evans *et al.*, 1961).

Hughes, T.E. 1954a. Some histological changes which occur in the gut epithelium of *Ixodes ricinus* females during engorging and up to oviposition. *Ann. Trop. Med. Parasitol.* 48:397–404.

Hughes, T.E. 1954b. The internal anatomy of the mite *Listrophorus leuckarti* (Pagenstecher, 1861). *Proc. Zool. Soc. London* 124:239–256.

Hughes, T.E. 1959. *Mites or the Acari.* Athlone, London.

Jeppson, L., H.H. Keifer and E.W. Baker. 1975. *Mites Injurious to Economic Plants.* Univ. of California Press, Berkeley.

Jones, B.G. 1950. The penetration of the host tissue by the harvest mite, *Trombicula autumnalis* Shaw. *Parasitology* 40:247–260.

Kaestner, A. 1968. *Invertebrate Zoology,* Vol. 2. Wiley (Interscience), New York.

Karg, W. 1971. Untersuchungen uber die Acarofauna in Apfelanlagen im Hinblick auf den Ubergand von Standarspritzprogrammen zu integrierten Behandleungsmassnahmen. *Arch. Pflanzenschutz* 7:243–279 (cited by Krantz and Lindquist, 1979).

Krantz, G.W. 1978. *A Manual of Acarology,* 2nd ed. Oregon State Univ. Book Stores, Corvallis.

Krantz, G.W. and E.E. Lindquist. 1979. Evolution of phytophagous mites (Acari). *Annu. Rev. Entomol.* 24:121–158.

Mitchell, R.D. 1955. Anatomy, life history, and evolution of the mites parasitizing fresh-water mussels. *Misc. Publ. Mus. Zool., Univ. Mich.* 89:1–28.

Mitchell, R.D. 1958. The musculature of a trombiculid mite, *Blankaartia acuscutellaris* (Walch). *Ann. Entomol. Soc.* 55:106–119.

Mitchell, R.D. 1962. Structure and evolution of water mite mouth parts. *J. Morphol.* 111:41–59.

Mitchell, R.D. 1964. The anatomy of an adult chigger mite *Blankaartia acuscutellaris* (Walch). *J. Morphol.* 114(3):373–391.

Nuttall, G.H.F., C. Warburton, W.F. Cooper and L.E. Robinson. 1908. *Ticks: A Monograph of the*

Ixodoidea. Part I. Argasidae. Cambridge Univ. Press, Cambridge, pp. 1–104.

Nuttall, G.H.F. and C. Warburton. 1911. *Ticks. A Monograph of the Ixodoidea.* Part II. Ixodidae. 105–348. Cambridge Univ. Press, Cambridge, pp. 105–348.

Prasad, V. and D.R. Cook. 1972. *The Taxonomy of Water Mite Larvae.* Mem. Am. Entomol. Inst., Ann Arbor, Michigan.

Reuter, E. 1909. Zur Morphologie und Ontogenie der Acariden. *Acta Soc. Sci. Fenn.* 36 (cited by Hughes, 1959).

Roshdy, M.A. 1962. Comparative internal morphology of subgenera of *Argas* ticks (Ixodoidea: Argasidae). 2. Subgenus *Chiropterargas: Argas boueti* Rouband and Colas-Belcour, 1933. *J. Parasitol.* 48(4):623–630.

Savory, T. 1977. *Arachnida,* 2nd ed. Academic Press, London.

Snodgrass, R.E. 1965. *A Textbook of Arthropod Anatomy.* Hafner, New York.

Treat, A. 1975. *Mites of Moths and Butterflies.* Cornell Univ. Press (Comstock), Ithaca, New York.

Woolley, T.A. 1969. The infracapitulum—A possible index of oribatid relationship. *Proc. Int. Congr. Acarol., 2nd, 1967,* pp. 209–221.

Woolley, T.A. 1979. The Chelicerae of the Gustaviidae. *In:* Rodriguez, J.G., ed. *Recent Advances in Acarology.* Academic Press, New York, Vol. 2, pp. 547–551.

Zakhvatkin, A.A. 1941. *Fauna of U.S.S.R. Arachnoidea.* Vol. 6, No. 1. Tyrogylphoidea (Acari). Transl. AIBS, Washington, D.C.

6

Musculature and Legs

A centipede was happy quite until a (mite) in fun
Said, "Pray, which leg goes after which?"
This raised his mind to such a pitch,
He fell distracted in a ditch, considering how to run.
                    —RAY LANKESTER, 1889

MUSCULATURE[1]

The muscles of mites are striated and each muscle appears to be made up of distinctive cells, although few studies have been made of body musculature and musculature of the legs is poorly understood (Legendre, 1968). Studies of mites parasitizing freshwater mussels and of *Blankaartia* are exceptions (Mitchell, 1955, 1962).

The musculature comprises two types: intrinsic and extrinsic. Intrinsic muscles are those that extend over the joints of leg segments and apparently are exclusively flexors in most mites (Fig. 6-1A), although a few exceptions have been noted in ticks, in higher Actinedida, and in some fur mites (*Listrophorus*; Hughes, 1954) (Fig. 6-1E). Intrinsic muscle

actions result in flexion. Elevation of the legs and protraction of the basal segments are different movements involving monovalent, bivalent, and rotatory actions (Hammen, 1970) (Fig. 6-1B).

Extrinsic muscles are those of the body and consist of dorsoventral, oblique, rotator, and elevator muscles; their positions correspond to their functions (Legendre, 1968). Origins and insertions of extrinsic muscles are on the body wall, on apodemes, or on endosternites. Most of these muscles are dorsoventral and modify the shape and contour of the body by regulating the turgor pressure of the body fluids. In most mites hydrostatic pressure causes the extension of the chelicerae, palps, and legs as well as modification of body shape (Treat, 1975; R.D. Mitchell, personal communication, 1977) (Figs. 6-2; 6-3C, D).

Muscles other than the extrinsic ones are associated with tubular organs. Some are found in the outer layer of Malpighian tubules of *Haemogamasus ambulans* (Young, 1968) and in the vaginal walls of ticks where they move

[1]Muscles are internal, but since they activate the legs and other external structures and at the present state of knowledge do not comprise an elaborate system like those described under *Internal Morphology*, they are considered here.

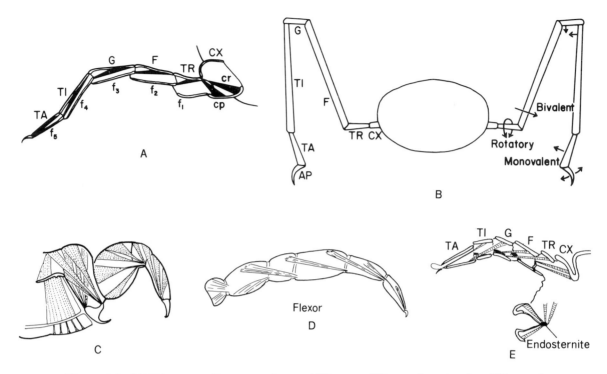

Figure 6-1. (A) Diagram of leg musculature (CX, coxa; CP, coxal protractor; CR, coxal retractor; F, femur, G, genu, TI, tibia; TA, tarsus) (drawing by Peter Eades; after R.D. Mitchell, personal communication, 1977); (B) movements of legs and leg segments (after Hammen, 1970); musculature of (C) gnathosoma of *Unionicola fossulata*, (D) leg I of same (after Mitchell, 1955); (E) leg musculature of *Listrophorus* showing flexor and extensor muscles and endosternite attachments (after Hughes, 1954).

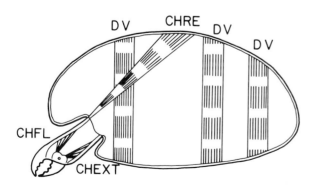

Figure 6-2. Diagram of extrinsic body musculature (CH FL, cheliceral flexors; CH RE, cheliceral retractors; DV, dorsoventral muscle bands) (after R.D. Mitchell, personal communication, 1977).

the endospermatophore upward (Balashov, 1972).

In the Gamasida the musculature does not show external evidence of attachment (as is visible in ticks). Dorsoventral muscles extend from interscutal plates to inguinal plates. Dorsoventral muscles also attach to the pygidial or opisthonotal shield and to the anal plate. Contraction of these latter muscles simultaneously opens the anus and compresses the gut. There may be other dorsoventral muscles that are incomplete in the podosomal area and these extend from the podosomal plate to the inner dorsal surface of an endosternite.

Muscles that operate the movable coxae are also attached to the edges of the endosternite

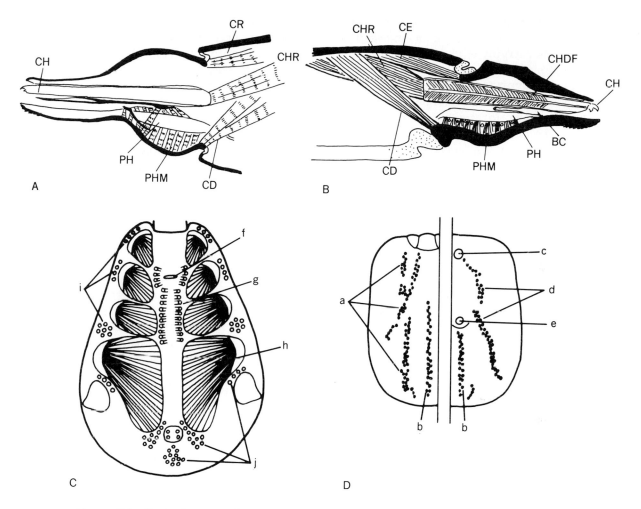

Figure 6-3. Capitular musculature of (A) *Hyalomma asiaticum*; and (B) *Dermacentor andersoni*. Body musculature of (C) *Dermacentor andersoni* and (D) *Hyaloma asiaticum* (BC, buccal cavity; CD, capitular depressor muscle; CE, capitular elevator; CH, chelicera; CHD, cheliceral digits; CHDF, cheliceral digit flexors; CHR, cheliceral retractors; CR, capitular retractors; PH, pharynx; PHM, pharyngeal muscles) (after Douglas, 1943; Balashov, 1972).

(Fig. 6-1E), as are the lateral tensor muscles from the body wall. Laelapids usually have flexor muscles of the gnathosoma attached to the anterior border of an endosternite. The flexors and elevators of the gnathosoma in the Parasitidae attach to the podosomal shield. In the Uropodina and Dermanyssidae, the mus-

cles that protract and retract the chelicerae insert posteriorly in the body near the anal region (Hughes, 1959).

A well-developed series of longitudinal muscles originates on the body wall dorsolaterad of coxae I in gamasids. These may cause depression of the outline of the body, but mainly function as tensors in support of the idiosoma. Such a function may be important in an engorged female or in one carrying eggs.

In ticks, four basic muscle groups are recognized and distinguished by location and function (Balashov, 1972). The groups of dorsoventral muscles are located by the external grooves in the skeleton of ixodid ticks and by the small discs on the surface of argasids. These muscles change the shape and size of the body and facilitate movement of the capitulum by altering hydrostatic pressure. They also function in oviposition and in defecation (Balashov, 1972; Hughes, 1959). The cervical groove lateral to the scutum and the scutolateral grooves mark the attachment sites of coxal muscles. The alloscutum has anterior and paramedian grooves reflective of the dorsoventral muscle attachments and the sheets of muscle that run between the caeca of the stomach. The lack of longitudinal muscles in ticks, found in nearly all other mites, accounts for the major masses of dorsoventral muscles (Hughes, 1959).

The musculature of soft ticks is similar to that found in hard ticks, but the muscles originate on small plates, usually arranged in rows. The dorsoventral muscles do not form sheets between the gut caeca (except in *Ornithodoros*).

Capitular elevator and depressor muscles, cheliceral retractor muscles, and a series of dorsal, ventral, and alate pharyngeal muscles are located in the capitulum of ticks (Figs. 5-3A, B; 6-3A, B). The musculature of the legs comprises heavy oblique muscle bands attached to the coxae and subcoxae. Flexor and extensor muscles in the legs of ticks are different from those of most other mites (Balashov, 1972).

Relatively little is known of the cytological and physiological aspects of the muscles of ticks. They are striated and the myofibrils are located centrally in the cells with a nucleated sarcoplasm around the periphery (Balashov, 1972).

The muscles of water mites are grouped on the basis of function: (1) coxal muscles, those that originate on the coxae and insert on the basal segment of the leg; (2) supportive muscles, those that originate on the body wall, coxae, or genital stirrup and insert on the transverse coxal ligament; and (3) dorsoventral muscles, those with both origins and insertions on the body wall (Mitchell, 1955).

The musculature of *Blankaartia ascoscutellaris*, a trombiculid mite, is divided into four discrete functional groups of muscles: those of the mouthparts, legs, genital field, and genital pore. The general body muscles form an integrated system with 31 pairs inserted on the body wall or associated sclerites. Dorsal, dorsoventral, and ventral muscles are distinguishable and function to extend the legs by turgor pressure and to maintain the general body contour and shape. Some of the dorsal muscles cause the dimples in the dorsal integument. The ventral muscles are heterogeneous. The length of leg muscles is related to the coxal apodemes, the size of coxal plates, and the articulations involved. Elevator, protractor, and depressor muscles are present in the legs and facilitate movement (Mitchell, 1962).

Little is known about the muscular system of Eriophyidae, but muscles are nonstriated (compared to the striated muscles of the Tetranychidae) (Witmoyer et al., 1972).

Muscles of the Astigmata are striated and consist of six basic groups: (1) muscles that move the mouthparts, longitudinal retractors that are fixed dorsally beneath the propodosomal shield, and rotators that bend the gnathosoma almost perpendicularly; (2) dorsoventral muscles attached to the exoskeleton and identified in location by the longitudinal depressions (rachidial ruts) on the dorsal exoskeleton; they flatten the body; (3) four pairs of longitudinal muscles extending from the propodosomal region to the level of coxae III; they shorten the body; (4) leg muscles, those that move the leg as a whole and are connected between the body wall and the endosternites and rotate the trochanters, and those that flex the segments of the legs (femur, genu, tibia, tarsus, and claws) (each flexor muscle stretching from one segment to the segment immediately preceding it in such a way that each segment of the leg contains parts of three muscles); (5) muscles of the digestive tract; (6) muscles of the genital tract (Zakhvatkin, 1959).

LEGS

Mite larvae are hexapod. Most nymphs and adults of Acari have four pairs of jointed legs. In mites with four pairs of legs, legs IV appear with the first nymphal stage. Eriophyidae have four legs in all stages after hatching, as the group name Tetrapodili implies. Particular lifestyles and habits affect the number of legs and the number of leg segments. Males of some Podapolipidae (*Podapolipus*) have only three pairs of legs; the female has only one pair. The nymphs of Demodicidae and other parasitic forms have fewer articles than the adults. Legs may be fused into double segments in some tarsonemids (*Schizocarpus*) and into legs each comprising a single segment in the parasite *Chirodiscus* (Hughes, 1959).

The jointed, movable limbs of arthropods, including mites, are made possible by the presence and local distribution of three main types of cuticle: solid sclerites, flexible arthrodial membranes, and hinges of ligaments composed of rubberlike resilin (Neville, 1975). The three types of primary articulation of legs were described earlier.

It is thought that the primitive acarine leg consisted of five segments. The six-segmented legs of Actinedid mites supposedly resulted from a secondary division of the femur (Mitchell, 1962). Each of the six movable segments of a leg is acted upon by a muscle inserted at its base (Fig. 6-1A). This means that the leg segmentation may have been misinterpreted or that there is a secondarily evolved muscle. The nomenclature of the leg segments should not be changed, however, until more is known about the musculature of the legs.

Other authors set the number of primitive leg segments at seven (Grandjean, 1954; Hughes, 1959; Evans et al., 1961; Hammen,

1970), but uncertainty exists concerning the homologies between the two groups Anactinotrichida and Actinotrichida. In the latter, Grandjean (1954) indicates that the leg is without a free coxa but has a trochanter, basifemur, telofemur, genu (also called patella), tibia, tarsus, and pretarsus including the apotele (ambulacrum). If the basifemur or telofemur remain fused the number of segments is six. The femora may be entire in the immatures and divided in the adults (*Bdella, Cunaxa, Palaeacarus*). Segmentation of the legs is compared in Figure 6-4.

Although the primitive number of leg segments may be six or seven (depending on the author) with arthrodial membranes between the articles, subdivisions may occur. In the Oribatida (e.g., *Palaeacarus, Acaronychus, Aphelacarus*) the femur may be divided into basi- and telofemur (Fig. 6-4F). Divided femora also occur in the Actinedida (e.g., Trombidiidae, Erythraeidae). Pseudoarticulation in legs is present in the femora of Rhagidiidae, Eupodidae, Anystidae, and Halacaridae. This division may also be present in the Gamasida. Subdivisions may occur in the tarsus where the basitarsus and up to 18 parts of the tarsus are found (Hughes, 1959). Primary leg segmentation (with muscles) and secondary segmentation (without muscles) in the legs of *Opilioacarus texanus* (trochanters 1 and 2, basi- and telofemur, basi- and telotibia, basi- and telotarsus, pretarsus and apotele (ambulacrum)) are noted by Hammen (1966).

Reduction of leg segments may include the incorporation of the coxae into the ventral surface of the podosoma. While coxae usually are movable in the Opilioacarida, Holothyrida, and many Gamasida, in some of the latter the coxae may be incorporated into the ventral surface. In some ticks coxae I may be slightly movable, but the other coxae are fixed in the ventral surface. Coxae of Astigmata and Oribatida are fused into the body wall and inward reflections of these segments form the internal apodemata for muscle attachments.

The Acari were primitively tridactyl (three-

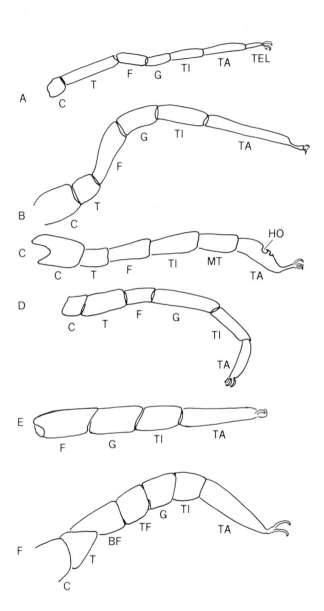

Figure 6-4. Diagrammmatic comparisons of leg segments of (A) Opilioacarida; (B) Gamasida; (C) Ixodida; (D) Actinedida; (E) Acaridida; and (F) Oribatida (A, ambulacrum; BF, basifemur; C, coxa; F, femur; G, genu; HO, Haller's organ; MT, metatarsus; T, trochanter; TA, tarsus; TI, tibia; TF, telofemur; TEL, telotarsus).

clawed) (Hughes, 1959; Evans et al., 1961). The description of *Protacarus crani* identified at least three bent distal setae on the legs. Variations of life-styles and morphology have led to differentiation into monodactylous and bidactylous conditions. Many of the Gamasida and the Actinedida have only lateral claws present. Loss of the lateral claws and retention of the middle claw (empodial claw) occur in some Actinedida, while the Astigmata and some Oribatida are monodactylous. Some of the specialized oribatids become secondarily tridactylous by the loss of the lateral claws and the subdivision of the medial claw.

The terminal ambulacrum (distal tarsal apparatus) of the leg is a remnant of the primitive apotele and consists of a basilar piece, claws, and empodia (pulvilli) derived from modified setae. Antagonistic tendons connect to dorsal levator and ventral depressor muscles in the tarsus and tibia, respectively, and move the ambulacrum. A pair of condylophores articulate with acetabula on either side of the basilar piece or freely articulate with the tarsus in Astigmata (Figs. 6-5A–C).

When the flexible membrane that connects the claws to the tarsus is elongated, it may become a pretarsus enclosing tendons that operate the claws and may end in a pulvillus or other padlike structure (empodium). Flaps that are associated with the distal end of the pretarsus may be lost or modified as holding structures in ticks and myrmecophilous or termitophilous species. In some instances (Macrochelidae and snout mites, Bdellidae) the whole ambulacaral apparatus is changed to a sensory function in leg I. One or more pairs of legs of male gamasids may be spurred (crassate) and used as claspers during copulation (Fig. 6-6D). Legs of Uropodina may be withdrawn into cavities or depressions in the venter (fovea pedales) for protection (Fig. 6-6E). In some Oribatida (Galumnoidea) sclerotized lateral flanges (pteromorphs) at the edges of the hysterosoma may be protective of the legs and used to cover them after withdrawal against the hysterosoma (Evans et al., 1961; Fig. 6-6F).

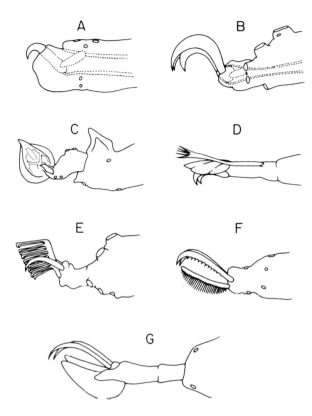

Figure 6-5. Tarsi and ambulacra of selected Acari: (A) *Tyroglyphus farinae* (after Grandjean); (B) Oribatida (after Grandjean, 1943); (C) Freyanidae (after Krantz, 1978); (D) Macrochelidae (after Krantz, 1978); (E) Tetranychida (after Baker et al., 1958); (F) Tydeidae (after Evans et al., 1961); (G) Ixodida (after Snodgrass, 1962).

Legs may serve functions other than locomotion. They may be antenniform in appearance and sensory, used to touch the substrate or to detect prey or other food. In the rake-legged mites (Caeculidae), spinelike apophyses on the paraxial surface of the front legs are used to capture and hold prey organisms (Fig. 6-8A). The legs of fur mites (e.g, Myobiidae, Listrophoridae) may be modified for clasping the hair of their host animals (Figs. 1-1, 6-8B, 6-9). Phoretic mites use the legs to clasp hairs or to hold to their transport hosts. The claws of legs I in some Rhynonyssidae are used as surrogate

chelicerae to pierce and tear the nasal tissue of their avian hosts (Krantz, 1978).

A number of mites exhibit modifications of the legs that function in a tactile way. Where this occurs, the palps are put to other uses, mainly raptorial or auxiliary feeding functions. Where the first pair of legs are modified as tactile appendages, they are raised above the ground and directed forward. The tarsal end of the leg may lack claws and be equipped instead with long setae (Fig. 6-7A, *Podocinum*). In others the segments of the legs are extremely long, attenuated, and arched in front of the mites in an exploratory manner (Fig. 6-7B, *Linopodes*). The first legs of Oribatids are tactile and carried above the substrate; they have long tibial solenidia for exploration and detection. Other mites (Gamasida, Actinedida) exhibit similar features (Lawrence, 1953).

While leaping powers are found in a number of insects, very few arachnids possess this capability, and among mites leaping is limited to a single family of Oribatida. The Zetorchestidae have legs IV enlarged in such a way that the mites can spring upward (Fig. 6-8D). Also, even though not directly involved with leg movement, a sailing, gliding action has been observed in the so-called large-winged oribatid mites (Galumnoidea) (Fig. 6-6F) that use the winglike pteromorphs (leg covers) as sails. Grandjean observed these animals gliding down from grass stems up which they had crawled.

In the Anactinotrichida, mites within the order Opilioacarida have nine primary leg segments with variable secondary segmentation. Two trochanters (basi- and telotrochanter) occur in legs III of *Opiliacarus texanus*, which is considered a primitive character (Hammen, 1966). The division of tibia I and all tarsi is a secondary development. Tibial division of leg I may include a basifemur and telofemur; the tarsus may have an acrotarsus and a pretarsus (=telotarsus 1, telotarsus 2) and pretarsus with an apotele (Hammen, 1970). This distal part of the leg is the ambulacral apotele or the ambulacrum (Fig. 6-6A) and usually consists of

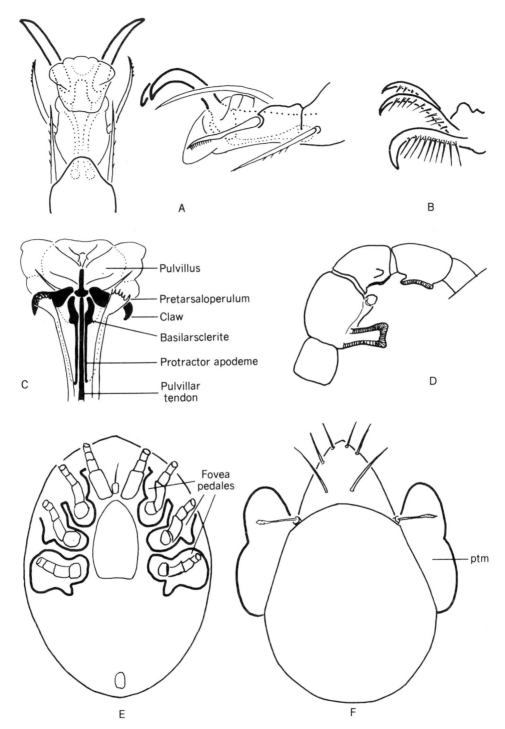

Figure 6-6. Tarsi and ambulacra of selected Acari: (A) Opilioacarida (after Hammen, 1966); (B) Pachygnathidae (after Grandjean, 1939); (C) *Haemogamasus* (after Evans and Till, 1965); (D) crassate leg of male *Pergamasus* showing enlarged apophyses for grasping female; (E) fovea pedales, recesses in venter for protection of legs (Uropodina, Gamasida); (F) pteromorphs, flexible extensions of the hysterosoma that protect the legs (Oribatida).

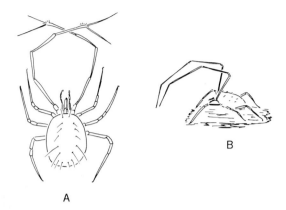

Figure 6-7. Exploratory use of legs of Acari: (A) *Podocinum* (Gamasida); (B) *Linopodes* (after Lawrence, 1953).

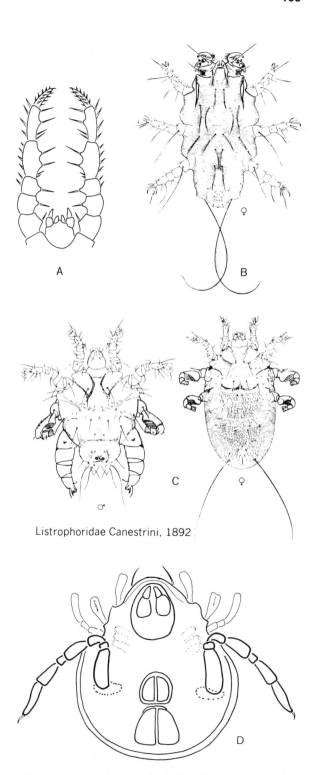

Listrophoridae Canestrini, 1892

paired claws and a padlike empodium. Single claws may occur as modified, clawlike empodia; an empodium may be in the form of a caruncle, be brushlike (rayed), or have suckerlike extensions. The stalk of the ambulacrum includes a basilar sclerite, sclerotized condylophores, and tendons to move the claws and/or empodium (Fig. 6-6C).

Coxae are free in Opilioacarida, Holothyrida, and Gamasida (Hammen, 1966); in a later publication (1977) Hammen claims that the Actinotrichida do not have coxae. In the more specialized mites of the Actinotrichida (e.g., *Alycus*) free coxae are not present and the legs exhibit six articulating segments: trochanter, femur, genu, tibia, tarsus, and apotele, without secondary segmentation.

The legs of ticks are usually divided into six segments: coxa, trochanter, femur, tibia, metatarsus, and tarsus (Fig. 6-4C). The coxae articulate with the body wall by means of membra-

Figure 6-8. (A) Legs I of rake-legged mites (Caeculidae, Actinedida); (B) legs I of fur-grasping mites (Myobiidae, Actinedida); (C) legs III, IV of fur-grasping mites (Listrophoridae, Astigmata); (D) jumping legs IV of Zetorchestidae (Oribatida) (after Baker et al., 1958).

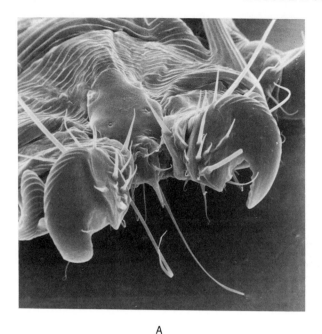

A

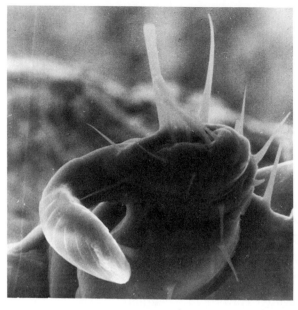

B

Figure 6-9. SEM of (A) *Myobia musculi* leg I; (B) male leg I of *Eadica brevihamata* (with courtesy and permission of Dr. F.S. Lukoschus).

nous connections, which allows some mobility in retraction and extension. The other articles of the legs are very flexible. Distally the tarsus ends in a pair or claws and a pulvillus (Ixodidae, Fig. 6-5G) or claws only (Argasidae). Haller's organ is a typical sensillum on the dorsum of the tarsi of both hard and soft ticks.

CHAETOTAXY

The chaetotaxy of legs is important in identification and classification. Likewise a review of Grandjean's analysis of leg setation (Norton, 1977) is important to the understanding of chaetotaxic systems and identification of setiform organs on the legs. A conceptual of parallel homology is employed to identify and orient setae of legs (Grandjean, 1961). If one envisions a mite with its legs perpendicular to its body, each leg would have an anterior side and a posterior side. These are designated respectively

the prime ['] and the second ["] sides. Thus, setae or organs on the anterior side of a given leg segment would be given the designation x['']; those on the posterior side would be x["]. Where mites have two pairs of legs directed anteriorly and the other two pairs directed posteriorly, conventionally structures closest to the sagittal plane of the body are paraxial (π); structures farther away from the sagittal plane are antaxial (α), which gives two possible positional designations to leg structure. In this way setae or setiform organs may have two positional notations (Grandjean, 1964) (Fig. 6-10A).

It is assumed that primitively the legs consisted of numerous small annuli, each supplied with a whorl of setae of some basic number (Grandjean, 1940). The modern segments and chaetotaxies resulted from fusions, numerical regressions of setal whorls, called verticils. The verticils of the genu, tibia, and anterior part of the femur may be distinctive, but are least evident on the tarsus, which implies origin from a

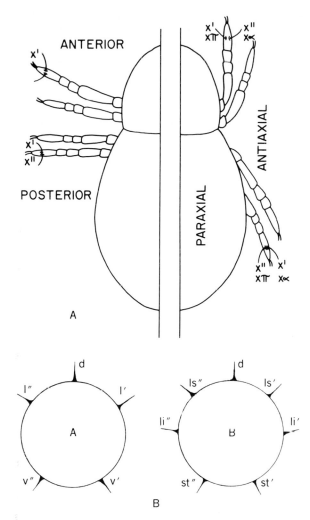

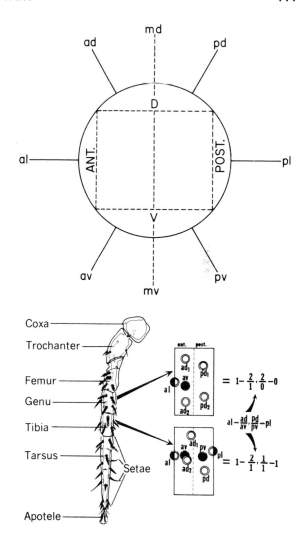

Figure 6-10. (A) Schematic orientation of legs for location and description of setae (after Grandjean, 1940); (B) two arrangements of verticils on legs (after Grandjean, 1940, from Norton, 1977).

Figure 6-11. (A) A system of setal positions on legs useful in nomenclatural and chaetotaxic studies (*Macrocheles plumosus*) (after Evans and Hyatt, 1963); (B) chaetotaxic formula expressed for the genu and tibia (from Krantz, 1970, 1978; after Evans and Hyatt, 1963).

small number of annuli (?four or more) (Norton, 1977). Verticils may include five or seven setae (Fig. 6-10B).

A system of setal nomenclature established for the Gamasida (Evans, 1963, 1972; Evans and Hyatt, 1963; Evans and Till, 1965) enables the characterization of setation on leg seg-

ments by a formula designating position of setal insertions related to the orientation of the leg segment (Fig. 6-11A, B):

anterolaterals (al) $\dfrac{\text{anterodorsals (ad)}}{\text{anteroventrals (av)}}$ $\dfrac{\text{posterodorsals (pd)}}{\text{posteroventrals (pv)}}$ posterolaterals (pl)

If a leg segment has two anterolaterals, two posterolaterals, three anterodorsals, three posterodorsals, one anteroventral, and one posteroventral, the letters translated into numbers would yield the following numerical formula:

$$\text{al} \; \frac{\text{ad}}{\text{av}} \; \frac{\text{pd}}{\text{pv}} \; \text{pl} = 2 \, \frac{3}{1} \, \frac{3}{1} \, 2$$

Chaetotaxy is important in the study of mites where the characteristic setal patterns of the idiosoma and leg chaetotaxy are distinctive for many species. Leg chaetotaxy is especially important in comparative studies [e.g., Cheyletidae, Fain (1979); Tydeidae, Andre (1981)].

REFERENCES

Andre, H. 1981. A generic revision of the Family Tydeidae (Acari: Actinedida). III. Organotaxy of the Legs. *Acarologia* 22(2):165–178.

Arthur, D. 1960. *Ticks,* Part V. Cambridge Univ. Press, London and New York.

Baker, E.W., J.H. Camin, F. Cunliffe, T.A. Woolley and C.E. Yunker. 1958. *Guide to the Families of Mites,* Contrib. No. 3. Inst. Acarol., University of Maryland, College Park.

Balashov, Y.S. 1972. *Bloodsucking Ticks (Ixodoidea)—Vectors of Diseases of Man and Animals,* Transl. 500 (T500). Med. Zool. Dep., USNMRU No. 3, Cairo, Egypt, U.A.R.

Cross, E.A. 1965. The generic relationships of the family Pyemotidae (Acarina Trombidiformes). *Univ. Kansas Sci. Bull.* 54(2):29–275.

Douglas, J.R. 1943. Internal anatomy of *Dermacentor andersoni* Stiles. *Univ. Calif. Publ. Ent.* 7(10):207–272.

Evans, G.O. 1963. Observations on the chaetotaxy of the legs in free-living Gamasina (Acari: Mesostigmata). *Bull. Br. Mus. (Nat. Hist.) Zool.* 10(5):277–303.

Evans, G.O. and K.H. Hyatt. 1963. Mites of the genus *Macrocheles* Latr. (Mesostigmata) associated with coprid beetles in the collections of the British Museum (Natural History). Vol. 9 (No. 9), pp. 327–401.

Evans, G.O. 1972. Leg chaetotaxy and the classification of the Uropodina (Acari:Mesostigmata). *J. Zool.* 167:193–206.

Evans, G.O. and W.M. Till. 1965. Studies on the British Dermanyssidae (Acari: Mesostigmata). Part I. External morphology. *Bull. Br. Mus. (Nat. Hist.)* 13(8):15–294.

Evans, G.O., J.G. Sheals and D. MacFarlane. 1961. *The Terrestrial Acari of the British Isles,* Vol. 1. Adland & Son, Bartholomew Press, Dorking, England.

Fain, A. 1979. Idiosomal and leg chaetotaxy in the Cheyletidae. *Int. J. Acarology* 5(4):305–310.

Grandjean, F. 1935. Les poils et les organes sensitifs portes par les pattes et le palp chez les Oribates. *Bull. Soc. Zool. Fr.* 60:6–39.

Grandjean, F. 1940. Les poils et les organes sensitifs portes par les pattes et le palp chez les Oribates. Deuxième partie. *Bull. Soc. Zool. Fr.* 65:32–44.

Grandjean, F. 1941a. La chaetotaxie comparee des pattes chez les oribates. (1e serie). *Bull. Soc. Zool. Fr.* 66:33–50.

Grandjean, F. 1941b. Observations sur les acariens (6e serie). *Bull. Mus. Nat. Hist. Natur.* (2) 13:532–539.

Grandjean, F. 1943. L'ambulacre des Acariens (2e serie). *Bull. Mus. Nat. Hist. Natur.* (2), 15:303–310.

Grandjean, F. 1954. Sur les nombres d'articles aux appendices des Acariens actinochitineux. *Arch. Sci.* 7:335–362.

Grandjean, F. 1961. Nouvelles observations sur les oribates (3e serie). *Acarologia* 6(11):170–198.

Grandjean, F. 1964. La solenidiotaxie des Oribates. *Acarologia* 6(3):529–556.

Hammen, L. van der. 1965. The Oribatid family Phthiracaridae. IV. The leg chaetotaxy of *Phthiracarus anonymum* Grandjean. *Acarologia* 7(2):376–381.

Hammen, L. van der. 1966. Studies on the Opilioacarida. I. Description of *Opilioacarus texanus* and revised classification of the genera. *Zool. Verh.* 86:1–80.

Hammen, L. van der. 1968. Stray notes on the Acarida (Arachnida) I. *Zool. Meded.* 42(35):261–280.

Hammen, L. van der. 1970. La segmentation des appendices chez les Acariens. *Acarologia* 12(1):11–15.

Hammen, L. van der. 1977. The evolution of the coxa in mites and in other groups of Chelicerata. *Acarologia* 19:12-19.

Hughes, A.M. 1961. The mites of stored food. Ministry Ag., Fish and Food. *Tech. Bull.* 9 (NSI). pp. 5-198.

Hughes, T.E. 1954. The internal anatomy of the mite *Listrophorus leuckarti. Proc. Zool. Soc. London* 124:239-257.

Hughes, T.E. 1959. *The Mites or the Acari.* Athlone, London.

Krantz, G.W. 1970. *A Manual of Acarology.* Oregon State Univ. Book Stores, Corvallis.

Krantz, G.W. 1978. *A Manual of Acarology,* 2nd ed. Oregon State Univ. Book Stores, Corvallis.

Jeppson, L.R., H.H. Kiefer and E.W. Baker. 1975. *Mites Injurious to Economic Plants.* Univ. of California Press, Berkeley.

Lawrence, R.F. 1953. *The Biology of Cryptic Fauna of Forests.* A.A. Balkema, Amsterdam.

Legendre, R. 1968. La nomenclature anatomique chez les Acariens. *Acarologia* 10(3):411-417.

Mitchell, R.D. 1955. Anatomy, life history and evolution of mites parasitizing fresh-water mussels. *Misc. Publ. Mus. Zool., Univ. Mich.* 89:1-28.

Mitchell, R.D. 1962. The musculature of a trombiculid mite *Blankaartia ascoscutellaris* (Walch). *Ann. Entomol. Soc. Am.* 55:106-119.

Neville, A.C. 1975. *Biology of the Arthropod Cuticle. Zoophysiology and Ecology,* Vol. 4/5. Springer-Verlag, Berlin and New York.

Norton, R. 1977. A review of F. Grandjean's system of leg chaetotaxy in the Oribatei and its application to the Damaeidae. *In*: Dindal, D.H., ed. *Biology of Oribatid Mites.* Syracuse State University, Syracuse, New York, pp. 33-62.

Snodgrass, R.E. 1962. *Anatomy of Arthropoda.* McGraw-Hill, New York.

Treat, A. 1975. *Mites of Moths and Butterflies.* Cornell Univ. Press (Comstock), Ithaca, New York.

Witmoyer, R.E., L.R. Nault and O.E. Bradfute. 1972. Fine structure of *Aceria tulipae. Ann. Entomol. Soc. Am.* 65(1):201-215.

Woodring, J.P and C.A. Galbraith. 1976. The anatomy of the adult uropodid *Fuscouropoda agitans* (Arachnida: Acari), with comparative observations on other Acari. *J. Morphol.* 150(1):19-58.

Young, J.H. 1968. Morphology of *Haemogamasus ambulans.* I. Alimentary canal. *J. Kans. Entomol. Soc.* 41(1):101-107.

Zakhvatkin, A.A. 1959. *Tyroglyphoidea (Acari)* (translation of *Fauna of USSR Arachnoidea,* Vol. 6, No. 1). Am. Inst. Biol. Sci., (translated by A. Ratcliffe and A.M. Hughes). Washington, D.C.

FORM AND FUNCTION— INTERNAL MORPHOLOGY

7

Digestive System

La Vie, c'est un action chimique.
—LAVOISIER

Many mites are fluid feeders and have the distinctive arachnid feature of preoral digestion. Enzymatic secretions from the salivary (oral) glands or from the midgut liquify the food to be ingested. Such liquification is related to the types of mouthparts present, to the internal digestive system, and to the process of digestion itself. The external structures of the digestive system include the gnathosoma and its associated structures described previously. The internal organs of the gut vary between orders. Salivary glands usually are present as accessory parts of the alimentary tract. Digestion is primarily intracellular and defecation is variable.

In food habits, mites range from general feeders and to those with specific diets. Ecto- and endoparasitic forms feed on blood, hemolymph, or tissue and cellular fluids of their invertebrate and vertebrate hosts. Follicle mites eat cell contents, sebum (the secretions of sebaceous glands), and lema (the secretions of the Meibomian glands). Some ectoparasites ingest skin dander, feather fragments, or hair. There are microphytophagous forms, fungivores, macrophytophagous feeders, carnivores, and scavengers that ingest decaying plant and animal matter; a single species may feed upon both living and dead material. Necrophagous and coprophagous examples are also found in several orders.

Adaptations of the feeding function and internal organs correlate with these food habits. Acarines with chelate chelicerae consume a variety of foods. Those with needlelike (styletiform) mouthparts show a wide range of feeding activities. Chelicerae range from the tiny needles of Eriophyidae to the extremely long ones in other plant-feeders. The phytophagous spider mites (family Tetranychidae) have bases of the chelicerae fused to form an extensible stylophore from which the long, curved, styletiform chelicerae may be projected to pierce the host plant tissue on which they feed (Fig. 5-1). Plant feeding Tarsonemidae or the Stigmaeidae have tiny stylets by comparison, but not as small as those of the gall-forming Eriophyidae. Feather quill mites (Syringophilidae) pierce the quill wall with tiny, rather straight stylets to extract the fluid food from the feather of the avian host (Kethley, 1971; Casto, 1974). External feather mites (Analgidae and Proctophyllodidae), not injurious to the host, feed on tissues of broken quills (Radford, 1950) or fragments of feathers (Atyeo, 1960). Slime mites, like *Histiostoma* (Histiostomatidae) stand "ankle-deep" in the fluid of vegetable decay and skim the bacterial "goodies" from their soupy meal into the digestive tract (Fig. 5-1B).

Gamasine blood feeders like the Macronyssidae pinch the skin tissue of the vertebrate host with their chelicerae, break the capillaries at the surface of the skin, and cause a pool of blood to form from which they ingest their meal. Hard ticks lacerate the host tissue with razor-sharp chelicerae, insert the toothed hypostome for an anchor, surround it with some cement, then with salivary fluid and pharyngeal pumping action become engorged with blood (Fig. 7-1).

The common "chigger" or "red bug," a larval trombiculid mite, inserts its curved, saberlike chelicerae into the host's skin. As a result of the host's reaction to the salivary secretions, a stylostome (=histiosiphon)—a "drinking straw"—is formed. The chigger sits at the top of this tube until it has ingested its meal, then drops off the host and leaves the latter to deal with the irritating itch that results from the bite (Fig. 7-2A–F; Jones, 1950). A remarkably branched stylostome is formed in beetles by the the feeding of a trombidiid mite, *Teresothrombium susteri* (Robaux, 1974) (Fig. 7-2G–H).

In contrast to these fluid feeders, oribatid mites usually chew solid food with their massive chelate chelicerae until the particulate size is ready for ingestion. Yet, even with oribatids, variations of feeding apparatus occur. For example, in Gustaviidae, chelicerae with long shafts and tiny, saw-toothed ends (Fig. 5-18B) probably macerate woody materials or fungal food.

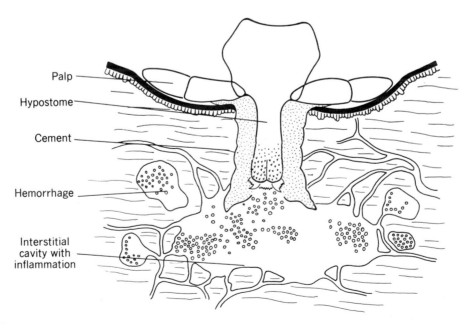

Palp

Hypostome

Cement

Hemorrhage

Interstitial cavity with inflammation

Figure 7-1. Tick feeding in mammalian host; mouthparts in place showing splayed palps, toothed hypostome surrounded by cement, infiltrating, hemmorhagic cavity of host, blood vessels and blood cells (after Balashov, 1972).

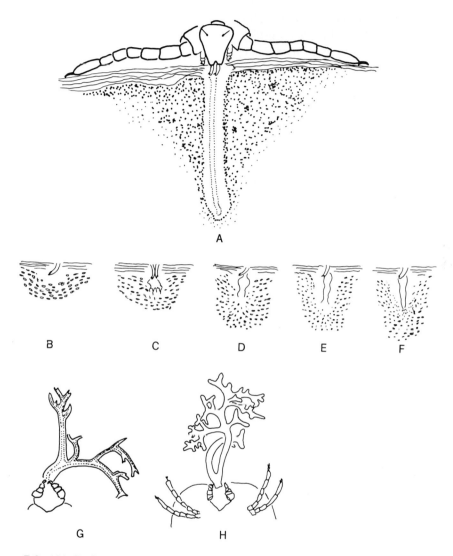

Figure 7-2. (A) Stylostome (=histiosiphon) and keratinization of skin around well-formed tissue canal caused by chigger bite (after Vitzthum); (B) initial skin reaction to physical irritation; (C) injection of saliva; (D) tissue canal forming; (E) injection of saliva; (F) suction arrested by protective tissue (after Jones); (G, H) stylostome of *Teresothrombium susteri* in interior of beetle (after Robaux, 1974).

The alimentary canal of mites generally comprises three main regions like the tract of other arthropods and is derived from similar embryonic anlage. The anterior foregut, formed from the stomadaeum of the embryo, includes the preoral buccal cavity, the mouth, and the sclerotized pharynx. The midgut, derived from the mesenteron or the middle intestine of the embryo, comprises the esophagus, the ventriculus (or stomach), and its attached caeca. In some instances (Oribatida) a crop (or ingluvies) may be interpolated between the esophagus and the ventriculus. (Because Oribatida feed on particulate matter rather than liquid food like most other mites, the crop acts as a temporary food storage organ in these beetle mites.) The hindgut, derived from the proctodaeum, consists of a short intestine, a rectal pouch, a rectal sac or rectal tube, a rectum, and an anus.

Salivary glands are usually associated with the anterior end of the digestive tract. While they are not directly parts of the alimentary canal, they usually empty into the preoral space of the gnathosoma and are important to digestion. They vary in number, size, and structure.

Typical also of many arthropods, the excretory organs or Malpighian tubules usually are connected to the digestive tract at the juncture between the midgut and the hindgut.

SALIVARY GLANDS

Morphology

All orders of mites have well-developed salivary glands, although these glands may differ in size, relative complexity, and origin (Woodring, 1976). Exceptions are found in some Astigmata (Prasse, 1969), but correlations cannot be made currently for the types of glands in the Anactinotrichida and Actinotrichida. The salivary glands of mites occur in the gnathosoma and idiosoma. Gnathosomal glands are usually simple and saclike. Some idiosomal glands are acinous or racemose, like tiny clusters of grapes, while others are more compacted; still others are distinctly tubular. The functions of these types of glands are related to (1) their form, (2) their alveolar types, and (3) the locations of their ducts, but little is known of the functions of many.

Ducts of the salivary glands are frequently lined with a cuticle and a system of spiral filum, threadlike structures that resemble the taenidia of tracheae (Fig. 7-3). This suggests that salivary glands and tracheae developed together where the spiral lining prevents collapse of the tubes. Not infrequently in dissections the

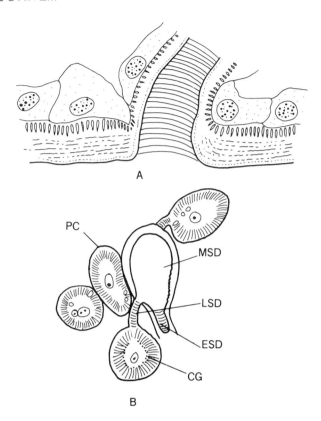

Figure 7-3. (A) Entrance of secondary salivary duct into main duct showing spiral filum or taenidia-like structures (after Balashov, 1972); (B) main, lateral, and efferent salivary ducts in *Argas persicus* (CG, coarse granules; ESD, efferent salivary duct; LSD, lateral salivary duct; MSD, main salivary duct; PC, peripheral cytoplasm) (after Chinery, 1964).

salivary ducts are confused with the tracheae, for at times the ducts and the tracheae coalesce.

The secretions of the salivary glands of mites vary with the structure the gland. Inasmuch as many mites are fluid feeders, it is thought that digestive enzymes produced by the glands contribute to preoral digestion. Some glands have a toxic saliva; anticoagulant agents in the saliva of ticks facilitate the ingestion of blood during feeding.

Opilioacarida–Holothyrida. Two infracapitular "salivary glands" occur in *Opilioacarus texanus*, each with a duct and a ductal orifice lateral to the mouth (Hammen, 1966). The glands themselves correlate with the "antennary glands" identified by With (1904) (cited by Hughes, 1959) (Fig. 5-4). Cheliceral, maxillary, and coxal glands are identified in Holothyrida. The two former are acinose (alveolar) with conical acini; the coxal glands consist of a sacculus, horn, and duct (Vitzthum, 1943).

Holothyrida have a pair of "maxillary glands" on either side of the pharynx, with large, conical cells around each acinus. Common ducts extend from the glands to the cheliceral space. Additional "maxillary" syncytial glands join by duct to the common salivary duct; the ducts have spiral chitinous thickenings like the taenidia of tracheae (Hughes, 1959).

Gamasida. The salivary gland of the gamasine *Glyptholaspis confusa* (with spiral lining) leads into the salivary stylus (siphunculus) adjacent to the corniculus (Hammen, 1964).

Racemose clusters of cells situated between and above the coxal glands comprise the salivary glands of *Fuscouropoda agitans* (Woodring, 1976) (Fig. 7-4A). Groups of smaller, inactive salivary cells lie close to the salivary duct and have a dense, uniformly basophilic cytoplasm. Secreting, functional salivary cells are much larger, are irregular in shape, and have clear cytoplasmic vacuoles. Their nuclei are large, basophilic and have distinct nucleoli. When secreting, the cells enlarge and elongate so that they are more distant from the duct than the nonsecreting cells. The ducts are paired and internally sclerotized with a thin cuticle. At the gland, the collecting duct is perforated irregularly so that each pore receives the secretion of a single salivary cell. The ducts extend laterally to the cheliceral sheaths and empty into the salivary stylets.

Gamasid salivary glands are small compared to those in ticks. Simple racemose types occur in *Echinolaelaps echidninus*, the spiny rat mite,

and *Haemogamasus ambulans* (Jakeman, 1961; Young, 1968). The small clusters of cells are at the distal ends of small ducts. The salivary cells of the spiny rat mite are hollow, are relatively thin-walled, and have peripheral nuclei (Fig. 7-4D). The salivary glands of *Haemogamasus ambulans* lie anterolateral to the synganglion and comprise a cluster of seven globular units at the end of a short duct (Fig. 7-4C). Acinous glands in the moth ear mite (*Dicrocheles phalaenodectes* are larger and more extensive than the glands in *Echinolaelaps* and *Haemogamasus* and have nine major lobes at the end of a relatively short duct (Treat, 1975) (Fig. 7-4B). The salivary duct of *Ornithonyssus bacoti* ends in a salivary stylus. The cells appear granular and slightly vacuolated (Hughes, 1949, 1959).

Ixodida. Salivary glands of ticks have been studied in greater detail than others. Those of argasid ticks are more compact than those in ixodids and are generally restricted to the area above coxae I–III. Size and shape are nearly equal in unfed and recently engorged soft ticks. Although the secretory alveoli of tick salivary glands comprise several types, argasid ticks exhibit a most primitive type—a secretory alveolus with a single type of cell.

The salivary gland of *Argas (Carios) vespertilionis* and *A. (O.) brumpti* have lobulated masses of alveoli in the ventroanterior part of the body cavity. Two types of alveoli are present: The anterior mass is larger and less compact than the more mesial, smaller, and compacted cluster of alveoli (Fig. 7-4E). The salivary duct extends from each gland to the capitular foramen. In *Ornithodoros papillipes* multiple clusters of alveoli are situated on a branched system of tubules.

Salivary glands of ixodid ticks are racemose (grapelike clusters) with many efferent ducts (Fig. 7-4F) in the ventroanterior part of the hemocoel. They rarely extend posteriorly beyond the level of coxae IV (Amblyommidae). Size of the gland is determined by the physiological state and is generally larger in feeding ticks.

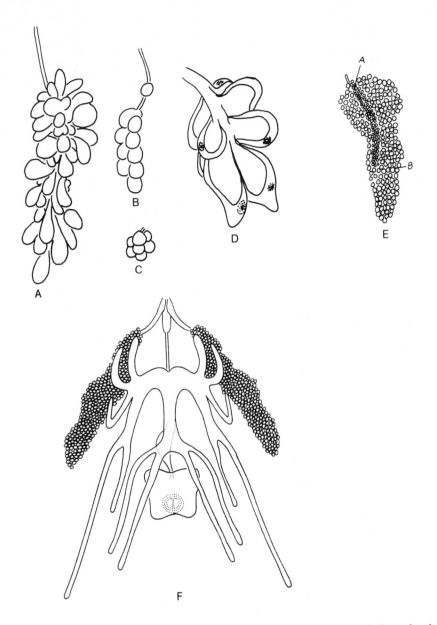

Figure 7-4. Salivary glands: (A) *Fuscouropoda agitans*; (B) *Dicrocheles phaelenodectes*; (C) *Haemogamasus ambulans*; (D) diagrammatic section of *Echinolaelaps echidninus*. (E) Two types of glands in *Argas (O.) brumpti*; (F) salivary glands and ventricular caeca of *Dermacentor andersoni*.

The ducts of the glands are epidermal in origin and are formed as deep invaginations of the body wall. Main ducts are paired and branch within the glands. Alveoli are clustered on these secondary efferent ducts. Main ducts show an internal spiral filum (Fig. 7-1); muscle fibers are absent (Balashov, 1972).

Histology of Tick Salivary Glands. Distinctions between types of alveoli in the salivary glands of ticks vary according to the author and the analytical techniques involved. As a result the descriptions that follow are not completely correlated as to types, but reflect the state of the research involved.

Two principal types of alveoli occur in the salivary glands of ticks: pyramidal and secretory (Balashov, 1972). The histology is heterogeneous, especially for the several types of secretory alveoli. The pyramidal type of alveolus, so called because of its shape, is found in the anterior part of the salivary gland and differs in size in various species. Originally this type of gland was assumed to consist of a single cell with several nuclei. Electron microscopy and other investigations show that the pyramidal alveolus is formed of a number of indistinctly separated pyriform or polygonal cells with their apices and efferent ducts directed toward the main salivary duct (Fig. 7-5). Pyramidal alveoli are characterized by a peripheral zone of fibrils perpendicular to the basement membrane and with mitochondria between.

Little change in structure or size occurs in the pyramidal alveoli from the time of differentiation at molting until the end of feeding. The alveoli lack inclusions of protein, mucoids, liquid globules, and secretory vacuoles. They also show relatively little RNA, but treatment with NET (nitroblue tetrazolium) demonstrates high activity of succinic dehydrogenase. The similarity of the fine structure of these alveoli with avian salt cells and similar enzymes used in sodium transport, suggests an osmoregulatory function for this type of alveolus (Balashov, 1972).

In *Argas* (*P.*) *persicus* and in other argasid species of ticks, the paired salivary glands extend posteriorly to the level of coxa III. Histologically, two types of alveoli are observed (Roshdy, 1972; Chinery, 1974). Type A is a narrow compact mass along the dorsomesial side of the gland. It is nongranular and is connected to the cuticular intima present in the ducts. Oval nuclei are observed in the alveoli, which appear to be syncytial. The cytoplasm is fibrillar with fine granular inclusions and is highly vacuolated toward the center of the alveolus. Watery and viscous solutions are secreted by two types of cells present. Neutral mucopolysaccharides, tyrosine, and firmly bound proteins may be associated with active feeding in argasids and may produce the active lytic component of argasid saliva. Special cells may be involved in the production of a sulfated anticoagulant. (In ixodids a similar secretion may be involved in tick paralysis, production of secondary cement for attachment, and tissue damage involving polymorphonuclear neutrophils (Moorehouse and Tatchell, 1966).)

Three types of cellular alveoli are identified in the salivary glands of *Dermacentor variabilis* (Coons and Roshdy, 1973) and similar alveoli occur in *Hyalomma asiaticum* (Balashov, 1972). One has a nongranular type of cell, the other two are granule-secreting forms. The alveoli are connected directly or indirectly to the main salivary duct or its branches and are innervated by branches of lobular nerves. The axons of these nerves frequently contain neurosecretory vesicles that enter the alveolus. Each of several cells that make up the nongranular alveolus has a basal region with membranous infoldings, mitochondria, and vacuoles. The cells of the intermediate region have fewer infoldings and more vacuoles. The cells of the apical region have few infoldings and mitochondria, some vacuoles, but many microtubules. Each of the two granule-secreting alveoli comprises at least three cell types interconnected by narrowed epithelial cells with microtubules and specialization at their membranal junctions with the alveolar cells. In the duct system a thin, valvular canal with a conical cuticular

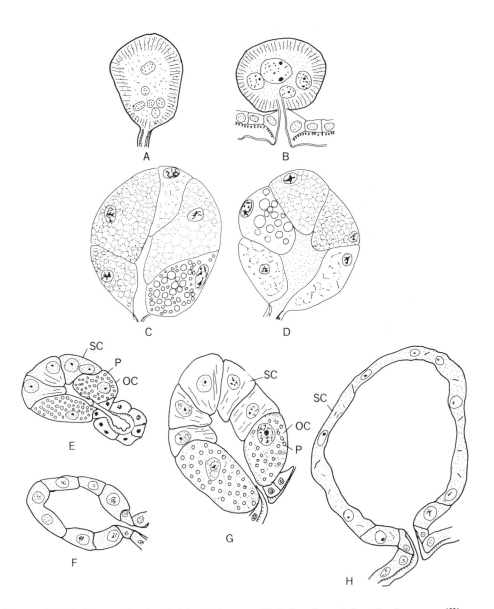

Figure 7-5. Salivary glands of ticks: (A) pyramidal alveolus of *Ornithodoros papillipes*; (B) same of *Hyalomma asiaticum*; (C) secretory alveolus of *O. papillipes*; (D) secretory alveolus of same releasing secretion into intralveolar lumen; Salivary glands of *Ixodes ricinus*; (E) alveolus replete with secretion at beginning of growth period; (F) empty alveolus at beginning of growth period; (G. H) empty alveoli at end of growth period (OC, orifice cells; SC, secretory cells; P, protein secretion). (After Balashov, 1972.)

guard valve lies adjacent to the microvillar apices of the alveolar cells. The valve is at the juncture of the alveolar duct and the lobular duct, which in turn joins the main salivary duct (Fig. 7-4F).

The subfamily Ixodinae (e.g., *Ixodes ricinus*) has a single type of secretory alveolus, comprised of seven to nine orifice cells at the base of the efferent duct (Fig. 7-5E–H). In recently molted ticks secretory substances are absent, but after postmolting is completed, the nuclei of the orifice cells enlarge and the cytoplasm becomes highly vacuolated; secretory cells also develop. When the tick attaches to the host, secretory activity begins and the alveoli grow rapidly. The vacuoles of the orifice cells enlarge, coalesce, and discharge the secretions, concurrently with the shedding of the apices of the cells of the alveoli. These materials are released into the intraalveolar cavity and the alveolus becomes a thin-walled vesicle (Fig. 7-5H). Subsequently, the alveolus collapses and the cellular structures degenerate.

The salivary glands in male *I. ricinus* are similar in structure to those of the female but in recently molted, unfed males the secretory alveoli are completely developed and replete with secretory materials.

A more complex structure of alveoli occurs in the subfamily Amblyomminae (e.g., *Hyalomma asiaticum*) than in Ixodinae (Balashov, 1972). Three pyramidal alveolar types are observed in females: Type I, pyramidal alveoli, are in the anterior part of the gland on the main efferent duct; Type II, alveoli of nearly spherical cells in the anterior two-thirds of the gland, empty into the primary branches of the efferent ducts; they shrivel up during secretory discharge; Type III alveoli make up the principal glandular mass and empty into the primary and terminal branches of the efferent ducts. A Type IV alveolus is found only in males. In the unfed ticks these alveoli are small vesicles composed of 7 to 10 cuboidal cells around a central cavity confluent with the efferent duct. No secretory activity is observed in this type (Balashov, 1972).

By electron microscopy various authors have found marked ultrastructural differences in cell types, secretory materials, and staining properties of the cells (e.g., Fig. 7-6).

Different species of ticks exhibit different arrangements of the types of alveoli and the types of cells involved. In addition to *a* through *d* cells, *e*, *f*, and *g* are noted, depending upon the staining properties of the cells and the successive stages of secretion (Balashov, 1972).

Saliva in Ticks. The complete histochemistry of tick saliva is not known even though cell types and salivary components are identified (Balashov, 1972; Roshdy, 1972). The composition of the saliva differs between the argasid

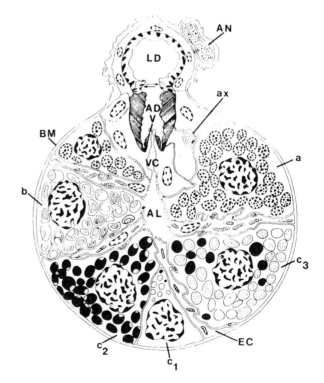

Figure 7-6. Reconstructed drawing of type II alveolus in unfed male *Dermacentor variabilis*. a, type *a* cell (AD, alveolar duct; AL, alveolar lumen; ax, axon); b, type *b* cell (BM, basement membrane); c1, c2, c3, different types of *c* cells (EC, epithelial cell; LD, lobular duct; VC, valvular canal) (after Coons and Roshdy, 1973).

and ixodid ticks. In argasids, a single secretion, probably including anesthetizing agents, is secreted by a single cell type. The components of the secretion are mucoproteins, glycoproteins, or possibly another complex of proteins and carbohydrates. Ixodid salivary glands, on the other hand, contain two different materials, protein and carbohydrate–proteins. Different materials are secreted from different gland cells in the secretory alveoli (Chinery, 1965a,b).

Protein secretion occurs in the vacuoles of orifice cells in *Ixodes ricinus*. Secretions in this and other ticks accumulate up to the end of postmolting development. The carbohydrate–protein secretions of the different types of cells in the Ixodinae and Amblyomminae resemble mucoproteins, glycoproteins, or possibly a complex near the neutral mucopolysaccharides.

Histochemical reactions of cytoplasm in salivary glands include carbohydrates, proteins, and RNA. Lipids are completely absent. Amounts of these materials vary between tick species, types of alveoli, and the types of secretory cells in the alveoli. Different substances and amounts come from the cytoplasm or from the secretory vacuoles and are identified by different staining techniques. The principal mucoproteins discharged are assumed to change during the process of secretion. Disintegration of the secretory cell membranes does not occur in argasids, in contrast to ixodid ticks, and secretions elaborated into the alveolar cavity are released into the ducts and into the salivarium. During and immediately after feeding, the ducts and salivarium are replete with these discharged materials. In unfed ticks the ducts and cavities are empty. The accumulation of the salivary materials and discharge of the same is asynchronous, so that cells are in different stages of secretion (Balashov, 1972).

Salivation in *Boophilus microplus* may be induced with cholinergic and adrenergic drugs such as noradrenaline (Edney, 1977). Little is known about the chemicals that stimulate the glands naturally.

Osmoregulation by Salivary Glands in Ticks. Ixodid ticks lack osmoregulatory coxal glands, and the elimination of water through the Malpighian tubules is negligible. Proteins, carbohydrates, and other macromolecular materials in the saliva of secretory alveoli suggest that the pyramidal alveoli function mainly as organs of saltwater exchange in these ticks. About two-thirds of the water and salts are eliminated during the feeding of *Ixodes ricinus* when blood from a vertebrate host is concentrated in the ventriculus (Lees, 1946, 1947). Excess NaCl and about 70% of the water ingested with the blood meal are eliminated through the saliva of the cattle tick *Boophilus microplus* (Tatchell, 1967). This is experimental evidence (using tritiated water) of the osmoregulatory function of salivary glands, at least in the ixodid ticks. The main osmoregulatory organs in the Argasidae are the coxal glands, so the salivary glands have little relation to this function. Fewer pyramidal alveoli in the salivary glands of argasid ticks compared to ixodids may be the reason for these differences.

Actinedida. Paired globular to elongated glands are found in the gnathosoma of the Pygmephorinae (Pyemotidae) (Cross, 1965). The salivary glands of *Eriophyes tulipae* Keifer, the curl mite, are simple, bulbous, paired glands that lie above the esophagus anterior to the synganglion. The salivary ducts connect with the buccal opening (Whitmoyer et al., 1972) (Fig 7-7A).

Simple salivary glands occur in the Demodicidae but in *Demodex folliculorum* they do not extend beyond the level of leg IV. In *D. antechini*, however, the salivary glands (and the synganglion) extend into the opisthosoma beyond the level of legs IV (Fig. 7-7B). The paired salivary glands of *D. folliculorum* are teardrop-shaped and consist of a single large cell surrounded by six to eight smaller, squamous cells with vacuolated cytoplasm and a few dark granules. Nuclei are large with a single nucleolus. Anteriorly the cells of each gland converge

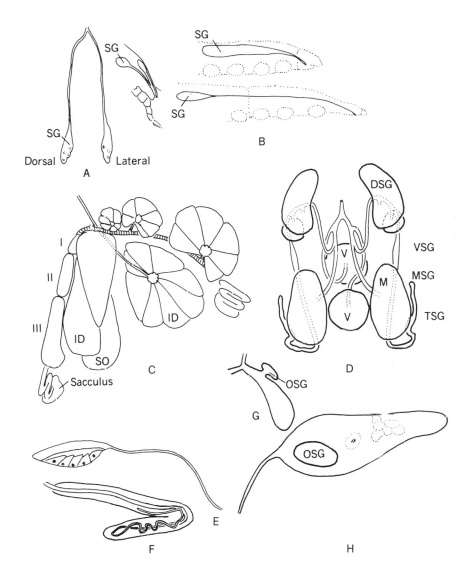

Figure 7-7. Salivary glands of Actinedida: (A) Eriophyidae, dorsal and lateral views (after Vitzthum, 1943 and Whitmoyer et al., 1972); (B) Demodicidae (*Demodex folliculorum* upper, *D. antechinus* below) (after Desch and Nutting, 1977); (C) Bdellidae (after Alberti and Storch, 1973); (D) Calyptostomidae, dorsal view (DSG, dorsal salvary gland; MSG, median salivary gland; TSG, tubular salivary gland; VSG, ventral salivary gland) (after Vistorin-Theis, 1977); (E) tracheal salivary gland, partially cut away and (F) tubular salivary gland of Tetranychidae (after Blauvelt, 1945); (G) lateral and dorsal view of oval salivary gland (OSG) and main salivary gland of *Trombicula autumnalis* (after Jones, 1950).

to form a solid process and connect with the sclerotized salivary duct that opens in the gnathosoma at the base of the stylets. The ducts enter through the foramen of the capsule and pass between the pharyngeal dilator muscles into the preoral cavity. Each duct has a crescent-shaped lumen for most of its length. In *D. antechini* the salivary glands are smaller than those of *D. folliculorum* (Desch and Nutting, 1977).

The podosomal gland complex of *Bdella longicornis* and *Neomolgus littoralis* (Bdellidae) has three exocrine glands and a coxal gland (Alberti and Storch, 1973) (Fig. 7-7C). The "salivary glands" are of two types. In one type the cells are filled with secretory granules. In the second type the granules are loosely distributed throughout the cytoplasm. A tracheal gland at the base of the chelicerae exhibits smooth ER, electron nontransparent granules, and a high degree of cellular interdigitation in contrast to the other types. The coxal gland found in association with these other glands comprises a sacculus and nephridial ducts similar to those of other euarthropods. The sacculus is lined with podocytes and the nephridial duct is lined with epithelial cells with extensive infoldings.

Calyptostoma velutinus has four paired acinous salivary glands and a single pair of tubular glands (Fig. 7-7D). The unpaired tracheal gland found in other prostigmatid mites is not matched, but the general arrangement of salivary glands found in trombiculids is present. The paired acinous glands include the dorsal, lateral, medial, and ventral salivary glands. Ducts from these glands join a common salivary duct to enter the buccal cavity. The paired tubular glands lie ventral of the dorsal, lateral, and medial glands. The saclike free end of this pair is looped forward beneath the median gland. Each duct of the tubular glands is confluent with the podocephalic canal before it joins the main duct from the salivary glands (Vistorin-Theis, 1977).

The silk-spinning apparatus of spider mites

has been described and interpreted in several ways, one of which is as a salivary gland (Blauvelt, 1945). Blauvelt reviews the literature on this topic and describes the details for the common red spider mite *Tetranychus telarius*. A pair of tubular silk glands arises above coxa I and extends posteriorly to the level of the ovaries where the glands double back on themselves. Nearly uniform in diameter, the tubular gland contains a ductlike lumen in the center that becomes convoluted toward the distal end of the gland. These tubular glands were considered to be homologous to the *glandes tubular* of Thor (1903). The cells of the glands are usually triangular in shape, clearly defined, but of variable size and arranged around the central lumen. The cytoplasm stains darkly. The duct of this gland extends anteriorly posterior to the rostrum where it unites with the corresponding duct from the other side and joins the common silk gland duct (Fig. 7-7E, F).

A pair of rather large reniform glands lies in the anterior part of the propodosoma above the fat body and the synganglion and anterior to the ventricular caeca (Fig. 7-7F). The margins of these glands conform to the contours of these organs but are indented by the passage bundles of tracheae, cheliceral nerves, and dorsoventral muscles. Numerous tracheae evidently pierce the glandular tissue. The cells of the reniform glands are large, elongated, and connected to each other at the region of the lumen. Different stages of secretory activity affect the histological appearance of the cytoplasm, which is usually of similar density throughout. At other times the distal cytoplasm of the cells stains very darkly and shows an irregular network with groups of darkened granules. The duct of the gland is narrow and short. It extends forward and ventral to the dorsal surface of the rostrum where it unites with the duct from the tubular silk gland.

An unpaired gland on the median line occurs in *T. telarius*. Comparable to the gland described by Thor (1903) for many of the prostigmatid mites and the "azygous salivary gland" (Michael, 1896), the gland is bilobed posteriorly. Anteriorly the gland lies in the posterior cleft of the mandibular plate. The cells have moderately staining cytoplasm and nuclei with a light peripheral ring. The duct of this gland extends forward between the peritremes, where it is embedded in sclerotized supports, and terminates near the base of the spinae. It discharges its secretions on the dorsal surface of the cheliceral stylets. It is assumed that this gland has retained its enzyme-secretory function (Moss, 1962).

Trombicula autumnalis exhibits two pairs of glands. A relatively large pair of elliptical glands extends posterolaterally to the margin of the anterolateral ventricular caeca. The surface appears reticulate because of the large polygonal cells with spheroid nuclei and granular cytoplasm (Fig. 7-7G, H). Variations in the pigmentation are related to feeding. In nonfeeding larvae these cells stain much darker.

The main salivary duct is composed of a nonstaining cuticle and arises from the anterior of the gland. This duct extends forward between the pharyngeal muscles and opens into the buccal cavity.

A second pair of salivary glands (oval salivary gland) is pressed against the anterior face of the larger glands so much that it is difficult to distinguish between them except by careful dissection. The structure of these oval glands is not well-defined (Jones, 1950).

Five pairs of glands comprise the salivary apparatus of the North American chigger, *Trombicula alfreddugesi*. These glands lie in the propodosoma behind the gnathosoma and form a mass that covers the anterodorsal and lateral surfaces of the synganglion. They differ from each other in general anatomy as well as in histological and cytological structure, although details on the latter are incomplete (Brown, 1952). The dorsal to ventral sequence of the glands is as follows: (1) The dorsal gland lies beneath the scutum and extends from the sensillary area nearly to the atrium. Its duct comes from the ventral surface. (2) The lateral gland is pear-shaped (pyriform) and is placed ventrolaterally to the dorsal gland with the smaller and directed anteriorly. Its duct emerges from the ventral, anterolateral surface about a third of its length from the anterior end. Smaller ducts from intercellular lacunae are visible. (3) The median gland is triangular with rounded corners, the small end directed anteriorly. This front end lies under the scutum beneath the sensillary area; its posterior end is mesad of the lateral gland. Smaller ducts similar to those of the lateral gland join the main duct. (4) The tubular gland is elongated with a tortuous arrangement beneath the anterior part of the lateral gland, and over the synglanglion to end in a dorsally directed blind tip. The duct of this gland is continuous with the anterior of the lumen. (5) Last, the ventral gland is bilobed. The larger of the two lobes is dorsal and rests over the anterolateral surface of the synganglion. The smaller, ventral lobe extends along the esophagus anterior to this structure. The duct for this gland originates at the junction between the two lobes.

The only citation of the salivary glands of an adult trombiculid (*T. akamushi*) describes a single pair of trilobed glands and a tubular gland with ducts that empty into the subcheliceral space in a position similar to that in *T. alfreddugesi* (Brown, 1952). Larger prostigmatid mites have a single unpaired and six paired glands, but a reduced number in smaller mites (Vitzthum, 1943; Brown, 1952). In 1986 Michael found five pairs of glands in the water mite *Thyas petrophilus* and Bader (1938) found the full six pairs in *Hygrobates longipalpus*.

Nothing is known of the enzymatic effects of these secretions in adults, but the histolytic properties of the larval saliva are well known. The penetration of the stratum corneum layer of the host by the chelicerae results in erosion of the deeper skin layers because of the saliva (Jones, 1950).

The salivary glands of *Unionicola fossulata* (Unionicola) were compared (Mitchell, 1955) with the general descriptions of these glands for water mites (Bader, 1938). Anteriorly, in this water mite a small mass of cells, the tracheal gland, lies immediately posterior to the chelicerae and empties by a short duct into the postoral cavity (Fig. 7-8A, B).

Two pairs of large reniform glands comprise the main mass of glandular tissue. These consist of a smaller, ovoid anterior gland mediad of a larger gland. The cells of the small gland are conical and the smaller ends come together in a central cavity that is drained by a simple duct. Each cell shows a darker, granular region at the

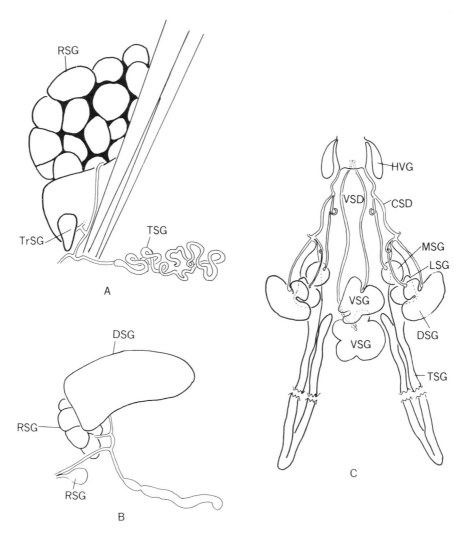

Figure 7-8. Salivary glands of water mites: (A) *Najadicola ingens*; (B) *Unionicola fossulata* (DGS, dorsal salivary gland; RSG, reniform salivary gland; TrSG, tracheal salivary gland; TSG, tubular salivary gland) (after Mitchell, 1955); (C) dorsal view of salivary glands of *Allothrombium lerouxi* (CSD, common salivary duct; DSG, dorsal salivary gland; HVG, hypostomal venom gland; LSG, lateral salivary gland; MSG, median salivary gland; VSD, ventral salivary duct; VSG, ventral salivary gland; TSG, tubular salivary gland) (after Moss, 1962).

tip; the nuclei of these cells are located in the basal part of the cells and contain a distinct nucleolus.

The larger, dorsolateral gland is irregular in outline and composed of cells that stain less darkly than the cells of the smaller gland; it exhibits coarse granules and a central vacuolated area. No distinct cavity for the reception of secretions is noted, but the gland is drained by a branched duct that joins the comparable duct from the other reniform gland to form a pair of common ducts that empty into the dorsal part of the post-oral cavity.

Tubular glands are present, but differ from the type described by Bader (1938). This simple, straight gland is closely appressed to the anterolateral caecum of the midgut. The ducts for this gland join the common duct from the reniform glands on each side.

The glands of *Najadicola ingens* (Unionicolidae) are similar in form and location to those of *Unionicola*, but with larger cells in the former. The reniform glands in this species are so compacted as to appear as one, but can be differentiated cytologically. Nuclei of the cells of these glands do not have distinguishable nucleoli (which may be artifactual). Granular and vacuolated cells comprise the histology of the gland. The cells are clustered around a small lumen that drains into ducts that join to form a common duct (Fig. 7-8A). The tubular glands are much longer than in *Unionicola* and much more convoluted; ducts join the common duct of the reniform gland before it enters the post-oral cavity. A saclike medial tracheal gland is present and has cuboidal cells in its walls and empties into the post-oral cavity as well (Fig. 7-8A).

It is postulated that the earliest trombidiform mites were predaceous, but restricted to a liquid hemolymph diet because of their mouthparts (Moss, 1962). Development of a salivary and digestive complex enabled these mites to utilize a greater variety of food. The salivary glands, where secretions are carried by ducts to the preoral cavity and injected into the body of the prey organisms, facilitated feeding.

It is inferred that saliva causes extensive histolysis and predigestion of muscle or other firm tissues. The liquified food then can be drawn up into the ventriculus by the pumping pharynx. The state of development of the salivary glands in such mites (e.g., *Allothrombium lerouxi*) thus enables the predator to ingest most of the tissues of the prey organism, leaving little but a collapsed and empty exoskeleton.

Descriptions of salivary glands for the higher trombidiform mites are available (Moss, 1962). In 1860 Pagenstrecher described paired reniform and tubular glands for a species of *Trombidium;* in similar species the common salivary duct and the blind cul-de-sac-type tubular gland are present. Five or six pairs of glands are described for *Thyas,* a water mite (Michael, 1895a) where the tubular gland was interpreted as a reservoir. Four paired glands and one unpaired gland occur in *Neomulgus litoralis* (Linnaeus) (=*Bdella basteri* Johnston) (Michael, 1896).

Salivary glands of *Allothrombium lerouxi* (Fig. 7-8C) are located in the gnathosoma and in the idiosoma (Moss, 1962). Although location and spatial relationships are fairly constant, there are individual variations in the shapes of the glands. The gnathosomal glands include two sets: (1) the hypostomal venom gland, a paired set of glands that lies slightly anterior to the epistome and extends along the wall of the basis capituli on either side; (2) a saclike cheliceral venom gland that extends from the base of the movable digit of the chelicerae into the basal joint. Both of these glands are thought to produce toxins.

Five pairs of idiosomal glands occur in different positions around the synganglion. Four of the pairs are relatively globular; the fifth is tubular (Fig. 7-8C). The dorsolateral, lateral, median, and ventral salivary glands have characteristic histological features that are different from the structure of the venom glands, but generally resemble each other. The histological structure of the tubular glands is strikingly different.

The largest of the idiosomal glands is the

dorsolateral, which is located below the integument at the level of the scutum and lies over the other idiosomal glands dorsally and laterally. Its ducts may loop before joining the common salivary duct. The smaller lateral salivary gland is located ventromediad. Its duct joins the common duct at the anterior end of the tubular gland. Medial salivary glands are the smallest of the idiosomal glands. These rounded bodies empty by ducts into the common salivary duct near the level of the anterior margin of the first coxa. The ventral salivary glands comprise a cluster of glands that lie in tandem and slightly overlap each other. The ducts from these ventral glands empty into the atrium independently of the ducts from the other glands.

The hollow, elongated tubular gland is somewhat broadened at its anterior end into a type of reservoir, but is of more or less uniform diameter posteriorly. It doubles back on itself within the idiosoma and its blind end approximates the posteroventral margin of the dorsolateral gland. It is homologous with the tubular gland in *T. alfreddugesi*. Its duct connects with the common salivary duct by an apparent valvular ampulla (Fig. 7-8C). The common salivary duct drains the lateral, dorsolateral, median, and tubular glands, and empties into the atrium. The ventral salivary gland is ducted separately to the atrium between the hypostomal venom gland and the common salivary duct opening. These ducts have internal, annular taenidia, which makes it difficult to identify and trace the salivary ducts of the higher trombidiids. Misidentification of tracheae as salivary ducts and the reverse condition have caused confusion (Jones, 1950).

The common salivary duct is modified within different groups of mites (Grandjean, 1938, 1944b, 1947; Jones, 1950). The common salivary ducts may be (1) completely internal (*Allothrombium, Trombicula, Cunaxa, Cheyletus, Smaris,* and *Odontoscirus*), or (2) internal for part of its length and then be exposed externally as an open or partially open trough on the lateral wall of the propodosoma. The trough may extend internally, combine with other salivary ducts, and open at the base of the chelicerae (e.g., *Anystis, Caeculus, Lordalychus, Pachygnathus,* and *Penthalodes (=Penthaleus)*). A single gland and a single duct were found in the genus *Cheyletus* (Moss, 1962; Hughes, 1959).

The common salivary duct also is described under different names by different authors (Moss, 1962). The term "podocephalic canal" relates to the external trough and duct associated with coxae I (Grandjean, 1938). This term is "imprecise and misleading" as well as "inappropriate and unacceptable" because of priority of other terms (Moss, 1962). The fusion of the basic cheliceral and pedipalpal segments to form the gnathosoma precludes consideration of discrete cephalic tagma as implied in Grandjean's terminology (Moss, 1962).

Speculatively, it may be that the glands expel their secretions at the initiation of feeding or over a continuous period (Moss, 1962). In *A. lerouxi* the venom glands are thought to expel their contents during feeding because there is no reservoir for storage of secretions. The expansion at the proximal (ducted) end of the tubular gland suggests a slight reservoir for storage of such materials. The presence of a valvular constriction at the junction of the duct with the common salivary duct also suggests a periodic release and mixing of such secretions with those of the other glands accessory to the common duct. The looping of the ducts and the influence of gravity on the flow of some secretions imply that capillarity during the feeding process is the main force by which secretions of the median and lateral glands flow to the mouth. While the secretions of the median and ventral glands are separated until feeding commences, materials from the tubular, lateral and dorsolateral glands may be stored in a common reservoir until use. The secretions of the tubular gland may also be carriers for smaller amounts of products from the dorsal group of salivary glands. Qualitative analysis of the secretions and additional investigation are necessary to reach complete conclusions (Moss, 1962).

Astigmata. The salivary glands of *Tyroglyphus farinae* consist of two pairs of acinous groups of cells arranged in tandem on each side of the body. The cells appear spongy with small granules, and include a large vacuole into which the nucleus is projected, covered by a thin layer of cytoplasm. Each of the glands opens into a common salivary duct (Fig. 7-9A) made up of thin-walled, flattened cells. The secretion of these glands is a mucus that lubricates the passage of food through the pharynx to the esophagus (Hughes, 1950).

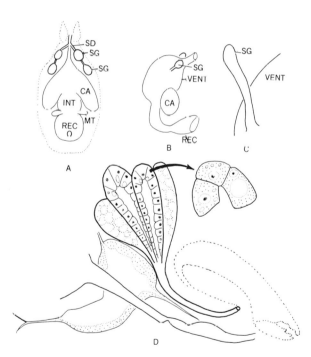

Figure 7-9. Salivary glands of (A) *Tyroglyphus fari-nae* (CA, caecum; INT, intestine; MT, Malpighian tubule; REC, rectum; SD, salivary duct; SG, salivary gland; VENT, ventriculus) (after Hughes, 1959); (B) *Hermannia arrecta*; (C) *Cepheus latus* (after Michael, 1883); (D) *Ceratozetes cisalpinus* (after Woodring and Cook, 1962).

The salivary glands in *Dermatophagoides farinae* (Pyroglyphidae), the house dust mite, are located dorsal to legs II in the idiosoma and are intimately intergrated with the large supracoxal glands. The salivary secretions are produced in acinous-appearing glands and pass by duct into the prebuccal cavity (Brody et al., 1972).

The salivary glands of *Listrophorous leuckarti* are "relatively enormous in size." The ducts of these glands open in the mouth cavity from which their secretions may be elaborated. Inasmuch as these mites scrape the scales from the hair of the host and feed on the epidermal secretions as well, the salivary secretions probably lubricate the food materials so they can be drawn in by the pumping pharynx (Hughes, 1954b).

Oribatida. The salivary glands (accessory glands or preventricular glands) of oribatids usually are conspicuously globose. At times they may be somewhat flattened or appear to be bilobed. The glands are situated at the anterior end of the ventriculus and ducted into the sides of this organ (Michael, 1884) (Fig. 7-9C). Their color varies from yellow to dark brown. The cells are loosely aggregated and somewhat delicate. Michael suggests that the glands secrete digestive fluids and found them occasionally filled with an orange-colored, oily secretion.

Salivary glands are present only in the adults of *Ceratozetes cisalpinus* and are derived from epidermal cells during the tritonymphal premolt period. Each gland comprises eight lobes. The lumen of each lobe empties into the common salivary duct. The common duct splits into a right and left branch, each of which empties mediad of the lingula. The cells of the glands form a single layer around the lumen of the lobe. Large nuclei and vacuoles are present. The saliva produced by these cells is thought to lubricate the food or possibly aid in digestion. It is not known why the immature stages lack these glands, especially since the food is similar for all stages (Woodring and Cook, 1962) (Fig. 7-9D).

ALIMENTARY CANAL

Just as the different gnathosomal features relate to feeding habits and the orders of mites, so the postoral organs of the alimentary tract show variations, but general features are consistent throughout the subclass. The tract has three principal regions although these may vary. The regions are as follows with the appropriate synonyms included parenthetically:

Foregut: stomodaeum

(anterior intestine; preintestine, pharyngoesophagus; ingluvies = crop in some)

Midgut: ventriculus

(= stomach, mesenteron, middle intestine; caecate in many)

Hindgut: proctodaeum

(posterior intestine, large intestine, colon, postcolon, rectal sac, rectal tube rectum)

Of the three main regions in the gut of mites, one type is described for the opilioacarid–gamasid–ixodid forms (Parasitiformes); a second type is indicated for Actinedida (Prostigmata, Trombidiformes); and a third type is identified for the Astigmata and Oribatida (Sarcoptiformes) (Baker and Wharton, 1952).

A similar system (attributable to Reuter, 1909) is based on the modifications of the caecate ventriculus and the proctodaeum in the major groups of mites. In this system the Opilioacarida and Gamasida represent one type of alimentary canal, and the ixodids exhibit a second type. The third type relates to the modifications that occur in some actinedid mites where there is no connection between the midgut and the hindgut nor an anus. The Astigmata and the oribatids comprise a fourth type where modifications and reductions of the ventriculus are noticeable, and where, in the

Oribatida, an ingluvies or crop may be present as a storage organ (Hughes, 1959).

Mouth

The preoral cavity of most mites is somewhat complicated. It is similar in structurally to the cavity of the Uropygi (Kaestner, 1968). The dorsal roof of the epistome (tectum capituli) is an extension of the anterior dorsal proterosoma. The epistome is fused laterally with the dorsal extensions of the pedipalpal coxae. The chelicerae are enclosed in the dorsal half of the preoral cavity. Below the chelicerae is a trough-shaped floor that extends forward as far as the labrum. The floor results from the fusion of the palpal coxal endites. Typically, the mouth is at the base of the labrum. The chelicerae are inserted above this floor and below the epistome, which allows for their protraction and retraction (Fig. 5-2). In some instances the gnathosomal connection to the prosoma may be membranous, which allows the whole tagma to be withdrawn within the prosoma into a cavity called the camerostome (Uropodina, Spelaeorhynchidae, Oribatida).

The mouth is crescentic in cross section in primitive arachnids, triangular in the primitive Actinotrichida, and H-shaped in the majority of the Acari, but variations may occur (Hammen, 1979) (Fig. 7-10A–C). The mouth opens into the buccal cavity and connects to the pharynx.

Pharynx

The suctorial pharynx or pumping organ is posterior to the mouth and buccal cavity. In the mites (as well as phalangiids and pseudoscorpions) the pharynx has an X-shaped lumen (Fig. 7-10C). The elastic properties of the pharyngeal walls and the shape of the lumen enable the opening to be tightly closed when the extrinsic muscles relax after contraction. Extrinsic musculature varies among mites. It is complex in some and reduced or lacking in others

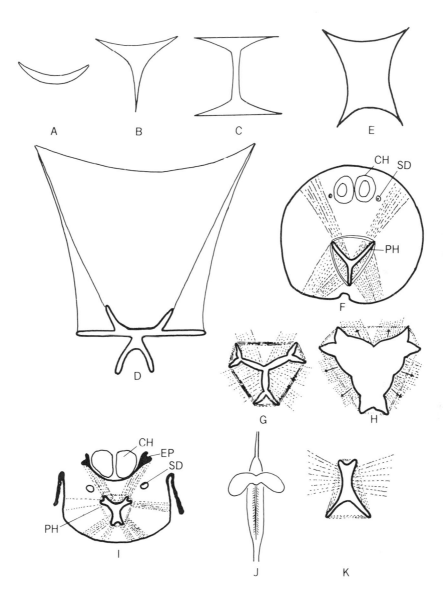

Figure 7-10. Diagrammatic cross section of the mouth in (A) a primitive arachnid; (B) primitive Actinedida; (C) the majority of Acari (after Hammen, 1968). Diagrammatic section of pharynx in (D) *Opilioacarus texanus*; (E) *Holothyrus coccinella* (after Hammen, 1965); (F) schematic section of pharynx of *Ornithonyssus bacoti* (after Hughes, 1959); (G) *Argas persicus*, contracted; (H) same, dilated (after Robinson and Davidson, 1913); (I) section of pharynx of *Ornithodoros papillipes* (CH, chelicerae; EP, epistome; PH, pharynx; SD, salivary duct) (after Balashov, 1972); (J) dorsal view of pharynx of *Dermacentor andersoni*; (K) section of same showing pharyngeal muscles (after Snodgrass, 1965).

(Gupta, 1979). The combination of pharyngeal elasticity and the powerful extrinsic musculature make this an efficient pumping organ in mites (compared to sucking stomach of spiders and amblypygids). The extrinsic musculature of the pharynx varies with the orders of mites; it is complex in some, reduced in others, and lacking in a few (Gupta, 1979). As a suctorial organ of some complexity, the pharynx is assumed to be derived from the stomodaeum of the embryo and because of its impervious chitinous lining, it has no absorptive function.

The pharynx of *Dermacentor* is an elongated, wedge-shaped sac without dorsal dilator muscles. The lumen is triradiate, narrowed above, and expanded ventrally with incurved walls (Fig. 7-10H, I). Seven pairs of winglike dilator muscles attach laterally; ventral dilators are also present. Each of these sets of muscles originates on the walls of the gnathosoma. In *Argas* and *Ixodes* the pharynx is wider dorsally and with a narrow ventral wall and dilator muscles that originate on the epistomial plate (Douglas, 1943; Arthur, 1946; Snodgrass, 1965). The pharynx in ticks contracts or dilates along its entire length and not by peristalsis as thought by earlier authors (Balashov, 1972; Gregson, 1967).

In the Eriophyidae the foregut has a pharynx with few cells, but with intrinsic and extrinsic muscles that attach to the apodemes of the prosoma. The esophagus passes through the synganglion to the midgut, which is a thin-walled sac. The hindgut is an expanded rectal sac connected by a small tube to the anus (Nuzzaci, 1979).

In the harvest mite, *Trombicula autumnalis*, the digestive tract comprises an oral recess, buccal cavity, pharynx, and a liverventriculus with caeca. The buccal cavity has no muscles attached, but the pharynx is equipped with radial muscles that provide the suction to pull in the food. Dorsal, ventral, and lateral dilator muscles are involved. Posterior to the pharynx, the slender, tubular esophagus, shorter than the pharynx, runs through the synganglion and connects to the ventriculus anteroventrally

through an esophageal–ventricular valve. As peristalsis fills the ventriculus with food, this valve prevents regurgitation into the esophagus. Ventricular caeca enable the distribution and storage of food from the main cavity of the ventriculus as it is filled. There is no hindgut, but an excretory organ connects to the anus (Jones, 1950) (Fig. 7-16B). The ventriculus and caeca are large and the excretory organ is median and dorsal in *Unionicola fossulata* (Fig. 7-16D). Calyptostomatidae have expanded ventricular caeca (Vistorin-Theis, 1977) (Fig. 7-16F).

Esophagus

The pharynx connects to a long, narrow tube, the esophagus, which is generally without a cuticular lining. It is assumed to be derived from the midgut and is not of stomodaeal origin (Fig. 7-11A). The esophagus invariably penetrates the synganglion of the nervous system like a pipe through a wall. It joins the ventriculus or stomach in different locations, but usually at a

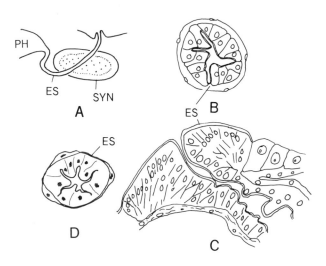

Figure 7-11. (A) Schematic drawings of pharynx (PH), esophagus (ES), and synganglion (SYN) of *Ornithodoros papillipes*; (B) cross section of esophagus of *Hyalomma asiaticum*; (C) same at junction with ventriculus (after Balashov, 1972); (D) schematic section of esophagus in *Ixodes* (after Arthur).

point on the ventroanterior surface of that organ.

In gamasid mites the pharynx may project into the stomach slightly from the ventral aspect. Histologically, the lining cells exhibit distinct nuclei. These cells are cuboidal in *Halarachne* and squamous in *Ornithonyssus* (Hughes, 1959). The esophagus of *Haemogamasus ambulans* has a smooth intima and is without folds and muscle layers (Young, 1968). In the spiny rat mite (*Echinolaelaps echidninus*) the esophageal walls are nonsclerotized. An esophageal valve prevents regurgitation of contents into the pharynx (Jakeman, 1961). A mesenteric origin is assumed for both the esophagus and the rectum in *Ornithonyssus bacoti*, which would account for the lack of a cuticular lining in these organs (Jakeman, 1961; Hughes, 1959). A cuticular lining is noted in the esophagus of the soft ticks *Ornithodoros papillipes* (Balashov, 1972) (Fig. 7-11A–D).

The esophagus in *Acarus siro* is an anterior extension of the midgut and is so sharply different from the pharynx that it is considered not of stomodael origin (Hughes, 1959).

Ventriculus—Stomach

The principal organ of the midgut is the ventriculus, which may exhibit lateral pouches or caeca. The digestive tract of mites is differentiated mainly by the morphological modifications of the ventriculus and its caeca and its connections with the rectum. Small size is correlated with the presence of simple, usually unbranched caeca in the mites (as well as in Ricinuleida, Pseudoscorpionida, and Opiliones). The number of caeca is greater in bloodsucking ticks; fewer caeca are typical of other orders.

Caeca are absent in the Opilioacarida and Holothyrida. Gamasid mites usually have two to three pairs. The ventriculus of the spiny rat mite comprises three central chambers and three pairs of lateral caeca (Fig. 7-12A). In *Haemogamasus ambulans* the ventriculus and its caeca are dimorphic for the sexes. Females show dorsal and ventral levels with differences

in the number and direction of the caeca. Males have a chordate-shaped ventral part of the ventriculus and lack the lateral caeca found in the females (Young, 1968) (Fig. 7-12B). The alimentary canal of other gamasid mites (*Ornithonyssus bacoti, Caminella peraphora, Urotrachytes formicarius, Fuscouropoda agitans*) are similar in many respects (Fig. 7-13).

The ventriculus is of endodermal origin in ticks. This section of the gut occupies most of the haemocoel and covers many of the other organs dorsally. Ixodids may have five or more pairs of primary caeca with smaller branches extending from these main trunks (Balashov, 1972; Sonenshine and Gregson, 1970). The general arrangement of seven pairs of tubular caeca holds for both argasids and ixodids (Fig. 7-14). Seven pairs of caeca occur in the embryo of *Ornithodoros moubata*, but all seven may not be formed initially in ixodids (Aeschlimann, 1958). The first pair of caeca may be delayed in development, which implies a secondary characteristic (Balashov, 1972). Five pairs of caeca are located anterolaterally in the region of the palps and legs I–III; two pairs are posterolateral. Variations in the configuration of the stomach and its caeca are numerous (Fig. 7-15).

The size of the ventricular caeca depends on the state of feeding: that is, they are distended and expanded when replete in the feeding animal; they may be reduced in size and even collapsed in nonfeeding or starving animals. Where the body is small and the stout legs are larger (Spinturnicidae) the caeca may extend into the legs (Hughes, 1959). The caeca increase the absorptive surface of the ventriculus in non-plant-feeding mites, but such expansions are not necessary in phytophagus mites because of predigested food.

Histologically, the walls of the stomach and caeca exhibit cuboidal cells on a distinct basement membrane. Larger, vacuolated cells are interspersed between the smaller cells. As the wall of the gut is extended, the cuboidal cells become continuous in extent and the larger cells are pinched off into the lumen of the gut where they lie as spherical bodies. When the

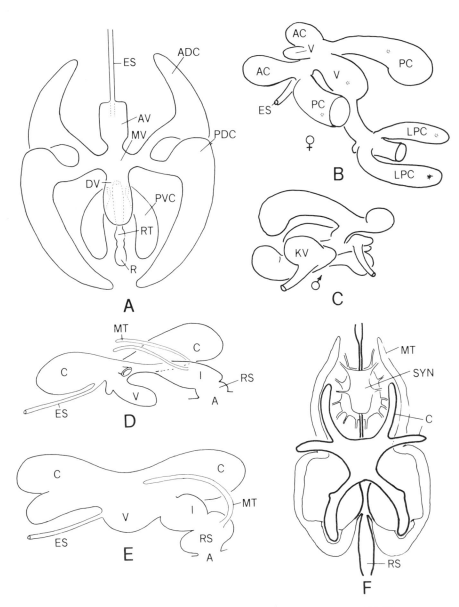

Figure 7-12. Alimentary canal in (A) *Echinolaelaps echidninus*; (B) *Haemogamasus ambulans* female; (C) same male (KV, chordateventriculus) (AC, anterior caecum; ADC, anterior dorsal caecum; AV, anterior ventriculus; E, esophagus; LPC, lateral posterior ventriculus, MV, middle ventriculus, PDC, posterior dorsal caecum; PVC, posterior ventral caecum; PV, posterior ventriculus; R, rectum; RT, rectal tube; **, mycetomes). Schematic lateral drawings of alimentary canals in Gamasida: (D) Parasitiformes; (E) Uropodina (after Hughes, 1959); (F) *Caminella peraphora* (after Ainscough, 1960).

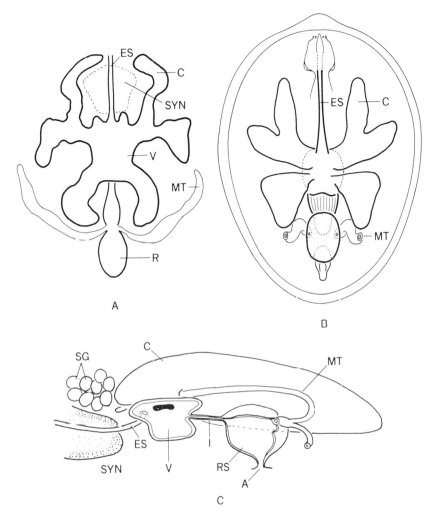

Figure 7-13. Alimentary canals in Gamasida continued: (A) *Urotrachytes formicarius* (after Vitzthum); (B) *Fuscouropoda agitans* (after Woodring, 1976); (C) *Ornithonyssus bacoti* (after Hughes, 1959) (A, anus; C, caecum; I, intestine; MT, Malpighian tubule; R, rectum; RS, rectal sac; SG, salivary gland; SYN, synganglion; V, ventriculus).

gut is distended with food, the cuboidal cells take on the appearance of pavement cells. The cells that are budded off into the lumen are thought to accumulate fecal materials in the form of small, dark spherules that are eventually eliminated in the feces along with the guanine secreted by the Malpighian tubules (Hughes, 1959).

The cells of the exterior of the ventriculus in the spiny rat mite are small and flattened. The inner cells are larger, more columnar, and somewhat vacuolar. Free cells in the lumen are assumed to be proliferated to supply digestive juices as in other mites. Oval, dark-staining bodies in the posterior median chamber of the ventriculus are assumed to be storage structures or possible production sites for enzymes (Jakeman, 1961).

Some "mycetome-like" bodies are found in the caudal tips of the ventral caeca and the mid-

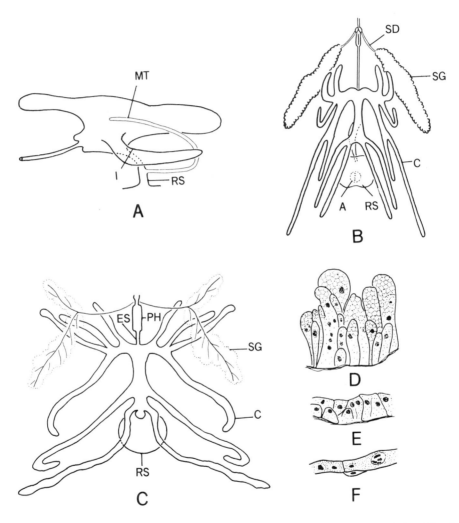

Figure 7-14. Alimentary canal in hard ticks: (A) schematic lateral view (after Hughes, 1959); (B) *Dermacentor andersoni*, dorsal view (after Douglas, 1943); (C) *Ixodes* sp.. Epithelilum of ventriculus; (D) anterior ventriculus; (E) midventriculus; (F) posterior ventriculus (after Arthur, 1960) (A, anus; C, caecum; ES, esophagus; I, intestine; MT, Malpighian tubule; PH, pharynx; RS, rectal sac; SG, salivary gland; V, ventriculus).

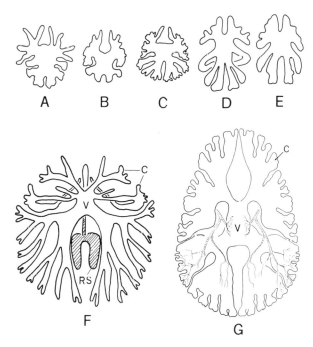

Figure 7-15. Schematic diagrams of the alimentary canal in soft ticks: (A–C) variations of ventricular caeca in larvae; (D) female of *Ornithodoros papillipes*; (E) female of *Alveonasus lahorensis*; (F) *Argas vespertilionis* (after Roshdy, 1961).

dorsal regions of caeca of female *Haemogamasus ambulans* (Young, 1968) these structures "consist of glomerates of tightly packed cells surrounded by a thin membrane." (Fig. 7-12B, C). Mycetomes are also described for *Ornithonyssus bacoti* (Hughes, 1959).

Intracellular digestion occurs in the ventriculus as evidenced by the vacuolated cells found in the organ (Krantz, 1978). In *Caloglyphus micophagus* the cells of the ventricular caeca appear ciliated except for the vacuolated cells among them (Rohde and Oemick, 1967). These mites eat dead and decaying insects.

The ventriculus in the actinedid mites varies within the order. Caeca are absent in the ventriculus of the wormlike Eriophyidae (Fig. 7-16A) and in the spider mites (Jeppson et al., 1975). In some actinedids (Parasitengona) the midgut ends blindly with few caeca present. There is no anus in some (Fig. 7-16B).

The armored predators in Nicoletiellidae (=Labidostommidae) exhibit a small pharynx and esophagus, a large ventriculus with massive caeca, a postcolon, a colon, and a rectum (Vistorin, 1979) (Fig. 7-16E).

The digestive system of Demodicidae is a simple, blind tube. The very narrow esophagus leads from the pharynx to the midgut. The midgut extends to the end of the opisthosomal cavity, where it ends blindly. The gut cells of *Demodex folliculorum* are of two types, all of which have a foamy apppearance. No lumen is present in the midgut, nor are excretory tubules present. Crystals (?excretory) do appear in the terminal, blind end of the midgut (Desch and Nutting, 1977).

The ventriculus in the Astigmata is relatively large and characterized by a single pair of short caeca (Hughes, 1961; Krantz, 1978) (Fig. 7-17A–D).

The digestive system of the house dust mites (*Dermatophagoides farinae, D. pteronyssus*) is divided into three major sections: foregut, midgut, and hindgut, each of which may be subdivided at least once based on the morphology (Brody et al., 1972). The anterior foregut begins with the pharynx, a muscular pump lined with cuticle and ribbed on the floor of the middle third. The pharynx is flattened and curves upward at its lateral margins. Dilator and constrictor muscles operate this organ.

At the juncture of the pharynx with the esophagus the digestive tube becomes plicated or folded as it is surrounded by the synganglion. Circular muscles extend between the plicae. The tubular areas between the folds are extensions of the hemocoel, but are separated from the nervous tissue by basement membrane. The space between the brain and the gut is filled with hemolymph and connective tissue. The plications give the esophagus an appearance of an eight-pointed star in cross section. The esophageal lumen is lined with an epicuticle, which may sluff off, and beneath that layer is the exocuticle with distinct pore canals. The esophagus emerges from the dorsal aspect of the synganglion and connects to the midgut.

The midgut of *D. farinae* is divided into two

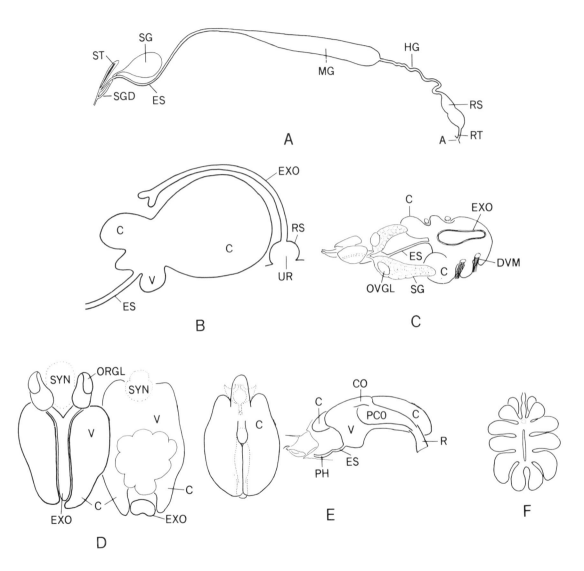

Figure 7-16. (A) Alimentary canal in *Aceria tulipae* (after Jeppson et al., 1975); (B) blind midgut of actinedid mite; (C) *Trombicula autumnalis*; (D) dorsal view of *Unionicola fossulata* (after Mitchell, 1955); (E) *Nicoletiella denticulata* (after Vistorin, 1979); (F) ventriculus and caeca of Calyptostomatidae (after Vistorin-Theis, 1977) (C, caeca; CO, colon; DVM, dorsoventral muscle; ES, esophagus; EXO, excretory organ; HG, hindgut; MG, midgut; OR GL, oral gland; OV GL, oval gland; PCO, posterior colon; PH, pharynx; R, rectum; RS, rectal sac; RT, rectal tube; SG, salivary gland; SGD, salivary gland duct; SYN, synganglion; U, uropore; V, ventriculus).

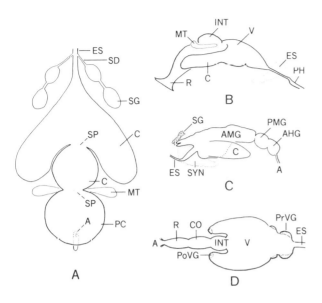

Figure 7-17. Alimentary canal in (A) *Tyroglyphus farinae* dorsal view (after Hughes, 1959); (B) *Tyrophagus dimidiatus* lateral view (after Vitzthum, 1943); (C) *Dermatophagoides farinae* lateral view (after Brody et al., 1972); (D) *Pterolichus nisi* dorsal view (after Vitzthum, 1943) (A, anus; AHG, anterior hindgut; AMG, anterior midgut; C, caecum; CO, colon; ES, esophagus; INT, intestine; MT, Malpighian tubule; PMG, posterior midgut; PRVG, proventricular gland; PoVG, postventricular gland; R, rectum; SD, salivary duct; SG, salivary gland; SYN, synganglion; V, ventriculus).

cells are cuboidal. Cells that line the caeca have numerous lysosomes.

A constriction separates the two regions of the midgut, and is evidently valvular since in living mites it is observed to expand and contract during the passage of food. Long microvilli are present in the lumen of the valve and in the surface of the posterior region of the midgut. A peritrophic membrane encases the food ball. In addition to the long microvilli, extensive rough ER and other organelles in actively secreting cytoplasm are typical of these posterior cells. Both the anterior and the posterior regions of the hindgut are lined with cuticle which has longitudinal folds in the anterior region. A mucus-type of lining is also present. Fecal balls are surrounded by peritrophic membranes and may contain parts of the cuticle of cannibalized mites. The cuticle of the posterior part of the hindgut has many pore canals. The posterior region of the hindgut terminates in the oval anus covered by the anal plates (Brody et al., 1972).

In oribatid mites the esophagus connects to the ingluvies (=crop) which acts as a storage organ for food. The crop leads to the ventriculus with its characteristic caeca. A pair of proventricular glands is also present (Fig. 7-18 A–F).

regions. The anterior region is about twice the size of the posterior region and has two caeca attached. The epithelium of the midgut varies. The epithelial cells of the anterior area are reduced in height and have few microvilli, mitochondria, and nuclei. Ventrally, however, there are two main rows of active cells that extend to the constriction at the beginning of the posterior region of the midgut. These cells have a single nucleus, short microvilli, rough ER, and large vacuoles. Cells that float freely in the lumen apparently originate from these ventral rows. The cytoplasm of these free cells is in a state of degeneration. Posteriorly, but still in the anterior region of the midgut, the epithelial

Midgut—Intestine

Usually a short intestine connects the ventriculus to the hindgut (rectum, rectal sac, rectal tube). Because the general arachnid proctodaeum is short, the rectum of many mites may be reduced in length as well. The proctodaeal derivation of the rectum implies that a cuticular lining is present and thus no absorption occurs, as in the stomodaeal pharynx.

In many of the acarines it is at the juncture of the midgut and the rectum that the Malpighian tubules are attached, if such are present. In others (some ticks and actinedid mites) the midgut ends blindly and the uropore is a separate opening (Fig. 7-16B).

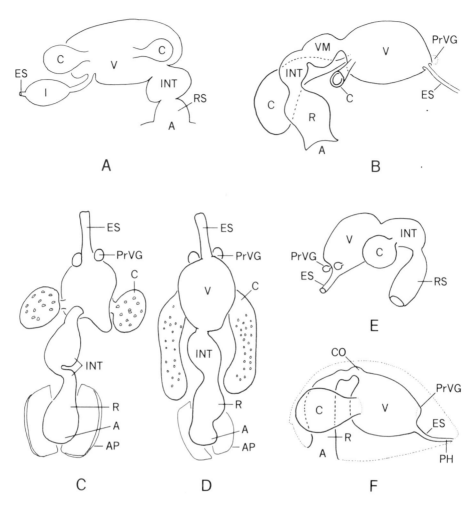

Figure 7-18. (A) Schematic of generalized alimentary canal of oribatid mite showing the ingluvies (=crop), ventriculus, and caeca from the lateral view (after Hughes, 1959); (B) lateral view of same with right caecum removed (after Bernini, 1973); (C) dorsal view of *Hoplophora magna*; (D) *Nothrus theleproctus*; (E) *Hermannia arrecta* (after Michael, 1883); (F) *Ceratozetes cisalpinus* lateral view (after Woodring and Cook, 1962) (A, anus; C, caecum; E, esophagus; I, ingluvies (crop); INT, intestine; PH, pharynx; PrVG, proventricular glands; R, rectum; RS, rectal sac; VA, anterior ventriculus; VM, middle ventriculus).

Hindgut

Various designations have been given to this part of the alimentary canal: hindgut (Winkler, 1888; Douglas, 1943; Wharton and Strandtmann, 1958), colon (Michael, 1892), rectal tube (Douglas, 1943; Jakeman, 1961), rectal sac (Ba-

lashov, 1972), and intestine (Young, 1968). The hindgut (rectal tube) of the spiny rat mite extends medially from the ventriculus to the rectum and may be deflected slightly by developing embryos in the hemocoel. A rectal valve is present at the juncture with the rectum (Jakeman, 1961). A similar form and location are described for *Haemogamasus ambulans*, but muscular sphincters are also observed at both the ventricular and rectal junctions. Hematin-filled cells occupy the lumen within 10 h after feeding. Then these clumps of cells are passed into the rectum (Young, 1968).

Rectum—Rectal Sac—Rectal Tube—Colon

The last chamber of the alimentary canal is the rectum. It connects with the anus and also receives the excretory tubes, if present, at its junction with the hindgut. The cells of this organ are columnar in the spiny rat mite and stain more deeply than the cells of the ventriculus or the hindgut (Jakeman, 1961). A chitinous intima is present, but may not be entire. In both the spiny rat mites (*E. echidninus* and *H. ambulans*) the rectum is swollen with fecal materials. When the latter mites have fed on blood, the contents of this organ are black.

The rectum in the soft tick *Ornithodoros papillipes* lies ventral to the rectal sac and has dilator muscles at its exit (Fig. 7-19A, B).

FEEDING HABITS

The feeding behavior of mites is classified into two principal categories: (1) functional specialization and (2) food type with overlapping of some categories. In the first instance mites may be polyphagous, digophagous, or monophagous. General categories and subcategories for the food type include:

Herbivorous

  phytophagous
  mycetophagous
  pollenivorous

Carnivorous (Zoophagous)

  predators

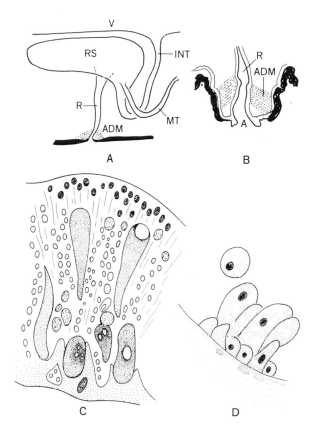

Figure 7-19. Schematic drawing of *Ornithodoros papillipes*: (A) rectum and rectal sac, rectum, dilator muscles; (B) enlarged view of same (after Balashov, 1972) (A, anus; ADM, anal dilator muscle; INT, intestine; R, rectum; RS, rectal sac). Digestive epithelium of (C) *Holothyrus gervaisi*; (D) *Ornithonyssus bacoti* (after Hughes, 1959).

  parasites
    hematophagous
    cytophagy, histophagy
    dander, skin, scale feeders
  cleptoparasites

Detritivores (saprophagous)

Omnivorous

The variety of specific feeding habits of the mites is indicated in the following list, but the list is not inclusive, nor are specific terms available for some of the types of feeding.

Algivorous = algophagous   Necrophagous
Carnivorous =zoophagous   Nectar feeders
Coprophagous   Nematophagous
Cytophagous   Pollenivorous
Detritovores   Parasitic–para-
                              sitoid–clepto-
                              parasitic
Hematophagous   Spore feeders
Fungivorous   Xylophagous
                              (tunnelers,
                              wood, punk)
Microphages
Mycetophagous (mycelia,
  spores, mildew, bracket
  fungi)

Eating habits of mites are varied and relatively few mites consume particulate food. Most of the food ingested is in a liquid state. Myrmecophilous forms eat the food of ants or feed on the ants themselves. Gamasids and sluggish uropodines feed on fungus and vegetable matter in manure, moss, and leaf mats; some feed on smaller animals. Uropodines are sluggish, slow-moving, and feed upon fungi and other vegetable matter. Four of the orders have species that feed on living plants, namely Gamasida, Actinedida, Astigmata, and Oribatida. Stigmaeidae and Tetranychidae exemplify mites that suck juices of plants. Fungal feeders are common among the Astigmata and Oribatida (Cloudsley-Thompson, 1958; Krantz and Lindquist, 1979).

Opilioacarida. Opilioacarids are carnivores. Fragments of millipedes and legs of phalangids, gamasids, and oribatids are found in the gut. Plant materials and quartz grains also occur. The food is eaten without liquification, which is exceptional among acarines. It is possible, however, that digestive enzymes are released prior to feeding inasmuch as fluid is observed in the capitulum during feeding (Grandjean, 1936b).

Large masses of pollen grains have been seen in the gut of *Opilioacarus texanus*. These mites are assumed to be nocturnal feeders for their hiding under rocks is a diurnal habit. Their feeding on pollen grains is a seasonal behavior because of the vernal flowering of the plants in semiarid regions. Pollen masses eaten were of single or multiple plant species (Hammen, 1966). Boluses of brown food also included appendages of unidentified arthropods, an evidence of predation (Cloudsley-Thompson, 1958). Evidently, these primitive mites are not as exclusively carnivorous as previously supposed. Fungal spores have also been postulated as food.

Gamasida. The food habits of the gamasid mites include predation, necrophagy, coprophagy, mycetophagy, consumption of pollen and nectar, and other methods. The larger predatory parasitoids feed on Collembola, acarid mites, and nematodes (Weis-Fogh, 1948). The slow-moving, nearly sedentary Uropodina are coprophagous in some instances (*Fuscouropoda marginata*). Others eat fungal hyphae and spores in the accumulated organic materials of compost heaps, leaf mats, and stored products like moist grain. Still others (*Prodinychus* sp.) eat living and dead animal matter (Hughes, 1959; Krantz and Lindquist, 1979). Predatory uropodines are rare (Krantz, 1978).

Among the predators of insects and mites that are plant pests, some members of the Phytoseiidae are important agents of biological control (e.g., *Typhlodromus* and *Amblyseius*). Other members of this family may feed on pollen, fungi, leaf sap, and powdery mildews. Species of the genus *Rhinoseius* (Ascidae) eat the nectar of flowers and may also feed on pollen. Some of the mites in another genus of this family (*Hoploseius*) scrape for food in the soft tissues of bracket fungi from the underside of the sporophore with modified chelicerae Hyphal and other soft tissues of different fungi provide food for species of the Ameroseiidae in stored products, bracket fungi, beetle galleries, and compost. Other species of this family seem to feed on nectar and pollen (Krantz and Lindquist, 1979).

Species of *Pneumolaelaps* (Laelapidae) are thought to feed on pollen in bees' nests because the chelicerae are similar to other pollen- and spore-feeding mites. Species of Parasitidae associated with North American bumblebees use a variety of nest substances as food, including mixtures of pollen and wax, at least in the deutonymph and adult stages (Richards and Richards, 1976).

The gamasids (and prostigmatid mites) are mostly fluid feeders. The presence of large salivary styli in many gamasine mites (e.g., *Glyptholaspis confusa*, Hammen, 1964) suggests preoral digestion, where the hornlike corniculi (=external malae) may act as restraining organs or hold the prey in the feeding process. They may act similarly to the rutella in the oribatid mites (Del Fosse et al., 1975). In those gamasid mites that are blood feeders, the corniculi are membranous and form a feeding tube with other mouthparts (Evans et al., 1961).

Ixodida. More information is available on feeding and digestion of ticks than in any order of mites. Feeding in hard ticks differs in several ways from the process in soft ticks.

Feeding in Hard Ticks. Once hard ticks have found a host they may crawl around on the body searching for a suitable site of attachment. What determines the exact site is not understood, but once located, the ticks spread their legs and elevate their bodies in a 45–60° angle to the surface of the host. By contraction of the dorso-ventral muscles, the turgor pressure of the interior is increased and the chelicerae are extended. The cheliceral muscles manipulate the cheliceral digits laterally as they cut through the skin to make a transverse wound. Simultaneously with the insertion, the palps are splayed out, which allows deeper penetration of the hypostome. Penetration extends through the stratified layer of skin within 3 to 5 min. The tick may withdraw from the wound and move to a neighboring site and repeat the process. Full attachment requires 10 min to an hour or more (Balashov, 1972). Lesions may occur at the feeding site (Tatchell and Morehouse, 1970).

When the capitulum is inserted and the palps are splayed on the surface of of the skin (Fig. 7-1), ixodid ticks secrete one of two types of salivary secretions. The first type is a viscous, milky fluid which completely surrounds the capitulum (Fig. 7-20) leaving a short cone on the surface of the skin. Inasmuch as the cement is nonantigenic, the adjacent host tissue does not become inflamed and this reduces reaction of the host to the mechanical penetration of the mouthparts. The solution solidifies into a type of cement sheath which adheres firmly to the skin of the host. The firmness of the sheath is shown by the imprint of the hypostome left on its walls when the cone is removed. The sheath provides anchorage for the mouthparts over lengthy periods of time.

The cement consists of two main components: (1) a cortex of carbohydrate–protein mixture stabilized by quinone tanning and SS linkages, and (2) an internum of lipoprotein (Fig. 7-20). The lipoprotein is secreted first and the cone material follows. Final rapid engorgement (after 48 h) of each stage is preceded by a secondary secretion of cement directly beneath the mouthparts which provides additional support. The cement layer also carries in between the corneum stratum and the remainder of the epidermis. The secondary deposition of cement fills the space left by the lysing of the epidermis where it adds to the attachment and keeps the tube clear of occlusion.

Once the mouthparts are firmly cemented into the host skin, the cheliceral digits are inserted into the secondary cavity and commence cutting again. With rocking motions of the mouthparts the hypostome is gradually pushed into the dermis. Ticks such as *Hyalomma* and *Ixodes ricinus* with long mouthparts may even

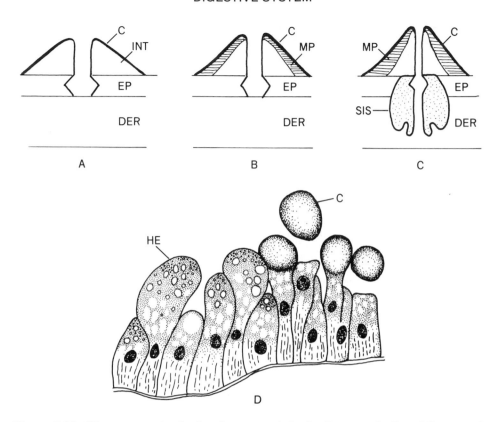

Figure 7-20. Three stages in the development of the feeding cone in *Boophilus annulatus*: (A) cone with cortex (c) and internum (int); (B) mixture of cortical secretion and lipoprotein (MP); (C) secondary internal secretion (SIS) into dermis (after Moorehouse and Tatchell, 1966); (D) composite of digestive cells of *Ornithodoros papillipes* (HE, hematin; C, cells shed into lumen) (after Balashov, 1972; drawing by Sonya Evans).

penetrate the subdermal area of the skin. Ticks with short mouthparts (*Dermacentor, Haemaphysalis, Rhipicephalus* and *Boophilus*) do not penetrate as deeply and the cement sheath may be restricted to a cone on the surface of the skin.

Deeper penetration results in the release of a second type of salivary secretion. This transparent liquid is injected into the wound, lyses the tissues in the hemorrhagic locus, and is an anticoagulant. Feeding consists of alternately ingesting blood, lymph, and lysed material from the hemorrhagic locus and injecting saliva into the tissues of the host by action of the muscular pharynx. Amounts and rates of sali-

vary secretion are different during different periods in the feeding process. Sucking may last 1 to 2, or 10 to 30 s and salivation approximately 1 or 2 s, each process having a time interval of from several seconds to several minutes. These separations of salivation and bloodsucking prevent a mixing of the fluids that move in opposite directions in the preoral cavity, which is both a salivary and blood channel. It is thought that the lower surface of the labrum fits in the lower wall of the pharynx and functions as a valve to prevent regurgitation. There may also be a correlation between size of chelicerae and size and development of retrorse teeth of the hyposotome (Arthur, 1951).

Histopathological studies show that the slight damage caused by mechanical penetration of the chelicerae and the hypostome is rapidly increased by the saliva. Necrosis of muscle fibers, blood vessels, and other tissues occurs together with inflammation due to the toxic and lytic ingredients of the saliva. In ticks with long capitula (e.g., *Ixodes*), severe inflammation accompanies the normal hemorrhagic syndrome.

In contrast to the heparinlike components of the saliva of leeches and bloodsucking Diptera, which inactivate thrombin and prevent fibrin formation, the anticoagulants of tick saliva are inhibitors of thrombokinase. Their action blocks the transformation of prothrombin to thrombin. These anticoagulants are thought to be glycoproteins or mucoproteins with a high thermostability (Balashov, 1972).

There is no marked disturbance of tick digestion even when fed on a wide variety of hosts (mammals, birds, reptiles). The ability of ticks to feed on blood from distantly related hosts is possible because of the stable chemical composition of the blood of terrestrial vertebrates. The blood of vertebrates is rich in protein and concentrated substances: hemoglobin, albumin, globulin, and cholesterol, which make up the basic components of the tick diet, and blood proteins and fats. Protein assimilation is of primary importance. Evidently, the serum albumins and globulins as well as globin are completely assimilated while most of the heme material of hemoglobin is lost with the feces. Some of the iron-containing heme material is necessary for tick development, however (Balashov, 1972).

Soft Tick Feeding. Our understanding of the feeding mechanisms of argasid ticks is incomplete because the subterminal position of the capitulum makes direct observations of the process very difficult. The hypothetical explanations given from analysis of the morphology indicate that a negative hydrostatic pressure is generated during the dilation of the pharynx as blood is ingested from the hematoma. Contraction of the pharynx results in positive pressure that pushes the blood into the midgut. (It has been noted that argasids exceed numerous bloodsucking insects in their capacity to ingest large amounts of blood with minimum irritation to the host.)

While observing the feeding behavior of *O. hermsi* on suckling mice, I found the nymphs feed to repletion in about 8 min, the adults in 12 to 15 min. Only small reddened areas of the skin result.

Soft ticks also have small, poorly developed hypostomal teeth; in some instances, the hypostome and teeth are vestigial. They are also intermittent feeders. Argasid chelicerae are proportionately larger, with sharp and effective cutting digits.

Argasid ticks attach to the host in a manner basically similar to ixodids. The cheliceral digits cut the skin, accompanied by a forward and backward movement of the capitulum that effects penetration into the wound. The legs are used to hold the ticks to the site of attachment, but are later relaxed. The palpi do not directly participate in the attachment.

Argasid ticks secrete a more fluid saliva that does not solidify as with some ixodids. The salivary secretions are released at irregular intervals, but the semitransparent material is released at the tip of the capitulum into the hemorrhagic area within the skin of the host. Drops of saliva are secreted alternately between periods of ingestion of blood from the hematoma in the skin.

Inasmuch as the soft ticks are intermittent feeders, take smaller meals and undergo faster engorgement than ixodids, they usually suck the blood only from small local hemorrhages of capillaries and an occasional larger blood vessel that is damaged by penetration. The rate of ingestion is faster than in ixodid ticks and exhibits a rhythm in feeding with rapid ingestion of blood alternated with short intervals where the sucking stops completely. It is not kown what causes the rhythm, but the action may facilitate the flow of saliva into the incision.

Three types of pathological changes are

caused by the feeding of ornithodorine ticks (Lavoipierre and Rick, 1955). First, the species in *Ornithodoros moubata* group are slow feeders and may engorge partially from superficial hemorrhages in 15 min or require 1½ h to engorge completely. Second, in the *O. coriaceus* group, the mouthparts are surrounded by small hemorrhages when the tick detaches. After detachment these areas enlarge several times and spread. They may also extend into the deep dermis and subcutaneous muscle of the host. In the third group, which consists of larger ticks (e.g., *O. savignyi*), feeding causes considerable hemorrhage also, but it is more localized than in the second group.

The anterior part of the gut of argasid ticks (buccal cavity, pharynx, esophagus) is lined with chitinous intima impervious to bacteria. The midgut is the location for absorption or for penetration by bacteria (e.g., *Borellia*), since the chitinous intima is absent.

Unfed argasid ticks exhibit a lining of a single layer of epithelium with basal nuclei. As feeding begins, the epithelium becomes multilayered with some enlarged, highly vacuolized cells. These cells are elongated distally and the expansion makes them appear clublike. Some of these cells may become detached and float in the lumen. Others evidently participate in digestion of crystalline hemoglobin. The bacteriocidal effect in the gut clublike cells following feeding is further evidenced by the presence of hematin and melanin, products of the digestion of hemoglobin, in the lumen for the life of the tick.

The feeding mechanics in *Boophilus microplus*, a one-host tick of cattle, involve short mouthparts, but the depth of penetration is identical in all life stages in spite of the time of attachment on the host. The penetration extends toward the base of the Malpighian layer at the margin of the dermis. Penetration may occur within 5 min of arrival on the host (Moorehouse and Tatchell, 1966).

Adults of the eastern dog tick, (*D. variabilis*) find a new host after molting from the nymphal stage and engorge with blood. An unfed female that measures 5 mm in length upon engorgement will become oval and turgid, about 13 mm long and 10 mm wide, and about 5 mm thick. The engorgement is accompanied by a flattening of the exocuticle and a stretching of the endocuticle.

Salivary Secretions and Water Balance in Ticks. During engorgement, large amounts of water are passed across the gut epithelium into the hemolymph and are returned to the host in the salivary secretions of the ticks. While the blood meal is concentrated, electrolytic and osmotic imbalance is avoided by returning excess water to the host (Tatchell, 1967). Ionic and water balance affect tick feeding in different ways (Kaufman and Phillips, 1973a,b,c).

Degree of concentration of the ingested blood more than doubles (e.g., a *Boophilus microplus* female increases from 10 mg to 200[+] mg) in host animals with low hemoglobin concentration. In those hosts with higher hemoglobin cocentrations in the blood there is a lower degree of concentration by the tick.

Concentration of the blood meal and water balance are related to the osmoregulatory–ionic functions of the coxal glands of argasid ticks.

Experimental work demonstrates that in some argasid ticks specific feeding responses occur when the amount of glutathion is reduced. Ticks respond to whole blood, serum, plasma, and hemolyzed red cells more readily than to fractions or treated plasma. It is presumed that feeding is mediated by two types of chemoreceptors situated on the mouthparts. The differences in experimental results indicate one type of receptor for glutathione (GSH), nucleotides, and ATP or DPNH; another type is specific for amino acids. Glucose is synergistic for both kinds of receptors (Galun and Kindler, 1968). Complete understanding of the feeding mechanism in argasid and ixodid ticks, according to these authors, will result from studies of the histology of the receptors and electrophysiological analysis of individual receptor cells.

The factors that induce feeding of *O. tholo-*

zani show that this tick imbibes saline containing reduced glutathione and glucose almost as readily as whole blood. The effect of these two compounds was synergistic when tested at concentrations found in blood, but glucose could not be replaced by fructose, galactose, arabinase, sucrose, and so on.

After prolonged starvation, marked feeding response was induced by isotonic saline containing glucose and one of the following amino acids: leucine, alloileucine; proline, valine, serine, alanine, phenaline, or glycine.

Solutions with ATP or DPNH in place of glutathione also induced maximal feeding response. L-glutamate and the SH-reagent, DTNB (5-5′-dithiobis-2-nitrobenzoic acid) inhibited feeding in the presence of glutathione, ATP, or DPNH. Addition of Mg^{2+}, Cd, Zn, or Ni prevented feeding on all solutions.

It is believed that feeding is mediated by at least two kinds of chemoreceptors: one specific for GSH, ATP, or DPNH, and the other for amino acids or glucose. The identification of several compounds that in certain combinations induce feeding in *O. tholozani* can account completely for the taste response to blood in chemical terms.

Blood Meal. Feeding in ixodid ticks usually consists of a single blood meal in each developmental stage, whereas argasid ticks feed intermittently in each stage. Digestion in ixodid ticks also differs from the process in the argasid ticks because in ixodids most of the hemoglobin and other blood proteins pass rapidly into the digestive cells from the lumen of the gut. The digestive cells of the midgut become hypertrophied and the lumenary content decreases. The blood remaining in the cavity of the gut undergoes similar changes in both argasids and ixodids. The hemoglobins may crystalize in conditions of low solubility, but remain in solution in conditions of high solubility. Hemoglobin concentration is correlated with oviposition and decreases markedly prior to the end of egg laying so that only traces of hemoglobin remain in the lumen.

In argasid ticks, the ventriculus and its caeca are associated with two functions: digestion and storage of food. In engorged ticks that have not discharged coxal fluid the ventriculus and its caeca are fully distended. After the discharge of coxal fluid the median and anterolateral caeca are less distended with blood than the anterolateral and posterolateral caeca. The posterolateral caeca act as the main storage reservoirs for undigested blood. They may remain distended for several months even though the ticks have not fed. Only in very emaciated ticks do the caeca become empty and collapse.

The blood meal is first digested in the ventriculus and in the proximal areas of the median and anterolateral caeca. Other caeca, especially the posterolateral pairs, essentially act as storage organs with thin walls around the blood meal without any digestive function (Balashov, 1972).

Digestive Epithelium. The epithelial cells of the ventriculus of argasid ticks vary in form and size depending on the physiological state the species of tick involved. The epithelial cells lie on a basement membrane surrounded by a reticulum of circular and longitudinal muscle fibers. These fibers do not form a distinct layer, but by contraction cause peristaltic movements of the ventriculus. Based on morphology, cellular inclusions, and function, the epithelial cells are of three types: digestive, secretory, and undifferentiated reserve cells, although the differentiation and relationships between these types is unclear. Reserve cells are thought to be the precursors of the other two types.

The proliferation of cells begins soon after ingestion of the blood meal in the immatures and just before molting. Adults at the close of feeding show mitotic divisions of the undifferentiated cells. It is possible that only one type of reserve cell in argasids differentiates into the digestive and the secretory cells. In one view (Tatchell, 1964) direct reciprocal transmutation of these cells is not likely. In another opinion a reversal of one cell type into another is possible when cells break off the apices containing a nucleus and the proximal part of the cells disintegrates. The apical part of such cells could account for the saliva-fast cells in unfed ticks (Balashov, 1972) (Fig. 7-20D).

The scheme of cells and digestion in larvae differs from the nymphs and adults. In the larva, the digestion and secretion cells are cuboidal and resemble each other. In unfed females of *Ornithodoros papillipes* the secretory and reserve cells predominate in the midgut, but may contain semidigested blood from a previous meal. Secretory cells are columnar with nuclei near their proximal ends and the cytoplasm contains numerous mitochondria and a fibrillar(?) endoplasmic reticulum mainly around the nucleus. Apical vacuoles do not stain with acid or basic dyes, but show positive PAS reaction for polysaccharides and glycogen granules. Reserve cells are cuboidal, much smaller, and tend to be overlapped or covered by secretory or digestive cells. The digestive cells are much larger than reserve cells, columnar, and expanded distally into the lumen of the gut. The nuclei are central in location and the cytoplasm contains vacuoles and other inclusions such as hematin granules and glycogen. Vacuoles are abundant in the apical ends of the cells and specific stains (azocarmine, iron hematoxylin, bromophenol blue, and benzidine) show proteins and hemoglobin. Fat vacuoles are also observed (with Sudan black or Sudan III, osmic acid). When the digestive cells become filled, they may detach into the lumen of the gut where they appear rounded like amebocytes. Lipids from the stroma of erythrocytes are assimilated into vacuoles of digestive cells simultaneously with the hemoglobin.

When fresh blood is ingested, the secretory cells are destroyed in large numbers. Their distal ends become swollen and break off or the whole cell is detached (Figs. 7-20D; 7-21). The digestive cells and reserve cells may be stretched so thinly that the wall of the gut becomes almost membranelike in thickness, resting on the muscular fibers.

Cells protrude into the lumen of the gut as the size of the lumen of the ventriculus and caeca diminishes. Digestive cells predominate shortly after ingestion of the blood meal; they are cuboidal at first, then become columnar. Digestive vacuoles form in the cytoplasm at the distal ends of the cells. Hemoglobin fills most of the apical vacuoles and is changed into hematin concretions. The intracellular digestion of hemoglobin is asynchronous in a single cell so that at a given time the cytoplasm may have large digestive vacuoles as well as the hematin end products. When the intracellular digestion of hemoglobin begins, new food vacuoles cease to appear. At the completion of digestion the digestive cells lack vacuoles and are filled with hematin material. Subsequently, these cells degenerate.

Pinocytosis is the main vehicle for absorption of the hemoglobin molecules and is assumed to occur when the vacuoles begin to appear in the digestive cells. The movement of the vacuoles through the cytoplasmic membrane cannot be explained by the usual cell-permeability processes (Balashov, 1972). Phagocytosis results in the uptake of hemoglobin and of erythrocytes in the blood meal. The apical ends of the cells appear to have microvilli, but electron microscopic evidence is needed for verification.

Differences in digestive epithelium occur between species of argasid ticks. In unfed *Argas persicus* the epithelial cells are much longer, and the proximal ends of the cells have fibrillar striations and are gathered into folds. Because the nucleated erythrocytes of avian blood resist hemolysis, they tend to be ingested whole by digestive cells, an example of intracellular digestion. Digestive cells and the lumen of the gut become so densely packed with hematin that it cannot be eliminated in *Ornithodoros*.

Each cell in the epithelium of ticks extends pseudopodia into the lumen of the gut. These pseudopodia engulf the materials previously ingested from the blood of the host. Intracellular digestion occurs within the vacuoles formed by these phagocytic cells (Andre, 1949).

Defecation in ticks involves the elimination

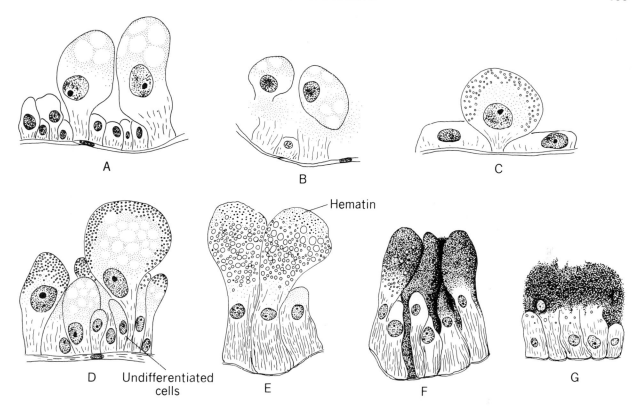

Figure 7-21. Digestive cells in *Ixodes ricinus*: (A) differentiated digestive cells; (B) cells shedding into lumen; (C) hypertrophied cell; (D) cells after first day of detachment; (E) cells before oviposition; (F) cells at end of oviposition; (G) cells after oviposition in dying female (after Balashov, 1972; drawing by Sonya Evans).

of waste products of digestion as well as the release of accumulated guanine from the Malpighian tubules. The release of these materials from the rectal sac occurs with definite cyclic regularity. In soft ticks the largest amount of feces is eliminated within a few days after molting; later, for nearly a week after a blood meal. Small fecal masses of guanine and/or hematin are usually released during the gonotrophic cycle. Similar elimination is found in larvae and nymphs, but stops a few days or several weeks prior to molting.

Adult ixodids eliminate the greatest amount of fecal material and guanine a few days after feeding. When adult females have detached from the host animal, defecation stops and the fecal and excretory materials are not eliminated; the female dies without defecation. Larvae and nymphs release maximum amounts of feces a few days after the end of feeding, but the process stops in engorged forms when the molting process begins and the epidermis separates from the old integument. The consistency of the feces depends upon the amount of water in the body and can be more liquid after feeding or solid in unfed animals (Balashov, 1972), which implies that the rectal sac reabsorbs water.

Because of the single blood meal ingested in each developmental stage of ixodid ticks, the cyclic changes of the epithelium of the gut are much more pronounced than in argasid ticks.

Such cyclic modifications result from the extremes of the feeding conditions, namely, a resting, essentially nonfeeding situation, then concentrated feeding activity prior to molting and egg laying. The physiological age of the unfed tick and the duration of the feeding period also influence these changes.

Digestion in ixodid ticks is much more intense than in argasids. The blood ingested in the first days of feeding is almost completely absorbed by the epithelial cells and they become hypertrophied. These swollen cells almost fill the lumen of the ventriculus and the caeca, where the rates of digestion are about the same. The ventriculus and caeca of engorged ticks are replete with blood and little change occurs prior to molting and oviposition.

Engorgement causes not only a general enlargement of the body, but also distension of the walls and flattening of epithelial cells. Within the first 24 h after the tick detaches, the cells become stretched and hypertrophied, then dome-shaped, clavate, or columnar. The process is indistinguishable from the differentiation of the reserve cells, which is signaled by the enlargement of the nucleus and the formation of RNA-rich fibrillar elements in the cytoplasm near the base of the cells. As assimilation of the food proceeds and molting or oviposition occur, the hypertrophied cells tend to fill the lumen of the gut. Differences in the complexity and the variety of changes occur in males, females, nymphs, and larvae, as well as between different species of ticks (Balashov, 1972) (Fig. 7-21).

Three different types of cells are described for the ventriculus and caeca of unfed ticks: undifferentiated reserve cells, degenerating digestive cells, and secretory cells (Roesler, 1934; Hughes, 1954a). Others assert that the gut wall has a single type of cell that varies in form and structure according to the particular phase of the digestive process (Chinery, 1964). Balashov (1972) maintains that the different types of cells in both argasid and ixodid ticks have a common origin and the undiffentiated reserve

cells are the common type (Balashov, 1972). He states:

> In unfed ticks, this cell type is most numerous and secretory cells are rare. Digestive cells filled with different types of inclusions form by reserve cell differentiation. In addition, secretory and digestive cells are apparently irreversibly specialized. When the functional activity cycle ends, the cells degenerate. Nevertheless, exceptions are not rare, since we repeatedly observed cells containing large secretory vacuoles to have hemoglobin-filled inclusions, i.e., secretory cells transforming into digestive cells. Thus, with definite reservations, we acknowledge the existence of three different cell types in the ixodid midgut: undifferentiated reserve cells, digestive cells and secretory cells similar to those of argasids. (p. 290)

The histology of caeca of unfed females differs mainly in the amount of food reserves within the cells that occur during feeding as the lumen of the gut fills and digestion proceeds. In *Hyalomma asiaticum* the structure of the ventricular epithelium in a newly molted female is similar to nymphs after molting with spherical, yellowish-brown, semitransparent food vacuoles in the cytoplasm and hematin inclusions. Undifferentiated reserve cells exhibit rounded nuclei close to the basement membrane. As the end of postmolting is reached, the hypertrophied cells collapse and disintegrate so that the histology of unfed female ticks is normal in appearance. Eventually the enlarged cells completely disappear, the ventricular and caecal walls become folded while hematin and cell debris accumulate in the lumen. Undifferentiated reserve cells appear unchanged.

In the ventriculus of the unfed adult female ixodids the epithelium is characteristically underdeveloped and lacks functional digestive cells. Those digestive cells that have been retained from the nymph act mainly as reservoirs of food and do not participate in the digestive cycle of the adult. The secretory cells begin to grow and enlarge as soon as the tick attaches to the host. This begins as an enlargement of the distal ends of undifferentiated reserve cells, from which the secretory cells are formed and extend into the lumen (Fig. 7-21). They are wedged between the undifferentiated cells and do not form a complete layer.

Increase in RNA is the first sign of change in the undifferentiated reserve cells and occurs even before liquid food has entered the gut, as shown by sections of ticks taken a few hours after attachment. The distal ends of the cells become greatly distended and vacuolated. Cell numbers increase rapidly as food is ingested and assimilated. Each cell (70–100 μm long) is usually columnar or clavate. The apical end of the cell is highly vacuolated; the basal end exhibits basophilic fibrillar structures thought to represent an endoplasmic reticulum. The RNA content of the cell increases with an increase in nuclear enlargement.

The entry of food into the gut first causes the secretory and digestive cells to disintegrate or to bud off into the lumen of the gut. Nucleated cell fragments appear like rounded hemocytes and later disintegrate into the lumen of the gut. Concurrently with this mixing, PAS-positive colloidal material is observed in the food mass. It is believed that the polysaccharide colloidal material is hemolysin (Hughes, 1954; Chinery, 1964). The budding off of the secretory cells is so extensive that only isolated undifferentiated cells remain on the basement membrane. The food thus appears to be separated from the hemocoel by only a thin membrane.

Liquid food that enters the gut is digested and assimilated immediately, but this is not accompanied in the first days of feeding by the cytoplasmic inclusion of food materials or digested products. Protein granules do appear in the periphery of the cytoplasm of digestive cells, however. Histological techniques are not sufficient to demonstrate finer details of protein breakdown, but growth of the tick integument, gonads, and other organs is indirect evidence of absorption of such materials. The absorption of hemoglobin is demonstrated by benzidine reactions and the identification of the brown-black hematin crystals that tend to clump together in larger cytoplasmic masses. The intracellular hemoglobin vacuoles formed after the host erythrocytes are lysed in the gut are easily recognized by staining techniques using axocarmine, iron, and hematoxylin and eosin. Digestive cells that are shed into the lumen may float among the contents of the gut and continue with intracellular digestion, shown by the accumulation of hemoglobin vacuoles and the hematin granules. Hematin-filled cells eventually disintegrate and release their contents into the lumen of the gut.

Rapid ingestion of blood expands the walls of the ventriculus. When the gut reaches its maximum expansion the engorged tick detaches from the host and the lumen is filled with undigested blood. The expansion of the caeca decreases slightly with the elimination of excess water from the food back through the saliva or through the coxal glands.

The ventricular cells of male and female *Ixodes* are similar. In nonfeeding males the changes in the epithelium of the gut are associated with the gradual utilization of the food reserves accumulated in the nymph. Recently molted males show many food inclusions in the cytoplasm of the gut epithelium, with indistinct cell boundaries. Intense secretory cell growth occurs after the male has attached to the host and before feeding begins. Upon ingestion of the blood meal, cells of the epithelium break away and mix with the blood and food granules accumulate rapidly in the digestive cells. The gut cells of male ixodid ticks have less hematin than those of females, which condition persists through the life of the males, although production of new digestive cells from the reserve cells continues. The gut wall of the males does not pass through the folding and "crypt" formation typical of females.

The digestive epithelium of larval ticks is flattened in the unfed stage. Upon feeding the cells enlarge and become dome-shaped and vacuolated. As the blood meal is ingested the gut expands and digestive cells demonstrate food inclusions in the cytoplasm. When the larvae detach from the host, mitotic divisions form some new cells that elongate into the gut lumen and show abundant food inclusions. Undifferentiated reserve cells persist into the nymphal stage and form the gut epithelium. Larvae of *Hyalomma asiaticum, Rhipicephalus turanicus,* and *Dermacentor pictus* have abundant food inclusions in the digestive cells.

Digestion in nymphs is similar to the same process in females. Food reserves are plentiful in the digestive cells that persist from the larva, but few nymphal digestive cells occur. The food reserves are almost exhausted in unfed, attacking nymphs. The dome-shaped digestive cells form as the nymphs feed and the cytoplasm of these cells becomes vacuolated. Before the nymph detaches, the gut wall has expanded greatly, mainly by proliferation of undifferentiated reserve cells. Cell division in these reserve cells continues from the beginning of the growth period through half of the molting period. There are, additionally, small, undifferentiated reserve cells between the enlarged digestive cells of the nymph that become the precursors to the secretory and digestive cells of the adult tick.

Ticks generally have a combination of intracellular and lumenary (extracellular) digestion. As explained earlier in the description of the feeding process of hard ticks (Ixodidae), a third type of digestion occurs. This is referred to as external (extragut, Balashov, 1972) because it is really the first phase of the digestive process and occurs outside of the tick body. The lytic effects of the tick saliva on the tissues of the host are similar to the process of digestion in spiders, where active proteases are injected into the body of the prey. The contents of tick saliva have not been investigated fully, but evidently include lysing agents. Also, the anatomy of the anterior part of the digestive tract in ixodid ticks prevents any regurgitation of digestive tract contents (Balashov, 1972; Gregson, 1967).

In the digestion of the blood meal, lumenary digestion is important because it involves hemolysis and the division of food components. The ingested blood mass in the ventriculus and caeca is uniformly deep red in color when stained with azocarmine shortly after hemolysis. The peripheral margins of the mass then change to a blue color signifying a further digestive breakdown. Evidently peristaltic mixing occurs and as digestion proceeds, distension of the gut wall decreases and the internal parts of the food mass are reached by the enzymes. This digestive activity is accelerated by the detachment and release of digestive and secretory cells into the lumen of the gut (Fig. 7-20D), which also increases the active absorptive surfaces available. Localization of the visible changes of the blood mass is assumed to imply parietal digestion (Balashov, 1972), a process that is also found in mammals and some other invertebrates. Structures resembling microvilli in the apical membranes of the gut cells together with highly alkaline phosphatase concentrations support the possibility of parietal digestion in ixodids, as well as argasids.

The final splitting of proteins and lipids is an intracellular process. Other aspects of digestion (external and lumenary) are subsidiary to final assimilation (Wright and Newell, 1964). The localization of enzymes (esterase and leucine aminopeptidases) occurs in *Argas persicus,* but weak acid and weak alkaline within the cells preclude action by aminopeptidase, which functions in an acid medium (Balashov, 1972).

Microscopic observations on the breakdown of hemoglobin in the cells indicate intracellular digestion because of the intense staining of the heme prosthetic group, but the process must still be considered more of a working hypothesis than actual knowledge. Microscopy does reveal that pinocytotic vesicles form, and that large drops of hemoglobin as well as whole erythrocytes may be phagocytized in the cells.

It is assumed that intracellular lysosomes discharge their enzymes into vesicles, which then become food vacuoles where splitting off of the heme and globin molecules occurs. The globin is assimilated and the prosthetic part—hematin granules—remains in the digestive cell until disintegration occurs.

Developing oocytes in *Boophilus microplus* are red-brown in color due to the accumulation of heme. The heme of the egg differs from the host hemoglobin. "The iron in this complex is trivalent and the protein differs from vertebrate globin. After hemoglobin splits, some ferric heme is believed to pass into the hemolymph through the basement membrane and to be bound to specific tick proteins. Thus a new type of iron-containing protein, called hemixodovin, is formed" (Brenner, 1959, cited by Balashov, 1972). This compound was found earlier by Wigglesworth (1943) in *Ixodes ricinus* and *Ornithodoros moubata* and is thought to be present in ticks of the genera *Argas, Hyalomma, Rhipicephalus,* and *Dermacentor.* Hemixodovin is present in the ventriculus during postembyronic development and becomes subject to intracellular digestion. Hematin and an unknown heme derivative are by-products of this process. Just how the hemixodivin functions is obscure, but it is suspected that it is used as a reserve protein during embryogenesis and as a source of energy for the larvae before feeding. It may also be used in the construction of enzymes that contain iron. The iron is not separated from the heme to form bile pigments in ticks as is found in *Rhodnius* and *Triatoma* (Wigglesworth, 1943), but traces of iron occur in digestive epithelium and fat bodies.

Argasid ticks store food reserves in the lumen of the gut mainly as undigested hemoglobin and perhaps some plasma proteins from the host. The posterolateral caeca are the main reservoirs and have little to do with digestion of the blood. The general digestive process is slow and the entire blood mass in the gut is not involved at the same time. The blood meal in *Ornithodoros moubata* gives specific precipitin tests for host blood proteins 6 months after the last blood meal. The hemoglobin inclusions in digestive cells are not reserves as is the case in ixodids. The more easily assimilated food reserves, glycogen, and fats, are refurbished slowly and decrease markedly in ticks that have been without a blood meal for a long time (Weitz and Buxton, 1953).

Ixodid ticks are different because the blood meal and any lysed host tissue ingested move rapidly from the lumen of the gut into the digestive cells and causes their hypertrophy. After feeding, these cells show an increased content of hemoglobin concurrent with the decrease in the food mass in the lumen of the gut. The hemoglobin content of these cells reaches its maximum toward the end of molting and when the undigested blood residues disappear. Stable fat reserves in vacuoles build as the hemoglobin decreases and then decrease in amount more slowly than protein residues. Eventually, the hemoglobin is transformed into fats.

The cells of the midgut not only perform the digestive functions, but act as reservoirs for the reserves of protein, carbohydrates, and fats. Some fat bodies in *Ixodes* and other genera also store lipid materials.

Actinedida. Prostigmatid mites represent a wide variety of feeding habits: they can be distinctive predators, parasitoids, definitive parasites, and fungivores. Algophagy and bryophagy are known. Phytophagy, however, is the major feeding practice of a number of major groups within the Actinedida (=Prostigmata). The majority of these mites use higher plants as food. Life cycles and mouthparts correlate with this condition (Evans et al., 1961). Each of the different types of feeders from predator to phytophage is related to the feeding structures present. The variations of the mouthparts of phytophagous prostigmatid mites are easily as pronounced as the differences between mouthparts of the prostigmatids and the others orders or mites (Krantz and Lindquist,

1979; these authors summarize many details of feeding in phytophagous mites.)

Phytophagous actinedids include Eupodoidea, Tydeoidea, Tarsonemoidea, Raphignathoidea, Tetranychoidea, and Eriophyoidea. Spider mites (Tetranychidae), false spider mites (Tenuipalpidae), and gall and rust mites (Eriophyidae) are exclusively plant feeders and comprise the families of greatest economic importance to plants in agriculture. Most of these plant feeders have either needlelike or styletiform chelicerae that can pierce the host plant tissue to suck cell contents or other liquid food. The phytophagous tarsonemid mites *Steneotarsonemus pallidus*, the cyclamen mite, is a serious greenhouse pest. The tiny stylets inflict serious injury to the upper surface of developing leaflets or young runners of the plants and cause malformation and death. Another species of this genus, *S. laticeps*, is responsible for damage to narcissus and other bulbs by feeding through the scales. Large populations may build up during the winter storage of bulbs and cause serious damage. Others of the plant-feeding genera and species feed on ornamentals, cereals, and ferns.

Algivorous and mycetophagous species occur in the genus *Tarsonemus*. *T. floricolus* feeds on mushrooms and tobacco, particularly when these materials are stored and have become moldy (Evans et al., 1961).

Mites of the family Pyemotidae have cheliceral styli that function independently of each other in protraction and retraction. The mouth is enclosed in an acuminate cone. Once the cheliceral stylets have cut a wound and withdrawn, the mouth cone is inserted and sucking begins (Krczal, 1959). The pharyngeal pump expands slowly, then constricts quickly to force the food into the gut. About 50 contractions per minute have been observed (Brucker, 1901). Bee mites feed in a similar way (Hirschfelder and Sachs, 1952).

Mouthparts of mites of the endeostigmatid family Nanorchestidae (Pachygnathoidea) are principally adapted to algophagous feeding in both aquatic and terrestrial environments, but these mites may be fungivorous as well (Krantz and Lindquist, 1979; Schuster and Schuster, 1977; Theron and Ryke, 1969; Grandjean, 1936b, 1939; Tragardh, 1909). They are thought to be the most ancient of the phytophagous groups and since their fossils date from mid-Paleozoic times, "one wonders whether or not the nanorchestids were adapted to feeding on green algae before the vascular plants arose in Mesozoic times" (Krantz and Lindquist, 1979).

The gnathosomal structures are peculiar in that a curved, sclerotized, pointed process extends forward between the chelicera and is assumed to be a piercing organ, but Grandjean (1939) suggested that it enabled the mites to jump by acting as a type of spring. The projection is really a tool for piercing the cells of the alga on which the mites feed, which agrees with Traghard's earlier assessment of the function of the mouthparts (Schuster and Schuster, 1977).

Limited information is available about the feeding habits of the Eupodoidea (most of them with needlelike chelicerae), but some members of the complex are known phytophages of grasses and herbs. Others possibly feed on mosses and lichens. Still others (*Eupodes* and *Linopodes*) are known to feed on fungi, including mushrooms. The family Rhagidiidae exhibit strongly chelate mouthparts and are known predators.

Penthaleus major (Duges), the winter grain mite, feeds on small grains and grasses and may infest legumes, ornamental flowers, cotton, and peanut plants. The feeding of the mites causes a silvering or graying of the tissue due to the extraction of chlorophyll. After several days, the tips of the leaves appear scorched and the plant may become stunted. In other situations the plant may die. Young plants are more affected by the feeding than older, more robust plants. The feeding action occurs as the result of penetration of the plants by the curved stylet of the chelicerae, and possibly the tridentate fixed digit as well (Jeppson et al., 1975).

The red-legged earth mite or "black sand mite," *Halotydeus destructor* (Tucker), is a crop pest in Australia, Cyprus, Zimbabwe, and South Africa. Clover, wheat, potatoes, tobacco, peas, and tomatoes are crops affected. Its feeding activities cause silver or white discolorations along the main veins of host plant leaves, which eventually results in a bleaching of the leaves, followed by a shriveling that appears as if the plants were scorched by heat. Young plants, particularly hothouse seedlings, may die quickly as a result of feeding injury. Examinations of the leaf tissue shows that the parenchymous cells of the palisade layer are empty, devoid of chloroplasts. Because these cells are absent, the space fills with air, resulting in the silvery appearance, and the epidermis of the leaf collapses with subsequent browning typical of the mite-feeding injury (Jeppson et al., 1975).

For a long time it has been thought that mites in the Tydeoidea were plant feeders, but conflicting observations show that Tydeidae are unspecialized feeders. *Tydeus californicus* (Banks) was assumed to be a predator of the citrus bud mite, yet is also identified as feeding on avocado and on aphid honeydew. Other tydeids cited by these authors are fungivorous and phytophagous, with body contents taking on the colors of the food consumed. *Pronematus* thrives on the pollen of cattail (*Typha*). Other populous species are assumed to feed on pollen blown by the wind onto trees (Krantz and Lindquist, 1979). A *Tydeus* species is found on grapes in Moldavia (Karg, 1971). *Lorryia formosa* is also cited by these writers as pests of the twigs of young citrus fruit in both Morocco and Brazil (Krantz and Lindquist, 1979).

The plant feeding observed in the Tydeidae appears to be a diversionary type derived from fungivorous habits and to be incidental rather than obligatory.

One would expect that tydeids which generally have fused cheliceral bases, reduced fixed cheliceral digits, and needlelike movable digits, would be in a position to adapt easily to obligatory feeding on higher plants; yet no group of tydeids has become specialized for obligate phytophagy. (Krantz and Lindquist, 1979)

Four distinctive morphological features of obligate phytophagous tydeids (as compared with predators) are (1) the absence of ambulacra on legs I, (2) striations on the dorsal idiosoma, (3) a different setal number and arrangement on the prodorsum, and (4) modifications of the gnathosomal structures (Malchenkova, 1967).

The unspecialized nature of the feeding of Tydeidae suggests that the family is in a transitional phase, evolutionarily speaking; that these mites may be tending toward phytophagy as a major feeding habit (Krantz and Lindquist, 1979).

Follicle mites (Demodicidae) feed on sebum (secretions from the sebaceous glands), cell debris, cell contents, and lema (secretions from the Meibomian glands). Other unique locations where secretions that comprise liquid food are most likely to be found include the sweat glands and the lymph glands of dogs (Nutting and Rauch, 1958).

The follicle mites have large salivary glands whose secretions may serve to liquify the contents of the cells on which they feed. The knife-like chelicerae puncture the cells of the host, and the contents, liquified by the salivary secretions and/or digestive enzymes, are sucked out. The liquification of the food by the chemical agents is necessary for passage of the food through the tiny mouth (Nutting, 1964). The larvae feed continuously but the deutonymphs do not feed. Females feed sporadically and intermittently. Males feed rarely. Nutting also records that mites move consistently to the supply of food—the sebum of the sebaceous gland (Spickett, 1961).

With respect to plant-feeding habits, probably no more diversified group exists than the Heterostigmata. In England one of the Pyemotidae, *Siteroptes graminum*, feeds on more than 30 species of grasses as well as wheat, barley, oats, and rye. The mites retard the growth of the influorescences by eating at the top of the stem within the sheaths. The influorescences are injured by the styletlike chelicerae and become distorted and silvered—which is described as "silver-tip." These mites may also be agents in the distribution of *Nigrospora oryzae*, a fungus that causes cob rot. Another species of this genus, *S. cerealium*, may spread the fungus (*Fusarium poae*) that causes bud rot of carnations (Jeppson et al., 1975). Mites of this latter species were reared successfully through several generations of agar cultures of fungus in the absence of green plant tissue (Krantz and Lindquist, 1979).

The snout mites, Bdellidae and Cunaxidae are voracious predators of other mites, and possibly insect larvae and eggs as well. Cheyletidae and Anystidae are predatory on similar organisms, particularly the insect and mite pests of plants.

The phytophagous spider mites characteristically have a stylophore and styletiform chelicerae. Thrusting action back and forth with lubrication from the salivary glands causes the chelicerae to penetrate to a depth of 100 μm (Jeppson et al., 1975). The long, sharp stylets are used to penetrate the plant cells as the mites feed. Cells of the epidermis of the plants lose the chloroplasts and the remaining contents tend to coagulate and turn an amber color. (As much as 60% of the chlorophyll may be lost.) Penetration of the cells of parenchymatous tissue of leaves is limited to the cells themselves, but where vascular bundles are affected, much damage may occur to adjacent bundles. Photosynthesis and transpiration are notably affected by the feeding damage of these mites. Destruction of epidermal cells may also cause morphological distortion of adjacent cells because of the excess growth required to fill vacant space. Damage to epidermal cells, to palisade cells and to spongy mesophyll cells differs with respect to the location and feeding of mites on upper or lower leaf surfaces. The stylets penetrate farther into cell layers from the upper surface than the lower surfaces of the leaves in all plant hosts. Injury to second palisade layers is greater in apple leaves than in plums.

Tetranychus (T.) urticae can suck the contents out of 18 to 22 cells per minute during feeding. The feeding pattern goes in a circle and leaves chlorotic spots on the apple leaves. Transpiration is greatly accelerated and the speckled leaf eventually falls off. The feeding of the mites directly affects photosynthesis by the extraction of leaf pigments, as shown by the large amount of these pigments in the feces of the mites. Bronzing also occurs because of damage to the mesophyll cells during feeding. It is not known, however, why different plants respond differently to feeding of the same species of mites. These variations are thought to be related to the chemicals injected by the spider mites during feeding (Jeppson et al., 1975).

Evidence exists that chemical toxins are injected into the plant tissue during feeding by spider mites. The nature of the chemicals is unknown, but the mechanism for injection involves flaplike structures at the apex of the rostrum adjacent to the stylets. Extension of these rostral flaps around the stylets enables fluids to be sucked up and toxins or viruses to be carried into the plant cells when the stylets penetrate the tissue (Jeppson et al., 1975; Tanigoshi and Davis, 1978).

The Eriophyidae are entirely phytophagous and are among the most highly adapted of the plant-feeding prostigmatids. Basically vermiform in appearance, these tiny gall, bud, blister, or rust mites live, feed, and reproduce in the microenvironments of the crevices under bud scales, in creases or depressions of leaf surfaces, or around the bases of petioles. They evidently ingest small amounts of cellular fluid by puncturing plant cells. Pedipalps, cheliceral guides, and stylets are involved in the feeding process, which is accomplished by the telescoping of the pedipalps and the penetration of the host tissue by the tiny stylets. Evidently the cheliceral stylets are inserted into the plant slowly enough that little mechanical damage results. The stylets remain in position while the mite feeds.

The comparative length of the stylets varies with the families involved and determines the depth of penetration into the host plant. The shortest of the stylets are in the gall formers. In the family Eriophyidae, the stylets are generally short. Nalepillidae and Rhyncaphytoptidae have proportionately longer ones. The auxilliary stylets are of unknown function, but are thought to be salivary ducts. The oral stylet enables the ingestion of nutrients because of the hinge attached to its pharyngeal pump.

The growth of galls and erinea are the means that eriophyids have of a continuous fresh food supply. These tissue growths sustain the mites by the development of succulent plant nurse cells, whose growth is stimulated in turn by the feeding of the mites. It is to be noted also that viral transmission can be accomplished by these mites because they feed on living rather than dead tissue (Jeppson et al., 1975).

Mites of the Parasitengona (Trombidiidae, Trombiculidae, Erythraeidae, water mites) are usually ectoparasitic as larvae. These mites may also have quiescent calyptostases (egg, prelarva, protonymph, tritonymph) so that the deutonymph is the only active nymphal stage. Deutonymphs and adults are free-living and frequently predaceous, particularly on insects. Such mite predators of small animals exsanguinate the prey organism and leave an empty shell. A nymph feeding on aphids holds the prey with the pedipalps, inserts the chelicerae, and sucks out the body fluids, discarding the empty skin. Larvae of *Balaustium florale* are evidently less parasitic and more predaceous in their habits, but all active stages feed on the pollen of buttercup (*Ranunculus acris*), English daisy (*Bellis perennis*), and common maple (*Acer campestre*) (Grandjean, 1946, 1947). The mites use their pedipalps to grasp and hold the pollen grain close to the infracapitulum. The needlelike chelicerae pierce the pollen grain. Within a few seconds the contents are sucked out, the shriveled, empty case is discarded for a second pollen grain, and the process is repeated (Evans et al., 1961). Larvae of one species feed to repletion on the leaf tissue of corn lily (*Veratrum* sp.) as well as other species that were predatory on pear psyllids and aphids in addition to eating pollen. Nymphs and adults of *B. aonidophagus* feed on red scale insects and red scale crawlers (*Aonidiella auranti*) by thrusting their chelicerae under the armor of the scale or the skin of other prey. These mites are cannibalistic in the captivity (when held in small containers) (Newell, 1963). Attacks on man by species of *Balaustium* are recorded: The mites normally fed on pollen from flowers in an orchard in British Columbia and secondarily attacked workers (Newell, 1963).

Mites in the family Trombiculidae are free-living predators in the upper surface of the soil most of their lives. Once during their lifetimes, however, they are parasitic, when the newly hatched "chigger," the larval stage, crawls on a host animal and attaches. Its penetration of the skin (Fig. 7-2) of the host enables it to suck the lysed tissues through the stylostome. If unable to engorge, it may reattach to another host. If engorged, it drops off the host, molts to a nymph, and feeds as a predator. Nymphs are never known to be parasitic. Changes to the predatory adult stage ensue. It is thus presumed that normally the mite feeds parasitically only once in its lifetime. The host may vary with the particular species of mite and with the opportunity for infestation. Hosts may include mammals, birds, reptiles, amphibians or arthropods, although the cases of chigger infestation of the latter two are fairly rare. Usually one host or a group of hosts is used by a single species of mite, although the range of hosts may be wide for some species (Harrison, 1953; Brown, 1952; Jones, 1950).

Mark–capture experiments and determination of the feeding times of trombiculid larvae show that chiggers are active only when dew is on the ground. Their rat hosts are nocturnal in habit, so the chiggers will have gone into hiding by 9 AM and not become active until after sunset. Whether the rat is released in the morning or the evening, therefore, it is unlikely to pick up any mites before sunset (7 PM). The rats are likely to be trapped in the earlier part of the night, after which their chances of picking up mites are considerably reduced, judging from the results of setting out rats in cages as bait. It seems fair, therefore, to allow that a chigger on a rat trapped within 24 hours of release has been central time 10 PM.

Facts indicate that most of the chiggers leave the rat during the night or early morning (Harrison, 1953).

Water mite larvae frequently parasitize aquatic insects, while the active deutonymphs and adults prey upon other aquatic arthropods or may feed on aquatic plants (Krantz and Lindquist, 1979). Adult water mites have very powerful salivary enzymes, for in a few minutes they completely drain mosquito larvae on which they feed and leave the empty skin. Extraoral digestion is essential to feeding because of the types of mouthparts involved (Bader, 1938).

Inasmuch as females of the genus *Hydrachna* use the chelicerae to incise sites in aquatic plants for egg laying, it seems likely that they may feed on the plant during this preoviposition activity, which may be a preadaptation to plant feeding for these mites (Krantz and Lindquist, 1979).

Schizeckenosy. Schizeckenosy is a term given by Mitchell and Nadchatram (1969) to a peculiar defecatory process in chigger mites. Because the higher trombidiform mites have no hind gut and the single, blind excretory tube opens through the uropore, no anus is present. Intracellular digestion in these mites implies that undigested residues accumulate in the cells of the gut until such cells are replete with fecal material. These cells break free and float in the gut lumen. In adults of the common scrub-itch mite, *Blankaartia acuscutellarus*, these floating cells are moved from the central lumen of the gut posterolaterally to a pair of special lobes or fecal pouches. When these storage lobes become replete with feces-packed cells, the gut opening to the pouch is closed. The filling of the fecal pouches appears to place the dorsolateral part of the body under high pressure, especially in the area immediately posterior to dorsalia 5 (identified externally as dimples in the integument). The body wall splits open by muscle action that moves the edges of the slit in opposite directions dorsally and ventrally. Subsequent depression by dorsalia 5 muscles extrudes the black fecal mass from the fecal pouches through the slit in the skin. The fecal pouches are different in size and not all of the pouch nor the cells are lost in this action. After expulsion of the fecal material the muscles relax and the integument closes and heals. Conspicuous scars may be observed on the skin of adult mites as mute evidence of previous defecation by schizeckenosy. Sometimes the fecal mass does not drop off and may be seen like a piece of dirt adhering to the surface of the dorsum. This phenomenon of gut schizeckenosy is observed in a number of species of chigger mites (Mitchell, 1958, 1964; Mitchell and Nadchatram, 1969) (Fig. 7-22).

Astigmata. The Astigmata exhibit a variety of feeding habits: they range from host-specific parasites to the more free-living forms. The latter are fungivorous, and/or graminivorous, saprophagous, and, very rarely, macrophytophagous (Zakhvatkin, 1959). It is also exceptional to find predators among these mites.

Closely related pairs of mite species (and insect species) feed on specific types of fungi (Sinha, 1968). A species in the pair may eat one variety of fungus, while the other species uses a wider selection of similar food. This selective mycophagy of some mites reduces the competition for food between species. Other fungivorous species (e.g., *Acarus siro* and *Lepidoglyphus destructor*) that may exist together in the same habitat eat different species of fungi and one mite cannot survive on the food of the other

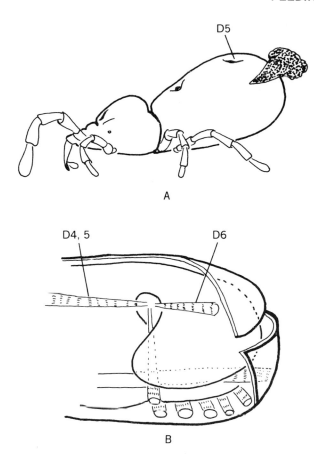

Figure 7-22. Schizeckenosy in adult *Balaustium acuscutellaris* showing (A) position of split in posterior body wall, dorsalia 4 and 5; (B) location of muscles involved in split (D4, D5, D6, dorsalia muscles) (after Mitchell and Nadchatram, 1969).

species. Because species of acarid mites (*Acarus, Tyrophagus*) may carry fungal spores into the gut and on their bodies that cause molding of the grain on which they feed, fungi and mites inadvertently cooperate to cause deterioration of the grain and thus perpetuate an abundance of food source for each, as well as damage to the stored product. Infestation of field-stored peanut pods by species of acarid mites (*Sancassania* and *Tyrophagus*) occurs. The mites carry spores of *Aspergillus* and cause injury to the cotyledons. Species of the genus *Sancassiana* can transmit spores of *Pythium* fungi (a pre-

ferred food) and thus increase the incidence of pod rot in peanuts (Shew and Beute, 1976).

Rhizoglyphus echinopus may feed on healthy bulbs and tubers as well as those attacked by fungi, and secondary injury may result (Sinha, 1968). These mites feed not only on the decaying tissue in affected bulbs, but extend their feeding to the healthy tissue of the plant (Michael, 1903).

Several instances of acarid mites damage living plants of greenhouse crops, such as winter spinach, and field crops, also tomatoes, cucumbers, and some cereals or grasses. Host preference and habitat selection are noted for *Tyrophagus neiswanderi, T. longior,* and *T. similis. Calvolia transversostriata* is cited as damaging to the leaf tissue of apples by its feeding habits in the spring. In the summer these mites change food preference to mycelia and sooty molds that develop on honeydew exudates of plant lice. Apple scab fungal spores were found by Nesbitt in the gut of these mites. *Cladosporium* fungus on filbert leaves and stems is an additional food source for these mites (Krantz and Lindquist, 1979).

Aquatic mites of the family Hyadesiidae feed on algae and seaweed found in their marine and estuarine habitats. Specific details of their feeding mechanisms are not known, but the normal color (olive brown) of their bodies suggests that they are mycetophagus.

Caloglyphus mycophagus eats dead and decaying insects (Rohde and Oemick, 1967). The cells of the lateral ventricular caeca appear ciliated except for the vacuolated cells in among them. The colon cells are characteristically tall, ciliated cells. The cells of the rectum vary from tall columnar cells to flattened cells with a brush border. Histiostomatidoe use modified mouthparts to sweep in microbial food (Hughes and Jackson, 1958; T.E. Hughes, 1953).

In the feeding process of *Listrophorus leuckarti*, the gnathosoma is moved back and forth by leg actions as the mite grasps the hair between the pedipalps. Salivary secretions lubricate the hair and the sharp points of the chelicerae scrape off the scales. The musculature of the gnathosoma and the chelicerae enable this to occur. Mites in a clasping action are unable to let go of the hair and may walk off the end (Hughes, 1954b).

It is generally thought that digestion in acarid mites is intracellular, especially in the ventriculus. Digestive epithelial cells of this organ absorb food into vacuoles. Hydrolysis occurs within the cells. Many of the cells are sloughed off or disintegrate, which releases the products into the lumen of the gut.

Extracellular digestion may occur in the posterior part of the ventriculus in *Dematophagoides farinae* Hughes. Such digestion provides for the assimilation of solid food materials (Brody et al., 1972). A peritrophic membrane may be formed to protect the delicate digestive epithelium and remains around the bolus of food. Like a sausage encasement, the membrane stays around the bolus as the fecal pellet forms and is passed to the rectum (Wharton and Brody, 1972). Other mites that ingest solid food likely form peritrophic membranes in a similar way (Krantz, 1978). An instance of unusual food is that where remains of lepidopteran wing scales have been found in the digestive tract of pyroglyphid mites (*Dematophagoides*). This type of food and feeding has not been noted previously (Treat, 1975).

Oribatida. A correlation exists between the structure of the chelicerae and the type of food eaten by the oribatid mites where the ratio of the length of the movable digit to the particular food consumed is involved (Schuster, 1956). A higher cheliceral ratio was observed in mites that feed on algae, pollen, and fungi, in contrast to macrophytophagous examples. Morphological features of the mouthparts correlate with feeding habits in other analyses (Grandjean, 1957b) where it is assumed that the subcapitular setae became hypertrophied to form the rutella and that these structures are important in the feeding process. Because of their position exterior to the chelicerae, these organs can be moved laterally as the chelicerae are extended and moved medially when the chelicerae are retracted. It is thought that the rutella could thus scrape off the external faces of the chelicerae and facilitate ingestion of the particulate food (Evans et al., 1961).

Oribatid mites feed principally on fungi or algae (Evans et al., 1961). D.A.C. Crossley (personal communication) and Luxton (1972) state that the oribatids usually feed in one of three major ways: mycophagy (microphytophages), saprophagy (macrophytophages), and panphagy (combinations). Phytophagy is identified as another type of feeding (Krantz and Lindquist, 1979).

Oribatids do not generally feed on higher plant material until it is partially decayed or fully saturated. If these mites do use higher plants for food, the feeding is usually confined to decayed tissue or parts in some stage of decomposition (Wallwork, 1958). The unspecialized feeders in this group include the Camasiidae, Caraboidae, and Hermanniidae. Box mites (phthiracaroids) eat decaying leaves, woody punk, and in some instances the fallen needles of conifers. Indeed, immatures of box mites (*Steganacarus, Euphthiracarus, Phthiracarus*) burrow inside pine needles (Jacot, 1939). Some species of *Steganacarus* seem to prefer the leaves of deciduous trees to pine needles. Other mites skeletonize partially decaying leaves, but do not feed on freshly fallen leaves. Phthiracaroids eat the mesophyll tissue of leaves (Hayes, 1967). It is also known that nymphs of *Hermannia* spp. eat moist, woody leaf material and tunnel in bark and deeper tissues (Wallwork, 1976; Woolley, 1960). *Rostrozetes flavus* prefentially eats decomposed outer sheaths of roots (Woodring, 1965). Some species of oribatids are predatory on nematodes, which contrasts to the nonpredatory feeding of most of their fellows (Rockett and Woodring, 1966). Gustaviidae exhibit special chelicerae for feeding (Woolley, 1979) (Fig. 5-18B).

Relatively few of the oribatids feed on living higher (green) plants. It is suspected that mosses are eaten by species of *Malaconothrus, Trimalaconothrus,* and *Achipteria. Achipteria* feeds on living moss as well as on fallen pine needles and fungi (Wallwork, 1958). The larvae and nymphs of *Minuthrozetes semirufus*, a ceratozetid, tunnel in the stems of grasses upon which their eggs are laid (Michael, 1884). Stems, sheaths, and rootlets of aquatic plants are invaded by species of *Hydrozetes* (Krantz and Lindquist, 1979). Experimentally, Woodring and Cook (1962b) fed fresh-cut green grass to *Scheloribates laevigatus*. The mites ate the food for several days (as evidenced by green fecal pellets), but it was not sufficient to sustain the culture.

Humerobates rostrolamellatus normally eats algae, but changes to feeding as a pest on the split fruit of cherries. *Perlohmannia dissimilis* may damage tulips, potatoes, and strawberries (Evans et al., 1961).

Striking examples of oribatids that feed on living green plants are the adults of a galumnoid mite, *Orthogalumna terebrantis* (=*Leptogalumna* of other authors). They feed on algae on the underside surfaces of water lettuce and other plants in South America. Females oviposit on water hyacinth (*Eickhornia*) leaves by excavating the surface to lay eggs. The larvae and three nymphal stages burrow in the leaves, where they feed on the parenchymal tissue. New adults emerge from within the leaves by cutting holes with their chelicerae (?and rutella) in the epidermis of the leaves (Wallwork, 1965). (It is interesting that there are about as many exit holes in the upper surfaces of the leaves as in the lower.) While these mites normally feed where the epidermis is broken on the leaves of the water hyacinth, they are capable of penetrating the leaf surface with the rutella as well as with the chelicerae. Mechanical injury to the epidermis caused by the feeding of snails or insects, or the internal burrowing of the larvae and nymphs of *O. terebrantis* may also result in openings that facilitate feeding (Del Fosse et al., 1975).

It was presumed that the oribatids were able to use only the microbial components of their food because they lacked the amylases necessary to digest the polysaccharide materials from the plant cells in their food. Indirect, contradictory evidence shows that some box mites are strictly xylophagous (Wallwork, 1958). They burrow in the heartwood of decaying stems and branches on the forest floor. Their fecal pellets show finely shredded woody tissue. Microflora in the gut of phthiracaroids that are assumed to be the source of enzymes for the hydrolysis of cellulose in the food. A fecal meal is required to establish the flora in the gut of the larva (C.J. Rohde, Jr., personal communication).

Oribatids secrete cellulase, xylanase, and a pectinase. They are microphytophagous and are able to macerate dead leaf tissue to the extent that the cellular surfaces are exposed to the enzymes and thus hydrolyzed. The macrophytophagous mites do, however, lack the enzyme capable of hydrolyzing trehalose, an important carbohydrate in fungi (Luxton, 1972). Of this condition Krantz and Lindquist (1979) state: "This may indicate a physiological adaptation for avoiding digestion of much of the microbial growth present in their food, and for egesting a more readily biodegradeable product replete with a complement of viable fungi. Such selectivity benefits the mites, since the vegetative material they consume is more readily digested when previously subjected to fungal action." Selectivity in feeding is emphasized when one realizes that as secondary decomposers oribatids digest more easily plant materials that are previously acted upon by fungi. Many types of fungi and spores are involved in the feeding process (Schuster, 1956).

Following a little more than half a century after Michael, modern studies on the feeding habits of oribatids involve laboratory rearing, direct examinations of litter and humus, and examination of gut contents to determine feeding processes (Riha, 1951; Wallwork, 1958). The feeding habits of 20 species of oribatids fall into a graded series from those with a wide variety of food materials to those which are restricted to a single food source: (1) nonselective feeders, (2) intermediate feeders, and (3) selective feeders. The food ranges through many types of plant materials in various stages of decomposition within the litter and humus. Fungi, algae, hemlock needles, yellow birch leaves, moss, fresh bark tissue, heartwood, decayed wood, fecal pellets of mites and snow fleas, as well as the carrion of dead mites and insects constitute the food of these mites.

Nonselective feeders in the litter include *Achipteria coleoptrata, Fuscozetes fuscipes, Scheloribates pallidulus, Peloribates* sp., *Oppia neerlandica,* and phthiracaroid mites. *Phthiracarus borealis* feeds on the tough plant tissue of hemlock needles and leaves of yellow birch, using both decaying and undecomposed materials. Decaying woody tissues are eaten by *Mesoplophora pulchra* and *Steganacarus diaphanum* along with some fungal spores and pollen grains. It is conjectural as to whether the latter are accidental gut contents or food deliberately selected.

The feeding habits of oribatid mites and the types of food eaten are correlated with the modifications observable in the chelicerae. Xylophagous mites, for example, have relatively large and powerful digits that can rip, tear, and macerate the woody tissue before it is ingested. The types of chelicerae are differentiated into chewing, grinding, tearing, and macerating forms (Wallwork, 1958) (Fig. 7-23). The capabilities of the chelicerae for comminuting food into particles are important to both the immatures and adults, as these capabilities influence the selection of food. Texture and moisture content of the food may also affect selection.

Selective feeders, such as *Heterodamaeus* sp., are representative of a specialized part of the forest floor community. They burrow or tunnel in the twigs and branches of hemlock and yellow birch. Some of the xylophagous species chisel out the burrows and tunnels, while others feed on the wood frass and fecal pellets produced. *Hoplophora magna* makes wellformed burrows in the heartwood of hemlock and yellow birch. The tunnel system is nearly filled with fecal pellets of finely shredded woody tissue. Species of *Pseudotritia* are more superficial in their tunneling, but their mas-

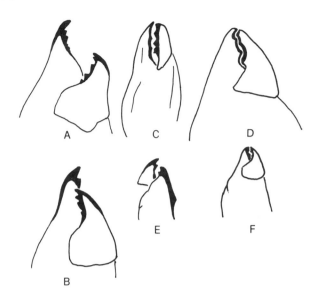

Figure 7-23. Types of chelicerae in Oribatida: (A) *Hermannia* (woodborer) open; (B) same closed; (C) *Phthiracarus;* (D) *Pseudotritia* (woodborer); (E) *Peloribates;* (F) *Achipteria coleoptrata* (after Wallwork, 1958).

sive, powerful chelicerae (Fig. 7-23) masticate woody tissue easily. Intermediate feeders include *Tectocepheus velatus* and a species of *Eniochthonius.* Particulate matter, decayed leaf materials, and occasional fungal spores were present in the latter.

The factors that permit or limit the choice of food are (1) the particle size, (2) the type of chelicerae, (3) the nature of the digestive system, (4) the condition of chemical decomposition of the food, and (5) the moisture content of the food. These factors interrelate in a number of ways and the relative importance of each factor also varies with different species and the life stages of the individuals. Intraspecific as well as interspecific differences also occur (Riha, 1951; Wallwork, 1958).

Oribatid mites play an important role in the decomposition of organic material and thus contribute to the formation of humus (Lebrun, 1979). Several observers record oribatids feeding on pupae of parasitic Hymenoptera, fly larvae, tapeworm eggs, and nematodes. Some oribatids demonstrate a predilection for nematophagy (*Pergalumna, Galumna, Fuscozetes,* and *Nothrus*) (Rockett and Woodring, 1966).

REFERENCES

Aeschliman, A. 1958. Development embryonnaire d'*Ornithodorus moubata* (Murray) et transmission transovarienne de *Borrelia duttoni Acta Trop.* 15:15-64.

Ainscough, B.D. 1960. The internal morphology of *Caminella peraphora* Krantz and Ainscough, with descriptions of the immature stages. Oregon State University Thesis, LD 4330, 53 pp.

Alberti, G. and V. Storch. 1973. Zur feinstruktur der "munddrusen" von Schnabelmilben. (Bdellidae, Trombiformes). *Z. Wiss. Zool.* 186(1/2):149-160.

Andre, M. 1949. Ordre des Acariens. *In*: Grassé, P. P., ed. *Traité de Zoologie.* Masson, Paris, Vol. 6, pp. 795-892.

Arthur, D.R. 1946. The feeding mechanism of *Ixodes ricinus* L. *Parasitology* 37:154-162.

Arthur, D.R. 1951. The capitulum and feeding mechanism in *Ixodes hexagonus* Leach. *Parasitology* 41:66-81.

Arthur, D.R. 1960. *Ticks,* Part V. Cambridge Univ. Press, London and New York.

Atyeo, W.T. 1960. A revision of the mite family Bdellidae in North and Central America (Acarina, Prostigmata). *Univ. Kans. Sci. Bull.* 40(8):345-499.

Atyeo, W.T. and N.L. Braasch. 1966. The feather mites Genus *Proctophyllodes. Bull. Univ. Nebr. State Mus.* 5:1-354.

Bader, C. 1938. Beitrag zur Kenntnis der Verdauungsvorgange bei Hydracarinen. *Rev. Suisse Zool.* 45:721.

Baker, E.W. and G.W. Wharton. 1952. *An Introduction to Acarology.* Macmillan, New York.

Baker, E.W., J.H. Camin, F. Cunliffe, T.A. Woolley and C.E. Yunker. 1958. *Guide to the Families of Mites.* Inst. Acarol., Dep. Zool., University of Maryland, College Park.

Balashov, Y.S. 1972. *Bloodsucking Ticks (Ixodoidea)—Vectors of Diseases of Man and Ani-*

mals, Transl. 500 (T 500). Med. Zool. Dep., USNAMRU No. 3, Cairo, Egypt, U.A.R. [published in *Misc. Publ. Entomol. Soc. Am.* 8(5):161-376 (1968)].

Bernini, F. 1973. Ultrastructural observations on the alimentary canal in Oribatidae (Acari). *Proc. Int. Congr. Acarol., 3rd, 1971,* pp. 67-71.

Bertram, D.S. 1939. The structure of the capitulum of Ornithodoros: a contribution to the study of the feeding mechanism in ticks. *Ann Trop. Med. Parasitol.* 33:229-258 (cited by Evans *et al.,* 1961).

Blauvelt, W.E.. 1945. The internal morphology of the common red spider mite *(Tetranychus telarius). Mem.-N.Y. Agric. Exp. Stn. (Ithaca)* 270:1-35.

Brody, A.R., J.C. McGrath and G.W. Wharton. 1972. *Dermatophagoides farinae:* The digestive system. *J. N. Y. Entomol. Soc.* 80(3):152-177.

Brown, J.R.C. 1952. The feeding organs of the adult of the "common chigger." *J. Morphol.* 91:15-51.

Brucker, E.A. 1901. Monographie de *Pediculoides ventricosus* Newport et théorie des pieces buccales des acariens. *Bull Sci. Fr. Belg.* (6) 35(4):365-451 (cited by Cross, 1965).

Casto, G. M. 1974. Entry and exit of syringophilid mites (Acarina: Syringophilidae) from the lumen of the quill. *Wilson Bull.* 86:272-278.

Chinery, W.A. 1964. The mid-gut epithelium of the tick *Haemophysalis spinegera* Neumann 1897. *J. Med. Entomol.* 1:206-212.

Chinery, W.A. 1965a. Studies on the various glands of the tick *Haemophysalis spinigera* Neumann 1897. *Acta Trop.* 22:235-266 (cited by Balashov, 1972).

Chinery, W.A. 1965b. Studies on the various glands of the tick *Haemophysalis spinigera* Neumann 1897. *Acta Trop.* 22:321-349 (cited by Balashov, 1972).

Chinery, W.A. 1974. Studies on the salivary glands of *Argas persicus. J. Med. Entomol.* 11(4):480-487.

Cloudsley-Thompson, J.L. 1958. *Spiders, Scorpions, Centipedes and Mites.* Pergamon, Oxford.

Coons, L.B. and M.A. Roshdy. 1973. Fine structure of the salivary glands of unfed male *Dermacentor variabilis* (Say) (Ixodidea: Ixodidae). *J. Parasitol.* 59(5):900-912.

Coons, L.B., R. Rosell-Davis and B. I. Tarnowski.

1986. Bloodmeal digestion in ticks. *In:* Sauer, R.J. and J.A. Hair, eds. *Morphology, Physiology and Behavioral Biology of Ticks.* Ellis Horwood, Ltd., Chichester, England, pp. 248–279.

Cross, E.A. 1965. The generic relationships of the family Pyemotidae (Acarina: Trombidiformes). *Univ. Kans. Sci. Bull.* 45(2):29–275.

Crossley, D.A. 1960. Comparative external morphology and taxonomy of nymphs of the Trombiculidae (Acarina). *Univ. Kans. Sci. Bull.* 40(6):135–499.

Del Fosse, E.S., H.L. Cromroy and D.H. Habeck. 1975. Determination of the feeding mechanisms of the waterhyacinth mite. *Hyacinth Control J.* 13:53–55.

Desch, C. E. and W. B. Nutting. 1977. Morphology and functional anatomy of *Demodex folliculorum* (Simon) of man. *Acarologia* 19(3):422–462.

Douglas, J.R. 1943. The Internal Anatomy of *Dermacentor andersoni* Stiles. *Univ. Calif. Publ. Entomol.* 7(10):207–272.

Edney, E.B. 1977. *Water Balance in Land Arthropods.* Springer-Verlag, Berlin and New York.

Evans, G.O. 1957. An introduction to the British Mesostigmata (Acarina) with keys to the families and genera. *J. Linn. Soc. London, Zool.* 43:203–259.

Evans, G.O., J.G. Sheals and D. MacFarlane. 1961. *The Terrestrial Acari of the British Isles,* Vol. 1. Adlard & Son, Bartholomew Press, Dorking, England.

Fain, A. and J. Bafort. 1964. Les Acariens de la Famille Cytoditidae (Sarcoptiformes). Descriptions de sept espèces nouvelles. *Acarologia* 6(3):504–528.

Fawcett, D.W., K. Binnington and W. P. Voigt. 1986. Cell biology of the ixodid tick salivary gland. *In:* Sauer, R.J. and J.A. Hair, eds. *Morphology, Physiology and Behavioral Biology of Ticks.* Ellis Horwood, Ltd., Chichester, England, pp. 22–45.

Furman, D.P. 1959. Observations on the biology and morphology of *Haemogamasus ambulans* (Thorell) (Acarina: Haemogamasidae). *J. Parasitol.* 45(3)274–280.

Galun, R. and S.H. Kindler. 1968. Chemical basis of feeding in the tick *Ornithodoros tholonzani. J. Insect Physiol.* 14(10):1409–1421.

Grandjean, F. 1936a. Un acarien synthétique: *Opilioacarus segmentatus* With. *Bull. Soc. Hist. Nat. Afr. Nord.* 27:413–444.

Grandjean, F. 1936b. Observations sur les Acariens (3 série). *Bull. Mus. Natl. Hist. Nat.* (2) 8:84–91.

Grandjean, F. 1938. Description d'une nouvelle prelarva et remarques sur la bouche des Acariens. *Bull. Soc. Zool. Fr.* 68:58–68.

Grandjean, F. 1939. Quelques genres d'Acariens appartenant au groupe des Endeostigmata. *Ann. Sci. Nat., Zool. Biol. Anim.* (11) 2:1–122.

Grandjean, F. 1944. Les "taenides" des Acariens. *C.R. Seanc. Soc. Phys. Hist. Natur. Genève,* 61:142–146.

Grandjean, F. 1946. Observations sur les Acariens (9e série). *Bull. Mus. Natl. Hist. Nat.* (2) 18:337–344.

Grandjean, F. 1947. Au sujet des Erythroides. *Bull. Mus. Natl. Hist. Nat.* (2) 19:327–334.

Grandjean, F. 1957a. L'infracapitulum et la manducation chez les Oribates et d'autres Acariens. *Ann. Sci. Nat., Zool. Biol. Anim.* (11) 19:233–281.

Grandjean, F. 1957b. Observations sur les Oribates (37e série). *Bull. Mus. Natl. Hist. Nat.* (2) 29:88–95.

Gregson, J.D. 1967. Observations on the movement of fluids in the vicinity of the mouthparts of naturally feeding *Dermacentor andersoni* Stiles. *Parasitology* 57:1–8.

Gupta, A.P. 1979. *Arthropod Phylogeny.* Van Nostrand-Reinhold, New York.

Hambdy, B.H. 1973. Biochemical and physiological studies of certain ticks (Ixodidea). Cycle of nitrogenous excretion in *Argas (Persicargas) arboreus. J. Med. Entomol.* 10(1):53–57.

Hammen, L. van der. 1961. Descriptions of *Holothyrus grandjeani,* nov. spec., and notes on the classification of the mites. *Nova Guinea, Zool.* 9:173–194.

Hammen, L. van der. 1964. The morphology of *Glyptholaspis confusa. Zool. Verh.* 71:1–56.

Hammen, L. van der. 1965. Further notes on Holothyrina (Acarida). I. Supplementary description of *Holothyrus coccilonella* Gervais. *Zool. Meded.* 40:253–276.

Hammen, L. van der. 1966. Studies on Opilioacarida. I. Description of *Opilioacarus texanus*

(Chamberlain and Mulaik) and revised classification of the genera. *Zool. Verh.* 86:1–80.

Hammen, L. van der. 1979. Comparative studies in Chelicerata I. The Cryptognomae (Ricinulei, Architarbi and Anactinotrichida). *Zool. Meded.* 174:3–62.

Harrison, J.L. 1953. Feeding times of trombiculid larvae. Malaysian parasites. X. *Stud. Inst. Med. Res. Malaya* 26:170–183.

Hirschfelder, H. and H. Sachs. 1952. Recent research on the acarine mite. *Bee World* 33(12):201–209.

Hughes, A.M. 1961. The mites of stored food. Ministry of Agr. Fish. and Food. Tech. Bull. No. 9, pp. 1–287.

Hughes, R.D. and C.G. Jackson. 1958. A review of the family Anoetidae. *Va. J. Sci.* 9(1):1–217.

Hughes, T.E. 1949. The functional morphology of the mouthparts of *Liponyssus bacoti. Ann. Trop. Med. Parasitol.* 43(3/4):349–360.

Hughes, T.E. 1950. The physiology of the alimentary canal of *Tyroglyphus farinae. Q. J. Microsc. Sci.* 91(1):45–61.

Hughes, T.E. 1953. The functional morphology of the mouthparts of the mite *Anoetus sapromyzarum* Dufour, 1839, compared with those of the more typical Sarcoptiformes. *Proc. Acad. Sci. Amst.* 56C:278–287 (cited by Evans et al., 1961).

Hughes, T.E. 1954a. Some histological changes which occur in the gut epithelium of *Ixodes ricinus* females during engorging and up to oviposition. *Ann. Trop. Med. Parasitol.* 48:397–404.

Hughes, T.E. 1954b. The internal anatomy of the mite *Listrophorus leuckarti* (Pagenstecher, 1861). *Proc. Zool. Soc. London* 124:239–256.

Hughes, T.E. 1959. *Mites or the Acari.* Athlone, London.

Jacot, A.P. 1939. Reduction of spruce and fir litter by minute animals. *J. For.* 37:858–860.

Jakeman, L.A.R. 1961. The internal anatomy of the spiny rat mite *Echinolaelaps echidninus. J. Parasitol.* 47:328–349.

Jeppson, L., H.H. Keifer and E.W. Baker. 1975. *Mites Injurious to Economic Plants.* Univ. of California Press, Berkeley.

Jones, B.G. 1950. The penetration of the host tissue by the harvest mite, *Trombicula autumnalis* Shaw. *Parasitology* 40:247–260.

Kaestner, A. 1968. *Invertebrate zoology,* Vol. 2 (trans. and adapted by H. & L. Levi) Wiley-Interscience, New York.

Karg, W. 1971. Untersuchungen uber die Acarofauna in Apfelanlagen im Hinblick auf den Ubergand von Standarspritzprogammen zu integrierten Behandleungsmassnahmen. *Arch. Pflanzenschutz* 7:243–279 (cited by Krantz and Lindquist, 1979).

Kaufman, W.R. 1986. Salivary gland degeneration in the female tick, *Amblyomma hebraeum* Koch (Acari: Ixodidae). *In:* Sauer, R.J. and J.A. Hair, eds. *Morphology, Physiology and Behavioral Biology of Ticks.* Ellis Horwood, Ltd., Chichester, England, pp. 46–54.

Kaufman, W.R. and J.E. Phillips. 1973a. Ion and water balance in the Ixodid tick *Dermacentor andersoni.* I. Routes of ion and water excretion. *J. Exp. Biol.* 58:523–536.

Kaufman, W.R. and J.E. Phillips. 1973b. Ion and water balance in the Ixodid tick *Dermacentor andersoni.* II. Mechanism and control of salivary secretion. *J. Exp. Biol.* 58:537–547.

Kaufman, W.R. and J.E. Phillips. 1973c. Ion and water balance in the Ixodid tick *Dermacentor andersoni.* III. Influence of monovalent ions and osmotic pressure on salivary secretion. *J. Exp. Biol.* 58:549–564.

Kethley, J. 1971. Population regulation in quill mites (Acarina: Syringophilidae). *Ecology* 52(5):1113–1118.

Krantz, G.W. 1978. *A Manual of Acarology.* Oregon State Univ. Book Stores, Corvallis.

Krantz, G.W. and E.E. Lindquist. 1979. Evolution of phytophagous mites (Acari). *Annu. Rev. Entomol.* 24:121–158.

Krczal, H. 1959. Systematik und Okologie der Pyemotiden Beitrage zur Systematik un Okologie Mitteleuropaischen. *Acarina,* Vol. I, Part 2, pp. 385–625 (cited by Cross, 1965).

Lavoipierre, M.M.J. and R.F. Rick. 1955. Observations on the feeding habits of Argasid ticks and of the effects of their bites on laboratory animals, together with a note on the production of coxal fluid by several of the species studied. *Ann. Trop. Med. Parasitol.* 49:96–113.

Lebrun, Ph. 1979. Soil mite community diversity. *In:* Rodriguez, J.G., ed. *Recent Advances in Aca-*

rology. Academic Press, New York, Vol. 1, pp. 603–613.

Lees, A.D. 1946. The water balance in *Ixodes ricinus* L., and certain other species of ticks. *Parasitology* 37:1–20.

Lees, A.D. 1947. Transpiration and the structure of the epicuticle in ticks. *J. Exp. Biol.* 23:379–410.

Luxton, M. 1972. Studies on the oribatid mites of a Danish beech wood soil. I. Nutritional biology. *Pedobiologia* 12:434–463.

Malchenkova, N.I. 1967. Klesch roda *Tydeus*—vreditel' vinograda v Moldavii. *Entomol Obozr.* 46:117–121 (cited by Krantz and Lindquist, 1979).

Michael, A.D. 1883. *British Oribatidae.* Ray. Society, London.

Michael, A.D. 1892. On the variations of the internal anatomy of the Gamasinae, especially in that of the genital organs and their method of coition. *Trans. Linn. Soc.* 9:281–324.

Michael, A.D. 1895. A study of the internal anatomy of *Thyas petrophilus,* an unrecorded Hydrachnid found in Cornwall. *Proc. Zool. Soc. London* 174–209.

Mitchell, R.D. 1955. Anatomy, life history, and evolution of the mites parasitizing fresh-water mussels. *Misc. Publ. Mus. Zool., Univ. Mich.* 89:1–28.

Mitchell, R.D. 1958. The musculature of a trombiculid mite, *Blankaartia acuscutellaris* (Walch). *Ann. Entomol. Soc. Am.* 55:106–119.

Mitchell, R.D. 1962. Structure and evolution of water mite mouth parts. *J. Morphol.* 111:41–59.

Mitchell, R.D. 1964. The anatomy of an adult chigger mite *Blankaartia acuscutellaris* (Walch). *J. Morphol.* 114(3):373–391.

Mitchell, R.D. and M. Nadchatram. 1969. Schizeckenosy; the substitute for defecation in chigger mites. *J. Nat. Hist.* 3:121–124.

Moorehouse, D.E. and R.J. Tatchell. 1966. The feeding process of the cattle tick *Boophilus microplus:* A study in host-parasite relations. Part I. Attachment to the host. *Parasitology* 56:623–632.

Moss, W.W. 1962. Studies on the morphology of the trombidiid mite *Allothrombium lerouxi* Moss. *Acarologia* 4(3):313–345.

Murphy, P.W. 1953. The biology of forest soils with special reference to the mesofauna or meiofauna. *J. Soil Sci.* 4(2):155–193 (cited by Evans *et al.,* 1961).

Newell, I.M. 1963. Feeding habits in the genus *Balaustrum,* with special reference to attacks on man. *J. Parasitol.* 49:498–502.

Nordenskiold, E. 1908. Zur anatomie and histologie von *Ixodes reduvius.* I. *Zool. Jahrb., Abt. Anat. Ontog. Tiere* 25:637–672 (cited by Douglas, 1943).

Nutting, W.B. 1964. Demodicidae—Status and prognosis. *Acarologia* 6(3):441–454.

Nutting, W.B. and H. Rauch. 1958. *Demodex criceti,* n. sp. (Acarina: Demodicidae) with notes on its biology. *J. Parasitol.* 44:328–333.

Nuzzaci, G. 1979. A study of the internal anatomy of *Eriophyes canestrini* Nal. *Proc. Int. Congr. Acarol., 4th, 1974,* pp. 725–727.

Prasse, J. 1969. Zur anatomie und histologie der acaridae mit besonderer breuchstigung von *Caloglyphus berlesei* und *C. michaeli.* III. Die Drusen und drusenahlichen Gebilde der Podocephalkanal. *Wiss. Z.-Martin-Luther-Univ. Halle-Wittenberg, Math.-Naturwiss. Reihe* 17:629–146 (cited by Woodring, 1976).

Radford, C.D. 1950. The mites parasitic on mammals, birds and reptiles. *Parasitology* 40:366–394.

Reuter, E. 1909. Zur Morphologie und Ontogenie der Acariden. *Acta Soc. Sci. Fenn.* 36 (cited by Hughes, 1959).

Richards, L.A. and R.W. Richards. 1976. Parasitid mites associated with bumblebees in Alberta, Canada (Acarina: Parasitidae; Hymenoptera: Apidae). II. Biology. *Univ. Kans. Sci. Bull.* 51:1–18.

Riha, G. 1951. Okologie der Oribatiden im Kalksteinboden. *Zool. Jahrb.* 80:407–450 (cited by Wallwork, 1958).

Robaux, P. 1974. Recherches sur le développement et la biologie des acariens "Thrombidiidae." *Mem. Mus. Natl. Hist. Nat., Ser. A (Paris)* 85:1–186.

Robinson, L.E. and J. Davidson. 1913. The anatomy of *Argas persicus* (Oken, 1818). Parts I, II, and III. *Parasitology* 7:20–48, 217–256, 297–317 (cited by Douglas, 1943).

Rockett, C.L. and J.P. Woodring. 1966. Oribatid mites as predators of soil nematodes. *Ann. Entomol. Soc. Am.* 59:669–671.

Roesler, R. 1934. Histologische, physiologische und serologische untersuchungen uber die verdauung bei der zeckengattung *Ixodes* Latr. *Z. Morphol. Oekol. Tiere* 28:297–317 (cited by Balashov, 1972).

Rohde, C.J., Jr. and D.A. Oemick. 1967. Anatomy of

the digestive and reproductive systems in an acarid mite (Sarcoptiformes). *Acarologia* 9(3):608–616.

Roshdy, M.A. 1961. Comparative internal morphology of subgenera of *Argas* ticks. I. Subgenus *Carios: Argas vespertilionis* (Latreille, 1802). *J. Parasitol.* 47(6):987–994.

Roshdy, M.A. 1966. Comparative internal morphology of subgenera of Argas Ticks (Ixodoidea, Argasidae). 4. Subgenus *Ogadenus: Argas brumpti* Neumann, 1907. *J. Parasitol.* 52(4):776–782.

Roshdy, M.A. 1972. The subgenus *Persicargus*. 15. Histology and histochemistry of the salivary glands of *A. (P.) persicus* (Oken). *J. Med. Entomol.* 9(2):143–148.

Savory, T. 1977. *Arachnida*, 2nd ed. Academic Press, London.

Schuster, R. 1956. Der Anteil der Oribatiden an den Zersetzungsvorgangen Boden. *Z. Morphol. Oekol.* 45:1–33.

Schuster, R. and I.J. Schuster. 1977. Ernahrungs - und fort -pglanzungsbiologische s tudien an der Milbenfamilie Nanorchestidae. (Acari, Trombidiformes.) *Zool. Anz.* 199:89–94.

Sen, S.K. 1935. The mechanism of feeding in ticks. *Parasitology* 27:355–368 (cited by Evans *et al.*, 1961).

Shew, H.D. and M.K. Beute. 1976. The role of soilborne mites in *Pythium* pod rot of peanut. *Proc. Am. Phytopathol. Soc.* 3:342.

Sidorov, V.E. 1960. Gut of argasid ticks as inhabitation environment of *Borrellia sogdianum. Zool. Zh.* 39(9):1324–1329. (Translation 638 (T638), NAMRU, Cairo).

Sinha, R.N. 1962. A note on associations of some mites with seedborne fungi from the prairie provinces. *Proc. Entomol. Soc. Manitoba* 18:51–53.

Sinha, R.N. 1964. Ecological relationships of stored-products mites and seed-borne fungi. *Acarologia* 6:372–389.

Sinha, R.N. 1968. Adaptive significance of mycophagy in stored-products Arthropoda. *Evolution* (Lawrence, Kans.) 22:785–798.

Snodgrass, R.E. 1965. *A Textbook of Arthropod Anatomy.* Hafner, New York.

Snow, K.R. 1970. *The Arachnids, An Introduction.* Columbia Univ. Press, New York.

Sonenshine, D. and J. D. Gregson. 1970. Contribu-

tions to internal anatomy and histology of the bat tick *Ornithodoros cooleyi. J. Med. Entomol.* 7(1):46–64.

Spickett, S.G. 1961. Studies on *Demodex folliculorum* Simon (1842). I. Life history. *Parasitology* 51:181–192.

Strandtmann, R.W. and G.W. Wharton. 1958 *Manual of Mesostigmatid Mites Parasitic on Vertebrates.* Inst. Acarol., Maryland Contrib. No. 4.

Tanigoshi, L.K. and R.W. Davis. 1978. An ultrastructural study of *Tetranychus mcdanieli* feeding injury to the leaves of "Red Delicious" apple. *Int. J. Acarol.* 4(1):47–51.

Tatchell, R.J. 1964. Digestion in the tick, *Argas persicus* Oken. *Parasitology* 54:423–440.

Tatchell, R.J. 1967. Salivary secretion in the cattle tick as a means of water elimination. *Nature (London)* 213:940–941.

Tatchell, R.J. and D.E. Moorehouse. 1970. Neutrophilus: Their role in the formation of a tick feeding lesion. *Science* 167:1002–1003.

Theron, P.D. and P.A.J. Ryke. 1969. The family Nanorchestidae Grandjean (Acari: Prostigmata) with descriptions of new species from South African soils. *J. Entomol. Soc. S. Afr.* 32:31–60.

Thor, S. 1903. Bemerkungen zur neueren Hydrachniden. *Nyt. Mag. Naturw.* 41:65–68.

Tragardh, I. 1909. Speleorchestes, a new genus of saltatorial Tromidiidae, which lives in termites' and ants' nests. *Ark. Zool.* 6(2):1–14.

Treat, A. 1975. *Mites of Moths and Butterflies.* Cornell Univ. Press (Comstock), Ithaca, New York.

Vistorin, H. E. 1979. Ernahrungsbiologie und anatomie des verdauungstraktes der Nicolettiellidae (Acari: Trombidiformes). *Acarologia* 21(2):204–215.

Vistorian-Theis, G. 1977. Anatomische untersuchungen an Calyptostomiden. *Acarologia* 19(2):142–257.

Vitzthum, H.G. 1943. Acarina. *In: Bronns Klassen und Ordnungen des Tierreichs*, Vol. 5, pp. 1–1011.

Wallwork, J.A. 1958. Notes on the feeding behavior of some forest soil Acarina. *Oikos* 9(2):260–271.

Wallwork, J.A. 1970. *Ecology of Soil Animals.* McGraw-Hill, London.

Weis-Fogh, T. 1948. Ecological investigations on mites and collemboles in the soil. *Nat. Jutland* 1:135–349 (cited by Evans *et al.*, 1961).

Weitz, B. and P.A. Buxton. 1953. The rate of digestion of blood meals of various haematophagous arthropods as determined by the precipitin test. *Bull. Entomol. Res.* 44:445–450.

Wharton, G.W. and A.R. Brody. 1972. The peritrophic membrane of the mite *Dermatophagoides farinae. J. Parasitol.* 58(4):801–804.

Winkler, W. 1888. Die Anatomie der Gamasiden. *Arb. z. Inst. Wien.* 7:317–354.

Whitmoyer, R.E., L.R. Nault and O.E. Bradfute. 1972. Fine structure of *Aceria tulipae. Ann. Entomol. Soc. Am.* 65(1):201–215.

Wigglesworth, V.B. 1943. The fate of haemoglobin in *Rhodnius prolixus* (Hemiptera) and other bloodsucking arthropods. *Proc. R. Soc. London, Ser B* 131:313–339.

Woodring, J.P. 1963. The nutrition and biology of saprophytic Sarcoptiformes. *Adv. Acarology* 1:89–111.

Woodring, J.P. 1965. The biology of five species of oribatids from Louisiana. *Acarologia* 7:564–576.

Woodring, J.P. 1976. The anatomy of the adult uropodid *Fuscouropoda agitans,* with observations on other Acari. *J. Morphol.* 150:19–58.

Woodring, J.P. and E.F. Cook. 1962a. The internal anatomy, reproductive physiology, and molting process of *Cerozetes cisalpinus* (Acarina: Oribatei). *Ann. Entomol. Soc. Am.* 55:164–181.

Woodring, J.P. and E.F. Cook. 1962b. The biology of *Ceratozetes cisalpinus* Berlese, *Scheloribates laevigatus* Koch, and *Oppia neerlandica* Oudemans (Oribatei), with a description of all stages. *Acarologia* 4(1):101–137.

Woolley, T.A. 1960. Some interesting aspects of oribatid ecology (Acarina) *Ann. Entomol. Soc. Am.* 53(2):251–3.

Woolley, T.A. 1969. The Infracapitulum—A possible index of Oribatid relationship. *Proc. Int. Congr. Acarol., 2nd, 1967,* pp. 209–221.

Woolley, T.A. 1979. The Chelicerae of the Gustaviidae. *In:* Rodriguez, J.G., ed. *Recent Advances in Acarology.* Academic Press, New York, Vol. 2, pp. 547–551.

Wright, K.A. and I.M. Newell. 1964. Some observations on the fine structure of the mid-gut of the mite *Anystis* sp. *Ann. Entomol. Soc. Am.* 57:684–693.

Young, J.H. 1968. The morphology of *Haemogamasus ambulans.* I. Alimentary Canal (Acarina: Haemogamasidae). *J. Kans. Entomol. Soc.* 41(1):101–107.

Zakhvatkin, A.A. 1959. *Tyroglyphidea (Acari)* (translation of *Fauna of USSR Arachnoidea,* Vol. 6, No. 1. (translated by A. Ratcliffe and A.M. Hughes). Am. Inst. Biol. Sci., Washington, D.C.

8

Circulatory System

Blood—the "milieu interieur."
—CLAUDE BERNARD (1813–1878)

Chelicerates generally have a well-developed circulatory system, as exemplified in the Merostomata, Scorpionida, and Araneae. The system is less well-defined in the pseudoscorpions, Opiliones, and Acari. The absence of a heart and blood vessels in the Tardigrada, Pauropoda, and most mites is due to small size. It is assumed that the ancestors of mites had a heart, even though the presence of a heart is not established in some extant forms. Simplicity in the circulatory system (as in other organs) may be primitive or secondarily simplified. In either case the conditions are correlated with size (Gupta, 1979).

Mites have an open (lacunar) circulatory system (Legendre, 1968) and the hemolymph (blood) bathes the internal organs, including the muscles of the body and legs. The hemocoel (blood cavity) is derived from a combination of the primordial blastocoel and parts of the schizocoelic spaces of the embryo and is thus a myxocoel. The principal organs are heteromorphic in origin (Beklemishev, 1969).

ORGANS

Most mites possess neither heart nor blood vessels (Andre, 1949), but all exhibit a hemocoel filled with blood in which a number of hemocytes (corpuscles) are suspended. A dorso-median heart exists in Holothyrida, Gamasida, and Ixodida; one or two pairs of ostia are present. The dorsal heart, when present, is generally described as flattened, of hexagonal or rhomboidal outline, usually with the extension of a single anterior aorta and without a posterior aorta. The heart usually is located anteriorly in the idiosoma (Evans et al., 1961). Unidirectional afferent valves are part of the ostia in the heart wall (Fig. 8-1A).

A heart occurs in *Pergamasus crassipes,* one of the more specialied gamasid mites (Young, 1968). The circulatory system of ticks is fairly developed, with a small, triangular heart and one or two pairs of ostia (Balashov, 1972). Ixodid ticks (*Ixodes ricinus, Hyalomma asiaticum*) show two pairs of ostia; a single pair is

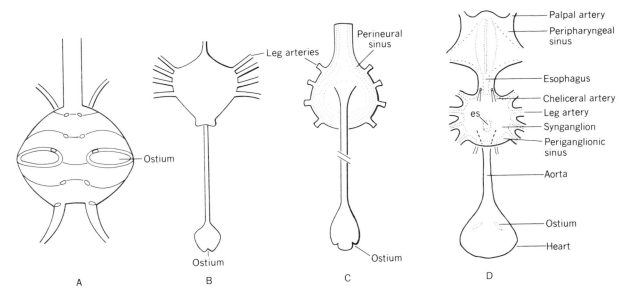

Figure 8-1. (A) Heart of gamasid mite showing two ostia (after Andre, 1949). Circulatory system of ticks: (B) heart, single ostium, perineural sinus, and partial leg veins of soft tick (after Balashov, 1972); (C) heart, two ostia, perineural sinus, and partial leg veins of *Dermacentor andersoni* (after Douglas, 1943); (D) heart, two ostia, periganglionic sinus, peripharyngeal sinus, and arteries of *Argas persicus* (after Balashov, 1972).

found in *Dermacentor andersoni* (Douglas, 1943) and in Argasidae (*Argas persicus*) (Balashov, 1972) (Fig. 8-1C, D).

The circulatory system of *Dermacentor andersoni* is more complex than in some insects and consists of some well-defined parts, namely, the heart, a dorsal aorta, perineural (periganglionic) sinus, a perioesophageal sinus, and the hemolymph. The heart is pyriform in shape (250 μm × 380 μm) with two simple ostia (valvular afferent openings) at the posterolateral margins. The heart is supported by five extrinsic muscles. Two muscle pairs insert laterally and a single muscle is attached to the posterior margin of the heart between the ostia (Douglas, 1943). *Argas persicus* is similar (Balashov, 1972; Robinson and Davidson, 1914).

The heart and the dorsal aorta, its anterior extension, have membranous walls. The aorta lacks intrinsic musculature, but structurally extends over the ventriculus and expands into the perineural sinus.

The perineural sinus invests the synganglion and extends around the four pairs of pedal nerves and into the coxal spaces of the legs. Its forward extension becomes the investment of the capitular nerves and the esophagus to form the perioesophageal sinus that serves the capitular region. This sinus opens at the posterior end of the pharynx and is called the peripharyngeal sinus (Balashov, 1972).

The circulatory system of the Eriophyidae is simply composed of various cells or of cellular organs lying in the hemocoel (Jeppson et al., 1975). Circulatory and respiratory systems are absent in *Demodex* (Desch and Nutting, 1977). In the trombidiform *Allothrombium lerouxi* the circulatory system is open, but a heart is absent.

CIRCULATION

Circulation of the hemolymph in mites is accomplished mainly by contraction of dorsoventral muscles, by movements of the body, and by the pumping action of a heart, if one is present. In *Dermacentor andersoni* two muscle pairs insert laterally and a single muscle is attached to the posterior margin of the heart between the ostia. Diastolic contraction is actuated by these extrinsic muscles. Systole results from the contraction of intrinsic muscles in the walls of the organ. The action of the heart is compared to "a clinching fist." Back pressure closes the valves of the ostia automatically and hemolymph is forced into the dorsal aorta and perineural sinus. The pulse rate is variable (20 to 128 per minute) and may cease for several seconds at a time (Douglas, 1943). The contraction of the dorso-ventral body muscles is the main force behind the extension of the legs and other appendages by increasing the hydrostatic pressure of hemolymph in the limbs and at the same time facilitating circulation. Flexor muscles in the body return the hemolymph from the extremities.

In Eriophyidae cells and cellular organs comprise the circulatory system and the hemolymph is circulated by body movements (Jeppson et al., 1975). It is suggested that in the higher trombidiform mites there is minimal hemocoelic circulation and that the midgut functions gastrovascularly. During schizeckenosy, when the integument splits and a lobe of the gut is lost, little fluid is eliminated from the hemocoel (Mitchell, 1964).

HEMOLYMPH

The hemolymph (blood) of arthropods generally, including mites, is a chemically complex fluid containing a number of different types of suspended cells or hemocytes. The hemolymph of Eriophyidae contains concentrically layered, ovoid granules (excretory) (Whitmoyer et al.,

1972); other mites may have like chemical inclusions. The extreme variability in the cells and in the chemical composition of the hemolymph is related to the cyclical events that transpire in the life of the mite, such as nutritional stages, growth processes, reproduction, and other metabolic activities. Amino acids, glucose, and lipids are found in the fluid part of the blood (Krantz, 1978). Few if any mites carry inherent respiratory pigments. If respiratory pigments are present (e.g., hemoglobin), they are carried in the hemolymph and not in the hemocytes; other pigments in the hemolymph may come from food (e.g., hematin in ticks from host blood).

The clear, colorless hemolymph of the soft tick *Ornithodoros kelleyi* occurs in the absence of a circulatory system and contains free-floating hemocytes (Sonenshine, 1970). The hemolymph in *Dermacentor andersoni* is a pale, slightly turbid, nearly colorless fluid with numerous large nucleated amoebocytes (Douglas, 1943).

HEMOCYTES

Hemocytes are the corpuscles suspended in the hemolymph. The number of hemocytes in the blood of arthropods varies proportionately with size. The average number is 75 000/mm^3 in Crustacea and Insecta. In Argasidae the number of hemocytes per unit volume of blood varies from 1000–2000 to 80 000–100 000 cells per microliter of hemolymph. In unfed ixodid larvae and nymphs the number of hemocytes is smaller than in engorged nymphs and adults (Clarke, 1973; Balashov, 1972), but note that "different hemocyte numbers in hemolymph smears may be partially explained by their temporary ability to become immobile and attach to connective strands and membranes" (Balashov, 1972, p. 191).

The acarines have all of the types of hemocytes found in scorpions and spiders, namely, prohemocytes, plasmocytes, granulocytes, spherulocytes, adipohemocytes, coagulocytes, oenocytes, and, questionably, cyanocytes. Some transitional forms of cells are also found. Basophilic hemocytes are known to contain glycogen (Gupta, 1979). Many hemocytes are mobile and have pseudopodia (especially phagocytes). Prohemocytes undergo mitosis without spindle formation (Dolp, 1970).

Hemocytes of *Ornithodoros kelleyi* have large, eccentric or central nuclei and a reticulated or vacuolated cytoplasm (Sonenshine, 1970).

It is difficult to classify the blood cells of different ixodid ticks by the appearance of the cells and the types of cytoplasmic inclusions. Prohemocytes are most prevalent (Tsvileneva, 1959, 1961a). Three types of hemocytes occur in the blood of *Boophilus calcaratus, Hyalomma detritum, H. anatolicum, Dermacentor dadhestanicus,* and *Rhipicephalus sanguineus* (Tsvileneva, 1961b,c). Balashov (1972) identifies the three types as trophocytes (20–25 μm) with large inclusions (Fig. 8-2A), basophilic phagocytes (14–18 μm) (Fig. 8-2B), and prohemocytes (5–6 μm). He also illustrates phagocytes (Fig. 8-2C) and oenocytes, a type of cell found in other arthropods (Tsvileneva, 1961b,c).

Numerous large nucleated ameboid hemocytes are present in the blood of *Dermacentor andersoni*. The cytoplasm of these cells is reticulate and contains dark granules identical in appearance with granules in the hemolymph itself along with larger eosiniphilic spherical elements that are apparently phagocytized by the corpuscles as similar elements are observed in the blood (Douglas, 1943) (Fig. 8-2).

Not all authors agree on the types of hemocytes. Krantz (1978) indicates that ticks exhibit three major types of hemocytes in the blood: small proleucocytes with big nuclei (5–7 μm), oval basophilic hemocytes (10–20 μm), and ameboid eosinophilous hemocytes (12–25 μm). Douglas (1943) illustrates hemocytes of *Dermacentor andersoni* (Fig. 8-3).

A classification of hemocytes has been made for four species of ticks in two genera from Ixodidae and Argasidae based on general structure and staining effects of Giemsa stain. The greatest concentration of hemocytes are in unfed ticks (Dolp, 1970). Hemocytes include relatively small, round prohemocytes (basophilic cells) (Fig. 8-4A), ovoid to fusiform plasmatocytes (Fig. 8-4B, C) of early and later stages, and larger, ovoid spherulocytes (Fig. 8-4D–F), usually with numerous purple-staining spherules. Spherulocytes are differentiated into two types based on the number and size of spherules present. Extreme pleomorphism does not occur in the plasmatocytes.

Numerous unclassified hemocytes are present in the hemolymph of *Allothrombium fuliginousum* (Moss, 1962). In *Caloglyphus berlesei* seven (Legendre, 1968) or eight

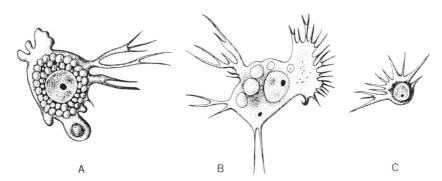

Figure 8-2. Hemocytes of *Boophilus calcaratus*: (A) trophocytes; (B) basophilic phagocytes; (C) prohemocytes (drawings by Sonya Evans; after Balashov, 1972).

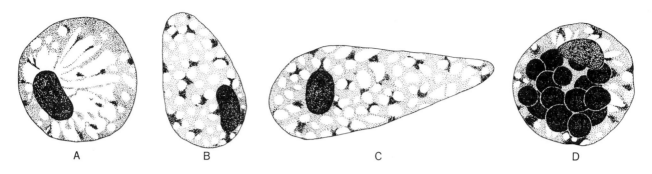

Figure 8-3. Hemocytes of *Dermacentor andersoni*: (A–C) unidentified hemocytes; (D) hemocyte with phagocytized eosinophilic bodies (drawings by Sonya Evans; after Douglas, 1943).

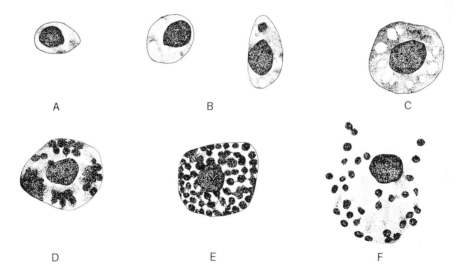

Figure 8-4. Hemocytes of Ixodidae and Argasidae: (A) prohemocyte; (B) early plasmatocytes; (C) advanced plasmatocytes; (D) early spherulocyte; (E) advanced spherulocyte; (F) ruptured spherulocyte (drawings by Sonya Evans; after Dolp, 1970).

(Kanungo and Naegele, 1964) different types of hemocytes occur, identified by a variety of techniques: (1) prohemocytes, (2) cystocytes, (3) granulocytes, (4) adipohemocytes, (5) spherulocytes, (6) amebocytes, (7) oenocytes, and (8) crystal cells. The prohemocytes are small spherical cells with a large nucleus and basophilic cytoplasm. Cystocytes are hyaline cells of unstable form that *in vitro* eject cytoplasmic threads like beaded strings in fresh wet preparations. Granulocytes are spherical with a rounded nucleus and eosinophilic granules in the cytoplasm, mainly around the nucleus. Adipohemocytes are rounded cells with large, eccentric basophilic nuclei. The cytoplasm contains many osmiophilic refractile droplets that stain with fat stains. Other nonosmiophilic cytoplasmic granules are also present.

Spherulocytes are rounded to ovoid cells with spherular basophilic inclusions. *In vitro,* these inclusions are nonrefringent, yellow in color, and variable in size. Amebocytes (apparently similar to plasmatocytes in other classifications) are the most abundant of the cell types in both fresh and fixed preparations. The nucleus is large and eosinophilic. Inasmuch as these cells are phagocytic, the acidophilic cytoplasm exhibits many variable inclusions during the phagocytic period; shape may be altered to ovoid or pyriform. The nucleus is large with distinct perinuclear granules. These amebocytes are more numerous and active during the quiescent premolting stage. In the protonymph they may migrate from the old to the new cuticle; their cytoplasm is granular during this period.

Oenocytes are large, conical cells with small, eccentric eosinophilic nuclei. At the tapered end is a long, flagellumlike process. Crystal cells are rounded without a discernible nucleus. Their hyaline cytoplasm contains birefringent, basophilic, variable-sized crystals.

Hematology of mites is in an early stage of development and the classification of acarine hemocytes is incomplete. Hematological studies of ticks are more numerous and more advanced than those of other orders.

For the present, it appears logical that developing prohemocytes (?=proleucocytes of Krantz, 1978) may be different functionally from plasmatocytes. Plasmatocytes and spherulocytes have differentiating features and sizes (which may intergrade), but functions remain obscure. Oenocytes, adipohemocytes, coagulocytes, and cyanocytes described as typical of acarine hemolymph (Gupta, 1979) require more study and comparison to determine their presence and differentiating features.

The limited summary in the table classifies the hemocytes of acarines with suggested possible equivalents from various authors. Uncertainties of equivalency and enigmatic status of cells are shown by a question mark.

	Gupta (1979)	Krantz (1978)	Balashov (1972)	Dolp (1970)	Kanungo and Naegele (1964)	Tsvileneva (1959)
Prohemocytes	*	*?	*	*	*	*
Plasmatocytes		*?	*?	*	*?	*
Granulocytes	*				*	
Spherulocytes	*	*?			*	*
Adipohemocytes	*				*	
Coagulocytes	*	*?				
Oenocytes	*				*	*
Cyanocytes?	*					
Crystal cells?						*

REFERENCES

Andre, M. 1949. Order des Acariens. *In*: Grassé, P.-P., ed. *Traité de Zoologie.* Mason, Paris, Vol. 6, pp. 794–892.

Balashov, Y.S. 1972. *Bloodsucking Ticks (Ixodidea)—Vectors of Diseases in Man and Animals,* Transl. 500 (T 500). Med. Zool. Dep. USNAMRU, No. 3. Cairo, Egypt, U.A.R.

Beklemishev, V.N. 1964. *Fundamentals of Comparative Invertebrate Anatomy,* Vol. 2. Izd. Nauka, Moscow.

Beklemishev, V.N. 1969. *Principles of Comparative Anatomy of Invertebrates,* Vol. 2. Univ. of Chicago Press, Chicago, Illinois.

Clarke, K.U. 1973. *The Biology of the Arthropoda.* Arnold, London.

Desch, C.E. and W.B. Nutting. 1977. Morphology and functional anatomy of *Demodex folliculorum* (Simon) of man. *Acarologia* 19(3):422–462.

Dolp, R.M. 1970. Biochemical and physiological studies of certain ticks. Qualitative and quantitative studies of hemocytes. *J. Med. Entomol.* 7(3):227–288.

Douglas, J.R. 1943. The internal anatomy of *Dermacentor andersoni* Stiles. *Univ. Calif. Publ. Entomol.* 7(10):207–272.

Evans, G.O., J.G. Sheals and D. MacFarlane. 1961. *Terrestrial Acari of the British Isles.* Adlard & Son, Bartholemew Press, Dorking, England.

Gupta, A.P., ed. 1979. *Arthropod Phylogeny.* Van Nostrand-Rheinhold, New York.

Jeppson, L.R., H.H. Keifer and W. Baker. 1975. *Mites Injurious to Economic Plants.* Univ. of California Press, Berkeley.

Kanungo, K. and J.A. Naegele. 1964. The Haemocytes of the acarid mite *Caloglyphus berlesei* (Mich. 1903). *J. Insect Physiol.* 10:651–655.

Krantz, G.W. 1978. *A Manual of Acarology.* Oregon State Univ. Book Stores, Corvallis.

Legendre, R. 1968. La nomenclature anatomique des les acariens. *Acarologia* 10(3):411–417.

Mitchell, R. 1964. The anatomy of the adult chigger *Blankaartia acusculellaris* (Walch). *J. Morphol.* 114:373–392.

Moss, W.W. 1962. Studies on the morphology of the trombidid mite *Allothrombium lerouxi* Moss. *Acarologia* 4(3):313–345.

Robinson, L.E. and J. Davidson. 1914. The anatomy of *Argas persicus.* Part 3. *Parasitology* 6:382–424.

Sonenshine, D.E. 1970. A contribution to the internal anatomy and histology of the bat tick *Ornithodoros kelleyi* Cooley and Kohls, 1941. *J. Med. Entomol.* 7(3):289–312.

Tsvileneva, V.A. 1959. Formed elements of the hemolymph in Ixodid ticks. *Dokl. Akad. Nauk Tadzh. SSR* 2:45–51 (cited by Gupta, 1979).

Tsvileneva, V.A. 1961a. Comparative histology of the blood and connective tissue. Formed elements of hemolymph in ixodid ticks. *Arkh. Anat. Gistol. Embriol.* 40:91–100 (cited by Balashov, 1972).

Tsvileneva, V.A. 1961b. Loose connective tissue of ixodid ticks *Arkh. Anat. Gistol. Embriol.* 41:79–88 (cited by Balashov, 1972).

Tsvileneva, V.A. 1961c. On the adipose body in ixodid ticks. *Dokl. Akad. Nauk Tadz. SSR* 4:57–60 (cited by Balashov, 1972).

Whitmoyer, R.E., L.R. Nault and O.E. Bradfute. 1972. Fine structure of *Aceria tulipae.* *Ann. Entomol. Soc. Am.* 65(1):201–215.

Young, J.H. 1968. The morphology of *Haemogamasus ambulans.* I. Alimentary Canal. *J. Kans. Entomol. Soc.* 41(1):101–107.

9

Respiratory System

L'Atmosphère, c'est un ingrédient vital pour la Vie.
—TYLER A. WOOLLEY

In its broadest terms the function of respiration includes the procurement of oxygen, its transportation from that dissolved in water or in ambient air into the body, and its subsequent intricate cellular and chemical utilization within the animal. Morphological features of this system are known for mites, but there is a paucity of information regarding the cellular and chemical aspects of respiration.

No single, uniform system for the exchange of gases exists in mites. Diverse methods correlated with variations of morphology and habitat are observed, but this diversity excludes the possibility of a single line of evolutionary development for the respiratory systems (Krantz, 1978).

The tracheal system in mites is an advance over the arachnid book lung and assists in the economy of internal moisture (Savory, 1977). It may also be that the replacement of book lungs by tracheae is related more to habits than to phylogeny and thus to the diversity of respiratory systems in the acarines (Levi, 1967; Levi and Kirber, 1976). Similarities exist between the tracheae of the Ricinulei and the Acari, yet the systems of these two groups differ from other arachnids (Ripper, 1931).

Tracheae are not always present in mites (Hughes, 1959), but when present, they may be simple or branched and open to the exterior through stigmatal or spiracular openings at different locations. The variations and complexities of these stigmatal features have been used more than any other anatomical system as a basis for systematics in the acarines. The use of stigmatal position in the classification systems as originated by Oudemans and employed in classificatory relationships is well documented (Hughes, 1959; Evans et al., 1961; Hammen, 1968b, 1973).

CLASSIFICATION OF RESPIRATORY ORGANS

A provisional classification of respiratory organs and stigmata of mites includes three series: (1) opisthosomatic, (2) genital, and (3) organs originating from integumental structures, for example, porose areas. The first series is opisthosomatic because of location and the assumption that the stigmata originated from lyrifissures. The arrrangement of the stigmata and the respiratory tracheae in Opilioacarida, Holothyrida, Gamasida, and Ixodida is considered homologous. The number of pairs of stigmata for this series varies, namely, four pairs for Opilioacarida, two pairs for Holothyrida, and a pair each for Gamasida and Ixodida. The second series includes the genital tracheae, which are exemplified in representatives within main orders of Actinotrichida, namely, Actinedida, Astigmata, and Oribatida. The genital tracheae are assumed to originate from accessory glands and are segmentally arranged. Because segmented accessory glands develop into trachealike organs in the pseudoscorpions, it is assumed that genital tracheae also originated from segmental accessory glands in the mites. Genital tracheae were discovered first in snout mites (*Cyta*, Bdellidae). More were observed later in the Glycyphagidae (Astigmata) and some of the inferior Oribatida (Fig. 9-10B).

The third category consists of (a) an axillary series, (b) superficial cuticular structures, and (c) hidden folds or depressions that serve the respiratory function. Examples of axillary stigmata include those found at the bases of the chelicerae (Actinedida), those at the base of the palps (*Eupelops*, Oribatida), and the coxal acetabula of the legs (many of the higher Oribatida). Superficial cuticular structures of supposed respiratory implication are sacculi, brachytracheae, and tracheae of the notogaster and legs of Oribatida. Areae porosae on the legs and notogaster also fit in this category. A few depressions of the cuticle (bothridia, anal folds, and apodemes) may have respiratory organs (e.g., platytracheae) in association (Hammen, 1968c) (Fig. 9-13).

In most orders of mites the process of respiration is effected by tracheae, sometimes well-developed and sometimes rudimentary; in other mites, the tracheae are absent. In mites, as in other arthropods, the tracheae are more or less complicated invaginations of the exoskeleton and their structure reflects this origin. The tracheae and tracheoles may have a distinctive spiraled intima of coiled taenidia composed of cuticulin and exocuticle (Witalinski, 1979) (Fig. 9-3). These taenidia vary in the number of turns in the coil and are sometimes difficult to observe. Inasmuch as the ducts of the salivary glands may have taenidia and in some instances salivary ducts may extend into the tracheal system, it has been suggested by several authors that the salivary and respiratory systems developed together. Thus Michael (1903) wrote: "If you find a duct with taenidia, it is probably a salivary duct." Tracheolar endings may be present, as in insects, but definitive information on these tracheoles is lacking.

The Acaridae, Tyroglyphidae, Sarcoptidae, Demodicidae, Eriophyidae, and larval ticks are examples of mites that have neither tracheae nor vestiges of identifiable respiratory structures (Andre, 1949). It is assumed that such mites respire through the integument or perhaps utilize the digestive tract in some way. Others may be anaerobic, or at least are aerobic at very low levels of oxygen tension.

SYSTEMATIC REVIEW

Opilioacarida. Adults in the Opilioacarida (Notostigmata) exhibit four pairs of simple stigmata arranged segmentally on the opisthosoma (*Opilioacarus segmentatus*) (Grandjean, 1935; Hammen, 1966). In *Opilioacarus texanus* (Chamberlain and Mulaik) the four stigmata are simple openings situated in a slightly crescentic arrangement (Fig. 9-1A). Each stigma opens into the trachea (Fig. 9-1F). The stigmata in *O. platensis* Silvestri (Fig. 9-1C) and *Adenacarus arabicus* are slightly different configurations (Hammen, 1969), yet similar.

Stigmata are absent in larvae; the number is variable in the nymphs, depending on the species. The stigmata are irregular, unprotected apertures in the leathery-appearing cuticle

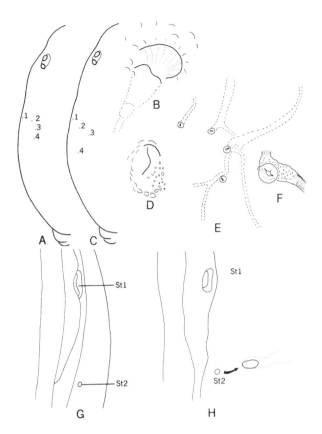

Figure 9-1. (A) Stigmata of *Opilioacarus texanus*; (B) enlarged view of stigma; (C) stigmata of *O. platensis*; (D) enlarged view of stigma (after Hammen); (E) stigma and tracheae of *O. texanus* (after Hammen, 1966); (F) enlarged view of stigma and tracheal trunks of *O. segmentatus* (after Grandjean, 1936); stigmata of (G) *Holothyrus coccinella*; (H) *H. grandjeani* (St1 stigmata 1; St2, stigmata 2) (after Hammen, 1961).

tinotrichida likely developed in a different way (Hammen, 1966).

Holothyrida. In the Holothyrida (Tetrastigmata) two pairs of stigmata are present anteriorly and dorsal to coxae III. Stigma I is elongate and has an open, interior vestibule. It lies in a lateral area of the integument adjacent to the lateral tectum (Hammen, 1961, 1965, 1966). Tracheae extend from the vestibule into the body. Stigma II is much smaller, is located posteriorly, and opens into a curved atrium confluent with a thin-walled diverticulum (Fig. 9-1G, H).

Gamasida. Tracheal trunks in adult Gamasida (Mesostigmata) extend internally from a single pair of external stigmata that open dorsolaterally near coxae III (or closer to coxae II or to coxae IV, depending on the families involved). The "middle" placement of these openings is reflected in the name Mesostigmata. Each stigma usually has an elongated trough, or groove, the peritreme. Most of the peritremes are long, more or less straight, and reach from the stigma to or slightly beyond the level of coxae I. Some may curl back upon themselves at the anterior end (Fig. 9-2A). In parasitic forms, the peritremes may be short or absent (Rhinonyssidae, Halarachnidae, Fig. 9-2I, J) (Strandtmann, 1960). Short, curled peritremes with slight branching are typical of some Zerconida (Fig. 9-3G, H) (Sellnick, 1958). In Uropodidae size and length of the peritreme are related to moisture of the habitat and the peritreme possibly acts as a plastron type of air-holding gill. Some gamasine families show peritremalia (postperitrematal plates) posterior to the stigmata (e.g., Digamasellidae), plates useful in identification (Fig. 9-2M–O).

In Parasitidae the stigma has a circular external orifice, a stigmal chamber, and an internal orifice that connects to a tracheal atrium; both the external orifice and the stigmal chamber have small spines on their surfaces. Branched, complex cuticular processes also occur. Neither of these processes is sensory. The

without any supporting plates, but are associated with the segmentally arranged lyrifissures. (Because of this location, Grandjean (1935) postulates that both stigmata and tracheae originated from lyrifissures by specialization.) The stigmata may be a part of a homologous series of opisthosomal origin in Parasitiformes (Anactinotrichida) (four pairs in Opilioacarida, two pairs in Holothyrida, and a single pair in Gamasida); stigmata in the Ac-

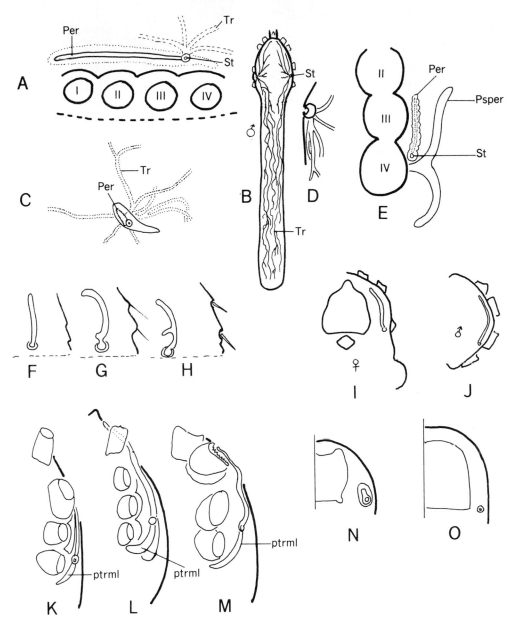

Figure 9-2. Schematic diagrams of the respiratory system in Gamasida: (A) general (after Evans et al., 1961); (B) female system *Halarachne zalophi* (after Vitzthum, 1943); (C) left stigmata, peritreme (Per), and tracheae (tr) of female of same (after Vitzthum, 1943); (D) enlarged stigmata (st) and tracheae of female *Orthohalarachne attenuata* (after Krantz, 1978); (E) coxae, peritreme, pseudoperitreme (Psper), and stigmata of *Aenictiques chapmani* (after Kethley, 1977); stigmata and peritremes of (F) *Parazercon*, (G) *Zercon*, (H) *Prozercon* (after Sellnick, 1958); (I) stigmata and peritremes of female *Periglischrus strandtmanni* and (J) male of same; (K) dorsal stigmata and peritreme of *Ptylonyssus*; (L) dorsal stigmata of *Sternostoma* without peritremes (after Strandtmann, 1960); peritremalia (ptrml) of (M) Digamasellidae, (N) Ascidae, (O) Laelapidae.

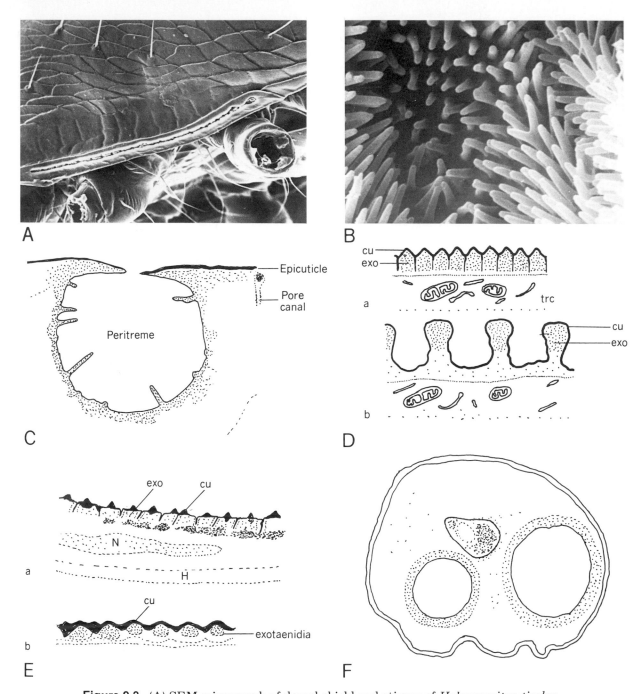

Figure 9-3. (A) SEM micrograph of dorsal shield and stigma of *Holoparasitus tirolensis* (450 ×); (B) SEM micrograph of stigmal chamber of *Parasitus viator* showing minute spines (4100 ×) (with permission of Witalinski, 1979); (C) section of peritreme of *P. viator* redrawn from micrograph; (D) drawings of sections of tracheal wall in insects (a) and mites (b) showing the cuticulin (cu) and exocuticle (exo) above the tracheal cell (tr c) (redrawn from TEM); (E) drawings of longitudinal sections of (a) trachea and (b) tracheole intima of *P. viator* showing cuticulin (cu) and exocuticle (exo) of taenidia (N, nucleus of tracheolar cell; H, hemocoel); (F) drawing of cross section of tracheolar cell with two tracheoles enclosed (redrawn from TEM section; after Witalinski, 1979).

groove of the peritreme is cuticular and circular in outline with inward-projecting spines (Fig. 9-3A–C). The stigmata are open, lacking a closing device, and the role of the peritreme is not understood (Witalinski, 1979) (SEM–TEM, Fig. 9-3-A, B).

Tracheae, stigmata, and peritremes are absent in all the larvae of gamasid mites,but protonymphs, deutonymphs, and adults have both stigmata and peritremes (Evans et al., 1961; Witalinski, 1979). For adults of the spiny rat mite (Jakeman, 1961), for *Parasitus* (Lawrence, 1953), for *Holoparasitus* (Witalinski, 1979), and for *Fuscouropoda agitans* (Woodring and Galbraith, 1976), tracheal trunks are associated with the stigmata and extend inward from the exterior opening (Fig. 9-4A). Stigmata of the female *Orthohalarachne attenuata* are lateroventral or ventral in position with the trachea extending into the podosomal region as well as into the elongated opisthosoma (Krantz, 1978). Simple stigmata and slightly sinuous

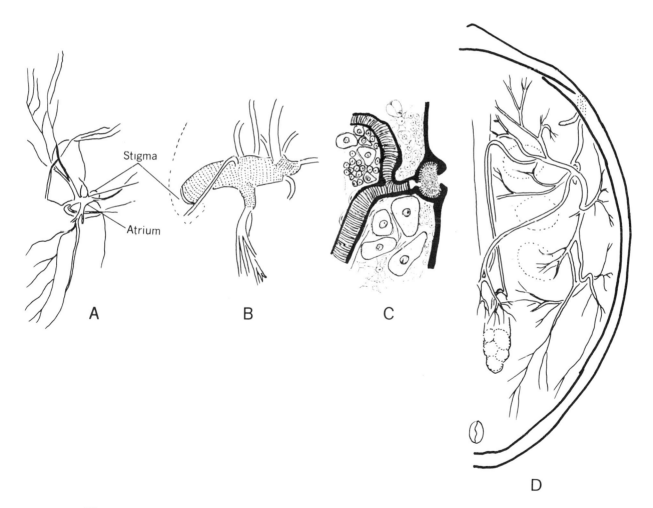

Figure 9-4. (A) Dorsal view of atrium and tracheae of spiny rat mite; (B) enlarged view of atrium showing taenidia (after Jakeman, 1961); (C) schematic section of stigmata and trachea of *Parasitus* (after Vitzthum, 1943); (D) drawing of dorsal view of tracheal system of *Fuscouropoda agitans* (after Woodring and Galbraith, 1976).

peritremes occur in the ectoparasitic Spinturnicidae (Tippetts, 1957).

In one of the trigynaspid families (Aenictoquidae) the real peritreme extends for about the length of two coxae, but a Y-shaped "pseudoperitreme" is observed laterad of coxae IV in *Aenictiques chapmani* (Kethley, 1977). The function of this structure is unknown (Fig. 9-2E).

Ixodida. The stigmatal entrances to the tracheae in ticks are different from most of all other mites and from the stigmata of most other arthropods. Nymphs and adults exhibit a single pair of stigmatal or spiracular plates adjacent to or posterior to coxae IV. (This placement "beyond" the middle position in the body is reflected in the name *Metastigmata* used by Evans et al. (1961).) These spiracular plates are diagnostic for the families and genera concerned, and are distinguished by their size, position, peritremal outline, and surface structure (Woolley, 1972). Larval stages lack stigmatal plates and tracheae. Cutaneous respiration occurs in larval hard ticks and argasid larvae may swallow air in respiration (Arthur, 1962).

Two stigmatal conditions exist in ticks. In soft ticks (Argasidae) the spiracular plate appears as a simple buttonlike, or flaplike fold of integument (Argas), or a conical projection *(Otobius)* (Fig. 9-5B) on the body. The size of the plate is small compared to that in hard ticks. In the buttonlike form there is a simple lidlike macula covering the ostium, which leads into the air system (the apex of the cone acts as the opening for others); when this hinged lid is raised the ambient air moves directly into the atrial cavity and thence into the tracheae (Roshdy, 1961; Balashov, 1972)

In adults and nymphs of *Ornithodoros kelleyi* and *Argas (Carios) vespertilionis* the system consists of paired stigmata, atria, tracheae, tracheal end cells, and tracheoles (Roshdy, 1961; Sonenshine, 1970). Larvae lack a respiratory system although atria and tracheae are described for larvae of *Argas* and *Or-*

nithodoros (Theodor and Costa, 1960). The stigmata are conelike in *Ornithodoros kelleyi* and buttonlike in *Argas C. vespertilionis*. In both they are located between coxae III and IV. Each of these stigmata consists of a padlike, elevated macula, a crescentic stigmata plate, and a crescentic slit, the ostium, between them. Discrepancies exist as to whether the stigmatal plate is porous, but several authors assume the plate to be like a sieve (Fig. 9-5C). However, it appears that the integument covering the stigmatal plate is a continuous thin procuticle and epicuticle, and that the ambient air is admitted to the atrium when the macula is lifted. Hydrostatic pressure of the hemolymph is the force that elevates the macula inasmuch as no muscles are present (Sonenshine, 1970).

The atrium in *O. kelleyi* is composed of a cuboidal epithelium covered with thin cuticle. The epithelium is continuous with the hypodermis and the cuticle is confluent with the linings of the stigmata and the tracheae.

Considerable variations occur in the numbers of tracheal trunks in different species of soft ticks. Major trunks range in number from three *(A. vespertilionis)*, to five *(A. brumpti)*, to eight (*O. kelleyi*). Three main dorsal and five main ventral trunks with their extensions occur in the latter (Sonenshine, 1970).

Hard ticks (Ixodida) have a larger stigmatal plate posterior to coxae IV. The surface of the plate has a sclerotized marginal frame, the peritreme, which takes a characteristic shape for genera, species and sometimes sex (Fig. 9-5D). A crescentic macula, usually placed within a somewhat eccentric pedicel, is surrounded by wrinkled integument. Peripheral to the macula the surface of the plate shows goblets or aeropyles (Fig. 9-6B). These air openings (called goblets in early descriptions) vary in number, size, and distribution. They have been used to differentiate descriptions of species. They may be circular or pyriform, some with straight underlying cuticular pedicels (*Haemaphysalis leporispalustris* (Fig. 9-6A, B) and some with stellate structures (Fig. 9-6C). The cuticular

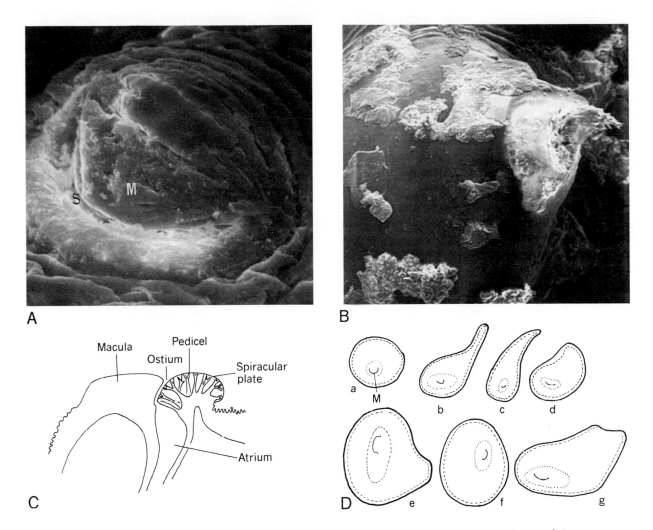

Figure 9-5. Stigmatal plates of soft ticks: (A) *Argas persicus* (S, stigmata; M, macula); (B) *Otobius megnini* (SEM micrographs). (C) Diagram of stigmatal opening (after Hoogstraal, 1956). (D) Shapes of stigmatal plates of hard ticks: a, *Ixodes ricinus*; b, *Rhipicephalus bursa*; c, *Rhipicephalus sanguineus* male; d, *R. sanguineus* female; e, *Dermacentor andersoni*; f, *D. variabilis*; g, *Amblyomma cajenennse* (a–d after Arthur, 1962).

pedicels or stellate pillars are extensions from the underlying hypodermis (Fig. 9-6D, E) (Woolley, 1972).

Scanning electron micrographs show aeropyles (air gates) that open into atrial chambers and subatrial chambers beneath the surface of the stigmatal plate (Woolley, 1972) (Figs. 9-6D, E). These aeropyles were assumed to be the functional openings for passage of the ambient air into the tracheae in the absence of a true ostium beneath the macula, "as has been claimed by previous writers who supposed that the chief or only route for gas exchange between the atrium and the ambient air was through the

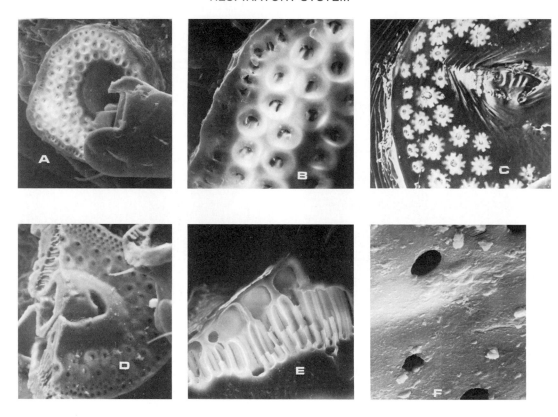

Figure 9-6. (A) Stigmatal plate of *Haemaphysalis leporispalustris* (300×); (B) enlarged view of aeropyles (1000×); (C) aeropyles of *Ixodes hexagonus* (SEM of Hinton); (D) cut stigmatal plate of *D. andersoni* (300×); (E) enlarged view of cut aeropyles and subatrial spaces of same (200×); (F) enlarged aeropyles of same (3000×).

so-called ostium . . . oils applied to the aeropyles readily flood the atrium, but do not penetrate into the atrium when applied to the ecdysial scar'' (Hinton, 1967). But the function of the aeropyles is not yet completely understood.

The muscular movement of the atrial walls and the lips of the ostium is thought to be responsible for the movement of ambient air into the tracheal system (Roshdy and Hefnawy, 1973). By histological studies these authors describe the stigmata of *Haemophysalis longicornis* (and other species) as an external plate, a middle subostial space, and an internal chamber that leads to tracheal trunks.

The cuticular macula extends into a tight conformation with the crescentic ostium and wedge-shaped ostial lip. The outer area of this lip is called the columella, and the inner extension of the cuticle, the stalk (Arthur, 1956). The edges of the ostium, the columella, and the stalk have a lumen that is confluent with the hemocoel and its hemolymph. A large cavity, the subostial space, surrounds the stalk.

The tracheal trunks leading from the stigmatal plate include (1) an anterior trunk to the legs, podosoma, and synganglion; (2) a median trunk to the salivery glands and reproductive organs; and (3) posterior median and posterior lateral trunks to the other viscera (Fig. 9-7A) (Arthur, 1962; Douglas, 1943). Taenidia line the tracheae (Fig. 9-7B).

The cuticle-lined atrial chamber leads to the tracheal trunks. Its ventral wall is flexible and valvular, in association with vacuolar connective tissue and hemocytes. An oblique muscle band inserts on the dorsal atrial wall and connects to the body wall. This muscle is used to-

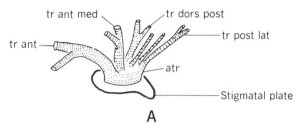

A

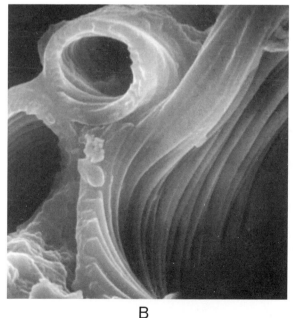

B

Figure 9-7. (A) Stigmata and tracheal trunks of *Dermacentor andersoni* (after Douglas, 1943) (atr, atrium; tr ant, anterior trachea; tr ant med, anteromedian trachea; tr dors, dorsal trachea; tr dors post, dorsoposterior trachea; tr dors lat, dorsolateral trachea); (B) cut trachae in *Ixodes ochotonae* (5000×) (SEM micrograph).

gether with hydrostatic pressure of the hemolymph to manipulate the opening.

Ticks treated with CO_2 died within 30 min when the ostial lip was separated from the edge (Roshdy and Hefnawy, 1973). In treatments with cyanide vapor ostial lips remained closed and ticks suffered no mortality from 30 min to 3 h. Roshdy and Hefnawy conclude that Hinton's idea of the nonfunctional ostium in hard ticks is in error and that the ecdysial scar may be an artifact. Their experimental results contradict the view that the macula is not a closing apparatus and that the aeropyles are functional "air gates" as proposed by Hinton (1967) and Woolley (1972).

If these pores were to communicate directly with the atrium, diffusion of cyanide gas through the pores would cause death of the tick. The numerous sensilla-like structures at the goblet bases observed in our study may be sensitive to CO_2 and other gases in the surrounding atmosphere. Response to these stimuli may regulate spiracular opening and closing. Experimental blocking of the porous part of the spiracular plate before gas exposure should reveal the actual function of these presumed sensory structures.

Actinedida. Most of the trombidiform mites (Actinedida) are tracheate, although exceptions do occur in Eriophyidae and Demodicidae (Jeppson et al., 1975; Nutting, 1964). The greatest number of morphological varieties of respiratory systems and locations of exterior stigmata are found in this order of mites. These variations are displayed in the locations of the stigmata and in the external manifestations of the peritremes. The main locations of the stigmata are (1) on or between the chelicerae or cheliceral bases, (2) on the noncheliceral parts of the gnathosoma, and (3) on the shoulders of the propodosoma (Fig. 9-8). It is evident that the types of feeding structures (e.g., the presence of a stylophore such as in the Tetranychidae) relate to this diversity of location (Evans et al., 1961). Also, in each of the major subgroups of this order, there may be mites in which genital

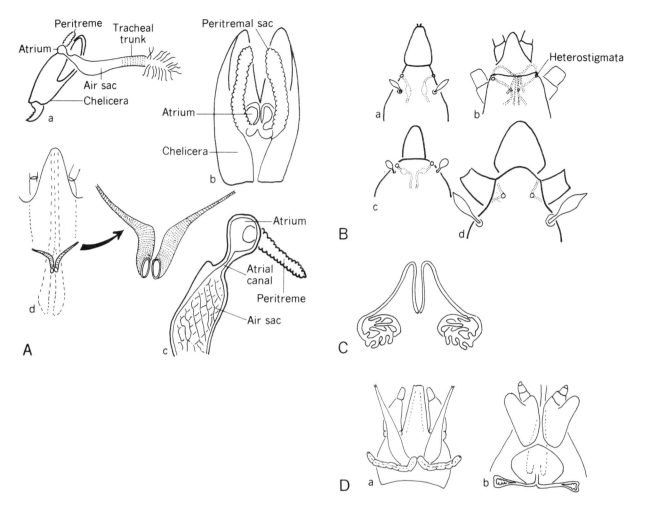

Figure 9-8. Respiratory systems in Actinedida: (A) cheliceral respiratory system of *Allothrombium fuliginosum*: a, diagram of chelicerae with atrium, peritreme, air sac, and tracheal trunk; b, dorsal view of peritremal sac, atrium, chelicerae (after Hughes, 1958); c, enlarged view of peritreme, atrium, and air sac; d, cheliceral stigmata of Erythraeidae (after Krantz, 1978); (B) stigmata on the "shoulders" of a, Tarsocheylidae; b, Heterocheylidae with heterostigmata; c, Tarsonemidae; d, Pygmephoridae; (C) peritreme of *Tenuipalpoides dorychaeta* (after Thewke and Enns, 1970); (D) emergent peritremes of a, Pterygosomidae, lizard mite; b, Harpyrhynchidae, bird mite.

tracheae are associated with the reproductive organs (e.g., Bdellidae, snout mites).

The peritremes are similar to those found in the gamasine mites, but are more diverse. They may be somewhat cryptic, slightly insunk integumental surface structures (Fig. 9-8B).

They may be slightly emergent above the surface of the body (Fig. 9-8C), or prominently raised in hornlike fashion (Fig. 9-8D). Peritremes may be entirely troughlike and smooth, or show a type of segmentation—a chambering—within them (Fig. 9-8C, D). Some of the

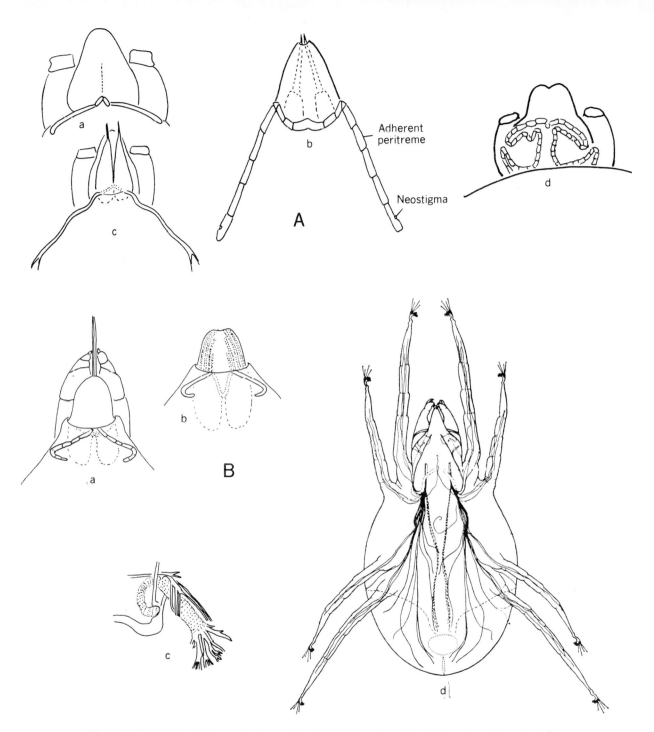

Figure 9-9. Actinedida systems continued: (A) chambered, adeherent peritremes of a, Rhaphignathidae; b, Cheyletidae with neostigmata; c, shoulder peritremes of Stigmaeidae; d, gnathosomal peritremes of Caligonellidae; (B) respiratory system of Tetranychidae (*Tetranychus telarius*): a, stylostome and peritreme; b, stylostome with air-filled grooves; c, sigmoid piece and accessory tracheal trunk; d, ventral view of tracheal system (after Blauvelt, 1945).

chambered peritremes are associated with definitive air sacs (Fig. 9-8A).

Conceptually the hypothetical ancestral prostigmatic (Actinedida) mite had two pairs of tracheal trunks extending from the body to the gnathosoma (Grandjean, 1938). One of the tracheal trunks was positioned above the other. The dorsal pair of trunks curved upward and opened between the bases of the chelicerae in the arthrodial membrane. The ventral pair opened appositionally below the cheliceral bases. Partial fusion of the cheliceral bases (e.g., *Retydeus viviparus*) results in the single median stigmata of the dorsal tracheae (=neostigmata of Grandjean, 1938). The ventral pair of tracheal trunks may open in the primitive location in some of the prostigmatid mites, but in the majority, the ventral tracheal trunks are lost and the dorsal pair remain in association with the cheliceral bases (Hughes, 1958). The stylophore of cheyletoids lacks a median division and the peritremes are usually external to the stylophore, humeral in position, and M-shaped with chambers (Fig. 9-9A).

Two tracheal trunks exist in *Cheyletus eruditus* and enter the gnathosoma in what is considered a primitive condition. The trunks end posteriorly at about the level of leg I and may show a small opening, the neostigmata (Hughes, 1958) (Fig. 9-8D); many fine tracheae extend to the internal organs. The tracheal trunks are supported by fused surfaces of the cheliceral bases, and because of the fusion, the styletlike movable digits are the only retractile parts. The air channels or peritremata are constricted at intervals by inner partitions that give the appearance of segmented chambers. The peritremes extend forward slightly and each ends in an elongated chamber in which is a neostigmata—a small lateral opening to the outside (Hughes, 1958).

In 1885, Berlese first recognized the similarities of Tarsonemidae, Scutacaridae, and Pyemotidae. He regarded the Tarsonemidi as transitional between the Oribatidi and the Sarcoptidi because of the trichobothria and sensilla. He thus (later) placed these families in the suborder Heterostigmata, a group in which only the females possessed stigmata.

Most male Pyemotidae differ from the females in the absence of a peritreme and all traces of the stigmata. The peritremes in females Pyemotidae are variously shaped and

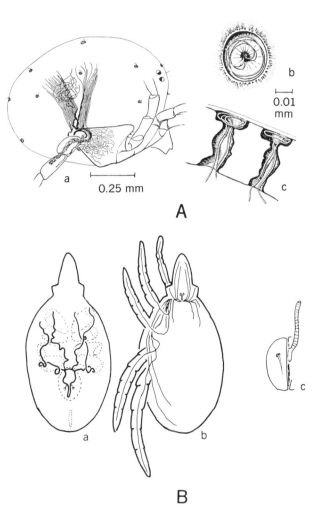

Figure 9-10. (A) Tracheal system of water mites: a, *Limnesia;* b, *Arrenurus* surface view of pit with tracheal loop; c, same, section of cuticle with tracheal loops in pits (after Mitchell, 1972); (B) tracheal systems of *Cyta latirostris*: a, genital tracheae; b, prostigmatic system (after Oudemans from Vitzthum, 1943); c, genital trachea of deutonymph of *Trachymolgus nigerimmus* (after Grandjean, 1935).

separated from the stigmata. "The shape, proximity and position of the peritremes are useful in classification at both the generic and specific levels" (Cross, 1965).

Species of *Cheyletiella* (an ectoparasite of some birds, rabbits, hares, dogs, and cats) show sexual dimorphism in the peritremes. Peritremes are usually larger in males than in females (Smiley, 1965, 1970). *Demodex*, a facultative aerobic mite, may live experimentally in distilled water, earwax, and machine oil for different periods of time (Nutting, 1965). Diffusion of oxidative metabolic materials of the host are used *in vivo* by these parasites, which lack anuses and respiratory and circulatory systems.

Genital tracheae are identified in *Trachymolgus nigerimmus* and *Cyta latirostris,* and genital air sacs are found in *Bdella semiscutata* (Bdellidae) (Figure 9-10B) (Vitzthum, 1943).

Peritremes are chambered in the spider mites (Tetranychidae). They begin at the base of the stylophore and overlay the anterior margin of the propodosoma. Integumental grooves on the surface of the stylophore may trap air (Blauvelt, 1945) (Fig. 9-9B).

No open respiratory system exists in water mites. Some water mites were assumed to have prostigmatic tracheal trunks, but numerous tracheae forming superficial networks and tracheal nets are observed. Each such trachea has a part of its length lying directly underneath the cuticle. One or both ends of the trachea turn into the body. Tracheal nets are absent in regions with heavy integumental sclerites, but muscles of the legs and mouthparts are well supplied. Special sets of tracheae in these areas meet the demand for greater gaseous exchange. In the water mite, *Najadicolu ingens*, tracheae are present as thin-walled tubes throughout the body. These are without taenidia and branching is uncommon. The proximal ends are attached to a sigmoid piece, but without a chamber (Mitchell, 1955, 1972) (Fig. 9-10 A).

Respiratory systems of chiggers include a pair of stigmata lateral to the gnathosoma.

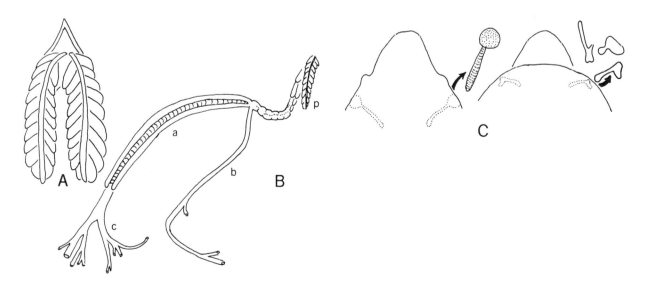

Figure 9-11. Respiratory system of *Podothrombium crassicristatum*: (A) superificial view of chambered peritremes; (B) a, anterior tracheal trunk with chitinous investment; b, lateral tracheal trunk; c, median tracheal trunk (p, peritreme) (after Feider, 1968); (C) pseudotracheae of *Grammolichus aureliana* showing location on idiosoma and several types from different species (after Fain, 1968).

Each stigma connects with an unbranched tracheal tube. The presence or absence of a respiratory system is a diagnostic characteristic for different subfamilies. No tracheae are found in *Trombicula alfreddugesi* or *T. splendens* (Trombiculidae) (Wharton, 1950; Brown, 1952).

The respiratory system of *Prothrombium crassicristatum* shows chambered peritremes connected to anterior, lateral, and median tracheal trunks (Fig. 9-11). The respiratory systems in the trombidial Actinedida are complex and vary with the evolutionary stadium and the taxa. These differences are useful in phylogenetic systematics and the taxonomic separation of some of the supercohorts (Feider, 1979).

Astigmata. The name of this order, Astigmata (a-/without), is indicative of the absence of respiratory stigmatal openings and implies that respiration takes place through the cuticle in the absence of tracheae. A pseudostigmatic slit above the base of legs I in many species was thought to be a functional spiracle, but this has been shown to be the external opening of a coxal gland (Zakhvatkin, 1959).

Pseudotracheae are found anterolaterally in some of the Glycyphagidae. They appear to be diagnostic for species (e.g., *Grammolichus*). Their function is unknown (Fain, 1968) (Fig. 9-11 C) (cf also Grandjean, 1937).

Oribatida. An older taxonomic name inclusive of the oribatids and the astigmatids is Cryptostigmata, which implies "hidden stigmata," located primarily in the leg acetabulae of the higher oribatids. Inferior oribatids evidently lack stigmata, but brachytracheae are present (Nothroidea, Phthiracaroidea) (Evans et al., 1961). Others (Cosmochthoniidae) have genital tracheae or extensive tracheae extending from the acetabula of legs I and II (Evans et al., 1961; Hammen, 1968c).

The main tracheae in adults of the higher Oribatida open in the acetabula of legs I and III and at the bottom of the furrow between the propodosoma and the hysterosoma (Fig. 9-12). Localized cuticular areas called sacculi and areae porosae are also found (Fig. 9-13). Sacculi

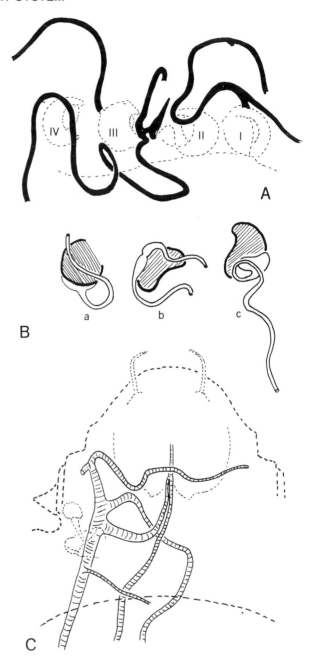

Figure 9-12. Respiratory systems in Oribatida: (A) main tracheae and acetabula of *Drymobates silvicola* (after Grandjean, 1934); (B) a,b,c, air sacs and tracheae of leg I and III of *Damaeus clavipes*; b, leg III of *D. geniculatus* (after Michael, 1884); (C) anterior tracheal trunks of *Neoliodes theleproctus* (prodorsum and trichobothrium in stippled outline) (after Grandjean, 1934).

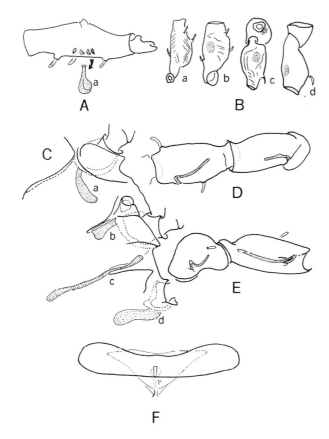

areae porosae are thought to be respiratory structures (Wallwork, 1969).

Brachytracheae occur in larvae and adults (Grandjean, 1934) and are associated with air sacs (*Neoliodes theleproctus*) (Vitzthum, 1943). Brachytracheae may be associated with the trichobothria or pseudostigmatic organs in oribatids, or may be connected to the coxal acetabula in both immatures and adults (Fig. 9-13C). Platytrachea are found in preanal folds of *Poroliodes farinosus* (Fig. 9-13F) (Grandjean, 1934).

Figure 9-13. Sacculi and areae porosae: (A) sacculi (air sacs) on femur I of *Platyliodes scaliger*; a, enlarged view; (B) areae porosae on a, femur I of adult; b, femur I of tritonymph; c, trochanter and femur IV of adult; d, trochanter and femur IV of tritonymph (after Grandjean, 1934); (C) brachytracheae of larva of *Neoliodes theleproctus*: a, brachytrachea I; b, apodematal II brachytrachea; c, brachytrachea at sejugal suture; d, brachytrachea III; (D) brachytrachea in leg III of protonymph; (E) brachytracheae of leg III of adult (after Grandjean, 1934); (F) platytrachea of *Poroliodes farinosus* associated with preanal folds (after Grandjean, 1934).

are small, distinctive pores on the dorsum of the hysterosoma opening from globose sacs beneath the integument. Areae porosae are larger areas on the dorsal surface of the hysterosoma as well as on some leg segments. These exhibit a papillate or a porose surface. Both sacculi and

REFERENCES

Andre, M. 1949. Ordre des Acariens. *In*: Grassé, P.-P., ed. *Traité de Zoologie*. Masson, Paris, Vol. 6, pp. 794–892.

Arthur, D.A. 1956. The morphology of the British Prostriata with particular reference to *Ixodes hexagonus* Leach. III. *Parasitology* 46:26–307.

Arthur, D.A. 1962. *Ticks and Disease*. Pergamon, Oxford.

Balashov, Y.S. 1972. *Bloodsucking Ticks (Ixodoidea)—Vectors of Diseases of Man and Animals*, Transl. 500 (T500). Med. Zool. Dep., USNAMRU No. 3. Cairo, Egypt, U.A.R.

Blauvelt, W.E. 1945. The internal morphology of the common red spider mite (*Tetranychus telarius*, Linn.). *Mem.-N.Y. Agric. Exp. Stn. (Ithaca)* 270:1–35.

Brown, J.R.C. 1952. The feeding organs of the adult common chigger. *J. Morphol.* 91(1):15–52.

Cross, E.A. 1965. Generic relationships of the family Pyemotidae. *Univ. Kans. Sci. Bull.* 45(2):29–275.

Daniel, M., V. Bozdech and C. Moucka. 1959. Vyskyt trudnika tukeveho (*Demodex folliculorum* Owen 1843) u lidi a jeho epidemiologie. *Cesk. Epidemiol. Mikrobiol. Immunol.* 8(1):-52–60 (Russian and English summaries) (cited by Nutting, 1964).

Douglas, J.R. 1943. The internal anatomy of *Dermacentor andersoni* Stiles. *Univ. Calif. Publ. Entomol.* 7(10):207–272.

Evans, G.O., J.G. Sheals and D. McFarlane. 1961. *Terrestrial Acari of the British Isles*. Adlard & Son, Bartholomew Press, Dorking, England.

Fain, A. 1968. Acariens nidicoles et detriticoles en Afrique au sud du Sahara. III. Espèces et genres nouveaus dans las sous famille Labidophorinae et Grammolichinae (Glycyphagida: Sarcoptiformes). *Acarologia* 10(1):86–110.

Feider, Z. 1968. La larve et la nymphe de *Podothrombium crassicristatum* (Acariformes: Podothrombiidae). *Acarologia* 10(1):29–43.

Feider, Z. 1979. The taxonomic importance of the respiratory system in *Trombidiformes. Acarologia* 21(2):239–248.

Fetscher, J. 1924. Beitrag zur Biologie der Acarusmilbe und zur Therapie der Acarusroude des Hundes. *Monatsh. Prakt. Tierheilkd.* 32(7–8):313–316 (cited by Nutting, 1964).

Grandjean, F. 1934. Les organes respiratoire secondaires des Oribates (Acariens). *Ann. Soc. Entomol. Fr.* 10(3):109–146.

Grandjean, F. 1935. Observations sur les Acariens (2e série). *Bull. Mus. Natl. Hist. Nat.* (2) 7:201–218.

Grandjean, F. 1936. Un Acarien synthetique. *Opilioacarus segmentatus* With. *Bull. Soc. Hist. Natur. Afr. Nord* 27:413–444.

Grandjean, F. 1937. *Otodectes cynotis* (Hering) et les prétendues trachées des Acaridiae, *Bull. Soc. Zool, Fr.* 62:280–290.

Grandjean, F. 1938. Retydeus et les stigmates mandibulaires des Acariens prostimatique. *Bull. Mus. Natl. Hist. Nat.* (2) 10:279–286.

Gupta, A.P. 1979. *Arthropod Phylogeny.* Van Nostrand-Reinhold, New York.

Hammen, L. van der. 1961. Description of *Holothyrus grandjeani* nov. spec. and notes on the classification of the mites. *Nova Guinea, Zool.* 9:173–194.

Hammen, L. van der. 1965. Further notes on the Holothyrina (Acarida). I. Supplementary description of *Holothyrus coccinella* Gervais. *Zool. Meded.* 40(38):253–276.

Hammen, L. van der. 1966. Studies on Opilioacarida (Arachnida). I. Description of *Opilioacarus texanus* (Chamberlin and Mulaik) and revised classification of the genera. *Zool. Verh.* 86:1–80.

Hammen, L. van der. 1968a. Introduction à la classification, la terminologie morphologique, l'ontogénie et l'évolution des Acariens. *Acarologia* 10(3):401–412.

Hammen, L. van der. 1968b. Studies on Opilioacarida (Arachnida). II. Redescription of *Paracarus hexophthalmus* (Redikorzev). *Zool. Meded.* 42(5):57–76.

Hammen, L. van der. 1968c. Stray notes on Acarida. *Zool. Meded.* 42(5):261–280.

Hammen, L. van der. 1969. Studies on Opilioacarida (Arachnida) III. *Opilioacarus platensis* Sylvestri, and *Adenacarus arabicus* (With). *Zool. Meded.* 44(8):113–131.

Hammen, L. van der. 1973. Classification and phylogeny of mites. *Proc. Int. Congr. Acarol., 3rd, 1971,* pp. 275–282.

Hammen, L. van der. 1977. Studies on Opilioacarida. IV. The genera *Panchaetes* Naudo and *Salfacarus,* gen. nov. *Zool. Meded.* 51(4):43–78.

Hinton, H.E. 1967. The structure of the spiracles of the cattle tick, *Boophilus microplus. Aust. J. Zool.* 15:941–945.

Hoogstraal, H. 1956. *African Ixodoidea. I. Ticks of the Sudan,* USNAMRU No. 1. Bur. Med. Surg., Cairo.

Hughes, T.E. 1958. The respiratory system of the mite *Cheyletus eruditus* (Schrank, 1784). *Proc. Zool. Soc. London* 130:231–239.

Hughes, T.E. 1959. *Mites or the Acari.* Athlone, London.

Jakeman, L.A.R. 1961. The internal anatomy of the spiny rat mite *Echinolaelaps echidninus. J. Parasitol.* 47:328–349.

Jeppson, L., H.H. Keifer and E.W. Baker. 1975. *Mites Injurious to Economic Plants.* Univ. of California Press, Berkeley.

Kethley, J.B. 1977. A review of the higher categories of Trigynaspida (Acari: Parasitiformes). *Int. J. Acarol.* 3(2):129–149.

Krantz, G.W. 1978. *A Manual of Acarology.* Oregon State Univ. Book Stores, Corvallis.

Lawrence, R.F. 1953. *The Biology of the Cryptic Fauna of the Forests.* A.A. Balkema, Amsterdam.

Levi, H.W. 1967. Adaptations of respiratory systems of spiders. *Evolution* 21:571–583.

Levi, H.W. and W.M. Kirber. 1976. An evoution of tracheae in arachnids. *Bull. Br. Arachnol. Soc.* 3(7):187–188.

Lindquist, E.E. and J.B. Kethley. 1975. The systematic position of the Heterocheylidae Tragardh (Acari: Acariformes, Prostigmata). *Can. Entomol.* 107:887–898.

Michael, A.D. 1884. *British Oribatidae,* Vol. 1.

Michael, A.D. 1903. Oribatidae. *In: Expedition Antarctique Belge, Résultats du Voyage du S. y. Belgica en 1897-1899,* Rapports Scientifique, Zoologie, Acariens, pp. 1-8.

Mitchell, R.D. 1955. Anatomy, life history and evolution of mites parastizing fresh-water mussels. *Misc. Publ. Mus. Zool. Univ. Mich.* 89:1-28.

Mitchell, R.D. 1972. The tracheae of water mites. *J. Morphol.* 136:327-335.

Nutting, W.B. 1950. Studies on the genus *Demodex* Owen (Acari:Demodicoidea). Ph.D. thesis, Cornell University, Ithaca, New York.

Nutting, W.B. 1964. Demodicidae—Status and prognosis. *Acarologia* 6(3):441-454.

Nutting, W. B. 1965. Host-parasite relations: Demodicidae. *Acarologia* 7:301-317.

Ripper, W. 1931. Versuch Einer Kritic der Homologiefrage der Arthropoden tracheen. *Z. Wiss. Zool.* 138:303-369 (cited by Gupta, 1979).

Roshdy, M.A. 1961. Comparative internal morphology of subgenera of *Argas* ticks. I. Subgenus *Carios: Argas vespertillionis* (Latreille, 1802). *J. Parasitol.* 47(6):987-994.

Roshdy, M.A. 1966. Comparative internal morphology of subgenera of *Argas* ticks (Ixodoidea, Argasidae). 4. Subgenus *Ogadenus: Argas brumpti* Neumann, 1907. *J. Parasitol.* 52(4):776-782.

Roshdy, M.A. and T. Hefnawy. 1973. The functional morphology of *Haemaphysalis* spiracles (Ixodoidae: Ixodidae). *Z. Parasitenkd.* 42:1-10.

Savory, T. 1977. *Arachnida,* 2nd ed. Academic Press, London.

Sellnick, M. 1958. Die Familie Zerconidae Berlese. *Acta Zool. Acad. Sci. Hung.* 3(3):313-368.

Smiley, R.L. 1965. Two new species of the genus *Cheyletiella Proc. Entomol. Soc. Wash.* 67(2):75-79.

Smiley, R.L. 1970. A review of the Family Cheyletidae (Acarina) *Ann. Ent. Soc. Am.* 63:1056-1078.

Sonenshine, D.E. 1970. A contribution to the internal anatomy and histology of the bat tick *Ornithodoros kelleyi* Cooley and Kohls, 1941. *J. Med. Entomol.* 7(3):289-312.

Strandtmann, R.W. 1960. Nasal mites of Thailand birds. *J. Kans. Entomol. Soc.* 33(4):129-151.

Theodor, O. and M. Costa. 1960. New species and new records of Argasidae from Israel. Observations on the rudimentary scutum and the respiratory system of the larvae of the Argasidae. *Parasitology* 50:365-386.

Thewke, S.E. and W. Enns. 1970. The spider-mite complex (Acarina: Tetranychoidea) in Missouri. *Monogr. Ser. Univ. Missouri Centr.* 1 1969:5-106.

Tippetts, T. 1957. Descriptions of a new *Periglischrus* from a bat, *Mormoops megalophylla senicula* Rehn, together with a key to the species of *Periglischrus. J. Kans. Entomol. Soc.* 30(1):13-19.

Vitzthum, H.G. 1943. Acarina. *Bronns Kl.* 5:1-1011.

Wallwork, J.A. 1969. Some basic principles underlying the classification and identification of cryptostigmatid mites. *In:* Sheals, J.G., ed. *The Soil Ecosystem.* Syst. Assoc. Publ., pp. 155-168.

Wharton, G.W. 1950. Respiratory organs of chiggers. *Proc. Entomol. Soc. Wash.* 52(4):194-199.

Witalinski, W. 1979. Fine structure of the respiratory system in mites from the family Parasitidae. *Acarologia* 21(3-4):330-339.

Woodring, J.P. and C.A. Galbraith. 1976. The anatomy of the adult uropodid *Fuscouropoda agitans* (Arachnida: Acari), with comparative observations on other Acari. *J. Morphol.* 150:19-58.

Woolley, T.A. 1972. The respiratory organs of ticks. *Trans. Am. Microsc. Soc.* 91(13):348-363.

Zakhvatkin, A.A. 1959. *Tyroglyphoidea (Acari)* (translation of *Fauna of USSR Arachnoidea,* Vol. 6. No. 1). Am. Inst. Biol. Sci., Washington, D.C. (translated by A. Ratcliffe and A. M. Hughes).

10

Excretory System

The best of all things is water.
—PINDAR, GREEK POET

Excretion implies the elimination of excesses of soluble and diffusible waste products of metabolism and includes water balance, so important in the osmoregulatory functions of the body. Excretion in mites is accomplished principally by Malpighian tubules, but also through salivary glands (in some ticks), single excretory tubules apart from the digestive tract (in some Actinedida), and by coxal glands and the hindgut (in, e.g., Oribatida). Descriptive terminology of the excretory organs of mites varies with authors so that the terms Malpighian tubules, excretory tubules, excretory sacs, excretory vesicles and excretory organs are employed in similar, but not always differentiating ways. In the Systematic Review that follows some of these terms are used interchangeably.

MALPIGHIAN TUBULES

Malpighian tubules are present in the Arachnida and in the Uniramia, although the origin of these tubules in the Arachnida is usually considered to be independent of such origin in the Uniramia. The tubules are absent in the related Merostomata, Crustacea, Tardigrada, Pentastomida, and Pycnogoinda (Gupta, 1979).

Many mites have discrete, paired Malpighian tubules similar in structure and location to those found in other arachnids, and connected to the digestive tract and into which excretory products are discharged. The tubules typically are closed at their distal ends, but have simple or complex loopings along their lengths to their proximal attachment between the midgut and the hindgut. Excretory products are eliminated with the feces through the anus. Primarily a single pair of tubules is present, but two pairs may occur or the tubules may be entirely absent. A secondary situation exists where expulsion of materials occurs through a single excretory vesicle unconnected to the digestive tract with an exterior opening (uropore) (Evans et al., 1961; Hughes, 1959; Vitzthum, 1943) (Fig. 10-4A).

The outer covering of Malpighian tubules (and possibly the single excretory vesicle) is a thin tunica propria (basement membrane) and muscle fibers covering an epithelium. The epithelium is composed of polygonal, finely granular cells with centrally placed nuclei, although the cell membranes are not always clear. The lumen of the tubule is reduced in a starved mite, but it becomes distended as feeding and digestion proceed and excretory materials accumulate. Distension of the lumen is also accompanied by vacuolation of the cytoplasm of the tubule cells (Balashov, 1972; Young, 1968) (Figs. 10-1, 10-3).

Contents

Materials eliminated by the Malpighian tubules vary according to the food consumed by the mites and these excretory substances differ in form and dimensions in the lumena of the tubules. Spherular concretions are composed of guanine (2-amino-6-oxypurine, a component of ribonucleic and deoxyribonucleic acids), a final product of nitrogen metabolism (Vitzthum, 1943; Baker and Wharton, 1952). Guanine has a low solubility and tends to precipitate while still in low concentration. It is a relatively insoluble, nontoxic product that can be stored by molting forms and handled by nonfeeding winter forms of the two-spotted spider mite (McEnroe, 1963). In ticks the guanine in the Malpighian tubules and rectal sac is in suspension or becomes a pastelike crystalline mass, which facilitates discharge without much water. During development, ecdysis, or in periods of starvation ticks receive little water from the environment and because of its low solubility the guanine accumulates in the Malpighian tubules. This accumulation prevents the concentration of the substance in the hemolyph from reaching toxic levels.

The guanine spherules tend to be white in color, differ in size, and are characterized by concentric layers or lamellae of varying complexity. They may be single, double, or with several spherules coalesced together. All guanine spherules are birefringent in polarized light. In living mites, this white material may be observed to move back and forth in the lumena of the tubules before it becomes mixed with the feces in the rectum prior to expulsion (Hughes, 1959; Balashov, 1972) (Figs. 10-1, 10-3).

Systematic Review

Excretory sacs, excretory tubules and Malpighian tubules occur in representatives of all the orders of mites except Oribatida (Woodring, 1973). As stated above, the number of excretory tubules varies in the different orders of mites. Opilioacarida, Gamasida, and many parasitic mites have a single pair. Holothyrida has two pairs. A single pair is found in Ixodida; their lengths and looped configurations are influenced by the feeding state of the tick. (In some ticks the lengths of the tubules may be twice that of the body.) In certain of the Actinedida (Tetranychidae, Trombidina) the midgut ends blindly without connection to the hindgut. In such instances a single, unpaired excretory vesicle is medial in position and extends posteriorly from its blind end in the hemocoel to a ventral or terminal uropore (Fig. 10-4A). Excretory tubules are absent in others of the Actinedida (Eriophyidae; Demodicidae). Small, simple Malpighian tubules may be present in the Astigmata, e.g., *Tyroglyphus farinae*, but are apparently lacking in most groups within the order. Excretory tubules are also absent in the Oribatida, where the hindgut assumedly functions in this process (Evans et al., 1961; Hughes, 1950, 1959), but where coxal glands occur (Woodring, 1973).

Gamasida. A pair of blind excretory tubules empties into the rectum of the spiny rat mite *(Echinolaelaps echidninus).* The tubules proceed from the attachment at the rectum the full length of the body and may extend into the coxae or trochanters of legs I. The anterior ends of the tubules are narrowed similarly to at the connection with the rectum. A tissuelike valve at this junction prevents backflow of materials (Jakeman, 1961).

The cells of these tubules are irregularly oval in shape with dark-staining nuclei and distinct nucleoli. As the guanine contents accumulate in the lumen, the cells become flattened. When not distended with excretory materials, three to five cells make up the circumference of the tubule. Contents of the tubules are moved along by peristaltic action of either the cells or a thin, outer tunic of muscle fibers (Jakeman, 1961).

The excretory tubules of *Haemogamasus ambulans* are similar to those of the spiny rat mite, and may extend into the coxae of leg I. The color of the contents is whitish, in contrast to the darker materials of the rectum, regardless of the food eaten. Movements carry the contents toward the rectum and a sphincter muscle controls release into the rectum. An external muscle layer covers the basement membrane surrounding the tubules. Peristaltic movements of the excretory tubules were observed in the embryos of the oviparous *Haemogamasus liponyssoides* (Young, 1968) (Fig. 10-1E).

The excretory tubules of *Caminella peraphora* K. and A. and *Fuscouropoda agitans* have placement similar to the that in gamasid mites (Figs. 10-1G).

Gamasid and actinedid mites lack guanine in the parenchyma, but the excretory organs are densely white-colored due to the accumulation of guanine spherules (Hughes, 1961).

Ixodida. More is known of the excretory system and its function in ticks than in any other order of mites (Fig. 10-2).

Argasidae. White guanine spherules are the major excretory product of third instar nymphs and adults of *Argas (P.) arboreus.* This material comprises 59.2 to 97.3% of the excreta by weight and 70 to 100% of the nitrogen excreted. Hematin constitutes 1.2 to 8.6% of the excreta by weight and 0.23 to 1.45% of the nitrogen excreted. No proteins or amino acids were detected in excreta by electrophotometric methods using pigeon hosts.

Excretion in soft ticks occurs in a definite cycle and the maximum quantity of the excretory products are produced 3 to 4 days after feeding. In the nymphs excretion stops almost completely 4 to 5 days before molting to the adult, but resumes thereafter. The excretory rhythm does not change seasonally, but the quantities of excretory products may vary with different seasons. Excretion of both guanine and hematin is affected by seasonal factors. The amount of guanine reflects the level of metabolic activity, which is lowest from August to December and highest from April to July. Large quantities of excreted metabolites are released during the latter period. The intrinsic cycle is not due to feeding, but cyclic metabolic rates are thought to be influenced by diapause. This in turn is related to the adaptation of this tick to the absence of the natural host *(Bubulcus i. ibis),* which disperses between late August and October and returns to the rookeries in late March and April.

The excretion of nitrogenous waste materials is under the control of gonotrophic rhythms of adults and the ecdysial cycles of immatures in the ticks as well as bloodsucking parasitic mites that feed irregularly. With few exceptions *(Ornithodoros moubata)* the products of digestion of the host blood, the disintegrated cells from the midgut, and the nitrogenous excretory products from the Malpighian tubules accumulate together in the rectal sac. While the tick is feeding, a large quantity of slightly altered host blood enters the rectum. The fecal material is thus a mixture of several substances and the proportional quantity of any one substance may change with the period of the life cycle involved (Balashov, 1972).

Elimination of excess water following engorgement and concentration of the blood meal by bloodsucking arthropods occurs through the Malpighian tubules in various insects and through the coxal glands in argasid ticks. Concentration of the blood protein is trebled by this process. During engorgement in *Boophilus*

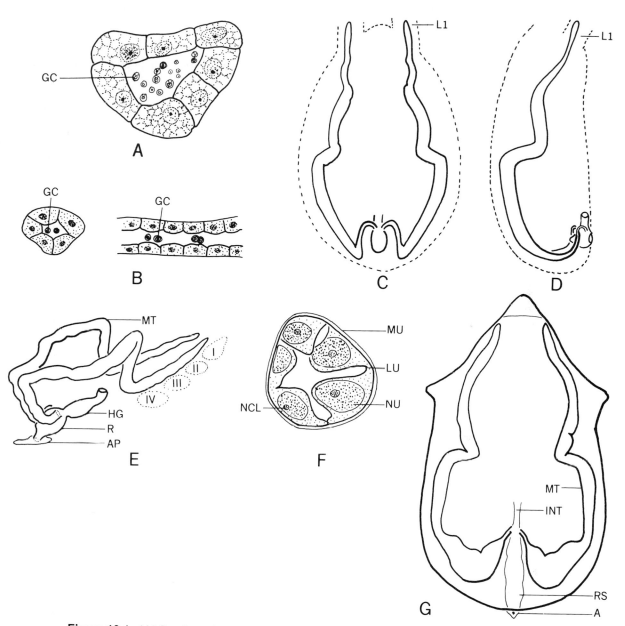

Figure 10-1. (A) Section of Malpighian tubule of *Hyalomma asiaticum* with vacuolated cells and guanine concretions (GC) in the lumen during first half of molting (after Balashov, 1972); (B) cross and longitudinal sections of Malpighian tubules of *Ixodes hexagonus* with guanine concretions (GC) in the lumen (after Vitzthum). Diagram of excretory system of *Echinolaelaps echidninus*: (C) dorsal view, (D) lateral view (L1, leg I) (after Jakeman, 1961). (E) Diagram of excretory system of *Haemogamasus liponyssoides* from lateral aspect, showing Malpighian tubules and juncture with midgut; (F) cross section of tubule of same showing muscle layer, nuclei and lumen (after Young, 1968); (G) excretory system of *Fuscouropoda agitans* showing the Malpighian tubules and juncture with rectum (after Woodring and Galbraith, 1976) (A, anus; AP, anal plate; HG, hindgut; INT, intestine; LU, lumen; MT, Malphighian tubules; MU, muscle; NCL, nucleolus; NU, nucleus; R, rectum; RS, rectal sac).

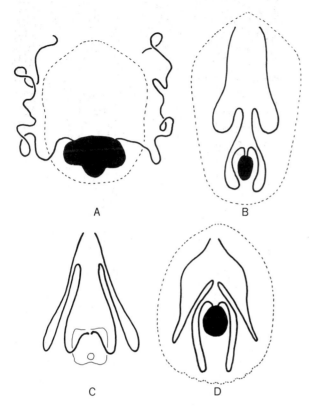

Figure 10-2. Excretory systems in ticks; Malpighian tubules and rectal sacs in (A) *Ornithodoros moubata* showing rectal ampulla (after Hoogstraal, 1956); (B) *O. papillipes* (after Balashov, 1972); (C) *Dermacentor andersoni* (after Douglas, 1943); (D) *Hyalomma asiaticum* (after Balashov, 1972).

microplus large amounts of water are passed from the gut of the tick through the epithelium into the hemocoel and returned to the host. "In the case of a tick engorging to a weight of 250 mg on a host with a hematocrit of 33 per cent . . . this would be in excess of 200 μl of water, representing 60–70 per cent of the host blood water." The determination of chloride concentrations of oral secretions, hemolymph, and whole engorged ticks also confirms the water excretory function of the saliva in these ticks. The oral secretions also remove other substances of host origin such as urea and histamines, which are assumed to be eliminated

through this same means. Tatchell concludes that in *B. microplus* and in other hard ticks the protein of the blood meal is concentrated, and imbalance of the electrolyte–osmotic system is avoided by this "remarkable adaptation of the salivary secretion" (Tatchell, 1967).

The white guanine spherules form in the Malphighian tubules. Hematin granules are characteristically black and often mixed with the guanine in the excreta. Fecal hematin results from intracellular digestion, and spectral analysis shows the black contains an iron protoporphyrin ring. Guanine excretion is low after feeding, but increases to more that 80% of the total excreta by weight.

Because ticks excrete iron as hematin, they probably do not digest or use the organic part of the hematin molecule. Enzymatic digestion of the blood meal stops after the removal of the globin part of hemoglobin in the blood (Hamdy, 1973).

Ixodidae. The Malpighian tubules of a recently molted female hard tick are replete with guanine spherules. Within 1 to 3 days after the molt the ticks eliminate the excreta that has accumulated during the premolting period when the tubule cells are cuboidal in shape. Subsequently, the tubule lumen shows a few spherules of small and medium size. Prior to attachment the unfed tick shows tubules with alternating white patches of guanine and darker empty patches in the lumen (Fig. 10-3).

Again, 1 to 3 days after attachment, the excreta accumulated during the nonfeeding period are eliminated and the cells of the tubule become polygonal with their apices extending into the lumen. Some of the cells show vacuoles in their distal ends. The diameter of the tubule decreases one and a half to two times and some cell division occurs, but not as much as in the premolting period. As the tick feeds to repletion some tubule cells continue to enlarge and develop rodlike striations at their apices. Other cells disintegrate partially or completely (Hamdy, 1977).

With the intensification of digestion in the feeding tick the amount of guanine deposited in the tubules exceeds the rate of discharge from the rectal sac, so accumulations occur in the local parts of the tubule. By the end of feeding, the tubule lumen is completely filled with floating guanine spherules and is the characteristic milky color. Distension of the tubule to three or four times the size of unfed ticks is accompanied by growth and proliferation of the cells of the tubules (Balashov, 1972).

After the hard tick detaches from the host, the tubules fill more rapidly with guanine; the diameter is 10 times greater and the walls more expanded than those of unfed ticks. The rectal sac is also enlarged and filled mainly with guanine. Both nymphal and adult stages of hard ticks excrete materials in addition to guanine. In decreasing order of volume the excretions are protein, a purine, guanine, and hematin (Hamdy, 1977).

The function of Malpighian tubules in larval and nymphal ixodids is similar to that of those in females except that the tubules do not fill with guanine. Larvae and nymphs periodically defecate during and after feeding. No defecation occurs during the period from premolting until ecdysis is complete because the rectal sac is not open to the outside; but the sac enlarges as it continues to receive guanine from the Malpighian tubules during this period. The rectal sac may be so enlarged just before the end of molting that it occupies most of the posterior half of the body cavity. The distension also causes the walls of the rectal sac to become membranous with scattered, flattened nuclei (Figure 10-3).

Distension of the Malpighian tubules is negligible in molting larvae and nymphs compared to that in the engorged female, but peristaltic movements carry the accumulated guanine spherules into the rectal sac. Transverse sections of tubules show that the increase in numbers of nuclei is one to two times in the larvae, three to four times in the nymphs, and five to eight times in the females.

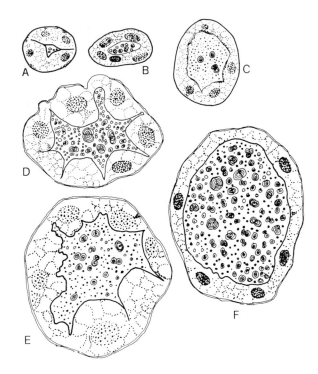

Figure 10-3. Sections of Malpighian tubules at various stages of guanine production in *Ixodes ricinus* showing the epithelial cells, nuclei, vacuoles, and muscular covering: (A) postmolting; (B) unfed for 1 year; (C) third day of attachment, packed with guanine; (D) completely engorged immediately after detachment; (E) before beginning oviposition; (F) before ending oviposition (after Balashov, 1972).

Actinedida. Relatively little is known of the excretory functions of the prostigmatid mites. Information on this function is better known in a few groups, but data are limited.

Careful observations have been made of the galls, erinea, and other places where colonies of eriophyid mites live. In none of these instances were evidences of excretions or fecal matter found. It is recognized that because the eriophyids have extremely small mouthparts and can feed only on predigested, liquid food in the upper surface of host plant tissue that they do not need to eliminate carbohydrates in the feces as do tetranychid mites (Jeppson et al., 1975). The excretory system in the grass mite *(Eriophyes tulipae)* is not well-developed. The metabolic waste products remain in the hemolymph in the form of concentrically layered, ovoid granules that resemble the granules common in the blood of many insects where waste products are found (Whitmoyer et al., 1972). They may also, like Demodicidae, be capable of eliminating metabolic wastes in the eggshells (Nutting, 1964) or in the spermatophore (Jeppson et al., 1975). In each of these instances the mechanisms for the release of waste materials would prevent fouling in the restricted and enclosed habitats of the eriophyid and demodicid mites. In the latter darkened "jack-stone-like bodies" (of Leydig) may be excretory products, possibly guanine. Since these bodies are incorporated into the ova prior to oviposition, this may be an interesting means of excretion (Nutting, 1964).

Chigger mites and other higher trombidiform mites lack a hindgut and what is called an anus functions as the external opening (uropore) of a blind, excretory tube (Fig. 10-4A). These mites exhibit a unique method of eliminating feces (and possibly excretory products) in the process of "schizeckenosy," the periodic splitting of the posterior body wall to allow fecal lobes to be released (Mitchell, 1964; Mitchell and Nadchatram, 1969). Peculiar excretory organs also are identified for Calyptostomatidae (Vistorin-Theis, 1977) (Fig. 10-4D).

Astigmata. Fecal material passing into the rectum of Astigmata is surrounded by a pertrophic membrane. Guanine particles are mixed with the feces and the pH of the gut is alkaline (7.0–8.0), which is assumed as evidence of an excretory function of that part of the digestive tract (Hughes, 1950). The Malpighian tubules join the tract at the junction of the colon and postcolon. They are usually short and devoid of granules in both the lumen and the cells; their histology appears similar to that of the rectal tissue. Because the guanine granules are absent in microscopic section and the thin-walled tubules do not appear white in the living

animal, it is assumed that they may have lost the excretory function and that this has been assumed by the rectum (Hughes, 1959).

Where phytophagous mites (e.g., *Histiosoma*) eat food of high protein content, the parenchyma tissue of the opisthosoma becomes densely packed with guanine bodies. *Acarus siro* fed on gluten-rich food show an increase in the number of guanine bodies. Such conditions suggest that excretion by Malpighian tubules is poorly developed (Hughes, 1961). Waste products of some acaridids may be retained in the fatty tissue or in the parenchyma of the body (Woodring and Galbraith, 1976).

Oribatida. Malpighian tubules are absent in the oribatid mites. The gut is assumed to function in excretion. Water and/or ion regulation by the gut cannot be discounted. Although uric acid or guanine crystals are observed in the excretory organs of the Gamasida, Ixodida, and the higher Actinedida, no such birefringent materials are found in the oribatids. Apparently nitrogenous waste products are secreted in soluble form and passed through the gut (Woodring, 1973).

COXAL GLANDS

Broad differences exist in the arrangement of coxal glands within the major divisions Anactinotrichida and Actinotrichida. The most primitive of the Acari, Opilioacarida, and Holothyrida possess typical arachnid coxal glands. In Opilioacarida the gland is tubular (With, 1904) or looped with a system of taenidia (Hammen, 1966) and opens into the precursor of the podocephalic canal (Woodring, 1973). The gland in *Holothyrus braveri* resembles the coxal glands in web-spinning spiders (Vitzthum, 1943) and has a sacculus and a looped labyrinth that opens on coxa I.

Little is known about the coxal glands of Gamasida, although one is described in *Echinolaelaps echidninus* (Jakeman, 1961). The gland

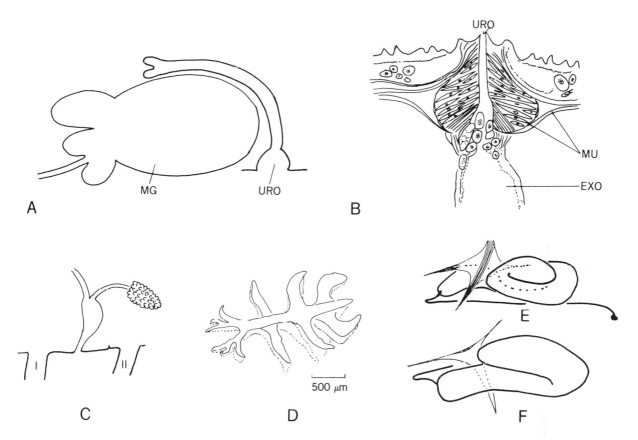

Figure 10-4. (A) Diagram of midgut of an actinedid mite with a single, unpaired excretory organ and uropore (URO; MG, midgut) (after Hughes, 1959); (B) diagram of section through uropore of *Hydrodroma despiciens* showing excretory organ (EXO), muscle (MU), and opening of uropore (URO) (after Vitzthum, 1943); (C) schematic of coxal glands of *Alveonasus lahorensis* showing the internal coxal pore and ducted accessory gland (after Balashov, 1972); (D) excretory organ of Calyptostomatidae, dorsolateral aspect (after Vistorin-Theis, 1977). Examples of coxal glands in Oribatida: (E) type A showing sacculus and labyrinth; (F) type B showing sacculus and labyrinth.

is composed of nine loosely packed large cells that surround a short, tubular atrium, so a typical sacculus and labyrinth are absent. A coxal gland with sacculus and labyrinth is described in a Uropodine (*Urotrachytus formicarius*) (Michael, 1894), but is atypical. The presence of coxal glands in other Gamasida is uncertain (Woodring, 1973).

The coxal glands in argasid ticks have been confirmed as osmoregulatory organs as well as ionic regulators. They retain ions while they eliminate water. The blood meal is thus concentrated without causing an ionic imbalance in the ticks (Edney, 1977). In some of the soft ticks the coxal glands have a thin-walled filtration chamber (?sacculus) and a complex of convoluted tubule (?labyrinth) (e.g., *Ornithodoros moubata*, Lees, 1946a; *Alveonasus lahorensis*, Balashov, 1972) (Fig. 10-4C). Hard ticks lack these glands (Saito, 1960; Till, 1961).

Coxal glands have not been described for the primitive actinedid mites, but inasmuch as the podocephalic canal is characteristic of many Actinotrichida (Actinedida, Astigmata, and Oribatida), a well developed coxal gland complex is also present. The podocephalic canal is external in primitive Actinedida but becomes an internal tube in the more advanced forms (Hammen, 1968). Glands resembling coxal glands are described for a few of the actinedids but other functions are associated with them (Woodring, 1973).

Evidently the Astigmata have lost the coxal gland secondarily, but accessory glands are present. What homologies exist between coxal glands and the podocephalic canal complex are not understood. The secretions of the supracoxal glands in these mites are a source of fresh water (Wharton and Furumizo, 1977).

Oribatida have typical arachnid coxal glands with a sacculus and a labyrinth, but of two types that correlate with the separation of oribatids into inferior and superior divisions. Type A coxal glands exhibit a labyrinth that coils clockwise in three 180° turns that make four parallel regions in the labyrinth. This type of gland occurs in all of the inferior oribatids (Fig. 10-4E). Type B glands exhibit one 180° turn, (Fig. 10-4F) in the shape of a hairpin, and are found in the superior oribatids. Internal chitinous supporting structures are different for each of these types of glands. It is assumed that the coxal glands function in osmoregulation. Filtration of the hemolymph is thought to occur through the walls of the sacculus by means of hemolymphic pressure and muscular action. Resorption of ions is assumed to occur in the labyrinth, which is noncollapsible because of the internal chitinous skeleton. Excretion may also also be a function of the coxal glands. An endocrine function in connection with molting is also ascribed to the lateral gland in some of these complexes (Woodring, 1973).

WATER BALANCE

To maintain physiological integrity mites must preserve the water content of the body and wa-

ter concentration within tolerable limits. Thermodynamically speaking, the body water pool responds as an open system in which simultaneous, independent influx and efflux of water are balanced during some increment of time. This equilibrium maintains body water content at levels that do not fall below or rise above vital upper and lower limits. Terrestrial species continually lose body water that must be replaced, and aquatic species continually gain water that must be eliminated if the amounts are to remain in balance (Arlian and Veselica, 1979).

The main avenues for uptake of water available to mites are imbibition through ingestion of free water or hydrated foods, the production of metabolic water, and the active and passive uptake of water from the ambient air or aquatic habitat through the general body surface or respiratory surfaces (Arlian and Veselica, 1979; Wharton and Devine, 1968).

Water loss from a mite to the air is transpiration; gain of water to a mite is sorption. The importance of these functions varies with the species. Water is lost by simple diffusion from the general body surfaces, respiratory surfaces, secretion of digestive fluids, defecation, excretion, production of pheromones or defensive fluids, reproductive products, or during oviposition. In a given period of time the number of molecules escaping (evaporating) and those condensing (sorption) will be in equilibrium with no net change occurring; that is, the rate of sorption and the rate of transpiration will be equal (Wharton and Devine, 1968).

In terrestrial mites, because of their small size, dominant problems are the conservation of water, the larger ratio of surface area to volume, and the relatively low water vapor density activity in their ambient environment. Transpiration from the integument can be eliminated by a waterproofed, waxy integument, a mechanism in which "arachnomorphs are conspicuously successful" (Edney, 1977).

Although the procuticle (exocuticle and endocuticle) is important in conserving water, the epicuticle appears to be most vital in preventing water loss. The epicuticle is a thin, nonchitinous layer, yet it is laminated into several distinct sublayers: a dense, homogeneous layer of protein in the epicuticle, a cuticulin layer, the outer epicuticle, and the outer membrane. Each of these sublayers is impregnated with wax and wax "blooms" may accumulate on the surface. A complex system of pore canals exists that apparently transports the wax from sites of production to the surface. These waxy coverings apparently constitute the principal barriers to water loss, for abrasive dusts, absorptive dusts, solvents, or detergents can significantly change the permeability of the integument (Hadley and Hall, 1980; Beament, 1959). Components of the exterior waxes include alkanes, wax esters, and sterols (Hadley and Hall, 1980).

Temperature and relative humidity affect water balance and survival of the spiny rat mite (*Echinolaelaps echidninus*). Low relative humidities result in high mortality of these mites because of excessive dehydration (Kanungo, 1964), but predessicated mites can absorb water from humid air (Kanungo, 1963). Temperature and relative humidity affect the water balance of females of this mite (Wharton and Kanungo, 1962).

Ticks resist dessication because of the waxy layer of the epicuticle. If exposed to increasing temperatures, water loss increases abruptly at certain "critical temperatures" as is the case in insects. Species that have higher critical temperatures are resistant to dessication within a broader range of temperatures. Resistance to dessication is also related to the type of habitat. The Argasidae are xeric animals. They live, peculiarly, in moist microhabitats in the xeric environment. Ixodidae are found in a wider variety of habitats and their ability to regulate water content aids in their dispersal.

Unfed ticks have the ability to take up water through the waxy epicuticle when exposed to higher relative humidity. Abrasion of the cuticle inhibits water uptake as it affects the hypodermal cells which appear to regulate the water imbibition (Cloudsley-Thompson, 1958). Humid air is the chief source of water for the unfed

hard tick, although water in the cuticle may be a source under experimental conditions. *Ixodes* does not drink and metabolic water is not considered to be significant in the water balance because a fasting tick can survive for limited periods at low humidities. In humid air these ticks have more efficient mechanisms for gaining water. Both fed and unfed ticks lose water rapidly by evaporation as well as through defecation and excretion after eclosion from the egg, after molting, and immediately after repletion and dropping from the host or through coxal glands (Lees, 1946a,b). It is paradoxical that the epicuticle can be the mechanism for water uptake in high humidity and the primary barrier to water loss as well.

In the two-spotted spider mite substantial quantities of ingested fluids are shunted from the esophagus to the hindgut for elimination, which facilitates digestion by concentrating the food in the ventriculus and controls the amount of water retained in the body. The two-spotted spider mite passes the equivalent of 25% of its body weight of water through its system in a 30-min interval (McEnroe, 1963).

Extraction of water vapor from the air by *Acarus siro* is governed by relative humidity. The higher the humidity, the greater is the amount of water taken up until saturation. An equilibrium is established between the humidity of the air and the water content of the animal. Sorption always takes place when the mites, in equilibrium with the ambient humidity, are transferred to a higher humidity. Longevity of these grain mites depends upon the loss of water and water loss is directly proportional to the saturation deficit of the air; longevity of the mites is indirectly proportional to the saturation deficit of the air, as long as the relative humidities are not below the critical equilibrium humidity of 71% relative humidity (Knulle and Wharton, 1964). Relative humidity influences feeding, water balance, and survival in house dust mites (Arlian, 1975a,b; Arlian and Wharton, 1974; Knulle, 1965a,b; Vannier, 1976; Wharton, 1963; Wharton and Richards, 1978).

REFERENCES

Ainscough, B. 1960. The internal morphology of *Caminella peraphora* K. & A., with descriptions of the immature stages. Oregon State University thesis LD 4330. Corvallis.

Arlian, L.G. 1972. Equilibrium and non-equilibrium water exchange kinetics in an atracheate terrestrial arthropod, *Dermatophagoides farinae* Hughes. Ph.D. thesis, Ohio State University, Columbus.

Arlian, L.G. 1975a. Water exchange and effect of water vapour activity on metabolic rate in the house dust mite, *Dermatophagoides. J. Insect Physiol.* 21:1439–1442.

Arlian, L.G. 1975b. Dehydration and survival of the European house dust mite, *Dermatophagoides pteronyssus. J. Med. Entomol.* 13:484–488.

Arlian, L.G. and M. Veselica. 1979. Water balance in insects and mites. *Comp. Biochem. Physiol. A.* 64A:191–200.

Arlian, L.G. and G.W. Wharton. 1974. Kinetics of active and passive components of water exchange between air and a mite, *Dermatophagoides farinae. J. Insect Physiol.* 20:1063–1077.

Baker, E.W. and G.W. Wharton. 1952. *An Introduction to Acarology.* Macmillan, New York.

Balashov, Y.S. 1972. Bloodsucking ticks (Ixodoidea)—Vectors of diseases of man and animals. *Misc. Publ. Entomol. Soc. Am.* 8(5):161–376.

Beament, J.W.L. 1959. The waterproofing mechanism of Arthropods. I. The effect of temperature on cuticle permeability in terrestrial insects and ticks. *J. Exp. Biol.* 36:391–422.

Cloudsley-Thompson, J. 1958. *Spiders, Scorpions, Centipedes and Mites.* Pergamon, Oxford.

Edney, E.B. 1957. *The Water Relations of Terrestrial Arthropods.* Cambridge Univ. Press, London and New York.

Edney, E.B. 1977. *Water Balance in Land Arthropods. Zoophysiology and Ecology,* Vol. 9. Springer-Verlag, Berlin and New York.

Evans, G.O., J.G. Sheals and D. MacFarlane. 1961. *The Terrestrial Acari of the British Isles,* Vol. 1. Adlard & Son, Bartholemew Press, Dorking, England.

Gupta, A.P., ed. 1979. *Arthropod Phylogeny.* Van Nostrand-Reinhold, New York.

Hadley, N.F. and R.L. Hall. 1980. Cuticular lipid biosynthesis in the scorpion *Paruroctonus mesaeinsis. J. Exp. Zool.* 212(3):373–379.

Hamdy, B.H. 1973. Biochemical and physiological studies of certain ticks (Ixodidea). Cycle of nitrogenous excretion in *Argas (Persicargus) arboreus* Kaiser, Hoogstral & Kohls (Argasidae). *J. Med. Entomol.* 10(1):53–57.

Hamdy, B.H. 1977. Biochemical and physiological studies of certain ticks (Ixodidea). Excretion during ixodid feeding. *J. Med. Entomol.* 14(1):15–18.

Hammen, L. van der. 1966. Studies on Opilioacarida. I. Description of *Opilioacarus texanus* and revised classification of the genera. *Zool. Verh.* 86:1–80.

Hammen, L. van der. 1968. Stray notes on the Acarida. *Zool. Meded.* 42(5):261–280.

Hoogstraal, H. 1956. *African Ixodoidea I. Ticks of the Sudan.* USNAMRU No. 1, Cairo, Bur. Med. Surg. Cairo, pp. 1–1101.

Hughes, A.M. 1961. The mites of stored food. Ministry of Agr. Fish. and Food Tech. Bull., No. 9, pp. 1–287.

Hughes, T.E. 1959. *Mites or the Acari.* Athlone, London.

Hughes, T.E. 1950. The physiology of the alimentary canal of *Tyroglyphus farinae. Q. J. Microsc. Sci.* 91(1):45–61.

Jakeman, L.A.R. 1961. The internal anatomy of the spiny rat mite, *Echinolaelaps echidninus. J. Parasitol.* 47:328–349.

Jeppson, L., H.H. Keifer and E.W. Baker. 1975. *Mites Injurious to Economic Plants.* Univ. of California Press, Berkeley.

Kanungo, K. 1963. Effects of low oxygen tensions on the uptake of water by dehydrated females of the spiny rat mite *Echinolaelaps echidninus. Exp. Parasitol.* 14:263–268.

Kanungo, K. 1964. Disappearance of blood from the gut of engorged *Echinolaelaps echidninus* (Acarina: Laelaptidae). *Ann. Entomol. Soc. Am.* 57(4):427–428.

Knulle, W. 1965a. Die sorptions und transpiration des wasserdampfes bein der Mehlmilbe (*Acarus siro* L.) *Z. Vergl. Physiol.* 49:586–604.

Knulle, W. 1965b. Equilibrium humidities and survival of some tick larvae. *J. Med. Entomol.* 2(4):335–338.

Knulle, W. and G. W. Wharton. 1964. Equilibriumn humidities in arthropods and their significance. *Proc. Int. Congr. Acarol., 1st, 1963,* pp. 299–306.

Lees, A.D. 1946a. The water balance in *Ixodes ricinus* L. and certain other species of ticks. *Parasitology* 37(1/2):1–20.

Lees, A.D. 1946b. Chloride regulation and the function of the coxal gland in ticks. *Parasitology* 37:172–184.

McEnroe, W.D. 1963. The role of the digestive system in water balance of the two-spotted spider mite. *Adv. Acarol.* 1:225–231.

Michael, A.D. 1894. Notes on the Uropodinae. *Urotrachytus formicarium. J. R. Microsc. Soc.,* pp. 289–313.

Mitchell, R.D. 1964. The anatomy of an adult chigger mite *Blankaartia ascuscutellaris* (Walch). *J. Morphol.* 114(3):373–392.

Mitchell, R.D. and M. Nadchatram. 1969. Schizeckenosy: The substitute for defecation in chigger mites. *J. Nat. Hist.* 3:121–124.

Needham, G. R. and P. D. Teal. 1986. Water balance by ticks between blood meals. *In:* Sauer, R.J. and R.J. Hair, eds. *Morphology, Physiology and Behavioral Biology of Ticks.* Ellis Horwood, Ltd., Chichester, England, pp. 100–151.

Nutting, W.B. 1964. Demodicidae—Status and prognosis. *Acarologia* 6(3):441–454.

Saito, Y. 1960. Studies on ixodid ticks. IV. The internal anatomy in each stage of *Haemaphysalis flava. Acta Med. Biol. (Niigata)* 8:189–239.

Tatchell, R.J. 1967. Salivary secretion in the cattle tick as a means of water elimination. *Nature (London)* 213:940–941.

Till, W.M. 1961. A contribution to the anatomy and histology of the brown ear tick *Rhipicephalus appendiculatus. Mem. Entomol. Soc. S. Afr.* 6:1–122.

Vannier, G. 1976. Principaux modes d'étude de la balance hydrique chez les Acariens. *Acarologia* 18(1):3–19.

Vistorin-Theis, G. 1977. Ernährungsbiologie Beobachtungung an Calyptostomiden (Acari: Trombidiformes). *Zool. Anz.* 199(5–6):381–385.

Vitzthum, H.G. 1943. Acarina. *Bronns Kl.* 5:1–1011.

Wharton, G.W. 1963. Equilibrium humidity. *Adv. Acarol.* 1:201–208.

Wharton, G.W. and T.L. Devine. 1968. Exchange of water between a mite, *Laelaps echidninus,* and the surrounding air under equilibrium conditions. *J. Insect Physiol.* 14:1303–1318.

Wharton, G.W. and R.T. Furumizo. 1977. Supracoxal gland secretions as a source of fresh water for Acaridei. *Acarologia* 19(1):112–116.

Wharton, G.W. and K. Kanungo. 1962. Some effects of temperature and relative humidity on water balance in females of the spiny rat mite, *Echinolaelaps echidninus. Ann. Entomol. Soc. Am.* 55:483–492.

Wharton, G.W. and A.G. Richards. 1978. Water vapor exchange kinetics in insects and acarines. *Annu. Rev. Entomol.* 23:309–328.

Whitmoyer, R.E., L.R. Nault and O.E. Bradfute. 1972. Fine structure of *Acaria tulipae. Ann. Entomol. Soc. Am.* 65(1):201–215.

With, C. 1904. The Notostigmata, a new suborder of Acari. *Vidensk. Medd. Dansk Naturh. Foren, Kjobenhavn,* pp. 137–192.

Woodring, J.P. 1973. Comparative morphology, functions and homologies of the coxal glands of oribatid mites. *J. Morphol.* 139:407–430.

Woodring, J.P. and C. A. Galbraith. 1976. The anatomy of the adult uropodid *Fuscouropoda agitans,* with observations on other Acari. *J. Morphol.* 150:19–58.

Young, J.H. 1968. The morphology of *Haemogamasus ambulans.* I. Alimentary Canal (Acarina: Haemogamasidae). *J. Kans. Entomol. Soc.* 41(1):101–107.

11

Nervous System

Natur nihil agit frustra.
—Sir Thomas Browne

The arthropod nervous system comprises supra- and subesophageal ganglia connected by circumesophageal commissures and usually an extended chain of paired ventral ganglia concomitant with segmentation. During ontogeny the formation of these neuromeres (ganglia) is associated with and directed by the coelomeres (Boudreaux, 1979b). The arthropod brain is generally distinguished by three parts (anterior to posterior): the protocerebrum, the deutocerebrum, and the tritocerebrum. The protocerebrum and deutocerebrum are thought to be derived from corresponding parts of the annelid brain. The deutocerebrum innervates the antennae of mandibulates. In the chelicerates, which lack antennae, the deutocerebrum is absent, but the protocerebrum and tritocerebrum remain.

The pattern of the arachnid nervous system is similar to that in other arthropods and embryonically demonstrates ganglia in each segment. The forward concentration of the nervous system during the ontogeny of arachnids,

however, results in a compacted, unsegmented mass in many. Absence of the deutocerebrum is attributed to the lack of antennae and a reduced function of the eyes. The subesophageal mass represents the fusion of segmental ganglia that moved forward during development. The number of ganglia varies in arachnids. Scorpions approach the primitive condition, with nine subesophageal, three prosomatic, and four opisthosomatic ganglia, a total of 18 when the cheliceral and cerebral ganglia are included. Spiders have five major ganglia. In the Pseudoscorpions, Opiliones, and Acari the subesophageal ganglia represent the fused posterior ganglia. No central nerve cord or ganglia are found in the opisthosoma (Savory, 1977). The consolidation of the nervous system into a single mass has proceeded further in mites than in any other group of arthropods (Hanstrom, 1928; Horridge, 1965). No organ system of the mites is more reflective of the loss of segmentation and consolidation than the central nervous system (CNS), even though the fusion of the ganglia is not as complete in larvae and nymphs as in adults. Little is recorded of the development, but embryonically the nervous system of mites comprises two main bands of tissue united by transverse commissures (Andre, 1949).

CENTRAL NERVOUS SYSTEM

The central nervous system of mites is a densely coalesced mass of nervous tissue from which the main nerve trunks extend. Preesophageal and postesophageal parts cannot be distinguished externally and circumesophageal connectives typical of other arthropods are absent (Eichenberger, 1970; Tsvileneva, 1965). (An exception to the compactness is cited for *Lardoglyphus konoi*; see Vijayambika and John (1975).)

The protocerebrum and ventral ganglia form the compacted mass of the central nervous system, the synganglion (or "brain") (Dogel, 1940) that is penetrated by the esophagus. Thus, the esophagus may be a general location marker for the boundary between the two main regions of this podosomatic nerve mass. The supraesophageal part of the synganglion (preesophageal or cephalic brain, Tsvileneva, 1965; Eichenberger, 1970) lies mainly dorsal to the esophagus. The subesophageal ganglion (postesophageal or body brain, loc. cit.) is generally the larger of the two regions and, with splanchnic ganglia, lies ventral to the site of penetration of the esophagus (Fig. 11-1). Neither of these major regions is well-defined externally, and with connecting commissures tightly incorporated or absent, the boundaries are indistinct. The true connectives and commissural links are open to question (Tsvileneva, 1965) (Fig. 11-1B–E). The narrowness of the esophagus is related to the reduction and loss of the commissures between the two areas of the synganglion, but some internal structures described below give indications of the locations of ganglia. (Horridge, 1965, reviews the historical papers related to the general anatomy of the acarine nervous system. The reader is referred to those volumes for details that are not described below.)

The dorsal part of the synganglion (=protocerebrum; =supraesophageal ganglia; =cephalic brain, Tsvileneva, 1965) consists of the paired cerebral, cheliceral, and pedipalpal ganglia and the stomodial pons. Also referred to as the brain (Eichenberger, 1970), the cerebral ganglia innervate the dorsopharyngeal muscles, the hypostome, the epipharynx, the eyes, and if present, sensilla in scutate mites. The chelicerae and cheliceral muscles are innervated from the cheliceral ganglia. Pedipalpal ganglia are located near the entry of the esophagus into the nervous mass and are connected by commissures to the cerebral ganglia. Nerves from the pedipalpal ganglia extend to the pedipalps and to the rostrum (Evans et al., 1961). If present, preoral commissures and postoral commissures are usually small. Optic nerves and optic centers may be present.

The ventral part of the synganglion [=tritocerebrum; =subesophageal–splanchnic ganglia; =body brain, Tsvileneva (1965)] has a dorsoventral zonality. The sensory region is usually ventral; the motor region is dorsal, but these zones are not as distinct as in insects. This part of the synganglion exhibits four pairs of pedal ganglia with nerve trunks that innervate the legs and the podosomatic musculature and opisthosoma (Legendre, 1968). Other nerves serve the gut, the reproductive tract, and additional internal organs of the idiosoma. One or two pairs of splanchnic ganglia (Oribatida and Astigmata) are assumed to represent the fused opisthosomatic ganglia of arachnids and to innervate the digestive and reproductive systems as well as the opisthosomal musculature (Evans et al., 1961). The nerve trunks that extend from the synganglion to the legs have a large number of conspicuous nuclei (Hughes, 1959). Rudiments of all nerves are present in the larval stage of ticks (Eichenberger, 1970).

Woodring and Galbraith (1976) state: "All Acari have four pairs of pedal nerves, a pair each of cheliceral and pedipalpal nerves, and

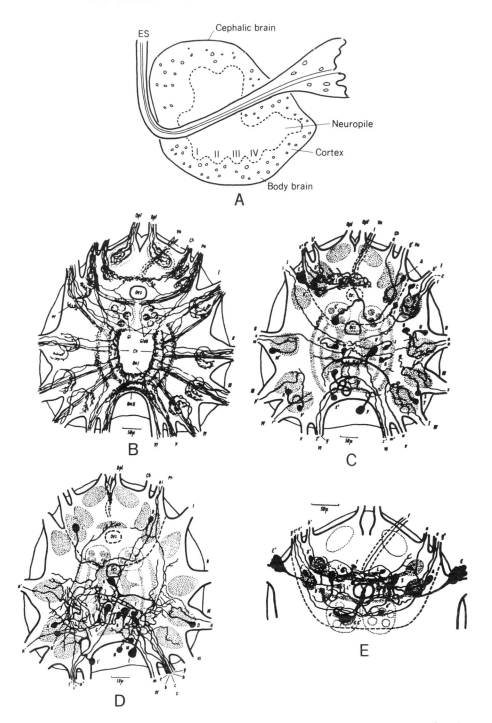

Figure 11-1. (A) Schematic section of the synganglion of *Ixodes hexagonus* showing two main ganglionic masses penetrated by the esophagus (ES); cortex peripheral, neuropile central (after Vitzthum, 1943); diagrams of nervous system of *Hyalomma sp.* (B) sensory neuropile synganglion; (C) ganglionic elements: motor and (part) associative-motor; (D) associative-motor elements of ganglion; (E) ventral region of cephalic brain (after Tsvileneva, 1965).

one to several pairs of stomodeal nerves. The number of additional nerves arising from both the sub- and supraesophageal ganglia, as well as the terminology of these nerves, among the suborders of Acari is highly variable.''

Main nerve trunks are described below, but it is noted that both sensory and motor nerve fibers are found in many of these trunks (Baker and Wharton, 1952). Association-motor neurons (polyvalent), characteristic of all chelicerates, are found in some ticks. Three types of nerve cells are distinguished in *Ornithodoros moubata*: motorneurons, association or control neurons, and neurosecretory cells (Eichenberger, 1970). The simplest reflex arc is bineural (Tsvileneva, 1965; Eichenberger, 1970).

PERIPHERAL NERVES

Most of the peripheral nerves of the synganglion are paired, although Michael (1895) found an unpaired esophageal (recurrent rostral) nerve and a pair of anterior nerves considered to be homologs of rostral nerves in spiders (Handstrom, 1919). In *Bdella* a group of nerve cells lateral to each eye is compared by Lang (1905) to a frontal organ. Most paired nerves are simple, originating in the synganglion and extending to their terminations. Some may have ganglionic swellings at their bases immediately remote from the synganglion. Such swellings are found in *Fuscouropoda agitans* and are considered as auxiliary ganglia (Woodring and Galbraith, 1976).

Little consistency exists in the literature for the names of the peripheral nerves of the synganglion. Apparently some authors have employed different terms for the same nerves, although some mites show enough variation that a particular label may not be completely applicable nor satisfactory. Homologies are difficult to establish at this point, but attempts have been made here to correlate the names and to use a consistent system that should partially reduce the confusion. At least, most comparable synonyms should be apparent. There will

still be some discrepancies, for there is not always agreement on the origin of the nerves and such must be established to identify homologies. Terminology has changed over time with increased investigations of the nervous system and older terms may not be as applicable as when initiated.

The following list of nerves summarizes most of the principal peripheral trunks that extend from the synganglion. Not all mites will have all of the nerves listed. In some instances the nerve trunk is peculiar to a given order of mites (e.g., tritosternal nerve in Gamasida; scutal nerve in *Allothrombium*, Parasitengona; rostral dorsal nerve in *Ceratozetes*, Oribatida). This list should identify the common nerve trunks and the possible synonymic relationships. The plural ending indicates that the nerves are paired; equivalent nerves are identified with an equal sign; nerves with questionable homology have a question mark before them.

?Cerebral nerve

Optic nerves = ocellar nerves, optic accessory nerves

Scutal nerves

Rostral nerve = esophageal nerve

?Dorsal nerve

Cheliceral nerves

Stomatogastric = stomodeal, esophageal

?Epipharyngeal

Pharyngeal nerves

?Hypostomal

Pedipalpal nerves

Tritosternal

Pedal or leg nerves I, II, III, IV

Accessory leg nerves

Ventral nerve = visceral = abdominal = intestinal

Splanchnic nerves

Gonadal = genital nerves

Rectal

Differentiation of the peripheral nerves for the anterior and posterior regions of the synganglion is illustrated by Binnington and Tatchell (1973).

Histology

The synganglion of mites has a histological organization similar to other arachnids. Externally it is covered by the mesodermal neurilemma, a thin sheath of connective tissue composed mainly of collagen fibers (Eichenberger, 1970; neural lamella, Coons et al., 1974). This sheath may have a single layer or several layers of cells. In some ticks homogeneous, finely granular, disorganized material makes up the sheath (Coons et al., 1974), which is thicker in ixodid ticks than in argasids. This external neurilemma covers the synganglion and may also extend over the peripheral nerves.

This covering is thought to act as a metabolic barrier and to have a role in the exchange of neurosecretions. It has been shown to have a special, mediating physiological role in insects, and inasmuch as the dorsal blood sinus in ticks supplies blood rich in nutrients to the central nervous system, it is assumed that the neurilemma similarly participates in the metabolic cycle of the system. The abundant nutrient supply is supplemented by the extensive tracheal network that ramifies through the nervous system and supplies abundant oxygen (as well as some support because of the chitinous structure) (Eichenberger, 1970).

In some instances (e.g., *Macrocheles muscaedomesticae, Ornithodoros moubata, Dermacentor variabilis*) an additional covering of glial cells (neuroglia), called the internal subperineurium, lies beneath the neurilemma (or neural lamella) and around the central core of the neuropile and fiber tracts (Obenchain, 1974b; Coons et al., 1974). Glial cells, nonnervous cells, and nonconductors of impulses are distributed throughout the whole synganglion. Though found in the layer of the subperineurium, they may also lie between the neurons and among the axons of the peripheral nerves. They may

have different forms depending on their location in the nervous system (the cortex or the neuropile), but are seemingly variants of a single basic type of cell (Eichenberger, 1970). They may occur as satellite cells, sheath cells, or act as fascicle border cells (Horridge, 1965).

It is hypothesized that these glial cells assist in the metabolic regulation of neural mechanisms based on the presence of secreted storage materials (e.g., mucopolysaccharides, glycogen) in the trabeculae of these cells. Intracellular spaces, or empty, membrane-bound vacuoles, mitochondria, microtubules, and free ribosomes may also be present (*Amblyomma americanum*, Coons and Axtell, 1971; Coons et al., 1974; Obenchain, 1974a). The glial cells may also function as a barrier to the penetration of ions (Horridge, 1965). They are also found within the neuropile. They originate either from cortex glial cells or from spiderlike glial cells that form within the neuropile. The latter type partially ensheath nerve fibers and surround tracheal elements (Coons et al., 1974).

Structural differences in the perineural glial cells are found between ixodid and argasid ticks. Large intracellular spaces or vacuoles are characteristic of argasids. These spaces are important in the trophic metabolism of the nervous system and are analogous to the glial lacunar system of insects. The nutrients stored in these vacuoles are passed to the neurons beneath the glial cells. Vacuoles and spaces are absent in the ixodid ticks, but large deposits of glycogen are present in the perineural glial cells. These cellular differences between argasid and ixodid ticks may be due to the remarkable differences in their feeding habits, which could in turn influence nutritional requirements and trophic reserves of the neural tissues (Coons et al., 1974). It is likely that these cells have a metabolic relationship with the neurosecretory cells of the nervous system. These latter cells are influenced markedly by the feeding activities of the ticks, so there could be corollary influence. The neurosecretory mechanisms involved in the nervous system will be discussed later.

The synganglion consists of two principal structural layers, the exterior cortex and the medullary interior or neuropile. The exterior cortex (also called the rind or cell rind) of the central nervous system is composed of the somas of nuclear-rich ganglionic cells, but the layer is not of uniform thickness; the rind is made up of many cells in some, fewer in others (Horridge, 1965). It is assumed that the concentration of these cells at specific locations may reflect the possible limits and locations of individual ganglia that have coalesced (Hughes, 1959; Obenchain, 1974a; Sonenshine, 1970; Woodring and Galbraith, 1976). Usually the cortex contains only cell somata, glial, and stem processes, but lacks dendrites, endings, and synapses.

The cell types of the cortex—the neurons—develop from neuroblasts of ectodermal origin. The cell bodies forming the cortex completely surround the neuropile in the preesophageal part of the synganglion in *O. moubata*. These neurons in the postesophageal part are found peripherally, ventrally, and laterally immediately beneath the neurilemma. They usually lie close together in the cortex and extend their processes singly into the neuropile. The cell body of the neuron consists of a nucleus with one or more nucleoli and the surrounding cytoplasm. The size of the nucleus and the amount of chromatin vary greatly. All neurons are unipolar, with a single process that bifurcates distally into an axon (efferent) and dendrites (afferent) that extend into the neuropile or fiber tract (Eichenberger, 1970; Tsvileneva, 1965). In *Acarus siro* the neurons are grouped in apparent ganglionic masses. In *Ceratozetes* they incompletely surround the neuropile (Hughes, 1959; Woodring and Cook, 1962).

Three functional types of neuronal bodies are present in the nervous system of many ticks (e.g., *Dermacentor variabilis; Amblyomma americanum; Argas arboreus*). These are motor or association-motor neurons, neurosecretory cells, and olfactory "globuli" cells in the first pedal ganglia (Coons et al., 1974; Ei-

chenberger, 1970; Ioffe, 1963; Tsvileneva, 1965).[1]

Type 1 neurons are common in all ganglionic centers. They vary in size (8–12 μm) and have large, chromatin-rich nuclei with relatively little cytoplasm. Free ribosomes, mitochondria, rough ER, Golgi, and lysosomes are found in the perikaryia. Some "dark" neurons (like those described for *Macrocheles muscaedomesticae*) are also found (Coons and Axtell, 1971). Type 1 cells correspond to motor or association-motor neurons and are abundant in the preesophageal part of the synganglion (Coons et al., 1974; Eichenberger, 1970). Motor neurons are considerably larger in size (12–16 μm) in the postesophageal part of the synganglion, and even larger (27–35 μm) in the abdominal ganglia. Nuclei have a loose chromatin network and a distinct nucleolus; the cytoplasm is abundant. The capability of some of these cells to conduct several impulses from various sources (i.e., gather impulses) is the basis Tsvileneva (1964) uses for calling them association-motor neurons (Eichenberger, 1970).

Type 2 are neurosecretory cells found among the other neurons. They are among the largest (5–15 μm) of the cells of the central nervous system. When actively secreting they measure 25–40 μm. The nucleus of such cells contains two nucleoli surrounded by loosely structured cytoplasm. Membrane-bound neurosecretory vesicles in the cytoplasm contain electron-dense materials. Mitochondria, rough ER, and lysosomes also are present.

The third type of neuron occurs in paired masses in pedal ganglia I of both argasid and ixodid ticks. These cells are fairly uniform in size and ultrastructure. They measure about 0.6 μm and typically contain fewer mitochron-

[1]The neurons of *Rhipicephalus sanguineus* are classified differently by Chow et al. (1972) as motor and ganglionic cells. The latter include neurosecretory cells, and spherical cells with a regular nucleus. The cells with irregular nuclei are evidently glial cells adjacent to neursecretory bodies (Coons et al., 1974).

dria, rough ER, and free ribosomes. The type 3 neuron is closely packed and exhibits a narrower glial sheath than the other types, or the sheath may be entirely absent.

Beneath the cortex is the medullary interior or neuropile, a tangle of dendrites, axons, and arborization fibers; but relatively little structural detail is known. The neuropile is the sole synaptic field, devoid or nearly so of ganglionic cell bodies (Horridge, 1965). Sensory and motor nerve fibers, glomeruli, and elements connecting the cephalic brain and the body brain are found in ticks. Two principal parts of the neuropile occur in the ixodid synganglion: (1) a preesophageal part connected with the cheliceral and palpal neuromeres (corresponding to the cephalic brain of other arthropods), and (2) a postesophageal section, called the body brain. Both motor and sensory nerves are found, but no association neurons are identified. The motor elements are positioned dorsally and their dendrites connect with various parts of the sensory neuropile in its ventral position. Motor-association neurons are present (Tsvileneva, 1965) (Fig. 11-1B–D).

The cephalic brain of ixodids exhibits signs of reduction of the protocerebrum. The optic ganglia are small and primitive. No mushroom bodies (corpora pedunculata or fungiform bodies) are present. The tritocerebrum has a well-developed stomodeal bridge with numerous links to other parts of the synganglion (Tsvileneva, 1965) (Fig. 11-1B–E).

Other anatomical features and known functions of the synganglion include the glomeruli, neurohemal organs, neurosecretory cells, and the great variety of sensilla found in the mites. Comparatively, much more physiological information on nervous function is available for ticks than for any other group.

In the neuroanatomical sense a glomerulus is a dense, usually spherical, tight knot of neuropile fibers. In *Boophilus microplus* glomeruli can be seen in whole amounts of the synganglion or in cut sections of the same. Both paired and unpaired glomeruli are found (Binnington and Tatchell, 1973). The glomeruli of *Hyalomma* are composed of knots of terminal branchings that are fairly large. They resemble the antennal glomerules of the brain in insects and other arthropods (Tsvileneva, 1965).

Globuli (?a type of glomerulus) may also be present elsewhere in the nervous system. These are clusters of small, unipolar, deeply staining neurons. Characterized by large nuclei and little cytoplasm, these groups of cells are thought to function as areas of higher association (Bullock and Horridge, 1965).

Neuropile tracts and structures resembling glomeruli are also found. Tracts extend from the pedal ganglia to the central neuropile. Transverse fibers suggest the possibility of a postesophageal commissure. Axons that enter the brain with peripheral nerves end in the corresponding ganglia or continue to a horseshoe-shaped central area around the esophagus. Some axons assumed to originate from Haller's organ enter the olfactory glomeruli from pedal nerve I. Other longitudinal tracts are demonstrated by methylene blue preparations (Binnington and Tatchell, 1973) (Fig. 11-2A).

Systematic Review

Holothyrida. Two anteroventral outgrowths on the subesophageal nerve mass, the corpora pedunculata (=organs of Thon), occur in Holothyrida and not in other orders. Their homology and function are not clear. Known as mushroom bodies in other arachnids, the corpora pedunculata have a densely staining neuropile associated with globuli cells. Other structural features of the nervous system in Holothyrida are obscure.

Gamasida. The Gamasida exhibit a generally bilobulate synganglion with the esophagus dividing or marking the region of the dorsal supraesophageal from the ventral subesophageal part (Hughes, 1959) (Fig. 11-2C, D). This bilobulate nature is less pronounced in some than in others.

Except for the dorsal surface of the brain, the neurilemma in *Fuscouropoda agitans* is

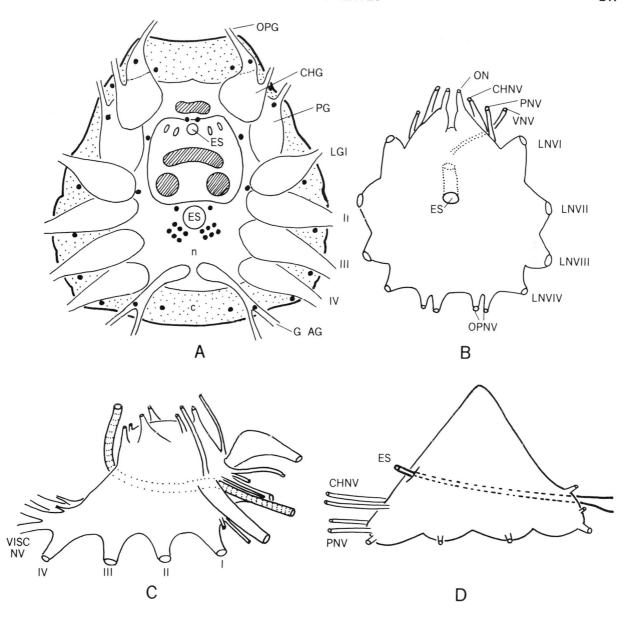

Figure 11-2. Schematic drawing of synganglion of *Boophilus microplus*: (A) view from the dorsum showing cortex (c), neuropile (n), ganglia, glomeruli (shaded), and neurosecretory cell groups (black dots) (CHG, cheliceral ganglion; OPG, optic ganglion; PG, palpal ganglion; ES, esophagus; GAG, genital-abdominal ganglion; LG I, II, III, IV, leg ganglia (after Binnington and Tatchell, 1973); (B) synganglion of *Hyalomma sp.* from the dorsal aspect (CHNV, cheliceral nerve; LNV, leg nerve; ON, optic nerve; OPNV, opisthosomatic nerve) (after Tsvileneva, 1965). Diagrams of synganglia in Gamasida: (C) lateral view of *Echinolaelaps echidninus* (after Jakeman, 1961); (D) lateral view of *Dichrocheles phaelenodectes* (after Treat, 1975) (CH NV, cheliceral nerve; ES, esophagus; PNV, palpal nerve).

hard to distinguish from fat body, coxal glands, and parenchyma tissue that is appressed to the synganglion. Pedipalpal and hypostomatic nerves arise from essentially one anterior root of the supraesophageal ganglion. Distinct oval swellings occur in the bases of the pedipalpal, pedal, and hypostomatic nerves. The swellings have a heavy layer of cortex and a thickened medulla of crisscrossed fibers in the neuropile, indicative of ganglia. (The cheliceral nerves and gonadal nerves lack this type of swelling.) Cross sections indicate weakly distinct glomeruli in the medulla of the synganglion, but they are not as well delimited as those found in ticks. The hypostomatic nerve is thought to be composed mainly of motor fibers. Cheliceral nerves have a thin glial sheath and innervate the internal cheliceral muscles as well as the retractors. A pair of thin, asymmetric stomodeal nerves leaves the posterolateral margin of the supraesophageal ganglion near the exit of the esophagus. This pair of nerves is thought to innervate the ventriculus and caeca.

Four pairs of thick pedal nerves arise from the subesophageal ganglion. One branch of each pedal nerve innervates the coxal muscle. Another main fork extends into the leg. A third branch, probably sensory, extends down to the tarsus. Robust gonadal nerves arise near pedal nerves 4 and are assumed to innervate the gonads as well as the muscles and other organs of the opisthosoma (Woodring and Galbraith, 1976).

The spiny rat mite, *Echinolaelaps echidninus*, has a synganglion from which several pairs of nerve trunks arise. The cortex of cells surrounds a fibrous neuropile composed of clearly defined ganglionic areas delimited by invaginations of the cortical layer and linked by commissures. Definitive nerve trunks extend from each of the ganglionic areas. A neurilemma or sheath covers the brain and is continuous over the nerve trunks (Fig. 11-2C).

Various authors divide the nerves of the synganglion into groups based upon their origin. Those nerves with origin in the supraesophageal ganglia innervate the gnathosoma and ventriculus. Nerves that originate from the posterior and ventral parts of the synganglion (subesophageal and splanchnic) innervate the legs and organs of the opisthosoma (Jakeman, 1961).

In *Dicrocheles phalaenodectes* the principal mass of the nervous system is the synganglionic mass or "brain" anterior to the ventriculus and behind the pharynx. The synganglion is a large, dorsally triangular mass bisected internally by the esophagus. The compact nature of this organ excludes any chain of ganglia and connectives, but fine nerve trunks extend to the appendages of other organs (Fig. 11-2D).

Ixodida. The fusion of the ganglionic elements of the tick nervous system leaves few if any traces of the external segmentation. In general the central nervous systems of argasids and ixodids are similar in structure. The differentiated development and mode of life of the hard ticks suggest that they have a better associated system and that cephalization of the medulla of the abdominal ganglia is more advanced (Eisenberger, 1970). The synganglion is ventral to the digestive tract and dorsal to the reproductive ducts. It lies at about the level of coxae I and II. It is enclosed in a sinus connected to the heart by the aorta. The hemolymph irrigates both the synganglion and the nerves extending from it (Tsvileneva, 1965).

As in other mites, the location of the esophagus divides the nerve mass into a dorsal and anterior supraesophageal part (cephalic brain of Tsvileneva), and a ventral and posterior subesophageal part (body brain, Tsvileneva, 1965). The dorsal part of the brain, which corresponds to the general arthropod encephalon, includes the protocephalon and the optic centers. In the eyeless species these optic centers are greatly reduced or there is an absence of the rudiments of optic ganglia. In addition, the cheliceral ganglia, which are coalesced in ixodid ticks, the stomodeal pons, and the palpal ganglia are included in this part. Paired optic, cheliceral, and pedipalpal nerves extend to their respective organs. An unpaired esophageal nerve from the stomodeal pons innervates the esophagus and pharynx. A network of ramifying tracheae is associated with the synganglion. Periganglionic blood sinuses invest the synganglion and the bases of the main nerves (Balashov, 1972; Eichenberger, 1970).

In ticks the supraesophageal part of the synganglion has paired cerebral ganglia fused in the midline. In ixodids, optic nerves extend to the eyes from these ganglia; such nerves are absent in the argasids. Areae porosae are also innervated from these ganglia. The cheliceral ganglia are ventral and slightly anterior to the supraesophageal ganglion. Nerves from the dorsal side of the cheliceral ganglia innervate both the internal and the retractor muscles of the chelicerae.

In hard ticks, pedipalpal ganglia are nearly level in position with the location of the esophageal entry into the synganglion; these ganglia are somewhat more ventral in some of the soft ticks, such as *Argas persicus,* where they are connected to the cerebral ganglia by short commissures. Inasmuch as these ganglia innervate the palps and the pharyngeal muscles, it is assumed that the nerves are mixed, with both sensory and motor fibers present (Hughes, 1959).

The major part of the synganglion comprises the subesophageal mass. Four pairs of pedal ganglia with essentially a metameric arrangement lie in sequence and give off four pairs of pedal nerves to the legs. The posterior end of the mass is a set of four abdominal ganglia that corresponds to the ventral nerve cord in other arachnids and is without traces of segmentation. Most of the idiosomatic organs are innervated by nerves from this region (Eichenberger, 1970; Balashov, 1972).

In *Dermacentor variabilis* a large nerve associated with the main salivary duct branches into smaller lobular nerves. The neural lamella around these nerves may have glycogen deposits. The individual axons of the lobular nerves are ensheathed by extensions of glial cells that have elongated nuclei (Coons and Roshdy, 1973).

Ticks evidently retain a simple bineural reflex arc because only motor and simple association motor neurons are found in the subesophageal part of the synganglion. Association neurons are absent in this posterior region, but are found within glomeruli in the supraesophageal part of the nerve mass. Glomeruli are present in the ganglia of leg I with connections to Haller's organ. The ganglia thus represent olfactory centers, as well as the location for spatial control of movement (Eichenberger, 1970).

Actinedida. The synganglion of *Tetranychus urticae* is fused into a compact mass as in other mites and surrounds the esophagus, which passes through the thickened part. Location of the synganglion is ventral and the mass is tilted upward at an angle of about 45°. Ventrally the mass is surrounded by the fat body and is also associated with the reniform silk gland and the main tracheal trunks. No external identification of ganglia is possible except by the location and emergence of the peripheral nerves. Sections of the central nerve exhibit ganglia beginning at the anterior apex with the optic ganglia and optic mass. Then in sequence posteriorly are the rostral, cheliceral, pedipalpal, and pedal ganglia, and the terminal abdominal ganglia (Fig. 11-3A–D).

Nerves of the supraesophageal part are both paired and unpaired. The optic and cheliceral nerves are paired. The rostral nerve is single, lies above the esophagus, and innervates the rostral muscles. The single stomodeal nerve arises from the posterior brain surface above the esophagus and extends to the ventriculus. Paired pedipalpal and pedal nerves arise from the narrower, posteroventral part of the synganglion (subesophageal). A single abdominal nerve extends from the posterior median abdominal ganglion and extends along the ventral surface beneath the gonads (Blauvelt, 1945).

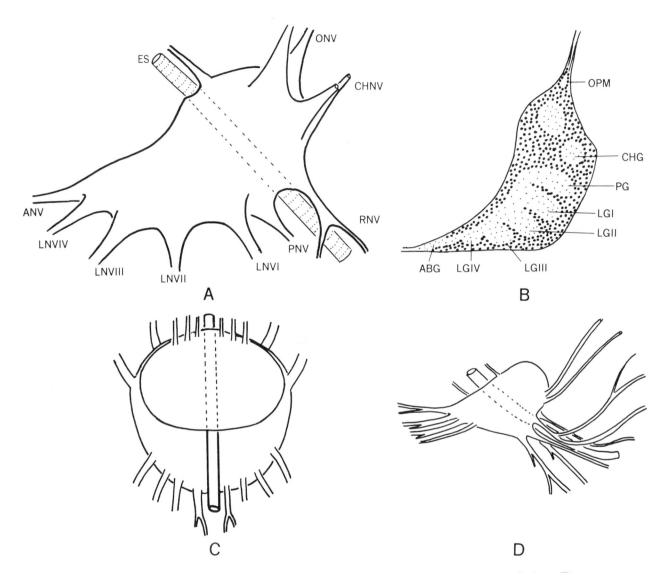

Figure 11-3. Diagram of the synganglion of *Tetranychus telarius*: (A) lateral view; (B) same, diagrammatic section (after Blauvelt, 1945). (C) Dorsal view of synganglion of *Trombidium fuliginosum* (after Michael, 1895); (D) lateral view of synganglion of *Allothrombium lerouxi* (after Moss, 1962) (ABG, abdominal ganglion; ES, esophagus; CENV, cervical nerve; CHG, cheliceral ganglion; CHNV, cheliceral nerve; LG, leg ganglion I, II, III, IV; LNV, leg nerve I, II, III, IV; OPM, optic mass; ONV, optic nerve; PG, palpal ganglion; PNV, pharyngeal nerve; RNV, recurrent nerve; VINV, visceral nerve).

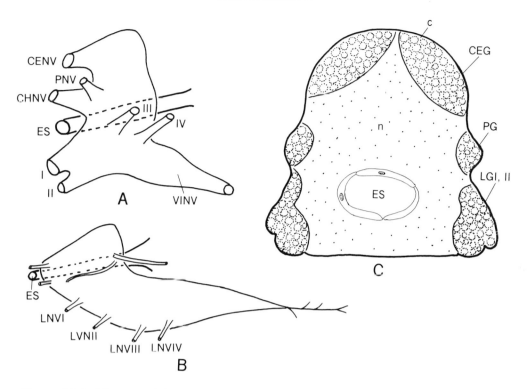

Figure 11-4. Diagram of the synganglia from the lateral aspect in (A) *Acarus siro* (after Hughes, 1959); (B) *Lardoglyphus konoi* (after Vijayambika and John, 1975). (C) Schematic section of synganglion of *Acarus siro* showing cortex (c) and neuropile (n) (after Hughes, 1959) (CEG, cerebral ganglion; CENV, cerebral nerve; CHNV, cheliceral nerve; ES, esophagus; LG, leg ganglion I, II; LNV, leg nerve I, II, III, IV; PG, pharyngeal ganglion).

The nervous system of Eriophyidae is composed of a single neurosynganglion located in the posterior cephalothoracic region. It consists of a central, membranous neuropile surrounded by numerous small nuclei. Some evidence exists "for lateral mesacon processes, particularly in the area of the extremities, but there is little evidence of a well-developed neural tube or canal extending posteriorly from the synganglion" (Jeppson et al., 1975).

Examples of the central nervous system in other mites are found in Figures 11-4 and 11-5.

SENSE ORGANS

Although mites have several kinds of sense organs, the physiological properties of these organs, with some exceptions, are not well known

(TREAT, 1975, P. 52).

Nearly all of the cuticular sensory structures on body and leg surfaces of mites are setalike and usually thigmotaxic, the primary function

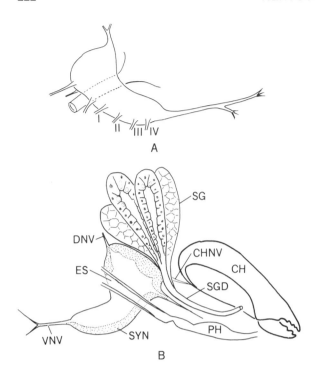

Figure 11-5. Diagrams of synganglion of *Ceratozetes cisalpinus*: (A) from the lateral aspect; (B) partial section with salivary glands (after Woodring and Cook, 1962) (CH, chelicera; CNV, cheliceral nerve; DNV, dorsal nerve; ES, esophagus; PH, pharynx; SG, salivary gland; SGD, salivary duct; SYN, synganglion; VNV, ventral nerve).

ascribed to most setae in the Arthropoda (Baker and Wharton, 1952). Some setae are chemosensory; others are of unknown function. These integumental, setal-type sense organs vary widely among the different acarine families and in some species. Five types are recognized: tactile setae, solenidia, famuli (microsolenidia), eupathidia, and trichobothrial sensilla.

The setae range through a variety of forms from plain and simple hairs to ornate palmate and plumose types. Scales and peg organs are found in a variety of forms. Specialized prosomatic trichobothria distinguish the Oribatida, but trichobothrial counterparts are found in several orders. Special sense organs of the scu-

tum of larval Trombiculidae are puzzling, but very diagnostic. These sensilla emerge, like other trichobothria (which they may well be), from pitlike or spheroid subcuticular receptacles and are innervated at the base. Flattened discs resembling the campaniform organs of other arthropods are called *sensilla auriforme*. Curious "erechthaesthetes" occur over the dorsal surface of water mites and the many neurons in these sensilla have distal processes that end in the terminal wall of small projections. Flattened sensory plates also occur in the genitalia of water mites (Horridge, 1965).

The tarsi of many mites are equipped with distinctive sensilla and hairs whose functions are not fully known or understood. In *Macrocheles muscaedomesticae*, a predator of eggs and first stage larvae of muscoid flies, there are distinctive setae and sensilla on tarsus I. The sensilla are distinguished as two types, a pointed form and a blunt, peglike form (sensilla basiconica). The number, size, and placement of these blunt pegs change during embryonic development as well as in stadial growth. Larvae have six, protonymphs have seven, and deutonymphs and adults have eight. There are porous and nonporous forms of the peglike sensilla, each with dendritic innervation in the lumen. The porous sensilla are similar to those found in Haller's organ of ticks. Olfactory and thigmotactic functions are ascribed to the porous and nonporous sensilla, respectively, although the latter lack a tubular base typical of mechanoreceptors. Amputation and behavioral experiments are the basis for these inferences (Coons and Axtell, 1973).

Setae on the legs of ticks are of two types. Long, slender, tactile hairs are found on the ventral aspects of the three distal segments of all the legs. These setae are pointed toward the tarsus and pulvillus. Each of these hairs has a fine, sensory process in the internal canal. Numbers of these tactile setae decline progressively from leg I to leg IV. Neither trichogen nor tormogen cells are visible. A second type of seta is present. These are short, straight hairs and are mainly dorsal and lateral in location on the legs. They differ little from the longer tactile setae except in size (Hess and Vlimant, 1986; Hirschmann, 1979; Lees, 1948).

In *Erythraeus,* a predaceous prostigmatid mite, long, spindle-shaped cells are attached to intersegmental membranes of the legs and are thought to be proprioceptors (Lees, 1948).

Haller's Organ

One of the most unusual of the cuticular sense organs in mites is Haller's organ, described originally by its author for a species of *Ixodes* near the close of the nineteenth century. It normally is found on the dorsal tarsus of legs I.

In *Opilioacarus segmentatus* the organ is singular and small, but well-developed in the surface of telotarsus I. In this primitive mite the organ contains two sensitive hairs or sensilla, the first of which is bacilliform (e) and partially extended from a pit. The second sensillum, Haller's organ (h), is claviform and is enclosed in a subdermal pouch or capsule (Grandjean, 1936) (Fig. 11-6). It is known in only one other order of the Acari, the Ixodida. Haller's organ has been compared (by Vachon) to tarsal organs in pseudoscorpions but is quite different (Arthur, 1956). Homologs, if present elsewhere in the mites, are not understood.

When a tick is in a questing position the

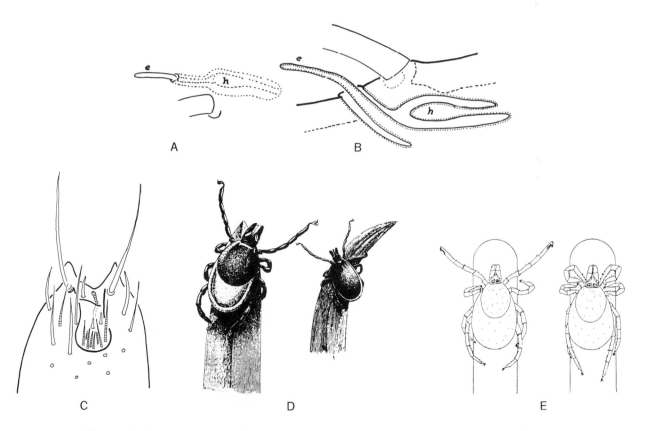

Figure 11-6. Diagram of Haller's organ in *Opilioacarus segmentatus* With: (A) from the dorsal view; (B) schematic section of the same (e, bacilliform sensillum; h, Haller's organ) (after Grandjean, 1936); (C) Haller's organ of *Holothyrus grandjeani* (after Hammen, 1961); (D) questing posture of male and female *Ixodes persulcatus* (after Pomerantzev, 1959); (E) unfed female in questing posture and at rest (after Lees, 1948).

front legs are held motionless aloft or are waved in the air at intervals for detection of the host. Even when walking the tick lifts its first legs above the substrate, for these legs are not usually involved in ambulation (Fig. 11-6D, E). Originally, Haller's organ was assumed to be an auditory organ equipped with otoliths. This idea was discarded for olfaction because both odor and body warmth of the hosts elicit directional and nondirectional responses by ticks (Lees, 1948; he reviews the history of these changes). More recent research shows that the distal pit of Haller's organ in ticks is a humidity receptor and the proximal capsule is olfactory in function. Mechanoreception is also noted in combination with chemoreception: Some ixodid ticks remain in ground burrows and respond to the vibrations of approaching host animals (camels, sheep), then crawl out to attack them (Foelix and Axtell, 1971, 1972; Zolotarev and Sinitsyna, 1965).

All genera of ticks possess a Haller's organ on the anterior dorsum of tarsus I, but differences occur from genus to genus and between species (Fig. 11-7C, D). Surface features, shape of the trough, variable closure of the capsule, types and location of sensilla, and other structural aspects of the organ are useful in systematic differentiation of species (Homsher and Sonenshine, 1975, 1977, 1979; Balashov and Leonovich, 1978; Roshdy et al., 1972; Woolley, 1972). Species of the genus *Ixodes,* for example, are differentiated by the presence of a closed or open capsule in the organ and the morphological variations are correlated phylogenetically with the nest-burrow parasitism typical of ixodine ticks (Balashov and Leonovich, 1978).

Figure 11-7. Schematic sections of Haller's organ to show the anterior pit and posterior capsule in (A) *Dermacentor* (after Arthur, 1960); (B) *Argas vulgaris* female (after Balashov, 1972); (C) Haller's organ of *Argas persicus* (SEM); (D) same of *Boophilus annulatus* (SEM); (E) semidiagrammatic representation of seven anterior pit sensilla (after Foelix and Axtell, 1972).

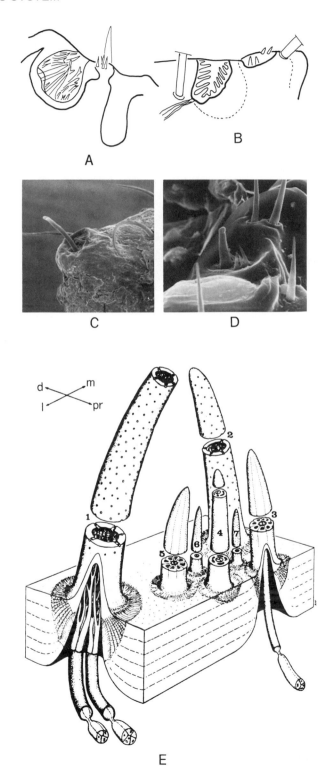

Variations between Haller's organs also differ in tick species common to desert and forest habitats (Zolotarev and Sinitsyna, 1965).

Investigations show that Haller's organ has a composite nature. In the majority of species two main sensory regions are present, each with characteristic sensilla. The first region consists of an anterior (distal) pit or trough in which sensilla basiconica are located. The shape of these sensilla is distinctive (Lees, 1948) and there are at least three types present (Arthur, 1956; Foelix and Axtell, 1972). The second sensory area is a posterior (proximal) capsule (or vesicle) within which are five to seven distinctively curved, bluntly pointed sensilla (four in the larvae) (Arthur, 1956). In some species the dorsal sensillum penetrates the opening of the capsule. In others it is straight or only slightly curved and may not emerge from the capsule (Arthur, 1956). Sensory cells underlie sensilla in both the pit and the capsule with dendrites from these cells extended into each sensillum (Lees, 1948; Arthur, 1956, 1962; Foelix and Axtell, 1972) (Fig. 11-6E).

From a dorsal view the surface of tarsus I shows four sets of setae in association with the pit (a) and the capsule (c) of Haller's organ. These are (1) distal (d, anterior to the pit), (2) lateral (l, at the sides of the capsule), (3) posterior (p, immediately proximal to the capsule), (4) medial (m, a transverse cluster of setae between the posterior setae and the proximal setae), and (5) the pair of proximal setae (p) (Axtell et al., 1973) (Fig. 11-8). (The posterior and medial setae constitute the "posterior bristle group" of Lees, 1948.)

Seven sensilla are typical of the anterior pit of Ixodidae in contrast to four in Argasidae. The lateral, posterior, medial, and proximal setae are in about the same relative positions in both families (Arthur, 1962; Roshdy et al., 1972). All sensilla of the anterior pit have thick cuticular walls and multiple innervation (Fig. 11-6E).

Studies of the ultrastructure of the sensilla show four types. The largest (d1, a1, a2,) exhibit thinner cuticular walls, but prominent cu-

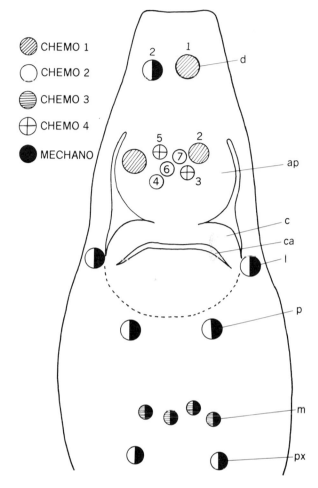

Figure 11-8. Schematic representation of the sensilla associated with Haller's organ of *Amblyomma americanum* and their functions (chemoreception, mechanoreception) (after Axtell et al., 1973) (ap, anterior pit; c, capsule; ca, capsule aperture; l, lateral; m, medial; p, posterior; px, proximal).

ticular pores 1200 Å in diameter. Each pore in the side of the sensillum contains a "plug." These sensilla are referred to as "large, plug-pore sensilla" (Foelix, 1972). Such plugged pores in sensilla are not reported for any other arthropods. A second type of small-pore sensillum (a3, a5) has longitudinal grooves on its surface. These pores are 100–200 Å in diameter and the sensilla resemble a spoke-wheel in cross section. They are similar to the sensilla coelo-conica on the antennae of grasshoppers. Size and structure of the pores is distinctive for a specific sensillum. Both types of pores could be assumed to protect against water loss or to act as valves (Fig. 11-9) (Axtell et al., 1973; Foelix, 1972; Foelix and Axtell, 1971, 1972). The third type of sensillum (a4, a7, a6) has a smooth surface without pores in the sides or the tip. Finely convoluted canals are found in the walls (*Argas (P.) arboreus*). A fourth type includes only the medial setae. These have small, internal, vase-shaped channels with slitlike apertures on the external surface of the sensillum. Because the sensilla of the pit are diversified, it is assumed that they are able to function as chemorecep-tors, hygroreceptors, and thermoreceptors, but electrophysiological proof is necessary to deter-mine specific functions for a given sensillum. Mechanoreceptive dendrites are absent even though the sensilla exhibit sockets (Foelix and Axtell, 1972) (Fig. 11-8).

It is not known if the plugs in the pores are movable, but the distance between the plug and the pore remains the same in all of the mi-crographs studied. Would the gap increase in ticks exposed to low humidities for long periods of time? We cannot tell at this point, but it is assumed that the material that suspends the plug is permeable to odor molecules because dyes penetrate. The dendrites in the fluid-filled lumen of the sensilla have endings that appose the inner tips of the plugs. It is not known, how-ever, whether diffusing molecules affect these dendritic endings directly or pass through the dendritic membranes to elicit response. The ex-terior coating of the sensilla appears to be anal-ogous to the mucus covering of the olfactory epithelia of vertebrates and to the covering of tarsal sensilla in *Macrocheles muscaedomesti-cae* (Foelix and Axtell, 1972).

The capsule may be closed over the top with just a simple slit opening, or have a more elabo-rate aperture, with fimbriated margin or stel-late-shaped opening, depending on the genera and species. It is not known if there is a closure mechanism involved. The capsule or vesicle is usually a spheroid depression in the tarsus. In-side the capsule are porous, olfactory sensilla (sensilla basiconica) as well as a number of non-sensory cuticular projections (pleomorphic se-tae). The thin-walled, porous capsular sensilla in ticks (and palpal sensilla in *Macrocheles muscaedomesticae*) resemble the olfactory sen-silla of insects (Slifer, 1970).

Seven elongated, blunt-tipped sensilla ex-tend from the basilar capsule wall and are di-rected toward the opening. These sensilla are unlike the tarsal bristles inasmuch as they lack sockets and are thin-walled, in contrast to the sensilla of the anterior pit with thickened cutic-ular walls. Again, as in the sensilla of the pit, the capsular sensilla are perforated and with plugged pores. Each lens-shaped plug is sus-pended within the pore by fibrous material (Fig. 11-9A, B).

Multiple innervation of the capsular sensilla is typical. The three to five dendrites in each sensillum extend from the proximal dendritic sheath (scolopale) into many distal branches (Fig. 11-8). Each branch has a microtubule in its center and is often apposed to a cuticular pore-plug. Nonsensory cell processes occupy the periphery of the lumen of the sensillum, but these processes do not possess the microtu-bules characteristic of the dendrites.

Dendrites of the sensilla exhibit a remark-able arrangement of the microtubules or ciliary structure. Receptors of vertebrates and inver-tebrates show ciliary structure of nine periph-eral double tubules, 9 + 2 or 9 + 0. Most ar-thropods show the 9 + 0 arrangement. In most ticks the sensilla show 11 peripheral double tu-bules without central elements, 11 + 0, al-though variations occur (9 to 12). *Argas* ticks have a 9 + 0; *Amblyomma* exhibit the 11 + 0 (Axtell et al., 1973; Foelix and Axtell, 1971; Roshdy et al., 1972).

The mix of chemo- and mechanoreceptors on the first tarsus of *Amblyomma americanum* has been identified by means of electron mi-croscopy (Axtell, 1979; Chu-Wang and Axtell, 1974). Four basic types of chemoreceptors oc-cur with differences in the walls and interior.

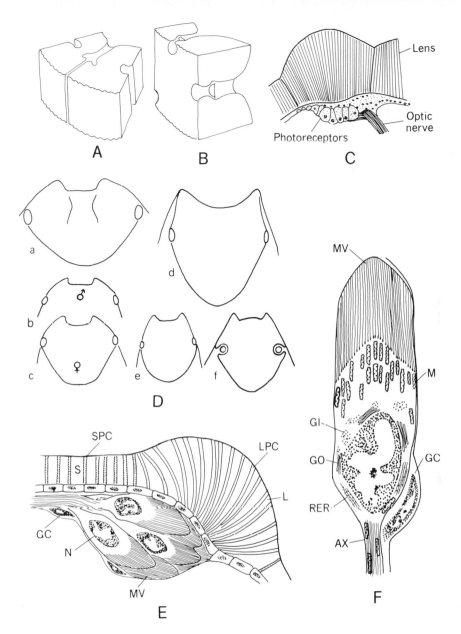

Figure 11-9. Types of pores in sensilla of Haller's organ: (A) without plug; (B) with conical plug in lumen of pore (after Foelix and Axtell, 1971). (C) Schematic section of eye of *Hyalomma sp.* showing lens, photoreceptor cells, and optic nerve (after Balashov, 1972); (D) eyes of Ixodidae, *Amblyomma americanum*: a, nymph; b, male; c, female; d, *Boophilus annulatus*; e, *Rhipicephalus sanguineus*; f, *Hyalomma asiaticum* (after various sources and Balashov, 1972); (E) schematic diagram of eye of *Amblyomma americanum* showing L, lens; LPC, lenticular pore canals; S, scutum; SPC, scutal pore canals; photoreceptor neurons with distal microvilli (MV), GC, glia cell; H, hypodermis; (F) photoreceptor neuron enlarged showing distal microvilli (MV) (M, mitochondria; GO, Golgi; Gl, glycogen; RER, rough ER; N, nucleus; A, axon; and adjacent glial cell (GC) (after Phillis and Cromroy. Reprinted with permission of the Editor, from the *Journal of Medical Entomology*, 13:685–698, 1977).

Mechanoreceptors are innervated usually with two dendrites; some show the presence of a tubular body. Some sensilla seem to have a combined function of contact chemoreception and mechanoreception (Fig. 11-8).

Beneath the capsule in the hemolymphic space is a large multicellular gland with presumed openings into the capsule cavity (*Amblyomma americanum*). Septate desmosomes connect adjacent gland cells. Occasionally a neurosecretory-type axon is enclosed within the basement membrane of the gland. The apical borders of the cells show numerous microvilli and a fibrous surface coating. Rough ER, Golgi, mitochondria, ribosomes, inclusions of lipids, and rosettes of glycogen characterize the cytoplasm of these cells. The glandular secretion is granular and homogeneous (Foelix and Axtell, 1972). Since the glandular apparatus opens into the capsule it is thought to compensate for fluctuating humidity and thus provide stability of the moisture level similar to such regulation in some insects (Leonovich, 1978).

Ocelli

All lateral eyes in the Arachnida are simple lens eyes with a cup-shaped retina. These differ in function from the faceted eyes of other arthropods in that light passes through the usually convex lens to different parts of the cuplike retina, depending on the angle of incidence. "It is possible that the precursor of the Pseudoscorpions, + Solifugae, and Opiliones + Acari + Ricinulei had only two pairs of lateral eyes, which are reduced in Opiliones, Ricinulei and Acari. It is difficult to tell about the mono- or polyphyletic origin of Acari (Hammen, 1970, 1973) . . . In some Acari this pair (median eyes) is fused into one" (Paulus, 1979). (The median eye is also called the frontal organ in water mites.)

Compound eyes are not found in the Acari, but ocelli are present in many mites. These are non-image-forming photoreceptors that re-

spond to light intensity. There may be a single median ocellus present, or one or two pairs of propodosomatic eyes. Water mites exhibit the most complex arrangement of ocelli among the acarines and may have a median and two pairs of propodosomatic eyes. Even when eyes are absent, mites respond to light intensity, usually by means of translucent integument over the synganglion (*Hydrozetes*, Oribatida). Changes in light intensity are detected directly by the brain and influence behavior. The simplest ocellus consists of a mass of pigment granules in association with the optic nerve. Pigments range in color from red to black, purple, and blue. A more complex eye will have a lenslike covering over the pigment or above the photoreceptive cells (Baker and Wharton, 1952).

Eyes in the Acari seem to have been influenced by regressive evolution. Eyes are said to be completely lacking in the Gamasida. The maximal development of eyes is in Endeostigmata and other prostigmates (Coineau, 1970). While the archetypal acarine had six eyes (Grandjean, 1958), in the Caeculidae, a good example of extant forms, the eyes are arranged in three groups, the lateral eyes and anterior eyes. Four lateral eyes in two pairs are present on each side in prominent tubercles near legs II. The third pair of eyes, which are considered primitive, are located anteriorly under the *naso*, or frontal protuberance (=epivertex). The naso is particularly well developed in primitive oribatids (Palaeacaroidea, Brachychthoniidae, Archeonothridae, and some Endeostigmata and Actinedida).

Coineau (1970) cites two examples of primitive anterior eyes, one in a spherical dome (Terpnacaridae, Sphaerolicidae), and the other in a reduced convexity under the naso (Pachygnathidae, Alicorhagidae). Two types occur in the Caeculidae, one with a single and one with a double ocellus.

Ocelli occur in many families of mites from the most primitive to the higher, more complex forms. Two pairs of ocelli occur in the very primitive Opilioacarida, but are entirely absent in the Holothyrida. The Gamasida also lack ocelli. Both major families in the Ixodida have ocelli, although they are more prevalent in the Ixodidae than in the Argasidae. One or two pairs of ocelli are present in some ixodid species and some are set in "sockets" (Fig. 11-9D). In others, a simple cuticular lens overlies the photoreceptor cells connected to the optic nerve (Fig. 11-9C) (Balashov, 1972). The scutal eyes in ticks are generally small, with unpigmented, convex lenses of hyaline integument superimposed above the hypodermis (Arthur, 1960). Located on the lateral margins of the scutum at the dorsum of the podosoma, the eyes are unlike those of other arthropod classes where the eyes are cephalic in position. Said to be "aberrant" and "improbable" as "efficient sense organs" (Horridge, 1965) they are neither aberrant nor improbable in function. They differ microanatomically from other arthropod eyes because of their simplicity and yet are efficient photoreceptors (Phillis and Cromroy, 1977).

The eye of *Amblyomma americanum* exhibits a slightly biconvex lens, well-developed photoreceptor neurons, and an optic nerve that connects directly to the optic lobes of the synganglion. The corneal lens of the eye is unusual because of what authors have called "striae" or "canals." As disclosed by electron microscopy, the microanatomy of the lens of this tick shows lenticular pore canals and microvilli in an arrangement that makes this eye unique. This is the first instance of an arhabdomate eye, wherein the lens has lenticular pore canals; all other arthropods exhibit eye lenses without these canals. This is a completely different adaptation of cuticular pore canals in the arthropods (Phillis and Cromroy, 1977) (Fig. 11-9E, F).

The eye of *A. americanum* is composed of 30–40 unipolar photoreceptor neurons in all stages (larva, nymph, adult), which persist as the eye enlarges during growth. The larval eye (with 20–30 neurons) is about one-fourth the size of the adult eye.

The lenticular pore canals converge in the area above the photoreceptor cells so that the lines inscribed on the longitudinal axis of the pore canals converge at a point in the microvillar region of the eye. The cuticular lens is roughly biconvex and the internal bulge of the lens is always proximal to the microvilli of the photoreceptor neurons. The pore canals are always perpendicular to the long axis of the photoneurons (cf. pore canals of scutum oriented in a dorsoventral axis). The lenticular canals (larger than the scutellar) exhibit a curvilinear path and are larger.

The photoneurons are connected to the optic lobes by a single neurite. These cells do not vary in size regardless of sex, age, or life stage of the tick.

Each photoneuron (retinular cell) is tipped with 7000 to 13 000 microvilli oriented perpendicularly to the path of light. The microvillar membrane is dome-shaped and the microvilli are independent and free within the glial investment. Intermediate cytoplasm lies between the microvilli and an area of mitochondria and intracellar channels. This is an area of intense metabolic activity and energy utilization, as noted in glycogen deposits and mitochondria. Coated vesicles, Golgi apparatus, and rough ER lie between the area of the microvilli and the nucleus. The axon of the neurite extends beyond the nucleus and has mitochondria. A glial sheath surrounds the base of the cell. The presence of microvilli distinguishes this eye as a rhabdomeric photoreceptor (Phillis and Cromroy, 1977) (Fig. 11-9F).

Many families in the Actinedida and water mites exhibit ocelli. An unusual dorsal location on the scutum of chigger mites is observed. *Peltoculus bobbianae* (Fig. 11-10C) illustrates this condition (Brennan, 1972). Trombiculid eyes generally have an elliptical lens, a discoidal mass of oily, red-pigmented material, and a dark-pigmented cup in the anterior eye of each ocular area (Jones, 1950) (Fig. 11-10D).

A few examples of ocelli are also found in the Astigmata and Oribatida. One of the more primitive oribatids, *Heterochthonius gibbus,* has three pigmented eyes, two lateral and one anteromedian. The latter is possibly a fused pair. This is quite different from the other Cosmochthoniidae (Grandjean, 1928).

Unpigmented lateral eyes are found in two places in another oribatid, *Eobrachychthonius sp* (Fig. 11-10E), namely, anteriorly at the base of the rostral hairs and laterally near the exobothridial setae. More or less distinct traces of unpigmented lateral eyes can be seen in other genera of the Brachychthoniidae (Trave, 1968).

The list below identifies most of the mite families that have one or two pairs of ocelli. Fused ocelli, ocular shields, ocular plates, and eye capsules are auxiliary features of some.

One Pair of Ocelli	Two Pairs of Ocelli
ACTINEDIDA	
Sphaerolichidae	Caeculidae
Labidostommatidae (+ pustule)	Adamystidae
	Bdellidae
Penthalodidae	Tenuipalpidae

Halacaridae (in ocular sheath)
Caligonellidae
Eupallopsellidae
Stigmaeidae
Tuckerellidae
Anystidae
Teneriffiidae
Erythraeidae
Trombiculidae (in scutum)

Tetranychidae
Cryptognathidae
Camerobiidae
Barbutiidae
Chyzeridae
Trombidiidae
Johnstonianidae
Eylaidae (ocular shield)
Hydryphantidae (ocular shield, eye capsule)

ASTIGMATA

Saproglyphidae

ORIBATIDA

Cosmochthoniidae (one fused)
Brachychthoniidae

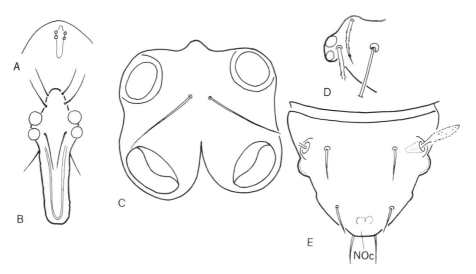

Figure 11-10. Ocelli of water mites: (A) ocelli on ocular plate of female *Limnochares crineta*; (B) enlarged view of ocular plate; (C) ocular plate and eye capsules of male *Rhyncheylais connexa* (after Cook, 1974); (D) ocelli of a trombiculid (*Peltoculus lobbianae*) (after Brennan, 1972); (E) naso ocelli (NOc) and lateral ocelli of Eobrachychthonius sp. (after Trave, 1968).

Palpal Setae

SEM studies confirm that no blunt, peglike sensilla are present on the palps of *Macrocheles muscaedomesticae*. The tined apotelic seta typical of gamasid mites is not set in a depression and is not innervated, but has a muscle and is thus movable. The other setae retain their spatial arrangement on the palp from the protonymph through the adult stadia. Evidence suggests that the palps are not involved in chemoreception in these mites (Coons and Axtell, 1973).

The palpal sensilla of ticks are located in a palpal organ on the ventral, distal tip of the fourth or ultimate segment of the palp. The sensilla are clustered in such a way that their arrangement appears diagnostic for the species involved. The sensilla are of the chemoreceptor type and resemble those of the capsule of Haller's organ, set in a small area of the cuticle, and innervated by sense cells beneath the cuticle. Other types of bristles, tactile setae, and short, straight hairs are also present (Lees, 1948) (Fig. 11-11A-D, SEM).

Other Integumental Structures

Some integumental structures in ticks with alleged sensory functions are ductlike and sometimes have complicated shapes. Such sensilla have been divided into "trichoid" and "tuft organs." The trichoid forms are thought to be proprioceptors and respond to cuticular strains. The tuft organs are presumed to be chemoreceptors and mechanoreceptors as well as being secretory (Arthur, 1960).

Areae porosae are thought to be sensory. Areae porosae are oval cribriform regions of the posterior dorsum of the basis capitulum in female ticks. They appear to comprise a compact system of ducts and are innervated (Douglas, 1943) and may function in pheromonal communication (Fig. 11-11E SEM).

Dorsal foveae are similar in structure to porose areas and are regarded as sense organs in

Dermacentor (Douglas, 1943), but are absent in *Ixodes* (Lees, 1948).

Peg organs found in mites are usually thin-walled and of two general types: (1) blunt or pointed on the surface, or (2) sunk in pits (*Ixodes*) (Horridge, 1965).

Three types of sensilla occur in the harvest mite, *Trombicula autumnalis*: (1) tactile sensilla, both plain and plumose, (2) peg organs, and (3) minute sensory rods confined to the first leg (Jones, 1950) (Fig. 11-11G).

Trichobothria

Bothrionotic sensilla (cuticular pits from which a hair emerges) are known in many arachnids, but the best examples found in the mites are the trichobothria of the Oribatida. Each of these paired organs on the propodosoma has a bothridium with internal helical curvatures and varied emergent hairs (Dahl, 1911). In other mites trichobothria are found on the idiosoma, palps, and legs. In *Bujobia* a simple hairlike trichobothrium is found on tibia I behind the solenidion (gamma). Legs IV of *Cyta latirostris* and *Trachymolgus* have trichobothria. Actinotrichid mites exhibit a core of actinochitin. The formation of the hair and acquisition of its function are related to the varied placement on the body (Figs. 11-11F; 11-12A-C).

Trichobothria are thought to respond to delicate movements of air, each sensillum oscillating in a single plane. A set of them determines the direction of a disturbance in the air and facilitates orientation to air currents (Pauly, 1956; Tarman, 1961). Astrotactic responses (to the light of the sun and moon) is also postulated (Avory, 1977, p. 26).

Slit Sense Organs—Lyrifissures

Slit sense organs (=lyriform organs or lyrifissures) are the most plentiful and enigmatic of the sense organs of arachnids (Barth and Seyfarth, 1972). They exist as single or compound organs on the surface of the body and legs (Savory, 1977). A single spider may have 3000 slit

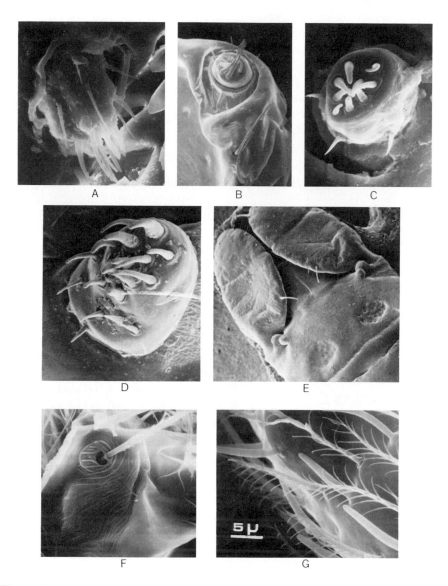

Figure 11-11. Palpal sensilla of ticks (SEMs): (A) *Haemaphysalis leporispalustris*; (B, C) *Amblyomma americanum*; (D) *Ixodes kingi*. (E) Areae porosae in *Ixodes kingi*; sensilla in *Eutrombicula splendens*: (F) sensillum and hairs; (G) plumose hairs and sensilla of legs (SEMs).

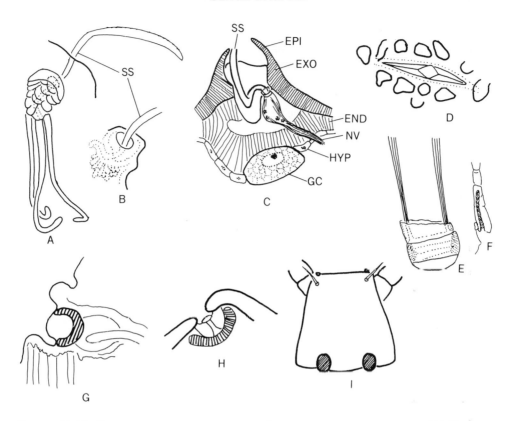

Figure 11-12. Sense organs. Trichobothria of (A) *Steganacarus magnum*; (B) *Nothrus sylvestris* (after Grandjean, 1934). (C) Schematic section of *Ceratoppia bipilis* (after Tarman, 1961); (D) lyrifissure of *Opilioacarus segmentatus* With (after Grandjean, 1936); (E) Claparede organ of *Damaeus onustus* (after Grandjean, 1955); (F) organ at base of leg I in larval oribatid (after Grandjean, 1946) (END, endocuticle; EP, epicuticle; EXO, exocuticle; GC, gland cell; HYP, hypodermis; NV, nerve; SS, sensillum); (G) campaniform organs in *A. Tydeus sp.*; (H) same in *Oppia coloradensis* Woolley (after Brody, 1969); (I) ampulliform organs in Heterocheylidae (after Krantz, 1978).

sense organs on the body surface. The functions of the lyrifissures are not understood, but they are postulated to serve as auditory organs, chemoreceptors, thermal receptors, hygroreceptors, or mechanoreceptors. With regard to the last function, they may be used in orientation, directional movements, position (kinesthetic), and ?proprioceptors, as well as in balance. In spiders these organs have been shown to respond to vibrations of the substrate, sounds in the air, movements of the body, and walking action (Savory, 1977). They may also function similarly to campaniform organs in insects (Pringle, 1955).

Many lyrifissures are found in transverse rows in the cuticle of Opilioacarids. They appear to be segmentally arranged. Holothyroida exhibit three to four pairs on the dorsum and others are lateral in position. Gamasines have lyrifissures in characteristic locations on the chelicerae and the sternal plate. These slitlike structures are found on legs of many mites (Grandjean, 1936; Figure 11-12D).

Claparede's Organ

Claparede's organ, or urstigma, is a small sense organ that is typical of the larval stage of some Actinedida and Oribatida. This small, usually peglike organ is situated at the base of coxae I. It is without a condyle or fossa at its base so it does not turn, but may be extended or retracted by tendons attached to its base. It varies in size within the Oribatida and is especially large in the inferior oribatids. It is an important sensory organ (Grandjean, 1955; Fig. 11-12E, F). The function is uncertain, but G.W. Krantz (personal communication) indicates that Alberti considers the organ osmoregulatory, as are the genital acetabula with which this organ is homologous.

Other Sense Organs

Brody (1969) found campaniform sensilla in the external bumps of the integument of *Tydeus*. These were thought to be sensitive to air pressure or to vibrations, or perhaps were responsive to the flexing of the cuticle at the surface. Each of these receptors consisted of a small dome, a thin membrane, and a nerve fiber (Fig. 11-12G). Another type of campaniform sensillum was found in the folds of the exocuticle of the oribatid *Oppia coloradensis* (Fig. 11-12H).

Ampulliform organs are found in the females of the Heterocheylidae (Fig. 11-12I). Located at the posterior margin of the propodosoma, these spherical structures are of unknown function. Immatures and males exhibit normal scapular setae in these locations. Heterocheylid mites are found under the elytra of passalid beetles. Whether they are parasitic has not been determined, but these organs may be related to their habit.

There may be many other sensory structures associated with the integument that are not delineated completely (George, 1963). Special pores, pits, clefts, and depressions exist in the integument and are innervated. Just how they function is not known, but they may be special sense organs (Baker and Wharton, 1952).

Biochemistry of Nerve Function

The discovery of acetylcholine (Ach), its enzymatic hydrolysis by cholinesterase (ChE), and its synthesis by choline acetylase (ChA) are examples of advances in the biochemistry of nervous function. Referred to generally as the "cholinergic system," the chemical steps involved explain the function of neurons and the synaptic responses associated with impulses. Although pharmacological and biophysical studies are taken to determine the functions of the system, biochemical investigations have provided most of the information available. Mehrota (1961) established the presence of Ach in *Tetranychus urticae* Koch (25 μg/g of mite), which is about 10 times greater than the amount reported for nervous tissue in vertebrates. Ach has not been reported for other species of mites. Cholinesterase (ChE) was reported in *Acarus siro* (Casida, 1955), but was not demonstrated at first in the two-spotted spider mite. Later investigations established the presence of this chemical (Mehrota, 1963; McEnroe, 1960). Choline acetylase (ChA) is found in *T. urticae* (Mehrotra, 1963).

All three components of a cholinergic system (ACh, ChE, ChA) are demonstrated in the Acari. Just how far-reaching these cholinergic materials are in nerve impulse transmission is unknown. Biochemical, biophysical, and pharmacological methods may extend the information available and disclose additional details about other transmitter substances (e.g., epinephrine, aminolatyric acid) in mites (Mehrotra, 1963).

NEUROSECRETION

A neurosecretory system comprises a group or groups of neuroglandular or neurosecretory cells that elaborate neurohormones (cf. neurohormonal or neurotransmitter substances released from nerve cell synapses). Functionally, neurosecretion is the elaboration or production of visible droplets of materials within the cytoplasm of neurosecretory cells. Neurohormones are the substances secreted and released from these cells at some distance from their target organs.

The phenomenon of neurosecretion is thought to be ubiquitous among the metazoans, and neurosecretory cells are assumed to be present in representatives of all invertebrates. Prominent and somewhat complex neurosecretory systems are found in the Arthropoda. Many chelicerates have well-developed and well-compartmentalized neurosecretory systems, but data are fragmentary (Gabe, 1955). In Arachnida the system is associated with the brain and other ganglia and has been correlated with molting and reproduction (in contrast to the neurosecretory systems of Insecta where the developing system approxi-

mates that found in Crustacea and neurosecretory cells send their axons to the neurohemal corpus cardiacum). In other Chelicerata, clusters of cells are found in the supraesophageal ganglia and in association with the ganglionic regions of the fused subesophageal mass. Variations reflect the differences between major groups (e.g., Merostomata an Arachnida) as well as between representatives of the arachnids that have been studied (Bern and Hagadorn, 1965). Acarines have neurosecretory centers in both masses of the synganglion (Gabe, 1966; Tombes, 1979) (Fig. 11-13).

Tracheate arachnids exhibit periganglionic sinuses in association with neuroendocrine cells, for example, the esophageal neurohemal organ. Such an organ is found in many arachnids, in ticks, and in some Actinedida (Oben-

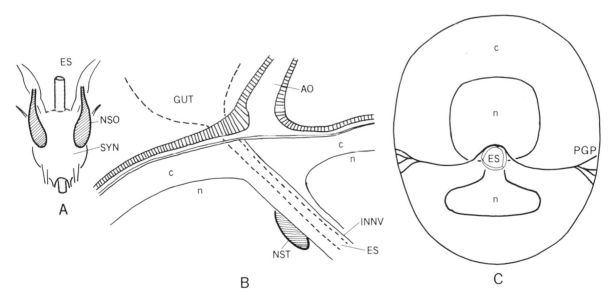

Figure 11-13. (A) Neurosecretory organs of *Fuscouropoda agitans* (after Woodring and Galbraith, 1976) (ES, esophagus; NSO, neurosecretory organ; SYN, synganglion); (B) neurosecretory tracts in *Boophilus microplus* showing cortex, neuropile, and neurosecretory tract (after Binnington and Tatchell, 1973) (AO, aorta; c, cortex; n, neuropile; NST, neurosecretory tract; IN NV, intestinal nerve); (C) diagrammatic section of *Boophilus microplus* showing neuropile, cortex, and paraganglionic plates associated with neurosecretory cells (after Binnington and Tatchell, 1973) (c, cortex; n, neuropile; ES, esophagus; PGP, pedal ganglionic plexus).

chain and Oliver, 1975). Ticks exhibit a complex neuroendocrine system in association with the periganglionic sinus (Roshdy et al., 1972).

Although the study of neurosecretion in the Acari is in its beginnings, a number of studies have elucidated the presence and function of neurosecretory cells. Such cells, as stated above, are usually located in part of the supraesophageal ganglia and the subesophageal ganglia; especially in the paraganglionic plates (Binnington, 1986; Binnington and Tatchell, 1973; Coons and Axtell, 1971; Gabe, 1955, 1966; Obenchain, 1974b; Obenchain and Oliver, 1975; Woodring and Galbraith, 1976; Tombes, 1979) (Fig. 11-13).

Neurosecretory cells differ from usual neurons in having more perinuclear cytoplasm. In all arthropods these cells are basically clusters in the ganglia and are associated with tracts of secretion-bearing axon fibers that terminate in close association with the vascular system and form a neurohemal organ. This organ is composed of specialized endings of the axons which store and release the secreted products into the hemolymph. By definition, the signs of secretion are found in cells that are otherwise definable as neurons. Described for the first time in argasid ticks as "plaques paraganglionaires" (Gabe, 1955), they are functionally identical to Schneider's organs I and II in Araneida, the X organs of Crustacea, and the corpora allata of Insecta (Eisenberger, 1970).

Neurosecretory products are frequently liberated at the site of their formation or accumulated much less rapidly than neurohemal substances. The accumulation of the secretion remains strictly extracellular. The travel route of neurohormones to receptor organs is often longer, since neurohormones are transported by humoral means along axons or the somas of secretory cells. Also, the physiological processes controlled by them are less rapid; their destruction or disintegration in the body is delayed for a certain time after their release (Gabe, 1955, 1966).

Neurosecretions apparently are stored temporarily in the neurohemal organs or dispersed

through the blood. In *Ornithodoros moubata*, secretory products of relatively indistinguishable neurosecretory cells are transmitted through their axons to peripheral bodies. The axons of the preesophageal part of the synganglion drain off the neurosecretions, but these axons cannot be traced in the dense meshwork of the neuropile. All that can be found, repeatedly, are accumulations of the neurosecretions in the ventral part of the synganglion. When the products leave the nerve mass they are either stored in retrocerebral endocrine organs or distributed by blood cells to other parts of the body. The neurilemma is thought to regulate exchange of neurosecretion from the synganglion to the "plaques paraganglionaires," but the physiological involvement is not fully clarified (Eichenberger, 1970).

The neurosecretory cells that give rise to this protocephalic neurosecretory pathway are arranged in four groups. Two are anteroventral and proximal to the cheliceral ganglia; two are posterodorsal and distal to the former. The pyriform, protocephalic neurosecretory cells are not much larger than the ordinary neurons (in two species of *Ornithodoros* they may reach a diameter of 20 μm). The cytoplasm has reduced ribonucleins, but contains a finely granular secretory product. The granules are heaped upon one another and are concentrated as masses in axon hillocks in the proximal part of their paths (Gabe, 1966).

Special cytological techniques in light and electron microscopy demonstrate visible droplets of "secretion" of neurohormones in the cytoplasm of neurosecretory cells. Selective staining facilitates detection (Gabe, 1955). The chemical nature of the products may differ, but a sulfur-rich protein is the ubiquitous component together with lipid and carbohydrate materials. The materials that stain are thought to be the carriers of the hormones rather than the actual regulatory agents themselves, and are usually of lighter molecular weight (Bern and Hagadorn, 1965; Gabe, 1955; Tombes, 1979).

Although the study of neurosecretion in the Acari is in its beginnings, a number of studies have elucidated the presence and function of neurosecretory cells. Such cells, as stated above, are usually located in part of the supra-esophageal ganglia and the subesophageal ganglia, especially in the paraganglionic plates (Binnington and Tatchell, 1973; Coons and Axtell, 1971; Gabe, 1955, 1966; Obenchain, 1974b; Obenchain and Oliver, 1975; Woodring and Galbraith, 1976; Tombes, 1979) (Fig. 11-13B, C).

Near the base of the pedipalpal nerves in *Fuscouropoda agitans* is a glandular structure of three to four enlarged cells. The narrow bases of these cells penetrate the medulla of the pedipalpal nerves close to their synganglionic connection. Although these large cells are widely variable in their appearance, the cytoplasm is consistently highly vacuolate. The cells are thought to be neurosecretory in function (Woodring and Galbraith, 1976); (Fig. 11-13A). Similar, paired, ductless glands are also described for *Echinolaelaps echidninus* (Jakeman, 1961) and for several families of gamasids (Coons and Axtell, 1971).

Various ganglia of *Ornithodoros moubata* have neurosecretory cells. Of the four types of cells identified in arthropods, types A and B are found in this tick. A-cells stain purple with PF and deep blue to blue-black with CHP. B-cells stain bluish green with PF and reddish with CHP (Eichenberger, 1970).

Neurosecretory production is greatly stimulated by food intake. A large amount of secretion is observed in the cytoplasm in engorged nymphs preparing for molting and in adults at the beginning of feeding. Upon completion of the molt or after engorgement, the amount of neurosecretion is reduced. Neurosecretions may also accumulate prior to diapause (e.g., *Dermacentor pictus*, Balashov, 1972).

Actively secreting cells are always present in all stages of the tick. B-cells continuously secrete whereas A-cells become particularly active after feeding. The blood meal is a positive factor in the elaboration of the secretions. Neurosecretory cells secrete more material 4 days after feeding than can be utilized; thus the neurosecretion stops and the materials are gradually used (Balashov, 1972; Eichenberger, 1970).

Fourteen to eighteen groups of neurosecretory cells are found in the protocerebral, cheliceral, pedipalpal, stomodeal, pedal, subesophageal, and opisthosomatic ganglia of *Dermacentor pictus* and *Hyalomma dromedarii* (Ioffe, 1964; Dhanda, 1967).

Neurosecretory vesicles are found in axons of nerves that connect to the salivary glands of *D. variabilis*, together with numerous microtubules and mitochondria (Coons and Roshdy, 1973). Significant neurosecretory centers are found in *Boophilus microplus*, another of the hard ticks (Binnington and Tatchell, 1973) (Fig. 11-2A).

Neurosecretory cells are present within the synganglion of Astigmata, for example, *Acarus siro*. Such cells are identified in the antero-dorsal aspect of the synganglion and are assumed to be involved in ecdysis and in the formation of hypopi (Hughes, 1964). Neurosecretory cells are not known at present in the Oribatida.

BEHAVIOR

Behavior embraces the sum total of the responses of animals to their external and internal environments. External environmental stimuli result in responses that involve the nervous system and perhaps chemically responsive organs or structures connected to the nervous system. Internal stimuli ellicit changes in various metabolic processes and the physiology of mites. Tactile setae and mechanoreceptors elicit specific responses.

Survival of animals in the environment involves these physiological, morphological, and behavioral responses. The responses result in relationships with the physical environment that ensure proper functioning and prevent injury or destruction of individuals and species.

The study of behavior among mites is in the forefront of investigations where disclosures of a complexity of activities boggle the mind. It is not possible to include in this writing the many types and details that are found in this phase of the biology of acarines. Examples and brief summaries of behaviors other than reproduction will have to suffice, together with literature citations that lead the reader to sources of information where this dynamic aspect of the acarological science is elaborated, (e.g., Baker and Delfinado-Baker, 1983).

Various behaviors are noted in the locomotion of mites. Oribatid mites use the long solenidia on the anterior pair of legs in a type of seeking, exploring behavior. In the Zetorchestidae (Oribatida) the last pair of legs is adapted for leaping. Various large-winged oribatids use the pteromorphs as sail planes for gliding down from grass stems. In the oribatids and in Caeculidae there is a peculiar "death feint" behavior where the mites stop locomotory movement and rest on the substrate; they remain motionless for some time and then carefully extend their legs and resume locomotion. This death feint (letisimulation) behavior probably is protective against predation. Water mites have extensive hairs on the legs with which swimming or suspension on the surface film of water is possible.

Opilioacarus either singly or in groups of two or three will remain on the undersurface of partially buried rocks, not in the soil. When walking, the front legs tap the substrate like antennae. If an object is encountered the mite retreats backward quickly. The legs are elevated in walking and the mite moves with great agility (Grandjean, 1936).

Spider mites have a behavior called "ballooning" by which distribution is effected. The mites spin a thread of silk and attach it to a leaf or a twig. Supported by this silken tie, the mite drops 2 to 5 in. and is suspended until a breeze breaks the thread and carries the mite away to a new location. Ballooning is usually the result of overpopulation or an unfavorable food supply (Jeppson et al., 1975).

The eriophyid mites, *Eriophyes ribis*, (which causes "big bud" on black currants) rear up on the tip of the opisthosoma and sway in the air. If an aphid, ladybird beetle, or other insect comes in contact, the mites are carried away. *Cheyletomorpha lepidoptorum* clings to hairs of moths with its chelate pedipalps and is carried by these lepidopteran hosts.

Feeding behavior varies among the orders. Predators, phytophages, zoophages, and saprophages differ specifically in the mechanisms by which food is obtained. In *Cheyletus aversor* four behavioral modes are noted: ambushing, moving, feeding, and resting. Predatory capture of the prey is more rapid than other movements and the responses depend upon the density of the predator population, interference between predators, length of time exposed, and level of successful search for prey. If prey density is high, instraspecific agonistic behavior occurs (Wharton and Arlian, 1972).

Orientation responses are definitively expressed. Hard ticks crawl to the distal tips of plants and extend the front legs in a host-seeking pose (Fig. 11-6D, E) that enables them to detect the host by means of Haller's organ. Before feeding and while at rest *Ixodes ricinus*, the sheep tick, may adopt a questing position with the first pair of legs extended into the air, or take an attitude of repose and fold the forelegs close to the body. While in the questing attitude or while walking, the forelegs may be waved back and forth in a manner similar to antennae. Ticks climb from the base to the tip of grass or rush stems, walk up and down the tip, and finally come to rest there. This is considered positive geotaxis (upward-turning near the tip) and kinesis (Lees, 1948).

Response to moisture in ticks is influenced by the physiological state. With water balance in the body normal, unfed ticks avoid high humidities, but rest with equal readiness in either moist or dry air. The negative taxis (avoiding response) is lost after dessication and a kinetic response occurs when the tick is active in dry air and comes to rest in more moist air. When the water is restored to the cuticle, negative taxis then recurs.

If unfed ticks are in an attitude of repose at

the end of grass stems and vibrations occur, they immediately assume the questing position. They cling readily to any object or animal that brushes against them. Most ticks will attach to the host if legs I or the palps are amputated; if both are lacking, none attach.

Temperature affects the behavior of ticks. If higher than 42°C. there is little avoiding response. They respond to gradients of air temperature and not to radiant heat. Temperature also affects their sense of smell, for within a temperature gradient they will respond rapidly at 37°C to odors of sheep wool and scents of dog, rabbit, cow, and horsehair.

Ticks respond to light by an immediate questing posture, especially if there is a sudden drop in light intensity (shading). Engorged ticks and newly molted nymphs seem to be strongly photonegative. The newly molted ticks avoid direct illumination, but this response diminishes with age (Lees, 1948). A circadian rhythm is involved in larval drop off in *Dermacentor variabilis* (Amin, 1970).

The rabbit tick (*Haemophysalis leporispalustris*) is highly host-specific. While adults feed almost exclusively on rabbits, some immatures will feed on birds. Larvae that leave the resting form of the rabbit during the day climb up on vegetation into a questing position. Experiments show that the selected position is at the height of the rabbit body and where the substrate has the greatest curvature. Such orientational behavioral limits the tick's spectrum of hosts to rabbits and ground-dwelling or ground-feeding birds (Camin and Drenner, 1978).

The bee tracheal mite (*Acarapis woodi*—Tarsonemidae) assumes an "ambush position" on the hairs of a dead host and remains in this position until stimulated by the tactile contact of a bee's hair (or a very fine needle). These mites also respond to vibration and positively to intermittent air currents, which aid them in locating the spiracle of the hosts. Wing vibration is less effective than the intermittent airstream. Penetration of the trachaea is by positive thigmotaxis. Orientation on the body of the bee

is by a sense of touch. Older bees with stiffer hairs offer greater resistance to penetration by the mites (Hirschfelder and Sachs, 1952).

The larval mites of the genus *Arrenurus* attach to specific sites on the odonate hosts and it has been suggested that this attachment is related to the way in which the larvae reenter the water after feeding on their hosts (Mitchell, 1959). These larvae encounter the membranes of the terminal abdominal segment where they attach first. If emergence of the host is arrested for half a minute or more after emergence has begun, the normal sequence of larval activity remains unaltered. If (by manipulation) the host's emergence is stopped earlier, the larvae will attach to other membranes. It appears that timing alone is responsible for site specificity and that whatever stimulus evokes the site selection behavior of the larvae is perceived during the first 30 s of host emergence (Mitchell, 1959).

The moth ear mite (Treat, 1975) accomplishes its orientation on its host in a remarkable way. From the boarding of a moth host by a fertile female to the dispersal of the fertile females of the next generation, the behavior is curious and distinctive. Peculiarly, the fertile females invade and occupy only one of the host's two ears. The advantage of the behavior is so that the other ear is intact and capable of detecting ultrasound; if an infested moth is captured by a bat, both the moth host and the ear mite colony are destroyed.

A fertile female of *Dicrocheles phalaenodectes* will respond to a new host by a temporary "freeze" or cessation of all movement. Then successive jerky movements and thrusts push her into the hairs and scales until she is hidden from view. This jerky progress continues until orientation is complete, after which she moves to explore the collar and shoulder regions of the host. The movements of the hairs through which she travels, as if through a "canebreak," identify location with the hairs bending in an indication of her direction. When she reaches the "crossroad"—a point between the two ears of the moth, a brief hesitation may occur, then she disappears in a tympanic recess. Her "housekeeping" is not begun immediately, but returning forays to the midpoint or "crossroad" have been interpreted as possibly laying a trail or posting a sign for other mites to follow. Her invasion of the tympanic and countertympanic membranes and engorgement follow, with the ensuing production of a colony of mites (Treat, 1975).

Sarcoptes scabei is normally assumed to be transmitted from host to host by contact and close association. Recent studies (Arlian et al., 1984) show that the mites have a sensory physiology and host-seeking behavior. These mites perceive specific host stimuli and actively move from host to host. They respond independently to both thermal stimuli and host odor. In close proximity both stimuli are effective; at greater distances host odor is more effective.

Feather mites are most interesting as objects of acarine behavior because of their obligatory ectoparasitic habit and occurrence in every avian order (except penguins and cassowaries). Each avian order has a specific acarofauna comprising a limited number of parasitic forms. Their "life-on-a-feather" results not only in morphological innovation, but behavioral differences because of their adaptations to the turbulent and hostile environment in which they live. Morphological variations differ between families as well as between genera in the same family. Specific species seek homes in definitive locations on the feathers, and because of this their behavior and morphology enables them to exist. Diverse morphology includes sexual dimorphism, less pronounced in females than males, but still reflected in the particular location on the host. Mites that live on the outer wing feathers, where the turbulence of the air is greatest, show greater size, more sclerotization, stronger coxal apodemes, dorsoventral flattening, and reduced dorsal chaetotaxy. Laterally placed legs and the behavioral ability to withstand the air turbulence and to reduce air resistance add to the adaptations. Behavioral adaptation coupled with the morphological features of males affect reproduc-

tion as well. Elaboration of a convex–concave posterior end with broadened lobes, lamellae, and setae, as well as copulatory discs of males and the lesser modifications of the female body facilitate successful copulation (Atyeo and Gaud, 1979).

Species of feather mites in the families Kramerellidae and Pterolichidae show specific site preferences on the feather of the greater sandhill crane. One species (*Gruolichus wodashae*) aligns along the shaft of the feather; another species (*Brephoscles petersoni*) chooses a site on the broad surface of the vane; two other species (*Geranolichus canadensis* and *Pseudogabucininia reticulata*) prefer a location at the distal tip of the feather (Atyeo and Windingstad, 1979). Similar preferential sites are described for species of *Trouessartia, Proctophyllodes, Analges,* and *Freyania* (from Dubinin, 1951; 1953; 1956; Evans et al., 1961) (Fig. 11-14).

Another group of feather mites, the Syringophilidae, inhabit the inside of the quills of feathers. They feed on the internal pulp and then insert their styletiform chelicerae through the calmus wall. These mites invade the umbilicus of developing feathers, where the newly formed females grow. As the females mature they show a single large egg inside the body. As the season progresses the larvae and nymphs appear and develop into the adult. Adult females emerge from mature quills and disperse to developing feathers. Mites transfer from the feathers of parent birds to their young during the diurnal brooding or roosting in the nest (Kethley, 1971; Casto, 1974a,b).

Defensive behavior may involve camouflage such as cerotegument or the retention of exuviae on the dorsum (Oribatida). Warning coloration or colors that blend more easily with the background (Trombidiidae) also are protective. There may be chemical defenses, not fully understood, that effect protection for some species.

Some of the complex behaviors include both social and reproductive responses. In the Cloacaridae, transmission of the mites from one turtle host to another is effected venereally

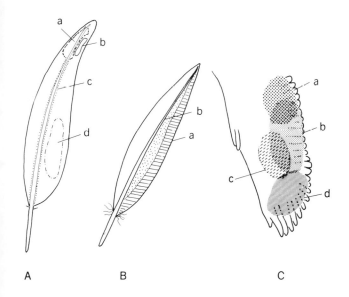

A   B   C

Figure 11-14. Site preference behavior in feather mites: (A) on 6th and 7th primaries of Greater Sandhill Crane; a, *Geranolichus canadensis*; b, *Pseudogabucininia reticulata*; c, *Gruolichus wodashae*; d, *Brephosceles petersoni* (after Atyeo, 1979); (B) primary feather of a duck: a, *Freyania largifolia*; b, *F. anatina*; (C) on hooded crow: a, *Trouessartia corvina*; b, *Proctophyllodes corvorum*; c, *Analges corvinus*; d, *Gubucinia delibata* (after Evans et al., 1961).

when the host mates. In a brief study of mites associated with bark beetles I observed that tarsonemid males remained close to the female imagochrysalis and often grasped the same and carried the female around until emergence. In *Dicrocheles phaelenodectes* a marriageable deutonymph is embraced by a prospective mate (Treat, 1975).

Phoresy (from the Greek, "to carry or to bear"), is defined by Farish and Axtell (1971), as "a phenomenon in which one animal actively seeks out and attaches to the outer surface of another animal for a limited time during which time the attached animal (called the phoretic) ceases both feeding and ontogenesis, such attachment presumably resulting in dispersal from areas unsuited for further development, either for the individual or its progeny." These

authors delete the implication of "transport" to a specific destination as a part of the definition. Treat (1975) devotes several chapters to this type of behavior and movement.

Mites of the genus *Coxequesoma* are known only from phoretic associations with army ants (*Eciton hamatus*). The dorsa of the mites are enlarged and expanded ventrally in a clasping mechanism for attachment to the coxae of the ant leg. This passive attachment is sufficiently strong to maintain the position on the leg even when the ant is preserved in alcohol (Elzinga, 1982) (Fig. 11-15).

Antennophorine mites arrange themselves symmetrically on the body of their transport ant host (*Lasius* (Fig. 11-16B). Phoretic deutonymphs (hypopi; see Chapter 12) orient to location and to hosts in specific ways (Fig. 11-16D).

In his review of mites associated with insects, Mostafa (1970a, b, c) relates that the first acarinarium (mite chamber) was discovered by Perkins (1899) on the basal segments of Xylocopine bees and some wasps. Many other authors since then have studied these hypopi and their commensal and phoretic behavior. Mostafa describes the behavior and ecology of such deutonymphs from wasps and other insects with details of their biology. Some saproglyphid mites have adjusted their developmental cycles to coincide with the cycles of their hosts. The rhythmic pulsations of the abdomen of a female wasp during oviposition may cause hypopi of *Vespacarus* to leave the body of the wasp and drop into a new uninfested nest or cell (Krombein, 1961). The orderly arrangement of hypopi in the acarinaria of *Ancistrocercus antilope* is an interesting and complex response in which these deutonymphs line up in orderly rows on males like soldiers awaiting transport. Female wasps are devoid of hypopi and acquire them in "venereal transmission" during mating. The deutonymphs then drop off into uninfested nests or cells (Cooper, 1954).

Mites from the families Mesoplophoridae, Oppiidae, and Oribatulidae (Oribatida) are phoretic on many insect species to which they attach to hairs by clasping mechanisms (Fig.

A

B

C

Figure 11-15. (A) *Coxequesoma umbocauda*; (B) same on leg of army ant (*Eciton hamatus*); (C) *Antennequesoma rettenmeyeri* attached to *Nonamyrmex* antenna (SEMs by permission of R.J. Elzinga).

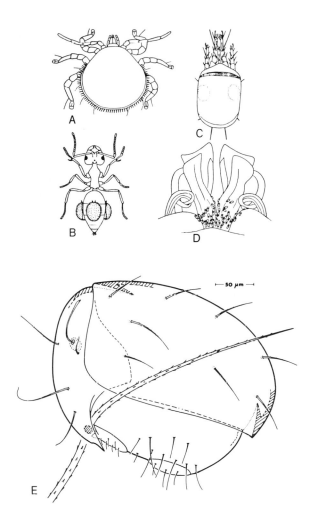

Figure 11-16. (A) *Neomegistus*, a flattened gamasine mite that lives permanently on the Natal Sirostreptid millipede, *Doratogonus setosus*; (B) antennophorine mites (*Trichocylliba*) symmetrically arranged on an ant (*Lazius*); (C) hypopus of *Caloglyphus*; (D) hypopi of same clustered on the sex organs of male Spirostreptid millipede (*D. setosus*) (after Lawrence, 1953); (E) *Mesoplophora* sp. showing method of attachment to an insect hair (after Norton, 1980).

11-16E). In *Mesoplophora* host hairs of suitable size and sites with relative absence of mechanical disturbances are the factors that determine passive transport. In passalid beetles these features are found in the ventrolateral area between the first and second thoracic segments. Symmetrical distribution of these mites on a phoretic host is an interesting phenomenon and ellicits questions as to how the mites determine that type of distribution, especially when when odd numbers of mites are present. Passive dispersal of oribatid mites by nest-building vertebrates is a common occurrence especially where specific microhabitats are shared in close relationships. Decaying stumps, subcortical locations with bark beetles, woody substrates, and decaying mats are sites where such phoretic associations occur (Norton, 1973, 1980).

Ascid mites (*Arctoseius cetratus*) are phoretic on mushroom sciarid flies (*Lycoriella*) and attach in rows on the pleural integument of the host in a stereotyped and symmetrical fashion (Binns, 1972). A new species of uropodid mite uses an anal pedicel in the deutonymphal stage for attachment to the phoretic host (formosan subterranean termite, *Coptotermes formosanus*). The attachment of the deutonymphs by means of these pedicels does not appear to affect adversely the movements or activities of the hosts. Behaviorly, one deutonymph attaches and faces posteriorly on the dorsum of the termite, which is interesting behavior. For an eyeless creature, this "sentinel" would have to possess other sensory devices for detection of changes (Phillipsen and Coppel, 1978).

REFERENCES

Amin, O.M. 1970. Circadian rhythm of dropping off of engorged larvae of *Dermacentor variabilis*. *J. Med. Entomol.* 7(2):251–255.

Andre, M. 1949. Ordres des Acariens. *In:* Grassé, P.-P., ed. *Traité de Zoologie.* Masson, Paris, Vol. 6, pp. 794–892.

Arlian, L.G., R.A. Runyan, L.B. Sorlie and S.A. Estes. 1984. Host-seeking behavior of *Sarcoptes scabei. J. Am. Acad. Dermatol.* 11(4):594–598.

Arthur, D.R. 1956. The morphology of Haller's organ in ticks. *Proc. Int. Congr. Zool., 14th, Copenhagen, 1953,* pp. 491–492.

Arthur, D.R. 1960. *Ticks,* Part V. Cambridge Univ. Press, London and New York.

Arthur, D.R. 1962. *Ticks and Disease.* Pergamon, Oxford.

Atyeo, W.M. and J. Gaud. 1979. Ptyssalgidae, a new family of Analgoid feather mites (Acarina; Acaridida). *J. Med. Entomol.* 16(4):306–308.

Atyeo, W.T. and R.M. Windingstad. 1979. Feather mites of the Greater Sandhill Crane. *J. Parasitiol.* 65(4):650–658.

Axtell, R.C. 1979. Tarsal sensory receptors of ticks. *Proc. Int. Congr. Acarol., 4th, 1974,* pp. 669–672.

Axtell, R.C., R.F. Foelix, L.B. Coons and M.A. Roshdy. 1973. *Proc. Int. Congr. Acarol., 3rd, 1971,* pp. 35–40.

Baker, E.W. and M. Delfinado-Baker. 1983. New mites (*Sennertia:* Chaetodactylidae) phoretic on honey bees (*Apis mellifera* l.) in Guatemala. *Int. J. Acarol.* 9(3):117–121.

Baker, E.W. and B.W. Wharton. 1952. *An Introduction to Acarology.* Macmillan, New York.

Balashov, Y.S. 1972. *Bloodsucking Ticks (Ixodoidea)— Vectors of Disease in Man and Animals,* Transl. 500 (T 500). Med. Zool. Dep., USNAMRU No. 3, Cairo, Egypt, U.A.R. [published in *Misc. Publ. Entomol. Soc. Am.* 8(5):161–376 (1968)].

Balashov, Y.S. and S.A. Leonovich. 1978. External ultrastructure of Haller's organ in ixodid ticks of the subfamily Ixodinae (Acarina, Ixodoidea) in connection with the systematics of this group. *Tr. Zool. Inst. Akad. Nauk SSSR* 77:29–36 (translation 1345 (T1345), NAMRU, Cairo).

Barth, F.G. and E.A. Seyfarth. 1972. Compound slit sense organs on the spider's leg. *J. Comp. Physiol.* 78:176–191.

Bern, H.A. and I.R. Hagadorn. 1965. Neurosecretion. *In:* Bullock, T.H. and G.A. Horridge, eds. *Structure and Function of the Nervous System of Invertebrates.* Freeman, San Francisco, California, Vol. 1, Chapter 7, pp. 353–429.

Binnington, K.C. 1986. Ultrastructure of the tick neuroendocrine system. *In:* Sauer, R.J. and J.A. Hair, eds. *Morphology, Physiology of Ticks.* Ellis Horwood, Ltd., Chichester, England, pp. 152–154.

Binnington, K.C. and R.J. Tatchell. 1973. The nervous system and neurosecretory cells of *Boophilus microplus* (Acarina: Ixodidae). *Z. Wiss. Zool.* 185:193–206.

Binns, E.S. 1972. *Arctoseius citratus* (Sellnick (Acarina: Ascidae) phoretic on mushroom sciarid flies. *Acarologia* 14(3):350–356.

Blauvelt, W.E. 1945. The internal morphology of the common red spider mite (*Tetranychus telarius* Linn.). *Mem.-N.Y. Agric. Exp. Stn. (Ithaca)* 270:1–35.

Boudreaux, H.B. 1979a. Significance of intersegmental tendon system in arthropod phylogeny and a monophyletic classification of Arthropods. *In:* Gupta, A.P., ed. *Arthropod Phylogeny.* Van Nostrand-Reinhold, New York, pp. 551–586.

Boudreaux, H.B. 1979b. *Arthropod Phylogeny with Special Reference to Insects.* Wiley, New York.

Brennan, J.M. 1972. A new genus and two new species of Venezuelan chiggers with eyes in the scutum. *J. Med. Entomol.* 9(1):16–18.

Brody, A.R. 1969. Comparative fine structure of acarine integument. *J. N.Y. Entomol. Soc.* 77(2):105–116.

Bullock, T.H. and G.A. Horridge, eds. 1965. *Structure and Function of the Nervous Systems of Invertebrates,* Vols. 1 and 2. Freeman, San Francisco, California.

Camin, J.H. and R.W. Drenner. 1978. Climbing behavior and host-finding of laravl rabbit ticks (*Haemaphysalis leporispalustris*). *J. Parasitol.* 64(5):905–909.

Casida, J.E. 1955. Comparative enzymology of certain insect acetylesterases in relation to poisoning by organophosphorus insecticides. *Biochem. J.* 60:487–496.

Casto, S.D. 1974a. Quill wall thickness and feeding of *Syringophiloidus minor* (Berlese) (Acarina: Syringophilidae). *Annals Ent. Soc. Am.* 67(5):824.

Casto, S.D. 1974b. Entry and exit of Syringophilid mites (Acarina: Syringophilidae) from the lumen of the quill. *Wilson Bull.* 86:272–278.

Chow, Y.S., S.H. Lin and C.H. Wang. 1972. An ultrastructural and electrophysiological study of the brain of the brown dog-tick, *Rhipicephalus sanguineus* (Latreille). *Chin. Biosci.* 1:83–92 (cited by Coons et al., 1974).

Chu-Wang, I.W. and R.C. Axtell. 1974. Fine structure of ventral and lateral tarsal sensilla of the hard tick, *Amblyomma americanum*. *Ann. Entomol. Soc. Am.* 67(3):453–457.

Coineau, Y. 1970. A propos de l'oeuil anterieur du nasi des Caeculidae. *Acarologia* 12:109–118.

Cook, D.R. 1974. *Water Mite Genera and Subgenera*, Mem. No. 21. Am. Entomol. Inst., Ann Arbor, Michigan.

Coons, L.B. and R.C. Axtell. 1971. Cellular organization in the synganglion of the mite *Macrocheles muscaedomesticae* (Acarina: Macrochelidae). *Z. Zellforsch. Mikrosk. Anat.* 119:309–320.

Coons, L.B. and R.C. Axtell. 1973. Sensory setae of the first tarsi and palps of the mite *Macrocheles muscaedomesticae*. *Ann. Entomol. Soc. Am.* 66(3):539–544.

Coons, L.B. and M.A. Roshdy. 1973. Fine structure of the salivary glands of unfed male *Dermacentor variabilis* (Say) (Ixodoidea: Ixodidae). *J. Parasitol.* 59(5):900–912.

Coons, L.B., M.A. Roshdy and R.C. Axtell. 1974. The fine structure of the central nervous system of *Dermacentor variabilis* (Say), *Amblyomma americanum* (L.), and *Argus arboreus* Kaiser, Hoogstraal & Kohls (Ixodoidea). *J. Parasitol.* 60:687–698.

Cooper, K.W. 1954. Venereal transmission of mites by wasps, and some evolutionary problems arising from the remarkable association of *Ensliniella trisetosa* with the wasp *Ancistrocercus antilope*. Biology of Eumenine wasps II. *Trans. Amer. Ent. Soc.* 80:119–174.

Dahl, F. 1911. Die Hohaare und das System der Spinnentire. *Zool. Anz.* 37:522–532.

Dhanda, V. 1967. Changes in neurosecretory activity at different stages in adult *Hyalomma dromedarii* Koch, 1844. *Nature (London)* 214:508–509.

Dogel, V.A. 1940. *Comparative Anatomy of Invertebrates*, Vol. 2. Leningrad (cited by Tsvileneva, 1965).

Douglas, J.R. 1943. *The Internal Anatomy of* Dermacentor andersoni *Stiles*. Univ. of California Press, Berkeley.

Dubinin, W.B. 1951. Feather mites (Analgesoidea). Part I. Introduction to their study. *Fauna U.S.S.R.* 6(5):1–363.

Dubinin, W.B. 1953. Feather mites (Analgesoidea). Part II. Epidermoptidae and Freyanidae. *Fauna U.S.S.R.* 6(6):1–411.

Dubinin, W.B. 1956. Feather mites (Analgesoidea). Part III. Pterolichidae. *Fauna U.S.S.R.* 6(7):1–814.

Eichenberger, G. 1970. Das Zentralnervensystem von *Ornithodoros moubata* (Murray), Ixodoidea, Argasidae, und seine post-embryonale Entwicklung. *Acta Trop.* 27(1):15–53 (translation in English; NAMRUT 419).

Elzinga, R.J. 1982. The genus *Coxequesoma* (Acari: Uropodina) and descriptions of four new species. *Acarologia* 23(3):215–224.

Evans, G.O., J.G. Sheals and D. MacFarlane. 1961. *Terrestrial Acari of the British Isles,* Vol. 1. Adlard & Son, Bartholomew Press, Dorking, England.

Fage, L. 1949. "Classe des Merostomaces." *In:* Grassé, P.-P., ed. *Traité de Zoologie.* Masson, Paris, Vol. 6, pp. 219–262.

Farish, D.J. and R.C. Axtell. 1971. Phoresy redefined and examined in *Macrocheles muscaedomesticae* (Acarina: Macrochelidae). *Acarologia* 13:16–29.

Foelix, R.F. 1972. Permeability of tarsal sensilla in the tick *Amblyomma americanum* (L.). *Tissue Cell* 4(1):130–135.

Foelix, R.F. and R.C. Axtell. 1971. Fine structure of tarsal sensilla in the tick *Amlyomma americanum* (L.). *Z. Zellforsch. Mikrosk. Anat.* 114:22–37.

Foelix, R.F. and R.C. Axtell. 1972. Ultrastructure of Haller's organ in the tick *Amblyomma americanum* (L.). *Z. Zellforsch. Mikrosk. Anat.* 124:275–292.

Gabe, M. 1955. Données histologiques sur la neurosecretions chez les Arachnides. *Arch. Anat. Microsc.* 44:351–383.

Gabe, M. 1966. *Neurosecretion.* Pergamon, Oxford.

George, J.E. 1963. Reponses of *Hemaphysalis leporispalustris* to light. Naegele, J.A., ed. *Advances in Acarology*, Vol. 1, pp. 425–430, Cornell Univ. Press, Ithaca, New York.

Grandjean, F. 1928. Sur un oribatide pourvu d'yeux. *Bull. Soc. Zool. Fr.* 53:235–242.

Grandjean, F. 1934. Les organs respiratoire secondaires des Oribates (Acariens). *Ann. Soc. Entomol. Fr.* 103:109–146.

Grandjean, F. 1935a. Observations sur les Acariens (2e série). *Bull. Mus. Natl. Hist. Nat.* (2) 7:201–218.

Grandjean, F. 1935b. Les Poiles et les organes sensitif portées par les pattes et le palpe chez les Oribatides. *Bull. Soc. Zool. Fr.* 60(1):6–39.

Grandjean, F. 1936. Un Acarien synthétique: *Opilioacarus segmentatus* With. *Bull. Soc. Hist. Nat. Afr. Nord.* 27:418–444.

Grandjean, F. 1946. Les poils et les organes sensitif portes par les pattes et le palpe chez les Oribates. (3e) *Bull. Soc. Zool. Fr.* 71:10–29.

Grandjean, F. 1955. L'Organe de Claparede et son ecaille chez *Damaeus onustus* Koch. *Bull. Mus. Natl. Hist. Nat.* 4:1285–292.

Grandjean, F. 1958. Au sujet du Naso et de son oeil infere chez les Oribates et les Endeostigmata (Acariens). *Bull. Mus. Nat. Hist. Natur.* (2), 30:427–435.

Gupta, A.P., ed. 1979. *Arthropod Phylogeny.* Van Nostrand-Reinhold, New York.

Hammen, L. van der. 1961. Description of *Holothyrus grandjeani* nov. spec., and notes on the classification of the mites. *Nova Guinea* 10(9):173–194.

Hammen, L. van der. 1970. La phylogenese des Opilioacarides et leur affinites avec les autres Acariens. *Acarologia* 12:465–473.

Hammen, L. van der. 1973. Classification and phylogeny of mites. *Proc. Int. Congr. Acarol., 3rd, 1971,* pp. 275–281.

Hammen, L. van der. 1979. Comparative studies in Chelicerata I. The Cryptognomae (Ricinulei, Architarbi, and Anactinotrichida). *Zool. Verh.* 174:1–62.

Hanstrom, B. 1919. Zur kenntnis des centralen Nervensystem der Arachnoiden und Pantopoden nebst schlussfolgerungen Betreffs der Phylogenie der genannten Gruppen. Inaug. -Dissertation, Zootomischen Inst. der Hochsh. zu Stockholm (Acarida, pp. 109–118) (cited by Blauvelt, 1945).

Hanstrom, B. 1928. *Vergleichende Anatomie des Nervensystems der wirbellosen Tiere unter Berucksichtigung seiner Funktion* Springer, Berlin, pp. 409–411.

Hess, E. and M. Vlimant. 1986. Leg sense organs of ticks. *In:* Sauer, R.J. and J.A. Hair, eds. *Morphology, Physiology and Behavioral Biology of Ticks.* Ellis Horwood, Ltd., Chichester, England, pp. 361–390.

Hirschfelder, H. and H. Sachs. 1952. Recent research on the acarine mite. *Bee World* 33(12):201–209.

Hirschmann, W. 1979. Geruchsorgan der Zecken. *Mikrokosmos* 6:176–177.

Homsher, P.J. and D.E. Sonenshine. 1975. SEM of ticks for systematic studies: Fine structure of Haller's organ in ten species of *Ixodes. Trans. Am. Microsc. Soc.* 94(3):368–374.

Homsher, P.J. and D.E. Sonenshine. 1977. SEM of ticks for systematic studies. 2. Structure of Haller's organ in *Ixodes brunneus* and *Ixodes frontalis. J. Med. Entomol.* 14(1):93–97.

Homsher, P.J. and D.E. Sonenshine. 1979. SEM of ticks for systematic studies: Structure of Haller's organ in five species of the subgenus *Multidentatus* of the genus *Ixodes. In:* Rodriguez, J.G., ed. *Recent Advances in Acarology.* Academic Press, New York, Vol. 2, pp. 485–490.

Horridge, G.A. 1965. The Arthropoda. *In:* Bullock, T.H. and G.A. Horridge, eds. *Structure and Function of the Nervous Systems of Invertebrates.* Freeman, San Francisco, California, Vol. 2, pp. 801–1165.

Hughes, T.E. 1959. *Mites or the Acari.* Athlone, London.

Hughes, T.E. 1964. Neurosecretion, ecdysis and hypopus formation in the Acaridae. *Proc. Int. Congr. Acarol., 1st, 1963,* pp. 338–342.

Ioffe, I.D. 1963. Der Bau des Nervenapparats von *Dermacentor pictus* Herm. *Zool. Ah.* 42:1472–1884 (English translation NAMRUT 64-19707).

Ioffe, I.D. 1964. Distribution of neurosecretory cells in the central nervous system of *Dermacentor pictus* Herm. *Dokl. Akad. Nauk SSSR* 154:229–232 (cited by Balashov, 1972; English translation NAMRU 3-T376).

Jakeman, L.A.R. 1961. The internal anatomy of the spiny rat mite, *Echinolaelaps echidninus* (Berlese). *J. Parasitol.* 47(2):329–349.

Jeppson, L.R., H.H. Keifer and E.W. Baker. 1975. *Mites Injurious to Economic Plants.* Univ. of California Press, Berkeley.

Jones, B.M. 1950. Sensory physiology of the harvest mite, *T. autumnalis* Shaw. *J. Exp. Biol.* 27:461–494.

Kethley, J. 1971. Population regulation in quill mites (Acarina: Syringophilidae) *Ecol.* 52(6):1113–1118.

Krantz, G.W. 1978. *A Manual of Acarology.* Oregon State University Book Stores, Corvallis.

Krombein, K.V. 1961. Some symbiotic relations between saproglyphid mites and solitary vespid wasps (Acarina: Saproglyphidae and Hymenoptera: Vespidae). *J. Wash. Acad. Sci.* 51:89–93.

Lang, P. 1905. Uber den Bau der Hydrachnidenaugen. *Zool. Jahrb. Anat.* 21:453–494. (Abstr. by R. Hesse, *Zool. Zentrlblt.* 12:574, 575).

Lawrence, R.F. 1953. *The Biology of the Cryptic Fauna of Forests.* Balkema, Capetown, Amsterdam, pp. 1–408.

Lees, A.D. 1948. The sensory physiology of the sheep tick, *Ixodes ricinus. J. Exp. Biol.* 25(2):145–207.

Legendre, R. 1968. Le nomenclature anatomique chez les Acariens. *Acarologia* 10(3):411–417.

Leonovich, S.A. 1978. Fine structure of Haller's organ in the tick *Hyalomma asiaticum* P. Sch. & E. Schl. (Parasitiformes, Ixodidae, Amblyommidae). *Entomol. Obozr.* 58(1):221–228 (translation 1343 (T 1343), NAMRU, Cairo).

Lindquist, E.E. 1975. Associations between mites and other arthropods in forest floor habitats. *Can. Entomol.* 107:425–437.

McEnroe, W.D. 1960. Cholinesterase in the two-spotted spider mite. *Bull. Entomol. Soc. Am.* 6:150.

Mehrotra, K.N. 1961. The occurence of acetylocholine in the two-spotted mite, *Tetranychus telarius* L. *J. Insect Physiol.* 6:180–184.

Mehrotra, K.N. 1963. Biochemistry of nerve function in Acarina. *Adv. Acarol.* 1:209–213.

Michael, A.D. 1895. On the form and proportions of the brain in the Oribatidae and in some other Acarina; with a note by E.M. Nelson. *J. R. Microsc. Soc.,* pp. 274–282.

Mitchell, R. 1959. Life histories and larval behavior of arrenurid water-mites. parasitizing Odonata. *J. N.Y. Entomol. Soc.* 67:1–12.

Mitchell, R. 1964. The anatomy of an adult chigger mite, *Blankaartia acascutellaris* (Walch). *J. Morphol.* 114(3):373–391.

Moser, J.C. 1981. Transfer of *Pyemotes* egg parasite phoretic on western pine bark beetles to the southern pine beetle. *Int. J. Acarol.* 7(1–4):197–202.

Moser, J.C. and L.M. Roton. 1971. Mites associated with southern pine bark beetles in Allen Parish, Louisiana. *Can. Entomol.* 103:1775–1798.

Moser, J.C., R.C. Wilkinson and E.W. Clark. 1974. Mites associated with *Dendroctonus frontalis* Zimmerman (Scolytidae: Coleoptera) in Central America and Mexico. *Turrialba* 24:379–381.

Moss, W.W. 1962. Studies on the morphology of the trombidiid mite *Allothrombium lerouxi* Moss. *Acarologia* 4(3):315–345.

Mostafa, A-R. I. 1970a. Saproglyphid hypopi (Acarina: Saproglyphidae) associated with wasps of the genus *Zethus* Fabricus. Part. I. *Acarologia* 12(1):168–192.

Mostafa, A-R. I. 1970b. Saproglyphid hypopi (Acarina: Saproglyphidae) associated with wasps of the genus *Zethus* Fabricus. Part. II. *Acarologia* 12(2):383–401.

Mostafa, A-R. I. 1970c. Saproglyphid hypopi (Acarina: Saproglyphidae) associated with wasps of the genus *Zethus* Fabricus. Part. III. *Acarologia* 12(3):546–582

Norton, R.A. 1973. Phoretic mites associated with the hermit flower beetle, *Osmoderma eremicola* Knoch (Coleoptera: Scarabaeidae). *Am. Midl. Nat.* 90:447–449.

Norton, R.A. 1980. Observations on phoresy by oribatid mites. *Int. J. Acarology* 6(2):121–130.

Obenchain, F.D. 1974a. Structure and anatomical relationships of the synganglion in the American dog tick, *Dermacentor variabilis* (Acari: Ixodidae). *J. Morphol.* 142:205–224.

Obenchain, F.D. 1974b. Neurosecretory system of the American dog tick, *Dermacentor variabilis* (Acari: Ixodidae). *J. Morphol.* 142:433–446.

Obenchain, F.D. and J.H. Oliver. 1973. Qualitative analysis of form, function and interrelationships of fat body and associated tissues. *J. Exp. Zool.* 186:217–236.

Obenchain, F.D. and J.H. Oliver. 1975. Neurosecretory system of the American dog tick, *Dermacentor variabilis* (Acari: Ixodidae). *J. Morphol.* 145:269–294.

Paulus, H.F. 1979. Eye structure and the monophyly of the Arthropoda. *In:* Gupta, A.P., ed. *Arthropod Phylogeny.* Van Nostrand-Reinhold, New York, pp. 229–383.

Pauly, F. 1956. Zur Biologie einige Belbiden (Oribatei, Moosmilben) und zur Funktion ihrer Pseudostigmatischen Organe. *Zool. Jahrb. Syst. (Oekol.), Geogr. Biol.* 84:275–328.

Perkins, R.C.L. 1908. Some remarkable Australian Hymenoptera. *Proc. Hawaiian Ent. Soc.* 2:27–35.

Philippsen, W.J. and H.C. Coppell. 1978. *Uroobovella formosana* sp. n. associated with the Formosan subterranean termite, *Coptotermes formosanus* Shitaki (Acari: Uropodidae—Isoptera: Rhinotermitidae). *J. Kansas Entomol. Soc.* 51:22–27.

Phillis, W.A., III and H.L. Cromroy. 1977. The microanatomy of the eye of *Amblyomma americanum* and resultant implications of its structure. *J. Med. Entomol.* 13(6):685–698.

Pomerantzev, B.I. 1959. Ixodid ticks (Ixodidae). *Fauna of U.S.S.R. Arachnida* 4(2):3–199. (Transl. by A. Elbl and G. Anastos. AIBS, Washington, D.C.)

Pringle, J.W.S. 1955. The function of the lyriform organ of arachnids. *J. Exp. Biol.* 32:270–278.

Roshdy, M.A., R.F. Foelix and R.C. Axtell. 1972. The subgenus *Persicargas* (Ixodoidea, Argasidae, *Argas*). Fine structure of Haller's organ and associated tarsal setae of adults. *A. arboreus* Kaiser, Hoogstral & Kohls. *J. Parasitol.* 58(4):805–816.

Savory, T. 1977. *Arachnida,* 2nd ed. Academic Press, London.

Sharov, A.G. 1966. *Basic Arthropodan Stock with Special Reference to Insects.* Pergamon, Oxford.

Sixl, W., E. Dengg and H. Waltinger. 1973. SEM investigations on ticks, with special regard to the sensillae, Haller's organ and apical palpal segment. *Proc. Int. Congr. Acarol., 3rd, 1971,* pp. 53–54.

Slifer, E.N. 1970. The structure of arthropod chemoreceptors. *Annu. Rev. Entomol.* 15(4):121–142.

Sonenshine, D.E. 1970. A contribution to the internal anatomy and histology of the bat tick *Ornithodoros kelleyi* Cooley and Kohls, 1941. II. The reproductive, muscular, respiratory, excretory, and nervous systems. *J. Med. Entomol.* 7(3):289–312.

Tarman, K. 1961. Uber Trichobothrien und Augen bei Oribatei. *Zool. Anz.* 167:51–58.

Tombes, A.S. 1979. Comparison of Arthropod neuroendocrine structures and their evolutionary significance. *In:* Gupta, A.P., ed. *Arthropod Phylogeny.* Van Nostrand-Reinhold, New York, pp. 645–667.

Trave, J. 1968. Sur l'existence d'yeux lateraux depigmentes chez *Eobrachychthonius* Jacot (Oribatei). *Acarologia* 10(1):151–158.

Treat, A.E. 1975. *Mites of Moths and Butterflies.* Cornell Univ. Press (Comstock), Ithaca, New York.

Tsvileneva, V.A. 1964. Architectonics of nervous elements in the brain of ixodid ticks (Acarina: Ixodidae). *Entomol. Obozr.* 44:241–257.

Tsvileneva, V.A. 1965. The nervous structure of the ixodid synganglion (Acarina, Ixodidae). *Entomol. Rev.* 44(2):135–142.

Vijayambika, V. and P.A. John. 1975. Internal morphology and histology of the fish mite *Lardoglyphus konoi* 3. Nervous System. *Acarologia* 17(1):114–119.

Vistorin-Theis, G. 1977. Ernährungsbiologisch Beobachtungung an Calyptostomatiden (Acari: Trombidiformes). *Zool. Anz.* 199(5–6):381–385.

Vitzthum, H.G. 1943. Acarina. *Bronns Kl.* 5:1–1011.

Wharton, G.W. and L.G. Arlian. 1972. Predatory behaviour of the mite *Cheyletus aversor. Anim. Behav.* 20:719–723.

Woodring, J.P. and E.F. Cook. 1962. The internal anatomy, reproductive physiology and molting process of *Ceratozetes cisalpinus. Ann. Entomol. Soc. Am.* 55(2):164–180.

Woodring, J.P. and C.A. Galbraith. 1976. The anatomy of the adult uropodid *Fuscouropoda agitans* (Arachnida: Acari) with comparative observations. *J. Morphol.* 150(1):19–58.

Woolley, T.A. 1972. Some sense organs of ticks as seen by scanning electron microscopy. *Trans. Am. Microsc. Soc.* 91(1):35–47.

Zolotarev, Y.K. and Y.Y. Sinitsyna. 1965. Chemoreceptive organs on the forelegs of ixodid ticks. *Vestn. Mosk. Univ., Ser. 6: Biol., Pochvoved.* 20(1):17–25 (translation 314 (T314), NAMRU, Cairo).

12

Reproductive System

And everything that creepeth upon the earth after his kind.
—GENESIS 1:25

A life history is not merely an assortment of adaptations but rather a system of adaptations turned to a particular environment.
—NICHOLAS C. COLLINS, 1973

INTRODUCTION

The variations in the reproductive systems of arachnids are unusual. The class is said to have three major types of gonads: (1) a ladder type (*Limulus*, scorpions), (2) paired saccular types (several orders), and (3) a single saccular type (pseudoscorpions) (Gupta, 1979, p. 528). The general arrangement of organs is distinctive for almost every group exemplified below.

Scorpionida: Network of tubules with longitudinal and transverse ducts in both males and females.

Pseudoscorpionida: Median ovary covered with follicles; median testes; paired oviducts and paired vasa deferentia.

Araneida: Two ovaries covered with follicles; two testes; paired oviducts and paired vasa deferentia.

Opiliones: Horseshoe-shaped ovary and U-shaped testis; anterior median oviduct; anterior median vas deferens; has a penis.

Acari: Paired, unpaired, single, and multiple gonads; single, paired, or unpaired oviducts and vasa deferentia; variations of accessory glands and genital openings; aedeagus present or absent.

The differences observed are related in part to the varied types of insemination, which are results of adaptation to terrestrial living. Spermatophores are common in the Pseudoscorpionida and Acari; such passive sperm transfer is considered primitive. Modifications of male pedipalps in spiders, legs III in the Ricinuelida and specialized chelicerae in Acari also facilitate this passive insemination. The intromittent organ, the aedeagus, and copulatory insemination that occur in Opiliones and in some Acari is an advanced condition (Savory, 1977).

Reproduction

The arrangement of the reproductive organs in mites is typically arachnidan in that, in general, they exhibit the paired saccular type of gonads (Legendre, 1968). Each member of the pair is derived from a coelomoduct that arises from mesodermal primordia and opens in the midline of the genital segment. Ovaries and testes are alike in that they may be fused, single, paired, or multiple. The eggs of the embryos of arachnids generally develop in pouches or pockets that project from the ovary into the hemocoel (Gupta, 1979, p. 578). Modifications have tended toward the unpaired condition with median gonads in both male and female systems. Single gonads and paired and single gonadal ducts are found. Accessory glands occur in both sexes but may be much more complex in the males.

Chromosomes

Extensive diversity occurs in the chromosomes; meiosis and sex determination in the different orders of arachnids and mites are no exception. Two types of chromosomes occur in the Acari, that is, monokinetic, with one centromere per chromosome, and holokinetic, having centromeres extended along the entire length of the chromosome ("diffuse centric") (Oliver, 1977). Well over 100 species of Acari observed in chromosome studies have demonstrated that mites in general have a low or moderate number of chromosomes. The diploid counts range from 3 to 36, but most diploid chromosome numbers are in the lower ranges (2 to 4). Relatively few organisms share this feature with mites (Oliver, 1964, 1977; Hansell et al., 1964; Oliver and Nelson, 1967; Oliver and Delfin, 1967).

Following is a brief summary of chromosome numbers (Oliver, 1964, 1966, 1971b; Oliver et al., 1973; Regev, 1974):

Gamasida	3–18
Ixodida	16–33
Terrestrial prostigmata	3–26
Hydrachnida	4–21
Astigmata	4–10
Oribatida	18–36

In *Ophionyssus natricis* the chromosome complement is n:9 and $2n$:18, a haplodiploid type of sex determination. Unexplained irregularities occur in both virgin and fecundated females (Oliver et al., 1963). Chromosomes numbers cited in Oliver (1971b) are:

	Male	Female
Gamasida		
O. bacoti	8	16
O. natricis	9	18
O. sylviarium	?	?
D. gallini	3	6
Typhlodromus occidentalis	3	6
T. caudiglans	4	8
Actinedida:		
Pyemotes ventricolor (?)	3	6
Siteroptes graminum	3	6
Raoiella indica	2	4

The chromosome numbers for *Cheyletus malaccensis* are two for males and four for females (Regev, 1974). Chromosome numbers are summarized by Oliver (1977) for species of ticks, water mites, Tetranychoidea, and some Gamasida. (This review is an excellent source of references on the subject of cytogenetics.)

In *Siteroptes graminum* two classes of eggs are found related to chromosome number. Fertilized eggs have a diploid number of six, which determines the female. Unfertilized (parthenogenetic) eggs, which are haploid with three chromosomes, develop into males (Cooper, 1937). The length of chromosomes is limited and controlled by the cross-sectional area of the metaphase spindle during cleavage. Each chromosome appears to shorten differentially during the prophases of both meiosis and cleavage (Cooper, 1939).

Sex Determination

Available reports show a wide range of sex determination extensive plasticity exists. The several types of mechanisms include XY, XO, and multiple sex chromosomal mechanisms; most hard ticks have XO. Variations of sex determination occur within the orders of mites and no particular type seems to be restricted to any order. In many species, the sex-determining mechanism is unknown at present (Oliver, 1977).

Species of both major families of ticks have recognizable sex chromosomes. In the Argasidae, species of *Argas* and *Ornithodoros* have an XX:YY sex determination. The male is heterogametic. In these instances the sex chromosomes are frequently several times longer than the longest autosomes. In the family of hard ticks, Ixodidae, sex is determined by different methods. Specific chromosome studies show that in *Dermacentor* the diploid number is 21 in the male and 22 in the female (Oliver, 1964).

Sex chromosomes determine the sex in *Argas columbarum* in which the male is also heterogametic. Theletoky is reported for *Ornithodoros moubata* and four species of Anoetidae (the haploid number is four in *Anoetus laboratorium*.) The most common method of sex determination is arrhenotoky (Oliver et al., 1963).

Species of *Ixodes* have XX females and XY males. Species of *Boophilus, Haemaphysalis, Hyalomma, Rhipicephalum, Dermacentor,* and *Aponomma* have XX females and XO males. Species of the genus *Amblyomma* show variations, that is, XX:XY, XX:XO, and some mechanisms of multiple sex chromosomes, namely, X1X1X2X2:X1X2Y, with the males being heterogametic. The sex chromosomes are identified by their large size.

A population of *Amblyomma moreliae* from the vicinity of Sydney, Australia, shows X1X1X2X2:X1X2Y while another population of the same species at Brisbane, 600 miles away, has the XX:YY sex determination system. Both of these systems are assumed to have developed from XX:XO types, since this is the most common mechanism that occurs in the family. Derivation was probably the result of reciprocal chromosome translocation in the Sydney form and a translocation with loss of the centromere of original sex chromosomes in the Brisbane population (Oliver, 1966).

In the Tetranychinae, sex determination is haploid–diploid (arrhenotokous); in the Bryobinae, reproduction is thelytokous; males are rare or unknown in many species (Huffaker et al., 1969). A "normal" sex ratio does not exist in tetranychid mites (Boudreaux, 1963). (See also Helle and Gutierrez, 1983; Helle and Wysoki, 1983; Potter, 1979).

Sexual Dimorphism

Mites are dioecious and frequently have sufficient sexual dimorphism for males to be distinguished from females by external features. Diagnostic features include smaller size in males, cheliceral modification, hypertrophied legs, tarsal or opisthosomal copulatory suckers, specific differences in sclerotization of epimera, coxae, ventral plates, or opisthosoma, size and location of the genital aperture, and differential setation. In relatively few mites a definite aedeagus (=penis) is a more or less pronounced male characteristic. Secondary sexual characteristics are manifest in Astigmata; sexual dimorphism is strongly or feebly demonstrated in oribatid mites. Bisexual forms are most common with about equal numbers of males and females. Laboratory inbred females of *Ornithonyssus bacoti* are more variable than males (Oliver and Herrin, 1974).

The presence of eggs, larvae, or protonymphs within the body of a mite is visual proof of the female sex, but differences in the position and sclerotization of the genital opening, internal vaginal sclerites or external genital apodemes, sperm induction pores, and copulatory bursa are also female characteristics. The greatly swollen hysterosoma (physogastry) of the Pyemotidae (Fig. 12-1) is a striking phenomenon that readily identifies a gravid female (Cross, 1965). In some instances an ovipositor (some Ixodidae, Endeostigmata, and Oribatida) identifies the female.

Reproductive openings for both sexes of

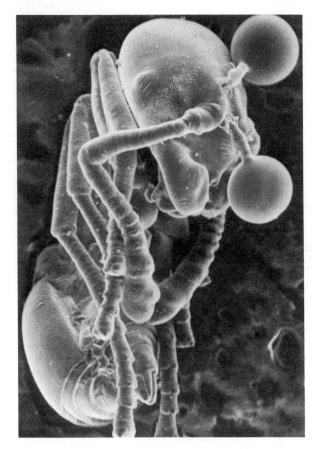

Figure 12-1. A. physogastric females of *Pyemotes tritici* on the head of *Solenopsis invicta* (SEM courtesy of USDA, W.A. Bruce, Savannah, GA).

which forms a sclerotized, chitinous intima within the genital atrium and/or vagina.

Slight sexual dimorphism occurs in the Opilioacarida (Hammen, 1977) (Fig. 12-2A, B). Setal locations and differentiation at the genital opening are major features. The several genital plates of female Holothyrida distinguish them from males, together with differences in sclerotization.

External dimorphism in the Gamasida is identified by chelicerae, legs, and gonopore. In

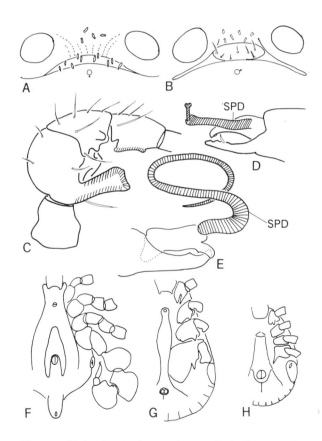

Figure 12-2. Genital openings of Opilioacarida, *Salfacarus legendrei*: (A) female; (B) male (after Hammen, 1977). (C) Crassate leg of *Pergamasus robustus*; male spermadactyls (SPD) of (D) *Typhlodromus reticulatus* and (E) *Pachylaelaps strigifer* ((C,E) after Vitzthum, 1943; (D) after Chant, 1955); ventral views of males of ticks: (F) *Margaropus winthemi*; (G) *Dermacentor occidentalis*; (H) *Amblyomma americanum*.

mites are usually on the venter between the coxae of legs II to IV, but the male genital opening in Tarsonemidae, Demodicidae, and Cheyletoidea may be dorsal. A more posterior and terminal genitoanal area occurs in several families of Actinedida (Labidostommatidae, Rhaphignathidae, Stigmaeidae, Caligonellidae, Cryptognathidae, Pomeranziidae, Tuckerellidae, Linotetranidae, Pterygosomidae, Tenerifiidae) and in the family Hemisarcoptidae in Astigmata.

In most females, the genital aperture is developed through an inflection of the body wall,

Dermanyssus gallinae, the males have tapering sides and the females are parallel-sided (Pound and Oliver, 1976). Chelicerae of gamasid males may be modified for sperm transfer. The spermadactyl (spermatophoral process) on the movable digit of male chelicerae is a sexually diagnostic feature, as well as a functional appendage for sperm transfer (Fig. 12-2D, E). Such males introduce sperm into sperm induction pores (podospermy) of females separate from the genital opening. Gamasid males that lack the spermadactyl introduce the sperm directly into the genital tract of the female (tocospermy). Such males (e.g., Parasitidae) may exhibit a small integumental boss or ridge on the movable digit of the chelicerae.

In the Gamasida, one or more pairs of legs may be modified for copulation. The crassate legs of Parasitidae, for example, have enlarged apophyses used in grasping the female during coition (Fig. 12-2C). The sternal shield is fused with the genital and ventral shields. The male genital aperture may be located anteriorly, ahead of the sternigenital shields, in the sternal region between coxae II, or more posteriorly between coxae III.

Most females show a distinctive genital shield (epigynial) posterior to the sternal shield. In parasitic gamasid females there may be a reduction of the genital and sternal shields or a complete loss (Fig. 4-20A). In some instances the genital opening is simple and unsclerotized. Some modifications of the epigynial shield arrangement occur in the trigynaspid forms (Fig. 4-20B). The latigynial and mesogynial shields and several variations of fusion of the genital and sternal or ventral plates are diagnostic for these families. Internal vaginal sclerites of different types serve to identify the females in some families (e.g., Macrochelidae). Claviform vaginal sclerites in a number of trigynaspid families are very characteristic. The genital opening and apodematal structures of the parasitic gamasid mites is reviewed by Fain (1963).

Sexual differences in the soft ticks (Argasidae) are indistinct and difficult to distinguish because males resemble the females. Sexual dimorphism in the hard ticks (Ixodidae) is more pronounced. Males show a complete scutum on the dorsum. In some males, enlarged coxal plates of legs IV (*Dermacentor occidentalis*) or the greatly enlarged legs IV (*Margaropus winthemi* (Fig. 12-2F), *Aponomma exornatus*) are very diagnostic and functionally important. Enlarged coxal plates and hypertrophied leg segments serve to prevent contact of the ventral surfaces of a copulating pair, a function attributable to anal shields as well. Removal of fecal materials during copulation occurs freely between the surfaces of the male and female. Were it not for this modification of the coxal plates and the enlarged legs, guanine of the feces would glue the sexes together inseparably during coition (Pomerantzev, 1959).

Female hard ticks have a partial scutum on the anterior dorsum; the alloscutum is pliable and facilitates engorgement. Female ixodids are also differentiated by two porose areas on the dorsal surface of the basis capitulum. In some female hard ticks, a slight, retractile ovipositor may be present.

Sexual dimorphism in the Actinedida ranges from relatively indistinct morphological features to hypertrophied, sometimes bizarre structural differentiation. This order has families in which the males exhibit separated anogenital openings, while those openings of the female are contiguous (Labidostommidae). The position of the female gonopore is of generic significance in the Ophioptidae. The opening is midventral at the level of leg II in *Afrophioptes* and posterior in *Ophioptes*. Tarsal claws may differentiate the male from the female (Halacaridae) or the contrast between the flagelliform legs IV of females and grasping legs IV of males may be diagnostic (Tarsonemidae). Peritrematal configurations over the stylophore separate females (more complex) from males in Caligonellidae. Spermathecal tubes of the female Eriophyidae are diagnostic at the familial level. Male palps are sexually diagnostic in water mites. Species of Myobiidae are differentiated by the configurations of the vulvar region of females and the aedeagus of males (Lukoschus and Driessen, 1970).

A distinct aedeagus occurs in many male Actinedida. In some, this intromittent organ is posterior (e.g., Iolinidae, Cheyletidae); in others the aedeagus is dorsal (Cheyletiellidae, Myobiidae, Demodicidae, Cloacaridae). Podapolipidae show variations of position of the aedeagus from posterior and terminal to an enormously hypertrophied dorsal aedeagus as long as the body (Figs. 12-15, 12-16).

The majority of the Astigmata are sexually dimorphic in both primary and secondary features. Within some species, the males show two types: (1) a homeomorphic form similar in structure to that of the female; and (2) a heteromorphic form characterized by enlargement of legs III or IV and hypertrophied secondary sex characters (Zakhvatkin, 1959). Four kinds of polymorphs are known in the males of the Acarida: homeomorphic, bimorphic, pleomorphic, and heteromorphic (Mahunka, 1979). Males frequently have an elongated, sclerotized aedeagus accompanied by tarsal or anal suckers. In one instance (Crypturoptidae), the sperm duct is evidently an intromittent organ (Krantz, 1978).

The males of stored products mites are smaller than the females and are further distinguished by the tarsal copulatory suckers on legs IV and the aedeagus (Hughes, 1961); some may also have copulatory suckers near the anus (Fig. 12-3C). The larger females frequently exhibit a terminal or posterior bursa copulatrix separate from the midventral genital opening (*Phyllostomonyssus conradyunkeri*). In the Gastronyssidae, the copulatory bursa is dorsal in position (Fig. 12-3D) (Fain, 1970a). Sperm transfer is made into this bursa (or sperm duct) rather than into the genital opening. Because the term "genital aperture" is misleading in the Astigmata and the opening is used mainly for oviposition, Krantz (1978) recommends the use of the term "oviporus" to make the differentiation. The oviporus may extend longitudinally or transversely. In longitudinal types, the

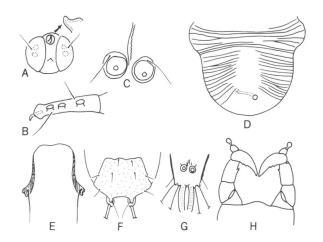

Figure 12-3. (A) Male aedeagus and (B) tarsal IV copulatory suckers of *Tyrophagus*; (C) anal suckers of male *Rhizoglyphus* (after Hughes, 1961); (D) dorsal copulatory bursa of female Gastronyssidae (*Phyllostomonyssus conradyunkeri*) (after Fain, 1970a); (E) female genital apodemes in Psoroptidae; (F) posterior of male *Chorioptes*; (G) *Proctophyllodes*; (H) hypertrophied legs II of Xolalgidae (after Atyeo, 1974).

opening may have sclerotized paragynial and epigynial elements associated with it. A pregenital sclerite, the epigynium, is found in both males and females. This sclerite may be separate or coalesced with the apodematal structure of adjacent coxae.

When a transverse genital aperture occurs in the female (e.g., Psoroptidae) paired genital apodemes may occur adjacent to the opening (Fig. 12-3E). Legs III of females are reduced and flagellate at the distal tip in contrast to the grasping legs III of the male. Some psoroptoid males have reduced legs IV and terminal lobulated opisthosomal projections (Fig. 12-3F). Legs IV are also reduced in Teinocoptidae and anal lobes are present. Legs II of the Xolalgidae are extremely hypertrophied and sicklelike at the distal end (Fig. 12-3H). Enlarged legs III or IV occur in males of some families together with anal suckers and the lobate posterior modifications of the opisthosoma (Fig. 12-3G) (Lobalgidae, Proctophyllodidae, Trouessartiidae, Alloptidae, Freyanidae, Gabucinidae, Listrophoridae, Analgidae).

In the Oribatida, body size, position of the rostral hairs, size of the genital aperture, together with the appearance of the ventral epimeral ridges are dimorphic features distinctive of sexes in *Epilohmannia* (Wallwork, 1962). A distinctive, protrusible female ovipositor and a male aedeagus (penis) are noticeable features that differentiate females and males among most beetle mites. Sexual dimorphism is recognized in the genera *Collohmannia, Epilohmannia, Pirnodus,* and *Parakalumma* (Mahunka, 1979). In addition to the presence of ovipositor and penis, in *Ceratozetes cisalpinus* the length and width of the genital plates are diagnostic for sex (Woodring and Cook, 1962).

Gynandromorphism—Sex Mosaics

Genetic abnormalities and asymmetrical development results in odd features in the bodies of mites. In *Caloglyphus dampfi* males are known where one side is heteromorphic, with large legs III, while the other side shows typical homeomorphic male features. Similar anomalies are recorded for some spider mites. Water mites may show abnormalities in the external reproductive apparatus. Many abnormalities are known among the ticks: hypertrophied leg structures, enlarged eyes, and secondary sex characters (Vitzthum, 1943). A phenomenon of particular interest is the condition of gynandromorphism, or sex mosaics. Intersexes are known among insects. Within the arachnids, gynandromorphism is known mainly among spiders, and the phenomenon is related to the behavior of the chromosomes. Temperature and gonadal infection by nematodes are cited as causes of this abnormality. Six arrangements of gynandromorphism are known (Savory, 1977).

Within the Acari, gynandromorphs are reported mainly from ticks. The frequency is relatively high in the genus *Amblyomma* (Rechav, 1977). In a field collection of *Amblyomma imitator* from Mexico, one gynandromorph occurred in 78 males and 14 females. The specimen was predominantly male on the left side and female on the right side (Fig. 12-4C, D) (Sundman, 1965). Most instances show the female characters on the left side and male characters on the right (Campana-Rouget, 1959). The opposite is cited by Vitzthum (1943) (Fig. 12-4A, B). A gynandromorph of *Amblyomma hebraeum* (Fig. 12-4E, F) from South Africa shows dorsal features of the female, but tegument and festoons of the male on the left side. The spiracles and legs III, IV on one side are of male character, but the genital opening is of the female type; the anus is typically male (Rechav, 1977).

Treatment of male *Dermacentor occidentalis* with apholateacetone solutions resulted in malformation, including two bilateral gynandromorphs. One had masculine and intersex characters on the right side. Males were attracted to these forms and evidently mated. One gynandromorph died without ovipositing; the other oviposited a small number of eggs (785). Larvae, nymphs, and adults were normal (Oliver and Delfin, 1967).

The various hypotheses for the development of gynandromorphs include partial fertilization of the developing egg, polyspermy with development of the fusion nuclei and accessory sperm (zygogenesis and androgenesis) in the same egg, elimination of sex chromosomes and nondisjunction of chromosomes, double fertilization in which one sperm fertilizes a polar body and another sperm fuses with the female pronucleus, two sperm fertilizing a binucleate egg, and somatic crossing over. Evidence supports each of these explanations, but other means could also be involved (Oliver and Delfin, 1967).

Lateral	Transverse		Crossed	
(1) M/F	(3) M	(4) F	(5) M/F	(6) F/M
	—	—	—	—
(2) F/M	F	M	F/M	M/F

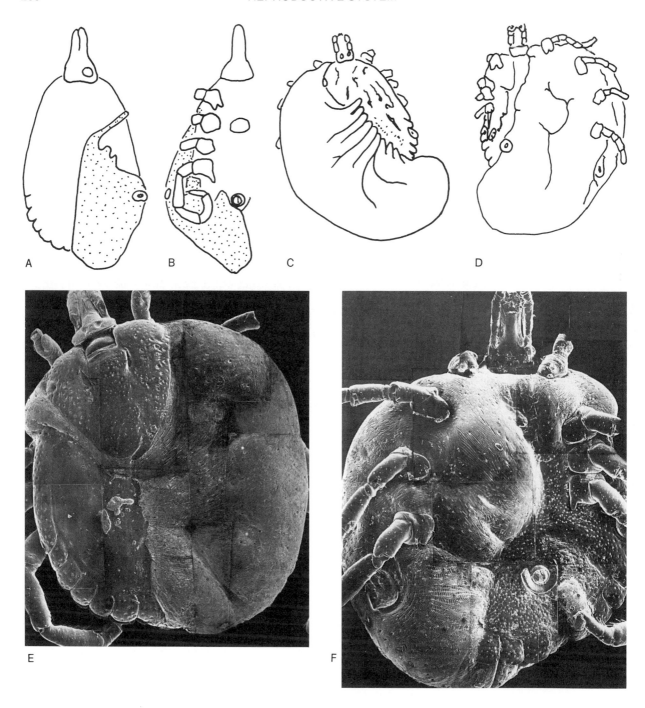

Figure 12-4. Gynandromorphs in ticks: (A) *Haemalaster neumanni*, dorsal; (B) ventral (after Vitzthum, 1943); (C) *Amblyomma imitator*, dorsal, (D) ventral (after Sundman, 1965); (E) *Amblyomma hebraeum*, dorsal, (F) ventral (SEMs by courtesy and permission of Y. Rechav, Tick Research Unit, Grahamstown, S. Africa).

THE MALE REPRODUCTIVE SYSTEM

Organs

The male system generally consists of testes, vasa deferentia, an ejaculatory duct, and associated accessory glands. Lower Gamasida and many Ixodida have unpaired testes. Testes are paired in Uropodidae (Woodring and Galbraith, 1976), in *Dermacentor andersoni* (Douglas, 1943), *Hyalomma asiaticum* and *Argas persicus* (Balashov, 1972), *Ornithodoros kelleyi* (Sonenshine, 1970), and the Acaridae (Rohde and Oemick, 1967). Multiple testes are found in some Actinedida; single testes are typical of Demodicidae and Eriophyidae.

Gamasida. Testes in the gamasid mites vary from fused medial structures such as those found in *Pergamassus crassipes* (Fig. 12-5A) and *Haemogamasus reidi* (=*ambulans* (Fig. 12-5E) (Hughes, 1959; Young, 1968) to bilobed forms with a tenuous medial connection. The paired, tubular condition is considered to be primitive (Hughes, 1959).

Germinal epithelium (germarium) is apparently limited in some mites to the apex of the unpaired testis (Young, 1968), and the histology changes abruptly at the vas deferens. In others (*Haemogamasus, Gymnolaelaps, Halarachne*), the germinal epithelium is uniform and continuous.

In some Uropodines, the testes are paired with a short vas deferens and a single accessory gland at the junction between the sperm duct and the ejaculatory duct (Hughes, 1959). Paired accessory glands are associated with the ejaculatory duct but little is known of their functions.

The testes of the uropodine *Leiodinychus krameri* consist of two lobes on each side of a massive accessory gland. The vas deferens leading from these paired testes extends to a substantial ejaculatory duct at the base of which are smaller accessory glands (Vitzthum, 1943).

The male reproductive system of *Fuscouro-poda agitans* (Fig. 12-5C) consists of a pair of testes, paired vasa efferentia, paired vas deferentia, seminal receptacle, a multilobed accessory gland, and the genital atrium, which acts as an ejaculatory duct. A definitive ejaculatory duct is not identified. About 60 to 80 spermatogonial cells enlarge and are surrounded by a membrane posterior to this germinal ring. Meiosis is synchronous so that all of the cells are in the same stage of reduction division within each cyst. As the cysts move posteriorly in the testis the mature sperm are located distally. When the cysts rupture, probably by hemolymphic pressure, the sperm are released into the lumen of the vas deferens and are moved by peristaltic action to the seminal vesicle. In this organ, the sperm are mixed with fluids from the rather complex accessory glands as the semen is extruded by compressions of the atrial wall. Hemolymphic pressure is assumed to force the sperm into the atrium and to cause the aperture to open (Woodring and Galbraith, 1976).

In contrast to the paired testes of Uropodina, the male gonads of gamasid mites such as *Pergamasus crassipes, P. viator,* and *Haemogamasus reidi* are single. In *P. crassipes*, the testis is fused and spherical with a germarium at its apical end (Vitzthum, 1943; Hughes, 1959) and in *P. viator* (Witalinski, 1975).

The male reproductive system in *Dermanyssus gallinae* has two semicircular halves that appear as a single, circular testis with a germarium (Fig. 12-5D). A thin vas deferens, a seminal vesicle, and accessory glands are also present. A single median accessory gland is flanked by a much larger, bilobed ventral accessory gland. The ejaculatory duct extends from the seminal vesicle to the gonopore (Pound and Oliver, 1976).

In *Haemogamasus reidi,* a nidicolous mite and ectoparasite of rodents in Europe, Asia, and North America, the male reproductive system is of the "ring-form" type (Fig. 12-5F). It consists of a cordate-shaped or pyramidal fused pair of testes with an apical germarium, two vasa deferentia joined to an ejaculatory duct and lobate accessory glands at the junction of

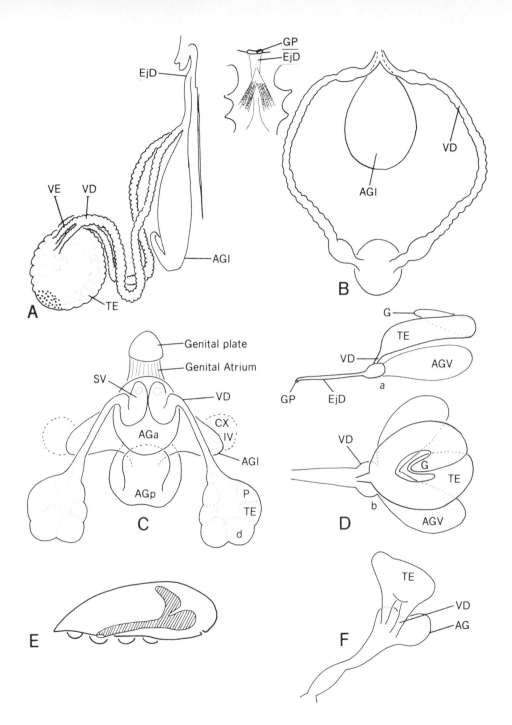

Figure 12-5. Male reproductive system of Gamasida: (A) *Pergamasus crassipes* (after Vitzthum, 1943); (B) *Pergamasus viator* (after Witalinski, 1975); (C) *Fuscouropoda agitans* (after Woodring and Galbraith, 1976); (D) *Dermanyssus gallinae* (after Pound and Oliver, 1976); (E) *Haemogamasus reidi* from lateral aspect, (F) enlarged (after Young, 1968) (AG, accessory gland; AGa, anterior; AGl, lateral; AGp, posterior; AGv, ventral; EjD, ejaculatory duct; G, germarium; GP, gonopore; SV, seminal vesicle; TE, testes, TEd, distal, TEp, proximal; VD, vas deferens; VE, vas efferens).

the vasa deferentia with the ejaculatory duct that extends to the genital opening. The epithelial histology of the vasa deferentia and ejaculatory duct is similar. The accessory glands are composed of large spongy cells; the glands are enclosed in a network of muscle fibers.

Large spermatogonia develop in the apex of the germarium at the tip of the testis. Each spermatogonium divides within an enclosing cyst that will contain 25 to 40 cells on maturity. The cysts are moved into the vas deferens, where the cyst membrane disappears and the cells become packed together. A secondary sperm cyst is formed in the vasa deferentia, and maturation occurs. The vasa deferentia serve both as sperm ducts and as seminal vesicles.

During mating, the sperm cysts pass into the ejaculatory duct, mix with fluid from the accessory glands, and are contained in a spermatophore. The spermatophore, formed in part by pressure of the dorsoventral muscles, is ejected and hardens on contact with the air. The length of the spermatophore varies (Young, 1968).

Ixodida. In ixodid ticks, the testes are paired, long, tubular forms with a complex set of accessory glands (Khalil, 1969; Arthur, 1962; Sonenshine, 1970; Roshdy, 1961), but in a number of instances the testicular tubes also have a small connection between their distal ends.

In argasid ticks (e.g., *Argas persicus*), the testis begins as an arched primordium in the larvae. The mass lacks evidence of genital differentiation, but consists of two principal cell types. Larger cells with vesicular nuclei are the primordial germ cells. Smaller cells form the seminal follicles and the mass is covered with epithelium and an acellular connective tissue. When the larvae have fed, the primordia of the testes enlarge, the vasa deferentia become more distinct, and the differentiation of the gonads continues through the nymphal stages (Balashov, 1972). Arrangements of the male organs are similar in *Ornithodoros kelleyi* (Sonenshine, 1970), *Argas persicus* (Fig. 12-7A) (Roshdy, 1962, 1966; Tatchell, 1962).

In the ixodid ticks, the testes are paired and elongated (Fig. 12-7B). They connect by narrowed vasa deferentia to the ejaculatory duct at the base of a complex, lobulated accessory gland. Variations of the lobes occur in different

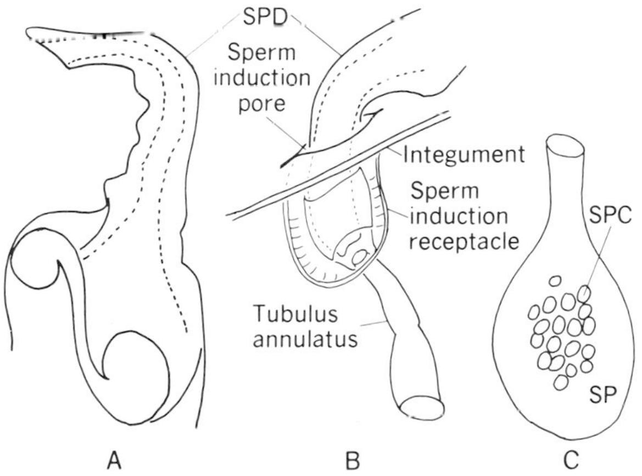

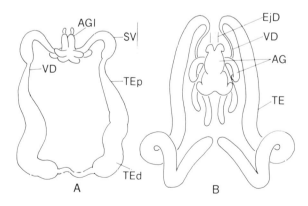

Figure 12-6. *Haemogamasus reidi*: (A) view of spermadactyl; (B) enlarged view, inserted into induction pore; (C) spermatophore (after Young, 1968) (SP, SPC, sperm cycsts; SPD, spermadactyl).

Figure 12-7. Male system of (A) *Argas persicus* (after Balashov, 1972); (B) *Dermacentor andersoni* (after Douglas, 1943) (AGl, accessory gland; EjD, ejaculatory duct; SV, seminal vesicle; TEd, distal testis; TEp, proximal testis).

species (*Dermacentor andersoni, Hyalomma asiaticum*) (Balashov, 1972; Douglas, 1943; Tatchell, 1962). Differentiated secretions of the lobes relate to the formation of the spermatophore.

Actinedida. Many Actinotrichida males (Tetranychidae, Acaridae) have a discrete, sepa-

rate pair of testes (Woodring and Galbraith, 1976) while others show an unpaired gonad (Anystidae, Halacaridae, Demodicidae, Eriophyidae). In others, for example, Bdellidae and Erythraeidae, the testes may be subdivided into a number of vesicles, and in some water mites, the testicular vesicles may differ from the common form.

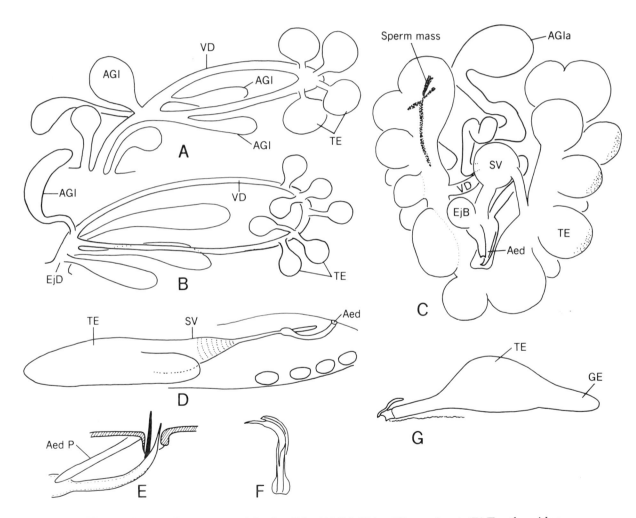

Figure 12-8. Male systems of Actinedida: (A) Bdellidae (*Neomolgus*); (B) Erythraeidae; (C) Trombidiidae (*Allothrombium fuliginosum*) ((A,B) after Hughes, 1959); (C) after Vitzthum, 1943); (D) male system of *Demodex antechini*; (E) view of aedeagus; (F) enlarged view of aedeagus (after Vitzthum, 1943); (G) *Eriophyes pini* (AGl, accessory gland; Aed, aedeagus; Aed P, aedeagus protractor E; B, ejaculatory bulb; EjD, ejaculatory duct; GE, germarium; PR, protractor; SV, seminal vesicle; TE, testis; VD, vas deferens).

The male system of *Neomolgus* (Bdellidae) may exhibit multiple testes and complex arrangements of accessory glands (Fig. 12-8A) (Hughes, 1959). The testes of male Erythraeidae are stalked, again with a ringlike vasa deferentia, but with different configurations of the accessory gland and ejaculatory duct (Fig. 12-8B) (Hughes, 1959). The testes in some Trombidiidae (*Allothrombium fuliginosum*) are lobulated but arranged in a horseshoe shape and connected to a spherical seminal vesicle by two short sperm ducts. Tubular and spherical accessory glands occur (Fig. 12-8C).

The testis in *Demodex antechini* is a single, U-shaped organ in the opisthosoma (Fig. 12-8D). It extends forward to a seminal vesicle (?sclerotized) that narrows anteriorly to become the vas deferens and connects to the curved aedeagus. On either side of the aedeagus is the aedeagal protractor that moves the organ dorsally out of the genital aperture at the level between legs I and II (Desch and Nutting, 1977). The pair of aedeagal protractors attach to a bulbous termination of the seminal vesicle and the beginning of the vas deferens (Fig. 12-8E). Eriophyidae males have a single testis, a simple vas deferens, and a genital opening (Vitzthum, 1943; Jeppson et al., 1975) (Fig. 12-8G).

Astigmata. The male system in cheese mites (Acaridae, etc.) (e.g., *Caloglyphus mycophagus*, Fig. 12-9A) consists of paired, pyriform testes, paired vasa deferentia, and a short ejaculatory duct (Rohde and Oemick, 1967). Accessory glands of various shapes attach to the ejaculatory canal (a single gland in Tyroglyphidae; two in Glycyphagidae). A copulatory organ, the aedeagus (or penis) lies in the genital slit of the male, partially hidden by overlying folds of cuticle (Fig. 12-3A). The penis is a stiff tube with supporting sclerites (apodemes) for muscle attachment. Sometimes an epiandrium, a chitinous platform, is arranged anterior to the genital opening. Associated with the penis are two pairs of genital suckers or acetabula. These can be extended outside the genital aperture. They

are thought to be chemoreceptors. Copulatory suckers are also found adjacent to the anal opening and on the tarsi IV (Fig. 12-3C).

Oribatida. The testes differ slightly in arrangement and size among families and between genera of beetle mites. They may be widely lobulate (*Cepheus tegeocranus*) (Fig. 12-9B), fused spheres associated with the seminal vesicle (*Belba geniculosa*) (Fig. 12-9D), or elongated sacs (*Nothrus theleproctus*) (Fig. 12-9C). Vasa deferentia are distinct and elongated in some, or reduced in size and integrated with the seminal vesicle. Separate accessory glands are present in some more primitive forms (*Cepheus tegeocranus*) (Fig. 12-9B).

Sperm

Arachnid sperm vary from the filiform (the most primitive type, found in scorpions) through a series of types to the immobile, encysted examples found in other orders (Gupta, 1979). The sperm of the Acari are specialized (Witalinski, 1979) and represent the highest type among arachnids (Fig. 12-10A) (Gupta, 1979). The mites possess fusiform, aflagellated sperm where the acrosome is single-layered, but the sperm are still motile. This motility, when present, is a new condition acquired secondarily by means of new organelles (Bacetti, 1979; Witalinski, 1979; Feldman-Muhsam, 1967). These organelles form a series of motile appendages replete with fibrils, but devoid of microtubules (Fig. 12-10B).

Synchronous meiosis in spermatogonial cysts is found in gamasid mites, uropodines, ixodids, and oribatids, but does not appear to occur in the Astigmata or in the Actinedid mites. "Synchronous cysts probably occur in all Anactinotrichida but in only one order of the Actinotrichida, the oribatids" (Woodring and Galbraith, 1976).

The spermatozoon of *Pergamasus viator* (Parasitidae) is rodlike with six morphologically different zones (acrosomal, mitochonrial, granular bodies, striated bodies, nuclear zone,

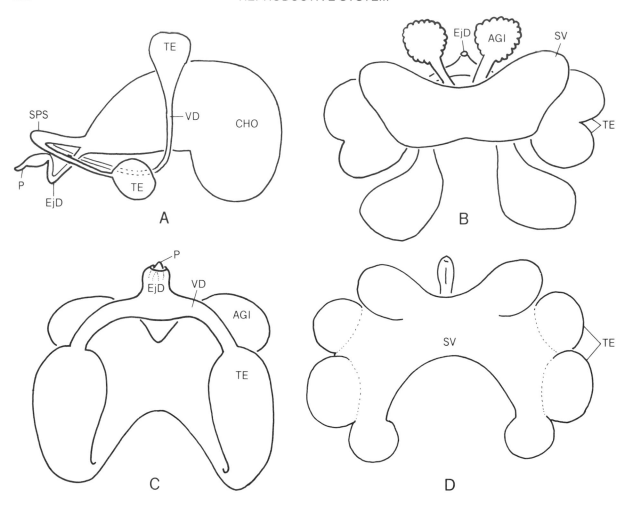

Figure 12-9. Male system of Astigmata and Oribatida: (A) *Caloglyphus mycophagus* from lateral aspect (CH O, chambered organ; EjD, ejaculatory duct; P, penis; SPS, sperm sac; TE, testis; VD, vas deferens). Male system of Oribatida: (B) *Cepheus tegeocranus*; (C) *Nothrus theleproctus*; (D) *Belba geniculosa* (after Michael, 1883) (AGl, accessory gland; EjD, ejaculatory duct; P, penis; SV, seminal vesicle; TE, testis; VD, vas deferens).

and retronuclear zone). Spermiogenesis involves the development of mitochondria, endoplasmic reticulum, and other intracellular modifications and constructions (Witalinski, 1979).

The production of sperm in *Dermanyssus gallinae* occurs within a circular testis (a fusion of two anterolateral testicular arms of the pro-

tonymphal stage). Haploid spermatogonial cells form haploid spermatids ($n = 3$) by mitosis. The process of spermatogenesis begins within an hour after the protonymph feeds, and continues through the adult. Feeding and nutrition are necessary for sperm production. Spermatids are rounded and have a reduction in cy-

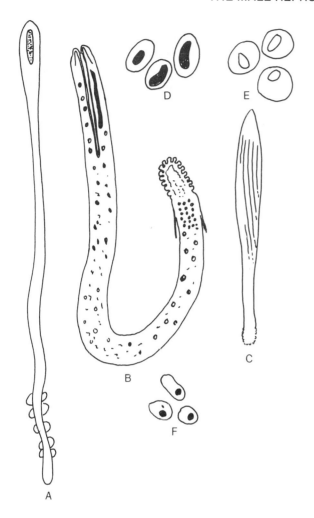

Figure 12-10. Spermatozoa of (A) Acarina (after Gupta, 1979); (B) *Amblyomma dissimilis* (after Bacetti, 1979); (C) Gamasida (after Vitzthum, 1943); (D) *Siteroptes graminum* (after Oldfield et al., 1970); (E) Eriophyidae (*Aceria cornutus*) (after Oldfield et al., 1970); (F) Oribatida (*Ceratozetes cisalpinus*) (after Woodring and Cook, 1962).

after the adult has molted (Pound and Oliver, 1976).

The spermatozoa of *Siteroptes graminum* are immobile with an ellipsoidal shape (Fig. 12-10D). The nucleus is flattened and somewhat concavoconvex in outline; it stains intensely with Feulgen and other types of nuclear stains. This simplified organization of sperm is similar to the type found in the Oribatida (Michael, 1883). Other authors describe the sperm of Erythraeidae, Trombidiidae, Anystidae, and the Hydrachnidae as aflagellated cells with short and thick wormlike tails (Cooper, 1939). Some authors may have described spermatids rather than mature sperm.

The sperm cells in Eriophyidae are about 2 μm in diameter and spherical in outline, with a dense, ellipsoidal nucleus (Fig. 12-10E). When transferred into the spermathecae of the protogynes, the cells are compacted, but retain the ellipsoidal shape of the nucleus (Oldfield et al., 1972).

The tubular testes of *Abrolophus rubipes* (Erythraeidae) produce spermatozoa in a dorsal part and store them in a ventral reservoir. The walls of the reservoir have glandular cells rich in ER and they secrete three types of proteinaceous secretory products, two of which are used in the construction of spermatophores.

The spermatozoa are spindle-shaped, with a threadlike nucleus. Mitochondria of the crista type surround the nucleus; smooth-walled cisternae occur peripherally. The sperm cells are covered by one of three types of secretory products from the reservoir of the testis; the other two products are used in the formation of the spermatophore. Spermatogenesis and delineation of the specific proteinaceous constituents of the testicular secretions comprise tyrosin, tryptophan, paraphenol, sulfhydryl forms, mucopolysaccharides, and lipids (Witte and Storch, 1973).

It is assumed that in oribatids the sperm or spermatids mature in the testes and are released as rather compact packets into the seminal vesicle (Fig. 12-10F). The cells of the seminal vesicle secrete a fluid in which the sperm is

toplasm. Although similar in appearance to spermatogonial cells, the spermatids exhibit dark-staining nuclei. Spermatocysts usually contain 18 to 36 cells, all in a synchronous state of division. Spermatogenesis begins in the protonymphal state and continues until 4 or 5 days distributed uniformly. The seminal vesicle also secretes the material from which the spermatophoral stalk is formed. This secretion is mixed with the other fluids. The mixture is forced into the vasa efferentia by hemolymphic pressure and moved along in the sperm ducts by peristaltic action into the ejaculatory duct and the penis. A sphincter in the seminal vesicles (at least in some species) controls the release of the mixture of semen and stalk materials into the sperm ducts. "The mixture of semen and stalk substance could not be propelled through the vasa deferentia by hydraulic pressure of the hemolymph because this would simply collapse the long, tubular vasa deferentia." The mixture is moved to the penis only when the male is ready to deposit a spermatophore. Hemolymphic pressure is the force that extrudes the penis, since only retractor muscles are attached to this organ (Woodring, 1970).

In *Ceratozetes cisalpinus*, sperm is inactivated and killed in both the males and spermatophores by temperatures of 35°C for 1 to 2 h. Even lower temperatures are thought to inactivate the sperm. The sperm sac on the spermatophore imbibes water quickly and often ruptures. In relative humidity of 50% or less the sperm sac shrivels within a day. It is also delicate enough that it can be dislodged or destroyed by a passing mite. Bacteria can penetrate the sperm sac and possibly enhance destruction. To assure fecundity, sperm needs to be fresh, not more than an hour old, when picked up by females (Woodring and Cook, 1962).

Sperm Transfer

Fertilization in the Acari is internal, which implies a transfer of sperm; but this is not always accomplished by direct, intromittent copulation. If the male possesses a spermadactyl on the chelicerae (Gamasida) (Fig. 12-6) (Krantz and Werntz, 1979), or an aedeagus (Fig. 12-15), the genital products may be introduced into induction pores, copulatory bursa, or directly into the genital tract of the female. Direct sperm transfer is known in the Ixodidae, some Actinedidae, and in most of the Astigmata. The second and most common mode of sperm transfer in Acari is an indirect or passive method that resembles the mechanism used by many other arachnids. It employs a sperm-carrier device, or spermatophore. Many Actinedida, a few Astigmata, and many oribatids use the spermatophore.

For a time it was believed that the principal method of sperm transfer in the gamasid mites was through the genital aperture of the female (Michael, 1892; Strandtmann and Wharton, 1958). Sperm induction pores were discovered later. It was then observed that the sperm were transferred into these auxiliary openings in the female by means of a hollow spermadactyl on the chelicerae of the males.

Sperm transfer is effected in two principal ways in the gamasid mites: tocospermy and podospermy (Athias-Henriot, 1970). Tocospermy is a condition where the tranfer is made directly into the genital tract of the female. Sperm are tranferred either directly or by means of a spermatophore. Spermadactyls are absent in males where tocospermy exists and the mites are haplodiploid in their genetic composition. This is typical of Uropodina, Parasitoidea, and some Ascoidea, for example.

The second type of transfer exemplified in Gamasida is podospermy, which involves secondary genital openings called sperm induction pores (Fig. 12-11). These induction pores are found in trochantera II, coxae IV, between coxae III and IV, and ventrolaterally on the surface of the idiosoma of the females involved. This type of sperm transfer is accomplished by means of hollow spermadactyls on the male chelicerae. These appendages are hollow auxiliary structures on the movable digit of the chelicerae. Sperm thus transferred into the induction pores passes into the genital tract of the female through the tubulus annulatus, ramus sacculus, and sacculus foemineus into the spermatheca of the female tract. Coxal insemination is suspected for *Dermanyssus gallinae*

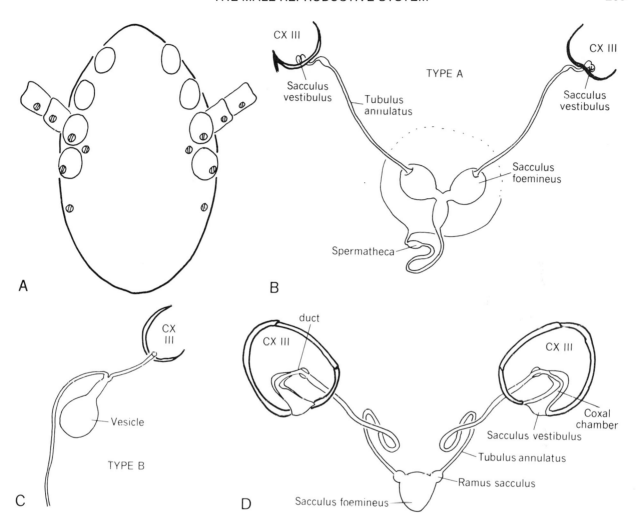

Figure 12-11. (A) Diagrammatic representation of location of sperm induction pores in Gamasida. Female sacculus complex in Gamasida: (B) *Macrocheles merdarius* (type A) (after Krantz, 1978); (C) *Amblyseus stramenti* (type B) (after Krantz, 1978); (D) *Macrocheles peltotrupetes* (after Krantz and Mellott, 1968).

(Pound and Oliver, 1976), and is known for phytoseiids (Dosse, 1959; Amano and Chant, 1978).

Two types of internal sacculi (sacculus foemineus) that connect the ducts to the genital tract of females exist. Type A exhibits a spermatheca between the sacculus foemineus and the end of the common duct leading to the vagina (Fig. 12-11A, B) (Krantz, 1978; Young,

1968). Eviphididae females are podospermous with an A type. Type B females have a vesicle in the duct near the induction pore (e.g., *Amblyseiius*) (Fig. 12-11C). Podospermy occurs in Eviphidoidea, Rhodacaroidea, Veigaiidae, Ologamasidae, and Otopheidomenidae. Some Acaroidea exhibit podospermy; others are tocospermous. Both tocospermy and podospermy occur in the Rhodacaroidae and Macronyssidae (Pound and Oliver, 1976).

Tocospermous sperm transfer is typical of ticks (Ixodidae) by means of the chelicerae into the genital tract of the female (Oliver et al., 1974).

Spermatophores

The second and most common mode of sperm transfer in mites is the indirect or passive method, in which the spermatophore is used. A differentiation is made for the spermatophores in the gamasid mites (Amano and Chant, 1978) (Fig. 12-12A, B).

The term ''endospermatophore'' delineates sperm packet material deposited in the spermatheca of the female. The term ''ectospermatophore'' refers to the encased sperm packet attached to the male chelicerae before injection into the female spermatheca. A similar designation is made for ticks (Feldman-Muhsam, 1967, 1986) (Fig. 12-12C–G). (See also Gladney and Drummond, 1971.)

In most other instances the spermatophore is stalked, pedicellated and erect (Fig. 12-14). It is secreted by the male inside the body and then extruded directly or through an aedeagus. The matrix of the spermatophore hardens on contact with air, but before hardening receives a sperm packet placed in varying types of receptacles and configurations.

Spermatophores are as varied in structure as the different groups of mites in which these sperm carriers are found. In many instances the sperm packet (Fig. 12-14D, F) resembles a rounded gemstone fork, placed in a ''pronglike'' setting like those on top of a ring with a spherical stone (Schuster and Schuster, 1966; Theis and Schuster, 1974; Oldfield et al., 1970, 1972; Sternlicht and Griffiths, 1974; Griffiths and Boczek, 1977; Jeppson et al., 1975; Oldfield and Newell, 1973).

The spermatophores may be transferred directly into the female (Ixodida; some Gamasida), but most frequently are deposited on the substrate of the habitat, on plant leaves, or in

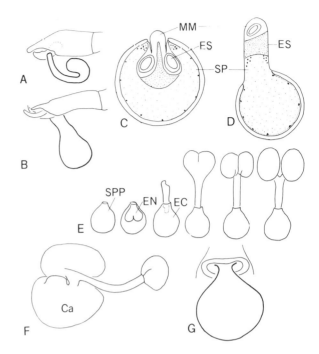

Figure 12-12. Spermatophores in Gamasida: (A) on movable digit; (B) on spermadactyl (after Evans et al., 1961). Spermatophores in ticks: *Argus persicus*, (C) unextruded, (D) extruded (ES, esterase secretion; MM, mucopolysaccharide–mucoprotein tip) (after Tatchell, 1962). (E) Consecutive stages of spermatophore formation and evagination in *Ornithodoros savigni* (after Feldman-Muhsam, 1967) (Ca, capsule; EC, ectospermatophore; EN, endospermatophore; SPP, spermiophore); (E) bulb invaginated to evaginated, (F) spermatophore enlarged, evaginated (Ca, capsule); (G) spermatophore attached in genital orifice (*O. moubata*) (after Nuttall and Merriam, 1911).

other places where the female mite is likely to find them. The methods of detection and introduction into the female vary as much as do the reproductive strategies related to copulation. Many gamasids and ixodid males use the chelicerae to place the spermatophore and its sperm packet into the female. In other mites, pheromonal substances may be the attractants by which the female is guided to the location, where she picks up the sperm packet herself, and abandons the stalked carrier. She may empty the sperm packet of its contents, leaving the case mainly intact (e.g., Eriophyidae; Oldfield et al., 1970, 1972). The pickup of the spermatophore in *Allothrombium fuliginosum* is described by Andre (1953) and illustrated in Figure 12-13 (after Robaux, 1974).

Spermatophores are known from many Actinedida (Bdellidae, Anystidae, Halacarida, Eriophyidae, Demodicidae, Erythraeidae, Trombidiidae, Trombiculidae), Hydrachnida, some Astigmata, and 13 families of Oribatidae (Oldfield et al., 1970, 1972; Sternlicht and Griffiths, 1974; Griffiths and Boczek, 1977; Jeppson et al., 1975; Oldfield and Newell, 1973). All bisexual species of oribatids produce free-standing spermatophores (Woodring, 1970).

The morphology of spermatophores is characteristic for families and genera, as well as some species. Spermatophores vary from simple, short vaselike forms (Acaridae) and tiny leaflike structures (Eriophyidae) (Oldfield et al., 1970) to elongated, more complex and sculptured forms (Bdellidae, Trombiculidae) (Fig. 12-14A, C, F) (Wallace and Mahon, 1976).

Spermatophores are produced by terrestrial and aquatic mites (Mitchell, 1958; Efford, 1966; Lanciani, 1972). The family of marine mites, Halacaridae, have produced spermatophores under laboratory conditions (Kirchner, 1967).

The spermatophores in the Eriophyidae consist of an enlarged base, a stalk, and an expanded apical head with a sperm sac. When isolated from spermatophores, virgin females of *Aculus cornutus* produce only males; when exposed to spermatophores, both males and females are produced. Males deposit between 16 and 35 spermatophores per day; the largest number deposited by one male was 614 (Oldfield et al., 1970; Oldfield and Newell, 1973). Spermatophores of *Eriophyes sheldoni* are glued to the substrate after emission. The sperm mass is supported on the flexible end of a stiff stalk (Sternlicht and Griffiths, 1974).

Summer form females (protogynes) of the peach silver mite (Eriophyidae: *Aculus cornutus*) find the spermatophores by specific tactile

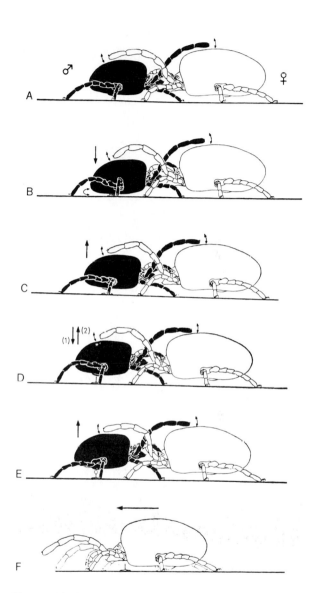

Figure 12-13. Diagrammatic representation of the placement of the spermatophore by male and pickup by female in *Allothrombium fuliginosum*: (A) preparatory phase; (B) primary emission of mucus by male; (C) phase of abdominal contractions; (D) secondary emission of mucous (1) by male and (2) stalk; (E) placement of spermatophore by male; (F) pickup of spermatophore by female (after Robaux, 1974).

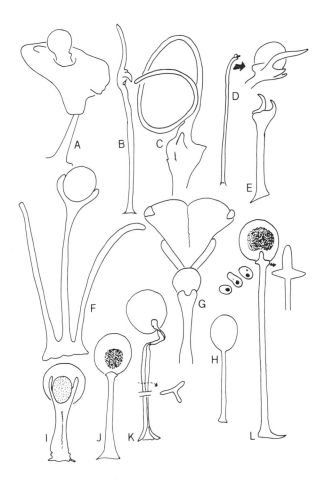

behavior using the legs. Such females apparently press the head of the spermatophore to release the sperm into the genital tract. The legs move rapidly at first and the mite is erect and motionless on its cauda during the transfer of sperm. Following the transfer, the female struggles slightly, with its legs moving, and when free, the spermatophore springs back to its position. Such changes suggest that the genital flap may grasp the spermatophore in the transfer. Such females that have visited a spermatophore produce both males and females. Protogynes that do not visit spermatophores reproduce the males only (Oldfield et al., 1972).

Species of the *Acarus, Lardoglyphus* and others produce spermatophores, which is characteristic of the genera. The sperm carriers are delivered into the bursa copulatrix of the female by means of a distinct aedeagus. The spermatophore is shaped into its final form by a sclerotized collar in the seminal receptacle. The final form of the spermatophore is not present in males because the aedeagus has a lumen much narrower than the main body of the spermatophore. It is assumed that the sperm material is fluid and is tranferred under pressure into the bell-shaped mold of the sclerotized collar. A single spermatophore is formed at each mating, so that repeated matings by a single female may be determined by counting the number of spermatophores present. In some mites (*Acarus siro*) the spermatophore breaks down in the receptacle; in others (*Lardoglyphus konoi*) the mass remains intact for later release of the sperm (Griffiths and Boczek, 1977).

Species of *Tyrophagus* do not exhibit a spermatophore, but the aedeagi in the species of this genus vary considerably, while the aedeagi of species of *Acarus* are similar.

The spermatophore in the Oribatida is composed of a noncellular, secreted stalk with a foot at the base. The stalk may show granules throughout its length. In damaeids, the stalk is sculptured somewhat like a triangular piece of extruded metal or plastic. At its apical end the stalk usually exhibits two short lateral sup-

Figure 12-14. Examples of stalked spermatophores: (A) Eriophyes (after Oldfield et al., 1970); (B) Bdellidae (*Bdella longicornis*) (after Alberti and Storch, 1976); (C) (*Bdellodes*) (after Wallace and Mahon, 1976); (D) Anystidae (*Anystus baccarum*) (after Schuster and Schuster, 1966); (E) Trombiculidae (*Hannemania*) (after Lipovsky et al., 1957); (F) Trombiculidae (after Evans et al., 1961); (G) Erythraeidae (after Witte, 1975); (H) Calyptostomidae (after Schuster, 1962); (I–L) Oribatida (after Da Fonseca, 1969; Woodring, 1970; Woodring and Cook, 1962); (K) *Damaeus quadrihastatus* (after Da Fonseca, 1969) showing ribbed stalk; (L) *Scheloribates laevegatus* showing blind filled spermatophore head with sperm mass (stippled), enlarged view of central and lateral supports, sperm cells (after Woodring and Cook, 1962).

ports and a median central support that penetrates the gelatinous vesicle surrounding the sperm sac (Fig. 12-14L). The sperm sac or sperm packet of spermatids or sperm is suspended in this spherical vesicle of clear material (Woodring and Cook, 1962; Woodring, 1970).

In the formation of the spermatophore males seek a smooth, dry surface free of feces. The male oribatid lowers its body and extrudes the aedeagus. A small part of the spermatophoral stalk is fixed to the substrate, then the mite's body is lifted upward and the stalk material is either drawn out between the penial sclerites or the upward motion of the body causes formation of the stalk. The penis is extended all of the time, so it appears that the former mechanism is used. The stalk hardens almost immediately in the air and the penis is retracted within the body. The male settles down on the stalk and the genital plates are nearly closed. The male retains this position for several minutes, then the body is raised again and discloses a complete spermatophore (Michael, 1883; Pauly, 1952, 1956; Woodring and Cook, 1962; Woodring, 1970).

When the female picks up the spermatophore, she raises her body high above the structure, opens the genital plates, and settles down with the plates closed. She remains for only a few seconds, visibly jerks upward, and moves away.

The number, size, and chemical composition of spermatophores varies among individual males and species of the oribatids. Productions begins about 10 days after emergence, becomes full at about 20 days of age, and may continue for a year or more. The maximum is 20 produced in 5 h, but this may vary with time and age of the male. Reduction to 20% of the number of spermatophores occurs in cultures where the females are absent (Woodring and Cook, 1962).

A very long penis is extruded by *Collohmannia gigantea* and a droplet of fluid is formed at the joint between the genutibia of leg IV. The females "nibble" on this material, but it is not known if the fluid contains sperm. No sperma-

tophore is observed in this oribatid (Woodring, 1970; Schuster, 1962).

A ribbed (triangular in section) stalk or pedicel occurs in the the spermatophore of *Damaeus quadrihastatus* (Fig. 12-14K) (DaFonseca, 1969). Structural detail of the head of the spermatophore is described by Fernandez (1981).

Aedeagus

Only males of the Actinotrichida have an extrusible aedeagus. A number of the families of Actinedida (e.g., Demodicidae, Myobiidae, Psorergatidae, Podapolipidae, Tetranychidae, Cheyletidae) and several families of the Astigmata (e.g., Acaridae, Chaetodaelylidae, Listrophoridae, Ewingidae, Proctophyllodidae, Kramerellidae, Pyroglyphidae, Gastronyssidae) have this intromittent organ for semen transfer during copulation. All other Acari exhibit passive, indirect transfer of spermatozoa through the spermatophores. The presence of an aedeagus, however, does not preclude the formation of spermatophores; in the Oribatida the aedeagus is present but is not used as an intromittent organ. All bisexual species of this order produce free-standing spermatophores (Woodring, 1970).

The greatest variety of structure and location of the aedeagus occurs in the Actinedida. The aedeagus is terminal in many actinedid males (Cheyletidae, Cheyletiellidae, Tetranychidae), but dorsal in others (Demodicidae, Myobiidae, Ophioptidae, Psorergatidae, Cloacaridae). Possibly the most highly developed aedeagus in the acarines is seen in the Podapolipidae, where the organ shows a range of small to large and terminal to dorsal locations different from those in Tarsonemidae and Pyemotidae (Fig. 12-15). R. W. Husband (personal communication) indicates that these mites, which are associated with insects, may show three types of mating positions that are related to the types of aedeagi present (Fig. 12-15K–M). In the retroconjugate position, the posterior ends of the male and female are together with

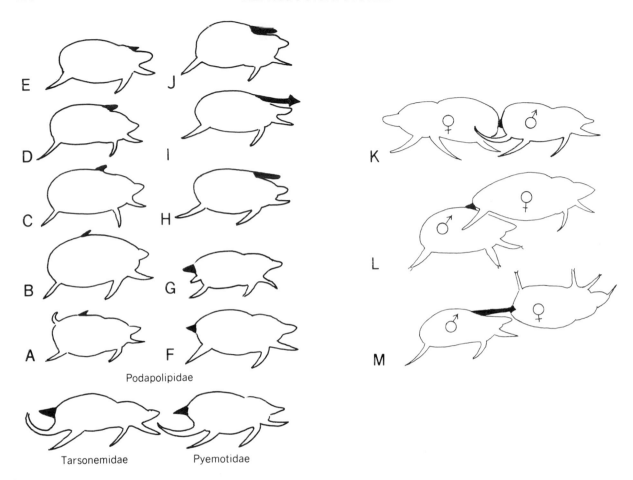

E

J

D

I

C

H

B

G

A

F

Podapolipidae

K

L

M

Tarsonemidae   Pyemotidae

Figure 12-15. Variations in the aedeagus (blackened) of Podapolipidae compared to Tarsonemidae and Pyemotidae. Podapolipid genera: (A) *Chrysomelobia*; (B) *Dorsipes*; (C) *Bakerpolipus*; (D) *Tetrapolipus*; (E) *Podapolipus*; (F) *Entarsipolipus*; (G) *Archipolipus*; (H) *Locustacarus*; (I) *Rhynchopolipus*; (J) *Podapolipus* (A–F, primitive; E–J, most modified). Three mating positions in Podapolipidae: (K) retroconjugate; (L) proconjugate; (M) inverted proconjugate (with permission of Dr. R.W. Husband).

the mites facing in opposite directions; a terminal aedeagus is involved. In the proconjugate position, the male and female face in the same direction with the male beneath; a small to moderate-sized dorsal aedeagus is characteristic. In the inverted proconjugate mating position, the female is upside down and the male right side up, but both are facing in the same direction.; the aedeagus is greatly hypertrophied (Fig. 12-16).

Podapolipid genera *Locustacarus* and *Rhyncopolipus* exhibit an enlarged aedeagus (Husband and Sinha, 1970) (Fig. 12-17). *Rhynchopolipus rhynchophori*, a podapolipid found under the elytra of the palm weevil, has a spearlike, dorsal aedeagus fully as long as the body of the male. In an unusual mating position the females hang upside down under the elytron of the weevil, firmly attached by strong chelicerae. The males crawl upright on the sur-

A   B

Figure 12-16. Aedeagus of *Rhynchopolipus rhynchophori*: (A) dorsal view of aedeagus in place; (B) enlarged arrowlike tip (SEMs by courtesy and permission of R. Husband).

face of the second pair of wings when the elytra are at rest and copulate with the larviform females. The anchor of the female chelicerae in the elytron of the weevil is strong enough to suspend both the male and the female during mating. The connection of the pair is probably also facilitated by the large aedeagus, which originates at the posterodorsal surface of the male, extends forward over the gnathosoma, and is partially held in place by propodosomal folds of the body. The spearlike terminus of the aedeagus probably assists in this holding attachment (Husband and Flechtmann, 1972; Husband, 1979) (Fig. 12-16A, B). In *Rhyncopolipus* the larval female also has an apparatus that clasps the aedeagus and the male has gnathosomal phlanges in which the aedeagus rests. An enlarged aedeagus also occurs in *Locustacarus trachealis* (Husband and Sinha, 1970).

One of the Bdellidae, *Neomolgus*, has an aedeagus with hooks at the terminal end (Vitzthum, 1943) (Fig. 12-17E). In *Pterygosoma sinaita* the aedeagus is forked (Jack, 1961) (Fig. 12-17H). Sexual dimorphism in the Myobiidae is distinctive at the species level. The males exhibit a distinct dorsal aedeagus that characterizes individual species. The hairs of the male genital opening in *Myobia micronydis* form a slide for the movement of the aedeagus in and out (Lukoschus and Driessen, 1970) (Fig. 12-17F). Dorsal aedeagi are also found in Cloacari-

dae, Ophioptidae, and Psorergatidae. In the Demodicidae, the aedeagus is dorsal, short, curved, and sclerotized. A distinctive aedeagal protractor muscle extends the organ out of the opening (Fig. 12-17L) (Desch and Nutting, 1977; Nutting and Rauch, 1958; Evans et al., 1961). Male Tetranychidae are differentiated at the species level by the shape and position of the aedeagus (Pritchard and Baker, 1955; Jeppson et al., 1975) (Fig. 12-17G).

While the variations of form and location of the aedeagus are not as diverse in Astigmata as in the Actinedida, aedeagi of males in the Acaridae, Glycyphagidae and other families of cheese mites are diagnostic for the species (Fig. 12-18A–C) (Hughes, 1961; Fain, 1969, 1970b; Vitzthum, 1943). The aedeagus in the Chaetodactylidae is forked internally (Fig. 12-18H) (Krantz, 1978). In the Ewingidae, the aedeagus is small with bell-shaped supports, and is covered by crescentic valves. Two pairs of minute genital acetabula are present (Yunker, 1970). Kramerellidae exhibit sickle-shaped aedeagi (Fig. 12-18G) (Krantz, 1978). In feather mites (Proctophyllodidae) the aedeagus may be recurved, extremely long, and accompanied by copulatory anal suckers (Fig. 12-18F) (Vitzthum, 1943; Atyeo and Braasch, 1966). The aedeagus of *Listrophorus leuckarti* lies in a pocket and is retractable, with a curved stylet extended from forked, sclerotized struts. Muscles attached to the struts cause its protraction

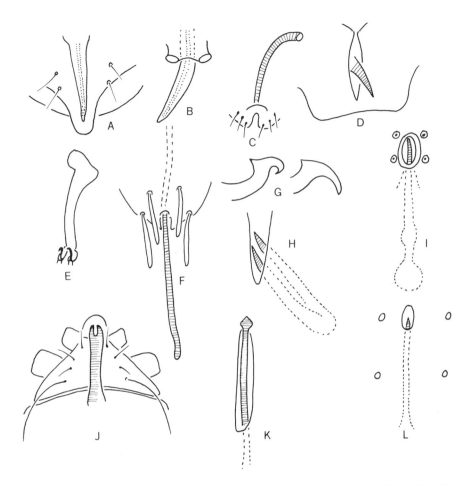

Figure 12-17. Variations of aedeagi in the Actinedida: (A) Cheyletidae (after Krantz, 1978); (B) Tydeidae (after Fain and Evans, 1966); (C) Cheyletiellidae (after Krantz, 1978); (D) Cloacaridae (after Camin et al., 1967); (E) Bdellidae (after Vitzthum, 1943); (F) Myobiidae (after Lukoschus and Driessen, 1970); (G) Tetranychidae (after Jeppson et al., 1975); (H) Pterygosomidae (after Jack, 1961); (I) Psorergatidae (after Evans et al., 1961); (J) Podapolipidae *Locustacarus trachealis* (after Husband and Sinha, 1976); (K) *Ophioptidae* (after Fain, 1964); (L) Demodicidae (*Demodex criceti*) (after Nutting and Rauch, 1963).

(Fig. 12-18E) (Hughes, 1954). The aedeagus in Pyroglyphidae and Gastroyssidae is pointed and encased in a sheath between the sclerotized margins of the genital opening (Fig. 12-18J). Anal suckers accompany this organ in the former family (Fain, 1970a).

Comparisons of the penises and types of ovipositors of 30 species of oribatids disclose two

types of aedeagi, weak and strong, and two types of ovipositors, short and long. Homologies indicate that primitive genitalic structure in these mites could function as either a penis or an ovipositor (Woodring, 1970).

The aedeagus of the Oribatidae is basically a double-walled oval cone into which the common vas deferens empties. Two types are exhibited,

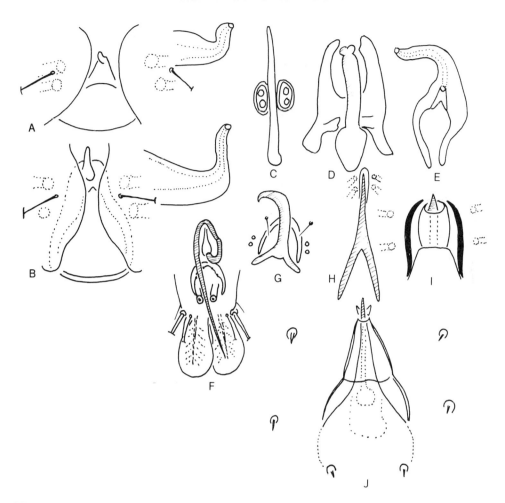

Figure 12-18. Aedeagi of Astigmata: (A) *Tyrophagus putriscentiae*; (B) *T. longior* (after Hughes, 1961); (C) Glycyphagidae (after Fain); (D) *Suidasia nesbitti* (after Hughes, 1961); (E) *Listrophorus leuckarti* (after Hughes, 1954); (F) *Proctophyllodes* (after Vitzthum, 1943); (G) Kramerellidae (after Krantz, 1978); (H) Chaetodactyllidae (after Krantz, 1978); (I) Pyroglyphidae (after Fain, 1965); (J) Gastronyssidae (after Fain, 1970).

a weak form and a strong form (Fig. 12-19A, B). The weak type is elongated, has sclerotized walls and a long central tongue, but lacks a cuticular bridge connecting the inner tube walls. The strong type of penis is shorter and thicker, with highly sclerotized walls. The aedeagus rests internally beneath the genital plates in the genital atrium (Grandjean, 1956; Woodring, 1970) (Fig. 12-19C).

THE FEMALE REPRODUCTIVE SYSTEM

Organs

The female reproductive system of mites consists of a single, fused ovary, a cluster ovary, or paired ovaries. Single and paired oviducts occur. Frequently, a distinct vagina is observed connected to the uterus or the genital atrium,

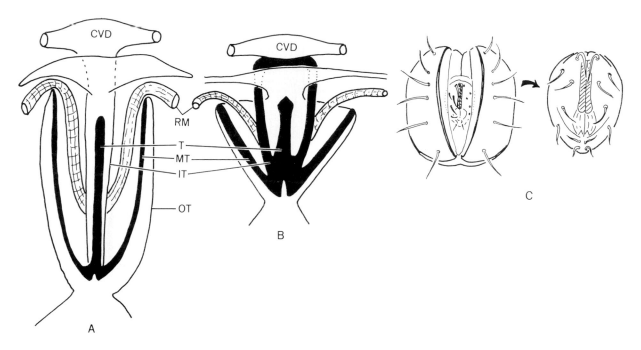

Figure 12-19. Aedeagi of Oribatida: (A) weak form; (B) strong form (after Woodring, 1970); (C) *Damaeus onustus*, aedeagus in place (after Grandjean, 1956) (CVD, common vas deferens; IT, inner tube; MT, middle tube; OT, outer tube; RM, retractor muscles; T, tongue).

or tied to the accessory glands. A distinct uterus may be present in some (Ixodida) but absent in others (Astigmata). In many gamasids a sperm induction pore leads by duct to an internal sacculus foemineus and thence to the main genital tract (Fig. 12-11). Females in one family of feather mites (Crypturoptidae) have a long, terminal external sperm duct (Fig. 12-20A).

Most often, eggs are laid directly without the use of an accessory organ, but a simple ovipositor is present in Opilioacarida; a partial one occurs in some Ixodida; several Endeostigmata (e.g., Pachygnathidae, Nanorchestidae) and some water mites show ovipositors. The beetle mites (Oribatida) have specialized ovipositors that differ among the families (Fig. 12-25) (Legendre, 1968; Grandjean, 1954).

Gamasida. In *Fuscouropoda agitans* (Fig. 12-20B) the female reproductive system consists of a small median ovary, paired oviducts, a median, muscular uterus, short vagina, seminal receptacle, accessory glands, and the genital atrium. Paired oviducts lead anteriorly from the ovary to a fused, medial junction with the uterus. The vagina is short, with a wrinkled cuticular lining. The seminal receptacle is egg-shaped, but is more or less a broad, posterior expansion of the genital atrium. The genital opening is covered by a large, ventral genital shield or plate.

Oocytes in the pyriform ovary appear as clusters of grapes; larger oocytes are peripheral. Funicular cells produce a stalk that flattens around the oocyte as the cell grows. Fine granular cytoplasm characterizes the primary oocyte. As growth occurs, the cytoplasm becomes reticular and yolk platelets accumulate. The nucleus and its nucleolus enlarge with an adjacent, hypertrophied mass of Golgi appa-

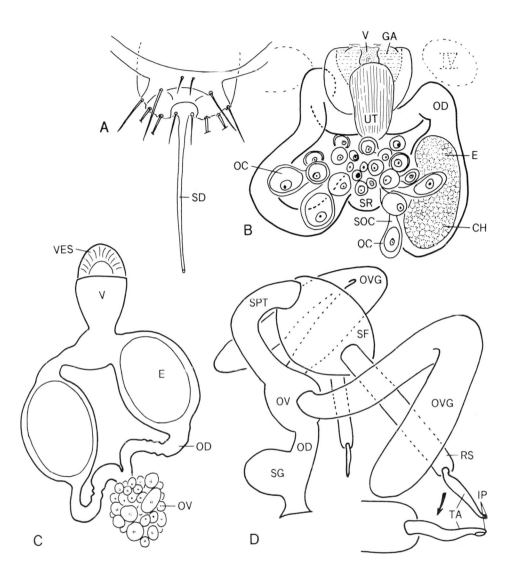

Figure 12-20. Female organs of Gamasida: (A) Sperm duct (SD) of female Crypturoptidae (after Gaud et al., 1972); (B) female genital organs of *Fuscouropoda agitans* (after
Woodring and Galbraith, 1976) (E, egg; CH, chorion; GA, genital atrium; OC, oocyte;
OD, oviduct; SOC, funicular stalk of oocytes; SR, seminal receptacle; UT, uterus; V,
vagina; female system of (C) *Leiodinychus krameri* (after Vitzthum, 1943); (D) *Haemogamasus reidi* (E, egg; IP, induction pore; OD, oviduct; OV, ovary; VES, vestibulum; V,
vagina; RS, ramisacculi; SF, sacculus seminus; SG, shell gland; SPT, spermatheca; TA,
tubulus annulatus; OVG, ovulogenous gland).

ratus. The single, largest oocyte is pushed into the ovarial lumen and into the oviduct by some unknown force, where it doubles in size, becomes filled with yolk platelets, and the chorion is formed. Mature females of *F. agitans* evidently lay a finished egg before another oocyte enters the oviduct.

The inner lining of each resting oviduct is a cuboidal epithelium underlain by circular muscle and a covering of peritoneum. The epithelial cells in the posterior half are active in the secretion of yolk; cells of the anterior half of the duct secrete the chorion. Peristaltic movements of the oviducal muscles move the egg into the uterus where it is stored temporarily. Compacted columnar cells make up the uterine epithelium which is covered peripherally by heavy bands of circular muscle that give the organ a longitudinally ribbed appearance. No secretion occurs in the uterus. The uterine muscles push the egg into the vagina (and possibly at least partially into the genital atrium).

The vaginal epthelium is composed of flattened cells with a pliable cuticular intima. No muscles are present in the vaginal wall, but the organ is supported by external muscles that prevent extrusion of the reproductive tract (prolapse) and aid in the passage of the egg into the genital atrium.

The squamous epithelium of the seminal receptacle is similar to and continuous with the epithelium of the vagina and the atrium. No musculature is present, but two bands of apodemal muscle insert on the posterior edges of the receptacle and probably are used to force semen from an enclosed spermatophore into the tract as an egg passes from the vagina into the genital atrium.

As an egg is forced from the uterus into the vagina and part way into the genital atrium, the seminal receptacle is compressed and semen is released onto at least one end of the egg. Hemolymphic pressure causes the release of the egg from the genital aperture after opening of the trapdoorlike genital plate by relaxation of the apodematal muscles attached (not by hemolymphic pressure). When the genital plate is lowered, the walls of the genital atrium are stretched and exposed and the egg is forced through the aperture to the exterior. Secretions of the vaginal glands are assumed to function in lubrication, activation of spermatozoa, and possibly fungicidal action (Woodring and Galbraith, 1976).

Leiodinychus krameri exhibits a fused median cluster ovary, paired oviducts, and a vagina that joins the genital opening (Fig. 12-20C) (Vitzthum, 1943).

The female of *Haemogamasus reidi* is podospermous and shows a slightly different arrangement of reproductive organs. The germaria (ovulogenous glands) are paired, elongated structures that lie free in the hemocoel and connect to a basal, spheroid ovary. A spermatheca extends dorsally from the ovary to the sacculus foemineus into which the sperm are introduced from the rami sacculi and tubuli annulatus. Each tubulus annulatus opens to the sperm induction pore in the ventral integument between legs III and IV. The oviduct extends from the spherical ovary to the genital opening. On its anterior base is an accessory shell gland (Fig. 12-20D) (Young, 1968). A comparable arrangement occurs in *Dermanyssus gallinae* (Pound and Oliver, 1976) and *Echinolaelaps echidninus* (Jakeman, 1961).

The spermatheca of female gamasid mites is sometimes used in the differentiation of species (*Typhlodromus*, Chant and Shaul, 1980; *Tropicoseius*, Dusbabek and Cerny, 1970).

Ixodida. Compared to the ovaries in insects, those in ticks are less complex, but are very efficient if one considers the number of eggs produced by a female ticks over 15 to 25 days (Brinton and Oliver, 1971a). Mating, oocyte production, and development are better understood for argasid ticks than for ixodids (Oliver, 1986).

The descriptions and figures of the reproductive tract of argasid females are similar (Fig. 12-21A, B). The ovary is single and U-shaped, joined to paired, convoluted oviducts with an expanded distal part posterior to the junction

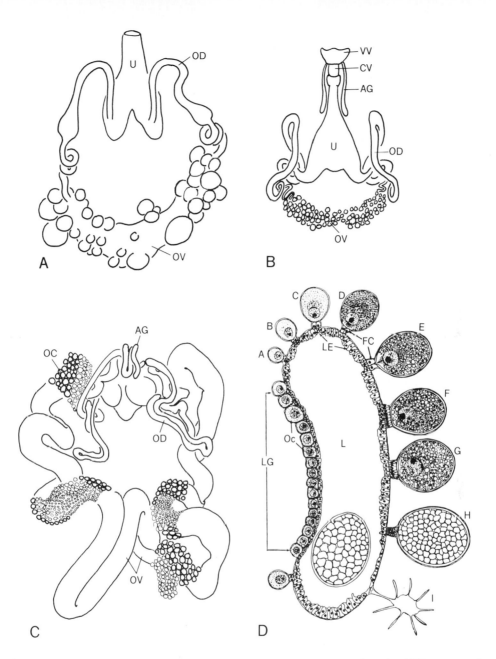

Figure 12-21. Female organs in Ixodida: (A) *Argas persicus* (after Aeschlimann and Hecker, 1969); (B) *A. ogadenus* (after Roshdy, 1966); (C) *Dermacentor andersoni* (after Brinton and Oliver, 1971b) (AG, accessory gland; CV, cervical portion of uterus; OC, oocytes; OD, oviduct; OV, ovary; SR, seminal receptacle; VV, vestibular portion of vagina; U, uterus); (D) ovarian development in *Dermacentor andersoni* (after Brinton and Oliver, 1971a) (Oc, undifferentiated oocytes in LG, longitudinal groove); oocyte development A,B,C,D around the ovary (FC, funicular cells; LE, lumenal epithelium); micropyle (arrow) in oocyte E; B–G, oocyte maturation; H, oocyte with vitellogenesis essentially complete; I, collapsed bag of tunica propria after ovulation (L, collapsed lumen of ovary).

with the distinct uterus. The expanded uterus extends to the vagina which it joins through a narrowed neck. A pair of tubular accessory glands is attached dorsally to the vagina (Arthur, 1962; Lees and Beament, 1948; Aeschlimann and Hecker, 1969). In *Ornithodoros papillipes*, the vagina is divided into cervical and vestibular parts on either side of the openings of the accessory glands (Balashov, 1972). Similar arrangements exist in *O. kelleyi, Argas ogadenus, Argas (Secretargas) transgariepinus,* and *Argas vespertilionis* (Sonenshine, 1970; Roshdy, 1961, 1963, 1966, respectively).

Vitellogenous materials incorporated in the eggs of *O. moubata* apparently come from different sources. The formation and transportation of these substances to the oocytes is accelerated by copulation. Penetration of the substances from the hemolymph to form yolk is aided by pinocytosis. Ribosomally rich cytoplasm in young oocytes is indicative of the active synthesis of protein. Golgi apparatus condenses the material and transports it to the ergastroplasm of the egg after the blood meal. Some of the proteins of the egg are extrinsic rather than intrinsic and involve pinocytosis, particularly the cuticular proteins that surround the egg (Aeschlimann and Hecker, 1969).

The female reproductive system in *Dermacentor andersoni, Ixodes ricinus,* and other hard ticks (Ixodidae) comprises a fused ovary and paired, convoluted oviducts attached to a uterus at the juncture with the seminal receptacle (Arthur, 1962; Brinton and Oliver, 1971a; Douglas, 1943) (Fig. 12-21C). The uterus extends to the vagina, on the dorsal surface of which is a pair of tubular accessory glands.

Oogenesis in Ticks. To better understand development and transovarial transmission of etiological agents of disease in ticks, Brinton and Oliver (1971b) studied the morphology and histology of ovarian development in *Dermacentor andersoni* (Fig. 12-21D). The ovaries in unfed nymphs are simple, tubular, horseshoe-shaped organs located behind the synganglion. Paired oviducts pass anteriorly into a vagina–

uterine complex. Accessory glands and a seminal receptacle are present.

The ovarian wall is made up of an external epithelium, interstitial cells lining the lumen, and oogonia. The ovary enlarges following nymphal engorgement and the subsequent change to the adult stage. This is the time of oogonial mitosis, a period when the length and diameter of the ovary increase.

When the adult stage is reached, the ovaries consist of oocytes, interstitial cells, funicular cells, and muscle cells. After molting, the ovaries of unfed females change by forming a longitudinal groove along the entire length. The groove does not function like the germarium of insect ovarioles, but has vitellogenic activities. By the time the groove is formed, all of the germinal tissue of the ovary has reached the stage of mature oocytes.

The interstitial cells become the lumenary lining and the oocytes remain essentially unchanged. A tunica propria of connective tissue is formed. Within 48 h after attachment of the female to a host, the oocytes migrate to the surface of the ovary and, covered by the tunica propria, protrude conspicuously into the hemocoel. They become attached to the ovarian wall by a pedicle of funicular cells and have 90% of their surface exposed to the hemolymph, which projection of the oocytes into the hemocoel gives the ovary a knobby appearance.

The development of the oocytes proceeds at different rates within the longitudinal groove, depending upon the time during feeding and after engorgement. During this period the ovary becomes twisted and convoluted. By the time oviposition is completed, the mature oocytes or eggs will have crowded into the lumen of the oviduct, leaving stumplike structures (the remains of the funicular cells) on the surface of the spent ovary.

Prior to oviposition, the eggs are probably primary oocytes inasmuch as no meiotic divisions were noted by these authors. Within 6 days after the completion of oviposition, mature oocytes are nearly totally absent from the ovary. Where thousands of oocytes were present on the surface of the ovary prior to ovulation, only a few immature cells remain together with the collapsed, convoluted "ghost-like" bags of tunica propria which formerly held the eggs.

No special germinal chambers or germaria occur in the tick ovary, nor vitellaria with trophocytes and follicular cells. Increased size of the ovary is the result of mitotic proliferation of oogonia and interstitial cells. It is not known what factors stimulate the development of the longitudinal grooves in the ovary and the growth of the oocytes, but these may result from certain hormones and unknown physical actions. The rapid enlargement of the oocytes in engorged females, when compared with the steady anabolic state during copulation and the period of rapid feeding after mating, is characteristic of the Ixodidae.

The intrinsic and extrinsic phases of oocyte development in *D. andersoni* overlap. The intrinsic or intraoocytic phase is already evident in unfed adult females and is characterized by heavy emissions of ribonucleoprotein from the nucleus, the coalescing of ribosomes with vesicles budded as saccules from dictyosomal cisternae to form dense vesiculate bodies, and the morphogenesis of mitochondria. These actions result in the formation of membrane-limited multivesicular bodies. The extraoocytic or extrinsic factors consist of micropinocytosis of hemocoelic fluids after the formation of a microvillar brush border around the oocytes within 4 to 6 days following attachment. Ingested fluids pass via the micropinocytotic tubes to membrane-limited amorphous reservoirs in the cortical ooplasm. These bodies subsequently fuse with dense bodies of intrinsic origin and form proteid yolk granules. At least four kinds of organelles contribute to the formation of the proteid yolk granules in *D. andersoni*: nuclear–nucleolus complexes, dictyosomes or Golgi bodies, mitochondria, and the plasmalemma brush border. The role of each is highly integrated during the oocyte development. The role of the smooth ER is evidently insignificant since this is sparse in the oocyte (Brinton and Oliver, 1971b).

Actinedida. Female systems vary in the Actinedida. The reduction of the number of organs in the female system is shown in *Bdella longicornis* (Linnaeus, 1758), where an unpaired ovary is connected to a single, somewhat convoluted oviduct (Fig. 12-22A). A seminal receptacle is connected to an expanded vagina some distance inside the vagina. The genital aperture exhibits genital covers (Vitzthum, 1943).

The genital tract of the female Demodicidae is a simple, single, tubular structure. A single, terminal ovary lies posteriorly in the hemocoel. The ovary connects to a tubular oviduct with an enlarged sperm storage area (seminal receptacle) immediately interior to the vulva (Fig. 12-22B) (Desch and Nutting, 1977). The single nature of the female reproductive system is also shown in the Anystidae, Halacaridae, and Eriophyidae (Hughes, 1959).

The female reproductive tract of Eriophyidae consists of paired ovarioles. The paired oviducts lead to the main genital canal and genital opening. Paired seminal receptacles or spermathecae are distinctive of the female and are attached to the oviduct near the genital opening. These spermathecae and tubes are diagnostic for subfamilies and species, and are useful in taxonomy (Jeppson et al., 1975).

As the ovariolar oocytes develop in the apical ovarioles, they pass downward into the oviducts, then into the main genital canal (Fig. 12-22E, F). Before reaching this canal, the oocytes accumulate nutrients (lipid bodies and yolk platelets) and nurse cells. The chorion and vitelline membranes become evident. A mature egg nearly fills the hemocoel of the female and is in position over the genital flap and ready to be laid (Vitzthum, 1943; Whitmoyer et al., 1972; Jeppson et al., 1975).

The reproductive system of *Allothrombium fuliginosum* (Parasitengona) is reminiscent of the tick system. The fused, single ovary is U-

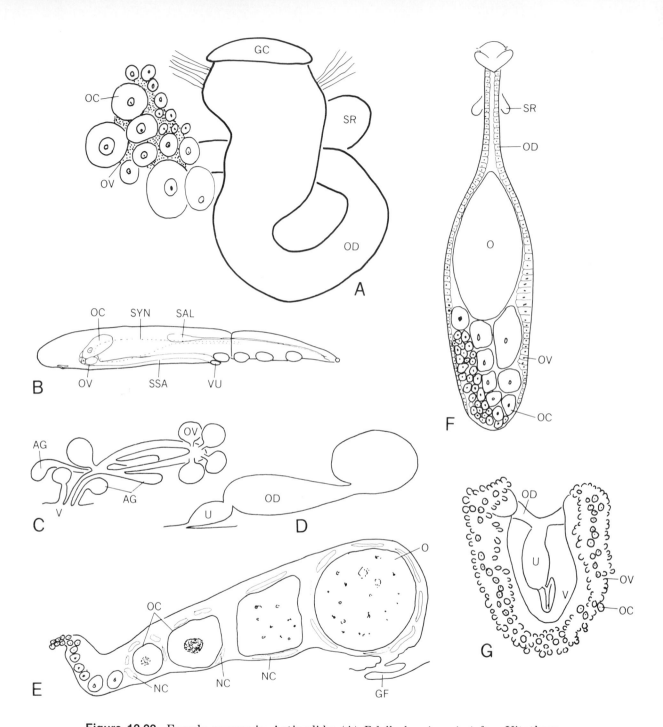

Figure 12-22. Female organs in Actinedida: (A) *Bdella longicornis* (after Vitzthum, 1943); (B) *Demodex antechini* (after Desch and Nutting, 1977); (C) *Neomolgus* (Bdelli-dae), (D) Anystidae (after Hughes, 1959); (E, F) sections of *Eriophyes tulipae* (after Jeppson et al., 1975) (AG, accesory gland; GF, genital flaps; GC, genital cover; NC, nurse cell; O, mature ovum; OC, oocyte; OD, oviduct; OV, ovary; SAL, salivary gland; SR, seminal vesicle; SSA, sperm storage area; SYN, synganglion); (G) Trombidiidae (after Vitzthum, 1943).

shaped with the eggs on the periphery. Paired, short oviducts connect to a broad uterus, which extends into the vagina, and in turn to the longitudinal genital aperture (Fig. 12-22G) (Vitzthum, 1943).

Astigmata. The female genital system of the cheese mites (Fig. 12-23I) consists of paired ovaries and oviducts which unite anteriorly in a common vagina, connected to the exterior genital opening. At the opposite end of the system is the copulatory pore or bursa copulatrix, an indented opening (Acaridae) or conical tube (*Ctenoglyphus, Glycyphagus*) (Fig. 12-23A, B),

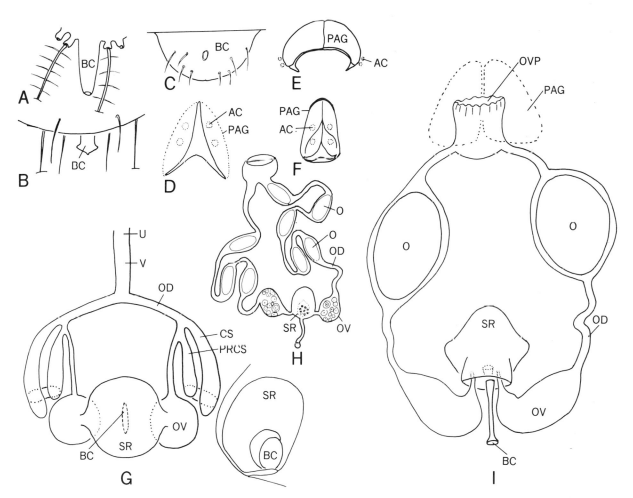

Figure 12-23. Female organs in Astigmata. Terminal bursa copulatrix (BC) in (A) *Ctenoglyphus* (after Hughes, 1959) and (B) *Glycyphagus domesticus* (after Hughes, 1959); (C) dorsal bursa in *Guanolichus* (after Krantz, 1978); female aperture or oviporus in (D) Acaridae, (E) Chortoglyphidae, and (F) Glycyphagidae (AC, acetabula; PAG, paragynial folds); female system of (G) *Caloglyphus mycophagus* (after Rhode and Demick, 1967) and (H) *Labidophorus platygaster* (after Vitzthum, 1943); (I) schematic system of Acaridae (after Griffiths and Boczek, 1977) (BC, bursa copulatrix; CX, chorionic sac; O, ovum; OD, oviduct; OV, ovary; OVP, oviporus; PRCS, prechorionic sac; SR, seminal receptacle; U, uterus; V, vagina).

or dorsal opening (Guanolichidae) (Fig. 12-23C), which is really the functional vagina inasmuch as it receives sperm or the spermatophore. The copulatory bursa leads from the exterior into a narrow canal with elastic walls, then to a sac-like seminal receptacle. Short canals connect this receptacle to the pyriform or ellipsoid ovaries in the system.

Thus two exterior openings in the female system are separated from each other: the posterior bursa copulatrix and the midventral genital aperture (oviporus), which is a longitudinal slit between epimera III and IV (Fig. 12-23D–F). A pair or more of longitudinal genital folds may cover the genital opening. Glycyphagid mites have a third, transverse fold that covers the aperture from behind. Females usually exhibit two pairs of internal suckers or genital papillae (=acetabulae) beneath the folds of the aperture. The anterior edge of the genital opening in females may also exhibit a crescentic sclerite, the epigynium, which is a strengthening structure for the opening. This structure may be fused with the sternal skeleton to form a perigenital or circumgenital ring that is characteristic of some families (Zakhvatkin, 1959).

The reproductive system of the female *Caloglyphus mycophagus* (Fig. 12-23G) consists of two spherical ovaries joined medially by an expanded, spheroidal seminal receptacle.

Oribatida. Females in the Oribatida usually exhibit fused, median ovaries. In some, the paired arrangement of the ovaries is evident, even though fusion is prominent. Paired oviducts extend from the ovaries to the vagina, sometimes with an intermediate expansion in each oviduct—the uterus—between the ovary and the vagina (Fig. 12-24A, B). Ova in the oviduct are large and filled with yolk. The proportionate size of the eggs of beetle mites is very large compared to the size of the oviduct and that of the genital opening. The eggs must be squeezed through a genital aperture that is perhaps a third less than the diameter of the egg. An ovipositor extends from the vagina for de-

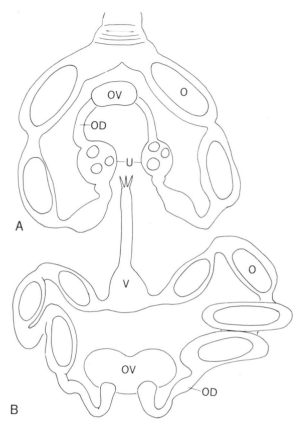

Figure 12-24. Female system in Oribatida: (A) *Damaeus geniculatus*; (B) *Xenillus tegeocranus* (after Michael, 1883) (O, ovum; OD, oviduct; OP, ovipositor; OV, ovary; U, uterus; V, vagina).

position of the eggs in the crevices of soil or litter.

TYPES OF REPRODUCTION

Oviposition

Most mites are oviparous. The eggs are variously shaped: ovoid, oval, spherical, elongated, or flattened. The egg case may be smooth, ridged, ribbed, or otherwise ornamented with striations, reticulations, bosses, or external protrusions (spines, hairs, clavate structures,

wax blooms, etc.). In some cases (e.g., Bdellidae) a protective silken cocoon is spun around
each egg (Alberti, 1973; Wallace and Mahon,
1972). In other instances (e.g., Tetranychidae)
the eggs are anchored to leaf surfaces with
silken guy wires. In most species the external
covering of the egg case is a waxy material vital
to the survival of the embryo. Ticks possess a
special organ (Gene's organ) by which this coating is applied; eggs without this protection des-
sicate and die very quickly.

The variously shaped eggs are frequently
disproportionately large for the size of the female, but because of their elasticity they can be
squeezed through a relatively small genital
opening without apparent damage (Krantz,
1978). Eggs of some mites are so large that

they cannot be deposited and the female dies,
after which the larva hatches.

The numbers of eggs laid by female mites
vary with the different orders. Several ticks
oviposit in clusters and the number of eggs laid
at one time is in the thousands. Female *Ixodes*
may deposit 10 000 eggs in one batch.

Oviposition may be accomplished in different ways. In plant-feeding mites and those that
live in decaying organic materials, the eggs are
usually laid at random on the food substrate,
but exposure or predation may reduce the number that hatch. Such situations are overcome
by multivoltinism and/or high fecundity (Solomon, 1945; Hussey et al., 1969). Where food is
limited and widely distributed in a fairly large
habitat, mites may be fairly selective for sites

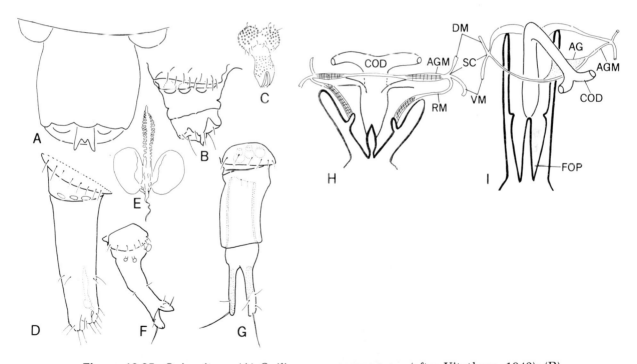

Figure 12-25. Ovipositors: (A) *Opilioacarus segmentatus* (after Vitzthum, 1943); (B)
Terpnacarus (after Grandjean, 1943); (C) *Hydrachna* (after Cook, 1974); (D) *Bdellodes*;
(E) ovipositor glands of same (after Wallace and Mahon, 1976); (F) *Nothrus sylvestris;*
G. *Eremaeus hepaticus* (after Grandjean, 1956a). Two types of ovipositors in Oribatida
(after Woodring, 1970); (H) short; (I) long (AG, accessory gland; AGM, accessory gland
muscle; COD, common oviduct; DM, dorsal muscle of ovipositor; FOP, finder of ovipositor; RM, retractor muscle of ovipositor; VM, ventral muscle of ovipositor).

of oviposition and use a protrusible ovipositor (e.g., Nanorchestidae, Oribatida).

Predaceous mites may lay the eggs in protective locations or where the hatched larvae (*Pergamasus quisquilarum*) will have easy access to the prey (Symphyla) (Berry, 1973). Parasitic mites use host tissue for oviposition. *Dicrocheles phaelanodectes* macerates the tissue of the tympanic cavity of the noctuid host before placing the egg in the cavity (Treat, 1975).

Brooding of clustered eggs occurs in sheltered "nests" of some cheyletid mites, but if the hatchlings remain in the area too long, they may be eaten by the mother (Summers and Witt, 1972; Krantz, 1978).

Soil-inhabiting mites oviposit eggs in locations that will be protective. Some of the Actinedida and most of the Oribatida use a protrusible ovipositor by which the eggs are placed in crevices or insterstices of the soil. Frequently the ovipositor is equipped with sensitive terminal "fingers" that enable the female to detect the proper site of oviposition (Fig. 12-25F, G). Some aquatic mites insert the eggs into the tissues of aquatic plants (Jeppson et al., 1975; Krantz, 1978) (See Figure 12.25a).

Ovipositors

The female ovipositor in mites comes in varying shapes, sizes, and complexities, but is not found in all orders. An ovipositor is present in the Opilioacaridae (Fig. 12-25A) but none is found in the Gamasida. A few ticks have a reduced ovipositor. Some of the Actinedida (e.g., Endeostigmata, Fig. 12-25B; Bdelloidea) exhibit this female organ. In the Bdellidae, special ovipositor glands accompany this organ (Fig. 12-25D, E). Limited numbers of female water mites have an ovipositor (Fig. 12-25C). The females of Oribatida commonly have a short, primitive type of ovipositor or an elongated, telescoped form (Fig. 12-25G–I) (Woodring, 1970).

Gene's Organ in Ticks

Female ticks in both Argasidae and Ixodidae have an egg-waxing organ that prevents dessication of the eggs after they are laid. This unique structure, Gene's organ, is not found in other mites (Lees and Beament, 1948), nor among other animals (Arthur, 1962). The organ, located beneath the anterior scutum of female hard ticks, consists of two to four lobes and a stalklike or balloonlike base that is capable of eversion. In ixodid ticks, the slitlike opening of the organ is in the camerostomal fold between the anterior scutal margin and the posterior margin of the basis capituli. In the nonscutate argasids the gland opens into the camerstomal depression. The broad basal stalk expands posteriorly into either two main horns (Argasidae, Amblyomminae) (Fig. 12-26A, B) or into four horns (Ixodidae, Ixodinae) (Fig. 12-26C). The horns may have singular glands or multiple folds of glandular tissue and may be greatly expanded. Internally there is a thin cu-

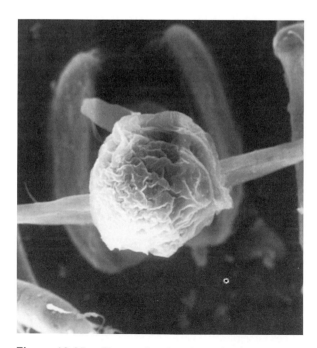

Figure 12-25a. Egg and ovipositor of *Eremaeus* sp. SEM.).

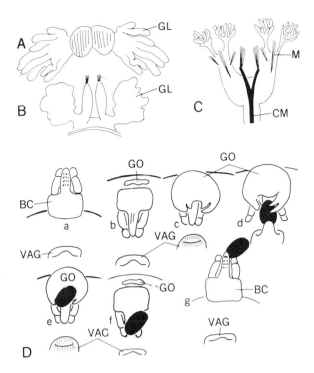

Figure 12-26. Gene's organ in (A) *Argas persicus* (after Robinson and Davidson, 1914); (B) *Argas (S.) trangariepinus* (after Roshdy, 1963); (C) Hyalomma asiaticum (after Balashov, 1972) (CM, cord of muscle attachment; GL, glandular part; M, muscle); (D) function of Gene's organ in *Ixodes ricinus*: a, ventral view of capitulum; b, capitulum bent ventral by exposing Gene's organ; c, Gene's organ expanded; d,e, expanded organ surrounds egg; f,g, coated egg attached to palps and capitulum prior to oviposition (BC, basis capitulum; GO, Gene's organ; VAG, vagina).

ticular lining, apart from the external cuticle, but composed of endo- and epicuticular layers that become thinner toward the horns. Fluid fills the space between the gland and the outer exoskeleton. Unfed ticks show no signs of glandular activity in this organ. The glands differentiate during feeding (Balashov, 1972). Protraction of the organ is evidently by means of hydrostatic pressure. Retraction is accomplished by specific muscles.

Glands that open into the areae porosae of female hard ticks function to lubricate Gene's organs as they are extruded in oviposition. If the areae porosae are covered or obliterated and this lubrication is absent, Gene's organs do not function properly or adhere to the capitulum (Feldman-Muhsam, 1963). (The history of Gene's organ and its function is reviewed by Lees and Beament (1948).)

Prior to oviposition, the capitulum of the female tick is directed anteriorly. Gene's organ is not visible and the genital aperture is a slitlike opening. When eggs are about to be extruded from the vulva, the tick bends the capitulum ventrad approximating the genital orifice. The organ of Gene is everted (from beneath the margin of the scutum in hard ticks) above the basis capitulum, whose contour follows. The horns of the organ are extended simultaneously with the prolapse of the vagina, which thus forms a tube, or type of ovipositor. As an egg leaves the genital canal, it is enclosed by the glandular parts of the organ, which alternately inflate and deflate coincidentally with palpal movements and manipulation of the egg. As a result, the egg receives a protective coating of waxy substance and Gene's organ is retracted. The capitulum returns to its normal position and the egg adheres to the dorsal surface of the hypostome or capitulum (Fig. 12-26D).

The eggs accumulate in a dense cluster or in a heap beneath the anterior region. If a female is disturbed during the process and the egg does not receive a waxy coating, it soon shrivels. Very few uncoated eggs hatch even in a humid atmosphere (Arthur, 1962; Lees and Beament, 1948; Snodgrass, 1965). They are highly susceptible to water loss and to the acquisition of water. When placed in distilled water, permeable, uncoated eggs of *O. moubata* may swell to nearly three times normal size before bursting (Lees and Beament, 1948).

Hatching

Reproductive types and different modes of

hatching in the oribatids include oviparity, ovoviviparity, parthenogenesis, and aparity (Strenske, 1949). The most common form of hatching is oviparity. Ovoviviparity is rare, if it exists at all (Woodring and Cook, 1962). Aparity implies that the larvae hatch within the dead female and are forced to eat their way out (Lipa and Chmielewski, 1966). The validity of aparity or ovoviviparity in oribatids is questioned, however (Woodring and Cook, 1962). Some of the higher oribatids, like Galumnoidea, exhibit aparity because the size of the eggs is extreme and too large for normal deposition (Jacot, 1933).

Ovoviviparity is exhibited in the Laelapidae. Viviparity is reported in females of Spinturnicidae, external parasites of bats. The Rhinonyssidae, nasal parasites of birds, are also viviparous (Mitchell, 1963). Several females of *Albidocarpus*, bat parasites (Chirodiscinae, Listrophoridae) have larvae inside their bodies and are termed larviparous (Pinichpongse, 1963a,b).

Ovoviviparity is noted for *Hirstionyssus staffordi* (Strandtmann and Hunt, 1951). In collections of this species from spotted skunks (*Spilogale* and striped skunks *Mephitis*) larval and protonymphal exuviae were observed within the females. Larvae were found within the uterus of female *Spelaeorhynchus praecursor*, an external parasite of bats (Dusbabek, 1970).

Physogastric females of the Pyemotidae (Fig. 12-1) may have young crawling within or upon the swollen idiosoma. While physogastry applies in several cases of the Heterostigmae, some pygmephorid females (*Trochometridium*) lay eggs in a normal way. Hatching occurs after development to the adult stage has been reached within the chorion of the egg (Cross, 1965).

Brooding of clustered eggs occurs in sheltered "nests" of some cheyletid mites, but if the hatchlings remain in the area too long, they may be eaten by the mother (Summers and Witt, 1972; Krantz, 1978).

Parthenogenesis

Most mites have the normal pattern of fertilization to produce offspring, but reproduction without fertilization also occurs (Oliver, 1971a). Beatty's (1967) definition of this agamic reproduction is:

> Parthenogenesis is the production of an embryo from a female gamete without any genetic contribution from a male gamete, and without eventual development into an adult. This deliberately wide definition includes the special case of gynogenesis, in which a spermatozoon does enter the egg and activates it into development, but does not exercise its second role of contributing genetic material.

The terminology identifying a parthenogenetic organism (parthenogenome; parthenogen; parthenote) and connotations vary with different authors (Oliver, 1971a). Gynogenesis is parthenogenetic development that is induced by some part of the mating act, such as the mechanical effect of coition, the influence of sperm or accessory materials, including the penetration of sperm into the egg, but lacking the fusion of the male genetic elements with the female gamete. Oliver designates the types of parthenogenesis (arrhenotoky, thelytoky, deuterotoky), the occurrence of parthenogenesis (occasional, obligatory, cyclical, facultative, paedogenetic), the cytological aspects (meiotic, ameiotic), and the relationships to taxa (biosystematics of parthenogenesis).

Lack of data on sex determination, development, and the cytogenetics of Opilioacarida and Holothyrida exclude these groups from identifiable agamic reproduction. It should be noted, however, that parthenogenetic species are found in all other major orders of the Acari.

The offspring of biparental union and fertilized eggs are always diploid females. Unfertilized eggs may develop in three different ways: arrhenotoky, thelytoky, and deuterotoky (also called amphoterotoky).

In arrhenotoky, haploid males develop from unfertilized eggs. This is the major type of parthenogenesis in some families, particularly among gamasine and actinedid mites (Filipponi, 1957).

Species that are arrhenotokous (haplodiploid) reproduce sexually and females have the maternal chromosome number through the addition of the male genome. Some species have unfertilized females that lay haploid eggs which develop into males. Still other species produce and oviposit haploid eggs following bisexual mating. The situation is analogous to thelytokous gynogenesis, but the mechanisms involved are elusive and not determinable from rearings alone.

Although most of the families of gamasid mites are probably arrhenotokous, thelytoky is present in many others. Both types of parthenogenesis are found in the Macrochelidae. Some species in this family are obligatorily thelytokous (Filipponi, 1964). No males are known from species of *Geholaspis, Veigaia,* or *Eulaelaps.*

Arrhenotoky appears to operate in closely related species in certain orders of mites. It appears to be quite common among the Gamasida. It is not reported from any of the ticks (Ixodida), nor from the Oribatida (Oliver, 1971b). Arrhenotoky is the predominant type of parthenogenesis in the Gamasida, especially in the families Dermanyssidae, Laelapidae, Macrochelidae, Macronyssidae, and Phytoseiidae. Combinations of situations are found within species of certain genera, however. Species of the genus *Macrocheles*, while mainly arrhenotokous forms, show only females in some species, females and males in others, and obligate thelytoky in still other species. Species of *Geholapsis* exhibit females only, while species of *Glyptholaspis* are arrhenotokous. Arrhenotoky is not reported from Ixodida or Oribatida (Oliver, 1971b).

One species of *Macrocheles* is capable of facultative parthenogenesis. A single virgin female may establish a new population after being transported to a new location on beetles (Filipponi, 1955). In 21 species of Macrochelidae, 19 species were shown to be arrhenotokous

and two species to be thelytokous (Filipponi, 1964). Males of *Macrocheles muscaedomesticae* are haploid and arrhenotokous. Some species of this genus, however, though morphologically identical, may exhibit different reproductive patterns in different geographic regions (Oliver, 1971b; Filipponi, 1964).

In thelytoky, all offspring from unfertilized eggs are females. Males are rare or absent, and if present do not mate. Thelytoky is known in all of the orders of the Acari except Opilioacarida and Holothyrida. The genetic mechanism of thelytoky is poorly understood even though it involves liberal amounts of heterozygosity. Such variety would be an advantage of heterosis and the adaptive aspects of polymorphism. The thelytokous species evidently use different chromosomal alternatives, but in known cases such species either (1) omit the first meiotic division (reduction) in gametogenesis, or (2) employ a method that restores the diploid number of chromosomes (e.g., fusion of the polar body nuclei, or fusion of the cleavage nuclei). Thelytokous parthenogenesis is rare in Gamasida, but is known in some ticks, in species of prostigmatid mites, and in oribatid mites. It is the only form of reproduction known in some geographic "races" or sibling species (Oliver, 1964, 1971b).

It is necessary for gynogenetic females to mate before oviposition can occur. This mating without penetration of sperm (sometimes with a male of another species) is a corollary of some types of parthenogenesis. Eggs not penetrated by sperm develop thelytokously.

Parthenogenesis is found in Ixodida (*Haemophysalis (Kaiserana) longicornis*), where germinal differentiation begins in the fed nymph. When parthenogenetic females mate with males from a bisexual race, bisexual offspring result. Females that have been inseminated deposit an equivalent or larger number of eggs than the uninseminated parthenogenetic females (Khalil, 1972).

The condition of thelytoky is rare in ticks. It may occur sporadically in a species that normally reproduces bisexually (XX:XY or

XX:XO system). In one of the soft ticks, *Ornithodoros moubata*, thelytoky occurs occasionally (Davis, 1951), is sporadic in species of the Ixodidae as well, and may be accompanied by morphological abnormalities. Most races of *Dermacentor variabilis* are nonparthenogenetic; some are parthenogenetic. Thelytoky is reported for *Amblyomma* and *Haemaphysalis* (Oliver, 1971b).

Polyploidy is also associated with thelytokous reproduction. In this condition, the animals have more than two sets of homologous chromosomes. Bisexual and parthenogenetic strains of *Haemaphysalis longicornis* (=*H. neumanni*) in Australia and Japan show diploid and triploid chromosomal numbers, respectively (Oliver, 1971b; Oliver et al., 1973).

The chief mode of reproduction in the Actinedida is by arrhenotoky, but thelytoky occurs in *Brevipalpus obovatus* (Piinacker et al., 1981). In the Tarsonemidae and Pyemotidae, rearing data confirm the haplodiploidy. Virgin females of *Siteroptes graminum* always produce males, while fertilized females give rise to offspring of both sexes. The sex ratio from these mated females is extreme (5% males to 95% females). Arrhenotoky is typical of Eriophyidae and many of their relatives, the spider mites. In the Tetranychidae, the more primitive subfamily, Bryobinae, has both arrhenotokous and thelytokous species. The subfamily Tetranychinae exhibits arrhenotoky exclusively, with chromosome numbers varying in the species of different genera ($n = 2$ to 7; $2n = 4$ to 14). A contributing factor to this situation is that there is more information regarding arrhenotoky in the spider mite family than in any other group.

Pterygosomidae exhibit arrhenotokous parthenogenesis, which may, in part, explain why an unusual sex ratio of only 5 males were found with 400 females in *Pimeliaphilus rapax*, an ectoparasite of scorpions (Beer, 1960). Arrhenotoky is also present and may be the chief type of reproduction in parasitic groups of Cheyletoidea and Anystoidea, including families such as Harpyrhynchidae, Anystidae, and Cloacari-

dae (Oliver, 1971b). Arrhenotoky is present in Demodicidae. *Demodex zalophi*, from the California sea lion, reproduces in this manner (Dailey and Nutting, 1979).

Parthenogenesis varies in the Astigmata. Species of Histiostomatidae (*Histiostoma fimetarium*) are arrhenotokous, but both thelytokous and arrhenotokous forms of other species may be found in the same habitat (Heineman and Hughes, 1969; Oliver, 1962). This family appears to be the only arrhenotokous family among the astigmatid mites.

Parthenogenesis is identified in a population of rose-colored tyroglyphid mites (Grandjean, 1941). The population (*Dondorffia transversostriata*) was observed to be in great abundance during spring and autumn. Collections were made of many adults, nymphs, and eggs. All were females, without exception. Grandjean's explanation of the phenomenon was that there had been a regression of the seminal receptacle in the females.

Deuterotoky (also called ampherotoky) is the form of parthenogenesis in which unfertilized eggs may produce males and females by agamic reproduction. This condition is known from the Astigmata (Hughes and Jackson, 1958; Heineman and Hughes, 1969) where it occurs in Histiostomatidae (=Anoetidae) (Hughes and Jackson, 1958) and in several ticks (e.g., *O. moubata, Amblyomma rotundatum*) (Davis, 1951). *Histiophorus numerosus* also exhibits this condition, according to Oliver (1971b). He cites Dubinina (1964) in the assertion that the condition occurs in the Labidocarpinae (Family Listrophoridae), where sexual and parthenogenetic reproduction may occur. Experiments of Nuttall (1913, 1915) induced parthenogenetic reproduction by immersing eggs of *Rhipicephalus bursa* in normal saline. All larvae thus produced died.

Depending on the species, sexual dimorphism is strongly or feebly demonstrated in oribatid mites. Bisexual forms are the most numerous, with males and females of about equal numbers in the population. Three major groups are separated on the basis of genital armature:

S, SP, and P. Group S is the most numerous and includes bisexual forms in about equal proportions (Brachypylina, Liodidae, Hermannidae, Palaeacariformes, and part of the Ptyctima). Group SP, in the superior oribatids, tends toward parthenogenesis, although in *Hydrozetes* the males are as numerous as the females. In *Scutovertex* the proportion of sexes varies. The third group, P, comprises the parthenogenetic oribatids. The presence of males is exceptional (100 females/1 male; 25 females/1 male). Diagastric oribatids and Macropylina exhibit this condition (*Epilohmannia, Pseudotritia, Parhypochthonius, Hypochthonius, Nothrus, Camisia, Tectocepheus*), which is considered to be an atavistic trait (Grandjean, 1941).

Parthenogenetic races may coexist with bisexual races, but the former dominate. In one special instance (*Platynothrus peltifer*) parthenogenesis was determined by raising an isolated tritonymph (July), which transformed into an adult (August) and laid eggs (September). The first larvae were observed on 26 September. By the 10th of November, 50 larvae or protonymphs were observed in the culture (Grandjean, 1947). The Nothroidea are also parthenogenetic (Grandjean, 1954).

Several questions have been raised about parthenogenesis and its atavistic character: Are the atavistic males more or less improbable in the development of eggs that become thelytokous? Does a regular mechanism intervene in the fertilization of females? Can a bisexual race exist concurrently with a parthenogenetic race that is more dominant? Grandjean's (1941) observations appear to support the parthenogenetic arrangement found in group P.

The anarthronotan oribatids are assumed to be commonly parthenogenetic (Trave, 1967). Of 113 adults of *Haplochthonius simplex*, 112 were females and 1 was male (Grandjean, 1946). All of the adults of *Trichthonius majestus* were females, many containing eggs (Marshall and Reeves, 1971). Mites of the species *Nothrus palustris* reproduce by thelytokous parthenogenesis (Lebrun, 1970). The phenomenon of par-

thenogenesis is found in *Dameobelba minutissima*, in a species of Nothrus, and is common in the Camisiidae and in the Eremaeidae (Grandjean, 1941, 1947, 1952, 1955). Parthenogenesis is found in *Oppia neerlandica*, a species in which males are not known (Woodring and Cook, 1962).

Paedogenesis

The phenomenon of paedogenesis occurs in a few insects, but was not reported for mites until recently (Baker, 1979). An undescribed species of *Brevipalpus* found on citrus leaves from the Dominican Republic developed paedogenetic protonymphs and deutonymphs. The normal life stages of this mite comprise the egg and prelarvae (calyptostase), larvae, protonymphal and deutonymphal stages, and the adult. In the instance cited, paedogenetic stages contained one, two, and three eggs each. Developing and fully formed larvae were also seen in some protonymphs and deutonymphs. Apparently the larvae broke through the hysterosomal cuticle in their emergence inasmuch as no genital opening is present in the nymphal stages. It is not known what causes the larva to develop into a normal reproductive form or into a paedogenetic protonymph.

DEVELOPMENT AND LIFE STAGES

Eggs and Their Development

The eggs of mites are ovoid, globular, sphaeroidal, elliptical, flattened, or bean-shaped (Treat, 1975). Eggs of Demodicidae, the follicle mites, are elongated, and in certain species assume the shape of the Y-shaped Meibomian glands of the host in which they are laid (Nutting, 1961). The surfaces of eggs may be glabrous, reticulate, ribbed, or roughened (Treat, 1975). They may be spined or have long, pedicellate–capitate projections (Bdellidae; Wallace and Mahon, 1972, 1976). Eggs appear colorless, transparent, opaque, pearly-white, or of pastel shades. Some are green, yellow, bright orange, or deep red (Jeppson et al., 1975; Treat, 1975). The color of an egg may change, and the surface may wrinkle upon contact with moisture (Jeppson et al., 1975).

Tetranychid females may lay 2 or 3, or perhaps up to 15 or 20 eggs per day, depending on the environmental factors and the strain of mites. A single female may lay 50 to 100 eggs, and a female *T. urticae* is cited as laying 312 eggs (Huffaker et al., 1969).

In Tetranychidae, the eggs may be suspended or tied with silken guy wires to the surfaces on which they are laid (Hazan et al., 1975a,b). Winter and summer eggs of spider mites differ in pigmentation and in placement on the host plant. The lid of the egg case of *Petrobia latens* is like "the top crust of a pork pot pie" (Evans et al., 1961).

Eggs of *Brevipalpus californicus* (Tenuipalpidae) are bright red and elliptical. The eggs are glued to the leaf surface of the host plant by a sticky substance. Eggs of the citrus flat mite, *B. lewisi*, are oval and pink. They are laid in the cracks and crevices of twigs, leaves, and the fruit of the host plant (citrus, walnut, ornamentals, grapes). Eggs of the privet mite, *B. obovatus*, are bright orange and elliptical when oviposited, but darken as hatching approaches (Jeppson et al., 1975).

The number of eggs laid by female ticks varies considerably. The minimum recorded for *Argas persicus* is 47, for *Rhipicephalus evertsi* it is 300, and for *Haemaphysalis leporispalustris* it is 59. Maximum numbers include 20 000 and 18 497 for *Amblyomma hebraeum* and *A. maculatum,* respectively (Manual of Livestock Ticks, USDA-ARS, 1965).

Two types of eggs are produced in the red-legged earth mite (*Halotydeus destructor*). The winter eggs are yellow to orange and are deposited in masses or a single layer. The masses are in contact with the soil or the underside of leaves. As the temperature increases, the oversummering eggs are formed. These eggs, which have a thicker shell than winter eggs, are retained within the body of the female, which serves to protect them from dessication. When the female dies, the eggs hatch (Jeppson et al., 1975).

The egg of *Acanthophthirium polonicus* (Myobiidae) is attached to the hair of its host, the pond bat (Fig. 12-27A), and appears to show the beginnings of limb buds (Pater-de Kort et al., 1979).

Fertilization and Cleavage

In *Siteroptes graminum*, the sperm enters the ovum during the first maturation metaphase of the egg and on the side opposite the site of maturation division. The copulation paths of the sperm may be different in separate eggs. Polyspermy is not known for this mite, although multiple entry of sperm is not precluded. The egg may swell at the point of entry and radial striations in the cytoplasm may delineate the pathway of penetration by the sperm. Sperm that penetrate the yolk become tripartite and bloated (Fig. 12-27B), and they stain more lightly. The deeper the penetration of this trilobed vesicle, the more ellipsoidal the sperm becomes in the principal axis of the egg. The nucleus becomes reticulated. The female pronucleus also elongates and increases in volume (Fig. 12-27C–E). The centroplasm of both male and female pronuclei that have approached each other at the equator of the spindle show striations along the principal axis of the egg. When both male pronucleus and the maternal karyomeres of the egg form a group of four elongated vesicles at the equator of the ovum, the process of fertilization is complete. Cleavage then follows and the specific details of mitosis ensue (Cooper, 1939).

During the early cleavage of the egg, each chromosome possesses an individual karyomere in which ultimately the chromosome condenses and disappears. Longitudinal division of the chromosomes is evidenced at late metaphase. An equatorial body is formed between the separating chromosomes, but this body degenerates in telophase. The chromosomes show parallel displacement during anaphase and

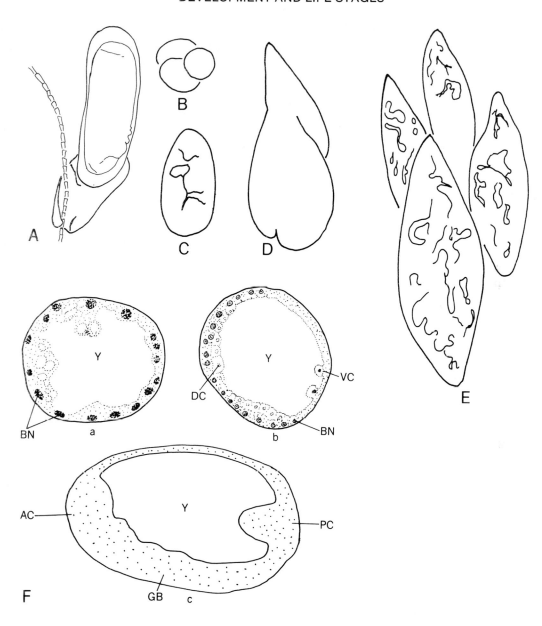

Figure 12-27. (A) Egg of *Acanthophthirium polonicus* (Myobiidae) attached to hair of bat host (after Pater-de Kort et al., 1979); (B–E) *Siteroptes graminum*, development of male pronucleus: (B) trilobed male pronucleus; (C) male pronucleus in copulation path; (D) male pronuclues shortly before reaching the equator; (E) smaller karyomeres of female pronucleus in gonomeric prophase of first cleavage (after Cooper, 1939); (F) embryogeny of *Acarus siro*: a, blastoderm formation around yolk; b, delamination (?) of cells and vitellophages; c, early embryo with anterior and polar caps (AC, anterior polar cap; BN, blastoderm nuclei; DC, delaminating cells; GB, germ band; PC, posterior polar cap; VC, vitellophage cells).

each chromosome becomes an elongated vesicle at telophase.

By the tenth cleavage division, karyomery and the formation of equatorial bodies have stopped and the mitoses of the early and late embryos and of the oogonia are normal (Cooper, 1939).

Embryogeny

The knowledge of mite embryogeny has remained at the same level for about 25 years. There is a paucity of literature on the subject, but mites appear to show a relatively primitive pattern of development. The eggs are centrolecithal and, as in other arachnids, the embryos develop in pockets or pouches that project from the ovary into the hemocoel (Gupta, 1979; Aeschlimann and Hess, 1984).

The majority of acarines exhibit superficial cleavage of the ovum. Total cleavage is limited to a few families (Tarsonemidae, Pygmephoridae, and Tetranychidae) (Krantz, 1978; Dittrich, 1965; Treat, 1975). A combination of initial holoblastic cleavage followed by superficial cleavage occurs in Demodicidae (Guilfoy and Nutting, 1987).

In intralecithal cleavage the nucleus of the fertilized egg divides synchronously, connected by reticuloplasmic strands and showing a cytoplasmic halo. Migration may occur prior to cleavage. The yolk does not shift. The cleavage nuclei continue to divide as cytoplasmic growth occurs (Aeschlimann and Hess, 1984). With this nuclear division and cell expansion the blastoderm is formed as an envelope around the yolk (Fig. 12-27F). Concurrent with nuclear division is the migration of cleavage nuclei to form special cells, vitellophages, which liquify the yolk so that it is available to the developing embryo (Krantz, 1978; Aeschlimann, 1958).

A superficial blastoderm forms around the yolk (Hughes, 1959). A thickened ventral germ band exhibits both anterior and posterior polar caps (Fig. 12-27F). The blastoderm is cuboidal. The germinal band does not invert as in spiders, but gastrulation is similar (Anderson,

1973). The germinal disc represents the telson of the embryo and determines anterior–posterior and dorsoventral orientations (Aeschlimann and Hess, 1984). The anterior cap begins the anlage for the future central nervous system—synganglion and nerves. As the ventral germ band develops, a ventricular area grows dorsally where a heart, if present, will be. Organogenesis is not understood (Aeschlimann and Hess, 1984). Gnathosomal and idiosomal areas are shaped and appendages formed. Segmentation is obscured in embryogeny, but traces appear. The chelicerae develop postorally in the embryo, but later migrate to a position anterior and dorsal to the mouth. Palps and limb buds develop (Jeppson et al., 1975; Treat, 1975; Hughes, 1959).

The cleavage and early development of *Demodex cafferi* show a graded series of germinal vesicles in the ovary (Fig. 12-28A). The smallest are thought to be oogonia; larger vesicles are previtellogenic oocytes. The initiation of vitellogenic growth occurs when the eggs enter the uterus (Fig. 12-28B). The germinal vesicle increases in size until the ovum reaches its mature size. During vitellogenesis, the chorionic membrane becomes prominent. The intrauterine, mature ovum exhibits a distinct nucleus (C), but is followed by a disintegration of the germinal vesicle and loss of an identifiable nucleus prior to oviposition. After oviposition, the first maturation division results in two equally divided daughter cells (D). Each daughter cell exhibits a central nucleus surrounded by a pocket of cytoplasm which extends as a network between the yolk bodies. A second maturation division results in four equally divided daughter cells (E). Nuclear cleavage occurs in a series of asynchronous differential divisions in which vitellophage (X) and blastodermal nuclei (Y) are differentiated (G, H). Proliferation and peripheral migration of blastodermal nuclei are accompanied by the growth of an internal cytoplasmic network. The vitellophage nuclei remain between the yolk bodies (I). The four-cell holoblastic condition appears to disintegrate with the degeneration of the inner cell parti-

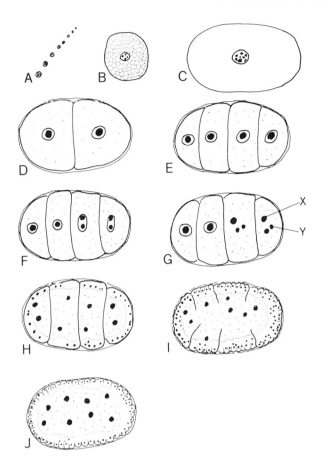

Figure 12-28. Schematic embryogeny of *Demodex cafferi*: (A) graded germinal vesicles in ovary; (B) vitellogenic growth on entering uterus; (C) intrauterine ovum of mature size; (D) two daughter cells after first maturation division; (E) four daughter cells after second maturation division; (F) nuclear cleavage; (G) asynchronous divisions result in vitellophage (X) and blastoderm (Y) nuclei; (H) peripheral migration of blastodermal nuclei; (I) four-cell holoblastic cleavage begins to disintegrate, superficial cleavage begins; (J) blastodermal layer of yolk free cells. (After Guilfoy and Nutting, 1987; Nutting, 1979).

tions. Further cleavage is superficial (J). The blastodermal layer of yolk-free cells is completed and the yolk bodies fuse to form a unitary yolk mass (W.B. Nutting, 1979, personal communication; Guilfoy and Nutting, 1987).

Time of development from egg to adult var-

ies from hours to a few days, to weeks, months, and years. Available food, temperature, humidity, and other factors affect the completion of the life cycle in populations of mites. Some species of Macrochelidae complete the cycle in 1.5 days (Axtell, 1969). One generation of *Dicrocheles phalaenectes* may be completed in 5 midsummer days (Treat, 1975). The wood tick, *Dermacentor andersoni*, requires 2 years for its life cycle (Hunter and Bishopp, 1911) while 4 to 5 years is required for the life cycle of *Ixodes uriae* (Eveleigh and Threlfall, 1975). (See also El Kammah et al., 1982, for the embryology of *Hyalomma dromedarii*.)

Generations per year and longevity are variable. In *Panonychis ulmi*, a multivoltine species of spider mite, eight generations may occur within a year (Jeppson et al., 1975). Univoltine mites occur in the soil, duff, litter, and some aquatic habitats (Hartenstein, 1962; Newell, 1973). Some species of *Tyrophagus* have a life span of less than 1 month (Krantz, 1978), while certain species of ticks live for several years (Balashov, 1972).

In some parasitic mites, there is a tendency to retain larval or nymphal features (Otopheidomenidae). Such a condition is referred to as "localized neoteny" (Evans, 1963).

The time of the life cycle of Australian Erythraeidae is identified through the following periods: egg (5–11 months); larva (1–3 weeks); nymphochrysalis (9–16 days); deutonymph (21–39 days), imagochrysalis (15–16 days); adult. This is a univoltine group and the cycle is annual (Lawrence, 1953, cites Southcott, 1946). *Allothrombium fuliginosum* passes through a prelarva (Fig. 12-29A).

In the life history of *Tetranychus cinnebarinus*, a spider mite pest of citrus fruit trees and truck crops in Israel, more eggs were deposited at 24°C and 38% relative humidity than at any other combination of temperature and relative humidity (Hazan et al., 1974).

Rapid development is typical of phytoseiid mites. *Phytoseiulus persimilis*, a specialized predator of tetranychids, develops to an adult within 4 to 5 days of oviposition; other species

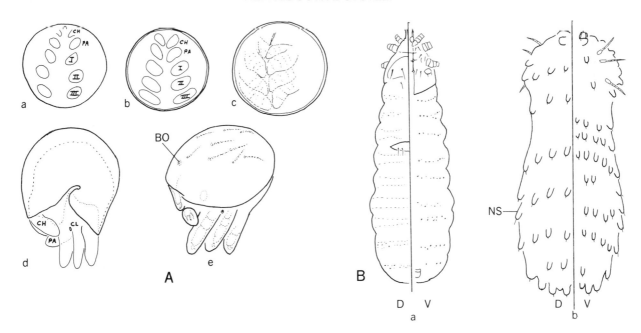

Figure 12-29. (A) Development of *Allothrombium fuliginosum*: a, ninth day of development; b, tenth day; c, twelfth day; d, hatching; e, larva (BO, bothridium; CH, chelicera; CL, Claparede organ; PA, palp) (after Robaux, 1974); (B) neosomy in a, Pyemotidae (after Cross, 1965); b, *Vatacarus* (after Audy et al., 1963) (D, dorsal; V, ventral; NS, neosomules).

require 5 to 7 days for development (Huffaker et al., 1969).

The developmental cycle of the straw itch mite, *Pyemotes tritici*, is completed in 5 to 7 days. Ovoviviparity in this species results in adult progeny and a high reproductive potential (ca. 250 adult offspring per gravid female). The population is 95% female, and these new females seek hosts immediately after birth (Bruce and LeCato, 1980).

Regarding embryonic homologies, Boudreaux (1979) states:

> The significance of variations in embryonic development among members of various taxa must be assessed with caution. Evolutionarily important mutations may affect any stage of development, either in delaying, advancing, suppression, or modification of expression, or derepression of the old gene codes, or in pro-

ducing totally new structures. . . . At any rate, the old law of recapitulation seems to be invalid. At best, the study of arthropod embryos may briefly allow the glimpse of the execution of gene systems that in the past controlled the primitive state of ancestral characters, and may be of limited usefulness in establishing homologies.

Neosomy

Neosomy is an unusual phenomenon exemplified repeatedly in the Acari. By definition it is a radical intrastadial metamorphosis that results in new body structures as a consequence of extensive cuticular growth (Audy et al., 1963, 1972). Neosomatic growth and development are phenomenal in chiggers of the genus *Vatacarus* (Trombiculidae) (Fig. 12-29B), a parasite in the lungs of sea snakes. Larvae of this genus increase 1500-fold or more and develop papilliform neosomules (external cuticular modifications) that enable the maggotlike larvae to move through the mucous of the host (Fig. 12-29B). Intranasal chiggers (*Fainiella, Cheladonia*) also exhibit neosomy in varying degrees. Chiggers parasitic on the wings of bats (*Riedelinia*) enlarge up to 750-fold and develop a neosomule that resembles a collar behind the gnathosoma. This collar (comparable to other adaptations of acarine parasites found on the wing or tail membranes of bats) facilitates attachment to the flying hosts (Audy et al., 1972).

Neosomy is identified in some males and many females of hard ticks (Ixodidae), but is limited to the larvae of soft ticks (Argasidae). Female *Ixodes ricinus* may enlarge 100 times. Gynandromorphs of ticks may show neosomy in the female part of this sex mosaic (Audy et al., 1972).

Several examples of neosomy are found in Pyemotidae (Fig. 12-29A). *Pyemotes ventricosus*, the grain itch mite, is an ectoparastic of insects. The female develops a greatly swollen opisthosoma and combines neosomy with *physogastry* and *tachygenesis*. Physogastry ("bladder belly") is a descriptive term applied to the swollen body of a female without reference to the developmental process. Such neosomatic females give birth to adult males and females without external ontogenesis through the nymphal stages (*tachygenesis*) (Audy et al., 1972). Physogastry enables the simultaneous production of large numbers of embryos. Development to the adult stage occurs within the mother and accompanies the hypertrophied growth of the female idiosoma (Moser et al., 1971).

Neosomy occurs in the plant parasite *Siteroptes graminum* (Audy et al., 1972). Following attachment to the host plant, the hysterosoma of the female swells with fluid to an enormous size— as much as 500 times the normal volume. Cooper (1937) indicates that the species is facultatively parthenogenetic (virgin females produce only males). Eggs develop within the hysterosomal sac, hatch, and grow into mature adults. Maturation of the brood results in a mass birth, rupture of the hysterosoma, and the death of the female.

Females of *Perperipes ornithocephala*, a nidicole of army ants (*Eciton*), enlarge to 10 times the volume of the unfed female (Fig. 12-1), but form neosomules of rough annular ridges that appear as segments and enable the mite to mimic larval ants. The neosomules are of two types: sclerotized mammilliform projections, or stellate concentrations of secreted materials (Cross, 1965; Audy et al., 1972).

Neosomy has been observed in water mites (larvae of *Hydrachna magniscutata*), some female Pterygosomidae (Actinedida), and female Teinocoptidae (Astigmata), all parasites of invertebrates or vertebrates.

Developmental Stages

In the developmental stages and life cycles of the Acari, one or more active immature stages usually occur between the egg and the adult. The stages of the life cycle vary with the order of the mites, the type of life-style (free-living, parasitic), distribution (phoresy and hypopi), and environmental factors.

The egg is usually laid by the female, but may be retained within the body until hatched. At oviposition it may pass through a developmental stage called the deutovum or prelarva. The larval stage is hexapod, a characteristic feature the acarines share with the Insecta and Ricinuleida. (An exception is found in the Eriophyidae where four legs are present in all stages.) Nymphal stages and adults are octopod (except the Eriophyidae and some parasitic forms). Nymphal stages vary in number in the different orders, but the normal sequence of stages is (prelarva), larva, protonymph, deutonymph, tritonymph, and adult.

Deutovum or Prelarva. It is assumed that none of the free-living mites passes through more than one active hexapod larval stage, but two stages are postulated in the phylogenetic history of the Actinotrichida (Grandjean, 1941, 1957). The first, a prelarva (=deutovum; calyptostase), is an inactive, preeclosion stage characterized by its distinct embryonic cuticle within the egg shell. Where it exists in mites and other arachnids, the prelarva is a primitive, nonfeeding stage (Krantz, 1978). It is postulated to have been an active stage preceding the larval molt (Grandjean, 1938). As an inhibited rather than a regressive form, it may be saclike, without legs or mouthparts, a condition called *calyptostasis* (Coineau, 1974a,b; Grandjean, 1946, 1947, 1962; Lions, 1967, 1973). Parasitiformes usually have no prelarvae, but may exhibit larvae with leg IV limb buds. Acariformes usually have a prelarval stage (David Walters, personal communication, 1986).

Phalangiacarus brosseti (Opilioacarida) (Fig. 12-30) has a primitive prelarval stage with well-developed and functional mouthparts and legs (Coineau, 1973; Coineau and Hammen, 1979).

A simple prelarva is noted for *Tetranychus linearius* (Grandjean, 1948). The calyptostatic prelarva in *Balaustium* (Erythraeidae) is found inside the oval, elongated egg (Grandjean, 1946). *Acanthophthirius polonicus* (Myobiidae), a parasite on the pond bat, has a similar prelarva (Fig. 12-33C) (Pater-de Kort et al., 1979). In some actinedid families (Anystidae, Bdellidae, Erythraeidae, Trombidiidae, and Trombiculidae) the stage exhibits simplified appendages and some body setae (e.g., *Allothrombium fuliginosum*) (Fig. 12-31E) (Vitzthum, 1943). Prelarvae are also typical of the oribatids *Trhypochthonius tectorum*, *Hermannia convexa*, and *Phthiracarus anonymum* (Fig. 12-38B, C) each of which exhibits a Claparede organ (Hammen, 1973). *Hoplophthiracarus pavidus* has up to four prelarvae inside the female, where the stage is clear, faintly colored, and with slight folds and ridges; vestiges of the chelicerae, palps, and legs I are present, but mainly as folds, and an urstigma is visible (Hammen, 1963). The box mite *Plesiotritia megale* (Euphthiracaroidea) also has a prelarva (Walker, 1964). The prelarva of *Camisia segnis*

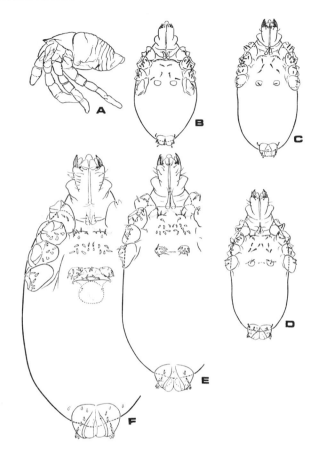

Figure 12-30. Life stages in *Phalangiacarus brosseti*: (A) prelarva, lateral and ventral aspect; (B) larva; (C) protonymph; (D) deutonymph; (E) female tritonymph; (F) adult female, ventral aspect (after Coineau and Hammen, 1979).

shows the anlage of the chelicerae, pedipalps, and leg I–III (Grandjean, 1936). In the higher Oribatida, the prelarva is saclike, lacks body setae, and may show bosses where the appendages develop (Evans et al., 1961; Lions, 1967, 1973).

Prelarvae may have three pairs of legs, mouthparts, and setae, but still may be inactive, an incomplete condition called *elattostasis* (Grandjean, 1957).

Prelarvae in the Acariformes are characterized by an *urstigma* (Claparede's organ) and a

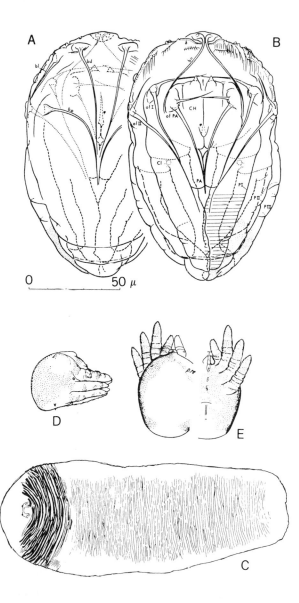

0   50 μ

Figure 12-31. Prelarva of *Eupodes strandtmanni*: (A) dorsal; (B) ventral (Coineau, 1976); (C) *Acanthophthirius polonicus* (after Pater-de Kort et al., 1979); (D) *Anystis*; (E) *Allothrombium fuliginosum* (after Vitzthum, 1943).

nonfunctioning stomodaeum. An anal opening, coxal glands, and distinct pretarsi may or may not be present (Krantz, 1978).

Eupodes strandtmanni, a viviparous species of mite, has an elattostatic prelarva (Coineau, 1976) (Fig. 12-31A). The prelarva of *Campylothrombium barbarum* Lucas also is similar, but shows the vestiges of ocelli as well (Robaux, 1970b). *Anystis* has a prelarva stage (Fig. 12-31D) (Vitzthum, 1943). The prelarvae are well-developed and active in Rhagidiidae and Adamystidae (Actinedida) (Ehrenberger, 1974; Coineau, 1979).

Larvae. Homeomorphic larvae are common among the orders of mites; that is, the larvae resemble the other stages in the life cycle; but some groups of the Actinedida have heteromorphic larvae that differ in morphology from the other stages. Heteromorphism may apply to one or more nymphs in the Acariformes (Hughes, 1959).

When the embryo is fully formed, it hatches by rupture of the egg case to become the hexapod larva. In some instances a deutovarial membrane is secreted by the blastoderm under the chorion and inside the shell. The space between the deutovarial membrane and the egg case is filled with fluid and contains some amebocytes. When the egg case is broken or ruptured, the deutovarial membrane encloses the larval stage with its limb buds and other structures. It also allows for expansion and growth after initial eclosion (Hughes, 1959; Jeppson et al., 1975). Passage from the deutovum or prelarval stage to the larval stadium may take a day and a half (Coineau, 1979) or two weeks (Robaux, 1971).

Eclosion from the egg of the larva of *Cheyletiella yasguiri* Smiley is effected by an egg-breaking structure. After oviposition, the female attaches the tapered end of the egg to a hair shaft of the host (2–3 mm above the base) by means of fine threads woven into a cocoonlike structure. The egg-breaking structure apparently is used similarly to such hatching spines in insects (Marchiando, 1981).

In the typical hexapod larva of Acari it is unusual for vestiges of legs IV to appear in the larval stage, although this has been observed in Opilioacarida (Coineau, 1973), and in embryogeny (Aeschlimann, 1958). The larval stage is characterized by little or no sclerotization, and the lack of external genitalia. Sclerotization, when present, is restricted to the podosomal region and ventral shields are weakly sclerotized, obscure, or absent (Krantz, 1978).

Larvae may be inactive and nonfeeding (Gamasida), voraciously predaceous (Cheyletida, water mites) or parasitic (Trombiculidae). In many instances, morphology makes the identification of larval stages difficult, although in Trombiculidae, the chigger mites, the species are identified predominantly by larval characteristics rather than by features of the adult (Brennan and Jones, 1959; Vercammen-Grandjean and Langston, 1971).

Little or no change of the initial form of the larva occurs in its development except where engorgement ensues. Intrastadial metamorphosis, *neosomy*, described earlier may greatly enlarge the cuticular growth and exterior structures (Audy et al., 1972).

Repression or abbreviation of the larval stage may occur because of environmental conditions. In *Siteroptes cerealium* (Pygmephoridae), the mating of sexually mature males and females results in larvae that molt into males and females (Krantz, 1978). Physogastric development of females may result in abbreviated development, as cited earlier.

When fully engorged, larvae of the Pterygosomidae, Trombidiidae, water mites, and other families enter into a resting stage called a *nymphochrysalis*. The epidermis retracts from the cuticle, muscle attachments are lost, and some tissue differentiation occurs. The nymphochrysalis may appear as an egg-shaped pupalike form (*Pterygosoma*) or exhibit limb buds (*Pimeliaphilus*). Upon rupture of the old cuticle, the nymphal stage emerges (Hughes, 1959).

Where the nymphochrysalis occurs, there is only one nymphal instar, followed by the *imagochrysalis*, another resting stage. The imagochrysalis (or telechrysalis) may, like the nymphochrysalis, show reduction in the appendages of the gnathosoma, and suckers may develop on the posterior ventral surface.

Nymphs. The interpolation of nymphal stages between larva and adult varies to a high degree among the orders. It is unusual for a single, free-living nymphal instar to occur in the life cycle, with the exception of the Ixodidae (hard ticks). A few parasitic forms (Pyemotidae, Pygmephoridae, Podapolipidae) develop directly from larva to adult without this intervening stadium. Many mites have two nymphal stages, but three nymphs are common in the life cycles of the Actinedida, Astigmata, and Oribatida.

The nymphal stages of mites are usually eight-legged. Differentiation occurs in the sclerotization of shields and plates of the idiosoma, setal patterns, genital acetabula, and leg setation. Some of these features are taxonomically important and identify the particular type of nymph as well as the major category involved.

Protonymphs. The first nymph following the larva is the protonymph. This is usually a free, active instar, but may or may not feed. It is normally found in and adapted to the environmental location where the other nymphal stadia will occur.

Deutonymphs. The deutonymph is the second nymphal stage in the life cycle and may differ from the adult only in size, the patterns of sclerotization, and setation (Krantz, 1978; Prasad, 1973). The deutonymphs of parasitic Gamasida do not feed (Evans and Till, 1965), nor do the phoretic deutonymphs of the Uropodina (Athias, 1975). Many deutonymphs of the Astigmata are heteromorphic, that is, they are completely different morphologically and behaviorally from preceding and succeeding stadia. Such heteromorphic nymphs, called hypopi (hypopodes, Treat, 1975) occur sporadically and may or may not occur in a specific generation, a condition called *facultative hypopody* (Krantz, 1978). These nymphs lack functional mouthparts and are principally phoretic stages.

Tritonymphs. The tritonymph is usually an active instar where it occurs, but may be a calyptostase. Where hypopi occur, in the Astigmata, the second homeomorphic nymph is considered to be a tritonymph because it is the third postlarval and preadult stage. Inasmuch as the tritonymph is found in relatively few groups of mites, the final molt to the adult stage usually occurs after the second nymphal instar (Krantz, 1978).

Life Cycles

Gamasida. In the Gamasida, the hexapod larva is almost always a nonfeeding instar. Dorsally, this stage frequently has an anterior podonotal shield and possibly a small posterior or pygidial shield. Ventrally, the sternal shield has three pairs of sternal setae and an anal shield is present. No stigmata or genital setae are present in the larva. Reproductive behavior differs between groups (Radinovsky, 1965).

Gamasida generally have two nymphal stadia between the larva and the adult: the protonymph and the deutonymph (Fig. 12-32).

The protonymph is octopod and dorsally may have either a large anterior podonotal shield and a small posterior pygidial shield, or a larger opisthosomal shield. Ventrally, the sternal shield exhibits three pairs of sternal setae. Stigmata and a pair of genital setae occur.

In the deutonymph, the dorsal shield may be entire, incised laterally, divided into two nearly equal parts, or have a large podonotal shield and a small pygidial shield. Ventrally, the sternal shield is attenuated posteriorly to a rounded point and has four sternal setae. The addition of the metasternal setae (sternal IV) is characteristic of the deutonymph and carries over into the adult. Genital setae are also present with stigmata.

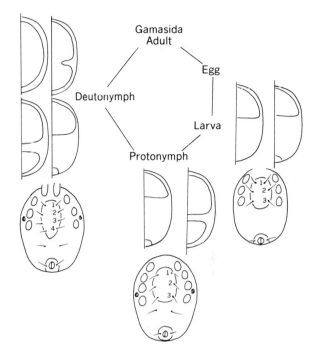

Figure 12-32. Life cycle of Gamasida showing schematic arrangements of dorsal plates and venter in larva protonymph and deutonymph.

Variations in the sclerotization and shield arrangements in the adults occur. The general pattern of shield development from larva to adult is one of increased sclerotizaton, yet some parasitic forms (Entonyssidae, Halarachnidae) show reduced shield development in the nymphal stages and incomplete or weakly developed shields in the adults (Evans et al., 1961; Prasad, 1973; Treat, 1975).

Free-living gamasid mites usually pass through four active stages in the life cycle: larva, protonymph, deutonymph, and adult. Parasitic gamasids may show specializations or modifications of the life cycle (Kinn, 1983). In a few of the Uropodina, the deutonymph may be phoretic and transfer to a new location by hitchhiking on an insect carrier. The phoretic nymph is attached to the insect by means of a pedicel called a styloproct that is inserted into the integument of the host.

Ixodida. In the life cycle of soft ticks, there may be as many as eight instars. In *Ornithodoros moubata* the tick emerges as a nonmotile larva (called "embryon eclos") and molts into an active nymph (N1) without a blood meal. Further nymphal stages (N2, N3, N4, N5) occur and become adults. Nymph 5 produces a few males and some Nymph 6, most of which develop into females. Unlike the larva, each nymphal instar molts into the next stage after taking a blood meal (Evans et al., 1961; Aeschlimann, 1958).

Investigations show that populations of *O. moubata* are made up of four races, each adapted to different food, temperature, and humidity requirements, but only slightly morphologically different (Clark, 1973).

Argasids feed rapidly and intermittently, and accumulate blood from the host animals as they are resting. This reduces the possibility of transport during inclement weather and keeps the host within feeding range (Hoogstraal, 1956). These several blood meals nourish the female as she deposits relatively few eggs at a time. The location of the argasids in the burrow, den, or house of the host, intermittent feeding, and small batches of eggs laid over time assist the fecundity of the argasids (Hoogstraal, 1956; see also Vogel, 1975).

Hard ticks have a single larval and a single nymphal stage between the egg and the sexually mature adult. Ixodids remain on the host for several days in each stage and slowly ingest large amounts of blood during engorgement. After this extended meal, females oviposit only once, but lay huge numbers of eggs over a period of several days. (Variations of life cycles of one-, two- and three-host ticks are detailed in the *Manual on Livestock Ticks* and Arthur, 1962) (Fig. 12-33).

Actinedida. In the Actinedida, the number of active instars ranges from none (e.g., Podapolipidae) to three (Endeostigmata, Bdellidae, Halacaridae) (Jeppson et al., 1975; Newell, 1973). This range in the number of active stages exem-

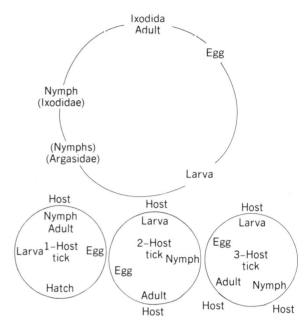

Figure 12-33. Generalized life cycle of Ixodida and diagrams of one-, two- and three-host life cycles in hard ticks.

plifies four rather distinctive life cycles (Evans et al., 1961) (Fig. 12-34A).

Some of the Tarsonemidae have only two stages following the egg, the larva, and the adult. Evidently, the nymphal stages are passed through in the skin of the larva. In pyemotid mites, both males and females may be produced within the body of the physogastric female without any external development before release of the final stage (Audy et al., 1972).

In *Eutrombidium rostratus* (Scopoli), which as larvae parasitize grasshoppers (Fig. 12-34B), an average of 4700 eggs are deposited per female over several weeks. The larvae hatch and engorge after attaching to a host insect. The life cycle includes the larva, (nymphochrysalis), nymph, (imagochrysalis), and adult. This mite is usually univoltine, with one generation per year, but one and a half generations per year are known, with an overwintering of the

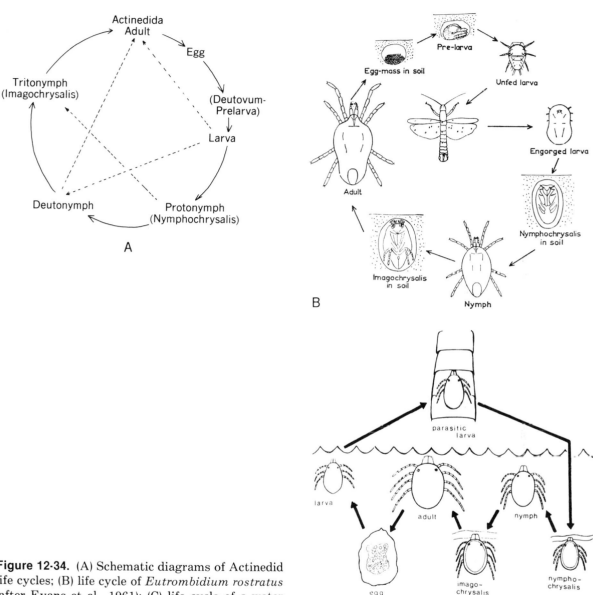

Figure 12-34. (A) Schematic diagrams of Actinedid life cycles; (B) life cycle of *Eutrombidium rostratus* (after Evans et al., 1961); (C) life cycle of a water mite (after Smith, 1976).

nymphs (Evans et al., 1961). Other groups of the Actinedida (e.g., Endeostigmata, Bdellidae, Tydeidae, Halacaridae) and water mites pass through three distinctive nymphal stages between the larva and the adult: protonymph, deutonymph, and tritonymph.

Typically, the Parasitengona (Trombidiidae, Trombiculidae, Erythraeidae, Hydrachnida) pass through only *one* active nymphal instar, the deutonymph. It is interspersed between two calyptostases, or pupalike stages: a nymphochrysalis (protonymph) and an imagochrysalis (tritonymph) (Fig. 12-34B, C). These inactive, *pharate* calyptostase forms are usually without appendages and leave *two* cuticles behind upon molting to the next stasis. The protonymph develops within the larval skin as a nymphochrysalis. The nymphochrysalis leaves the larval and the calyptostase cuticle behind. The imagochrysalis is interpolated between the active deutonymph and the adult. This imagochrysalis (tritonymph) emerges from the deutonymphal skin and the cuticle of the imagochrysalis (Newell, 1973).[1] In some Trombiculidae, all nymphal stages occur within the larval skin (Audy et al., 1972).

The universal presence of these calyptostases in the Parasitengona shows retention of structural regression and specialization (Newell, 1973), not necessarily evolutionary regression as suggested by Grandjean (1939b). Newell also correlates the numbers of nymphal stages of the Hydrachnida with the structural modifications of the groups concerned. In a later publication, Grandjean (1957) maintains that the calyptostases are the protonymph and tritonymph.

In *Podapolipus grassi* (Podapolipidae), the hatched larvae assume a "seeking position," much as do ticks, awaiting the presence of a host insect. Adults of this species are hexapod—essentially sexually differentiated larvae. Paedogenetic development of eggs occurs within the body of the mite. Eclosion occurs after the first molt, within 36 to 48 h after feeding. Males are produced by this process and mate with the parent. As many as 100 eggs adhere to the body of the female. Both sexes are produced, with males being larger than the paedogenetic eggs. Males are ectoparasitic on the female and may kill her. These stages are transferred from one host to another (Evans et al., 1961).

It should be evident that a broad diversity of life cycles occurs in the Actinedida. Because of the number of variations in the life cycles, selected examples are given.

Terpnacarus (Endeostigmata) has five stages of development: larva, protonymph, deutonymph, tritonymph, and adult. Such a life cycle, however, is not confirmed for *Lordalychus, Alicorhagia,* or *Sphaerolichus* (Grandjean, 1939).

The winter grain mite, *Penthaleus major,* has a life cycle of egg, deutovum (prelarva), protonymph, deutonymph, and adult. The red-legged earth mite, *Halotydeus destructor,* passes through four to five stages from larva to adult (Jeppson et al., 1975).

Pimeliaphilus podapolipophagus is a parasite of cockroaches. Females of this species lay eggs singly or in batches of 2 to 20. The larvae hatch from the eggs and pass through the (nymphochrysalis), nymph, (imagochrysalis), and the adult stages in the life cycle (Cunliffe, 1952; Evans et al., 1961).

Spider mites and their relatives (Tetranychoidea) usually have two active nymphal instars between the larva and the adult. In these forms there is an obligatory inactive period during development when the nymph does not move and the new cuticle and its myocuticular junctions are formed. This quiescent period ends with the splitting of the old cuticle and the eclosion of the next instar. Only *one* cuticle is left behind—that of the preceding instar (Jeppson et al., 1975; Newell, 1973).

The stages of the life cycle of Tetranychidae are egg, larva, protonymph, deutonymph, and adult. In some species susceptibility to environmental stress has been overcome by a quiescent stage or condition that tolerates the adverse situation in the habitat (Huffaker et al., 1969). Each nymphal stage and the adult stage is begun during periods of inactivity called protochrysalis, deutochrysalis, and teliochrysalis. The mite anchors itself to a leaf or to its silk webbing, contracts the legs and develops the new integument before molting to the next instar. Whether there is an increase in somatic

---

[1]The terminology for these calyptostases is reviewed by Newell (1973) and termed "elaborate and confusing." The simple terms *protonymph, deutonymph,* and *tritonymph* are proposed to reduce the confusion, but not all authors accept and use this terminology. (From REPRO 14)

cell numbers during the "chrysalis" stages is not known and no mitotic figures are found in stages beyond the larva. Size is increased by growth of cells, not by multiplication (Jeppson et al., 1975).

Spider mites develop rapidly and have short life cycles overall, but their fecundity is not necessarily high. They are noted for the rapidity with which infestations occur, which results from the large number of generations developed in a single season.

In the Demodicidae, the follicle mites, four stages of the life cycle are known. Spickett (1961) cultured *Demodex folliculorum* and calculated the life cycle to be about 14.5 days. The egg stage was approximately 60 h, the larval stage 36 h, the protonymphal stage 72 h, and the deutonymphal stage 60 h, while adults lived for up to 120 h. The sex ratio in this study was 1:1. Nutting (1964) observed 8 to 13 eggs in the ovary of *D. caprae*. If each egg was viable and developed up to 3 days, the life cycle would be longer.

By histological investigation, use of *in vitro* cultures and experiments, Spickett (1961) shows that more females develop than males and that the males of *D. folliculorum* live about half as long as the females. Males and deutonymphs are more tolerant of variable conditions than females. Also, males are found only at the opening of the follicle; hence, it is assumed that they spend most of their lives on the surface of the skin and enter the follicle for feeding and copulation. Fertilized females return to the follicle. It is not known how many eggs are produced, but the egg is large. Because only a few larvae are produced in culture, it is assumed that few eggs are produced by one female during her life. Two-, four-, and eight-cell embryos occur in *D. cafferi* (Nutting and Guilfoy, 1979).

After oviposition, the female moves back to the mouth of the follicle and dies, having completed the life cycle in approximately 120 h (14.5 days). About half of the mites in the mouth of follicle are dead, which means that the carcasses tend to plug the mouth of the fol-

licle and prevent fresh infections. This circumstance reduces the number of mites per follicle and prohibits overpopulation.

The life cycle begins as the fertilized female makes her way into the follicle and then into the associated sebaceous gland. She deposits an egg within 12 h. The larva usually hatches within 60 h after oviposition, feeds continuously in the gland for about 40 h, and then molts into the protonymph. The change in form usually occurs in the pilosebaceous canal of the follicle. The protonymph also feeds continuously, but is carried by the flow of sebum from the sebaceous gland toward the opening of the follicle. The legs of the protonymph are feebly developed, so they offer little effective resistance to the current of the sebum. The protonymph changes into the deutonymph in about 72 h. The deutonymph feeds for a short time and then crawls on to the skin surface, where it moves about randomly for about 12 h (but as many as 36 h), usually in darkness or in dim light. The deutonymph is evidently the stage of transfer and distribution on the skin. The deutonymph reenters the follicle after about 60 h to change into the adult. Females remain in the opening of the follicle until copulation occurs. They then move into the follicle.

Comparison of the morphology and life cycles of *D. criceti* and *D. aurati* show that males and females of the latter species are found in the same follicle. Ova of *D. aurati* are spindle-shaped; those of *D. criceti* are broadly oval. The life cycle of *D. aurati* occurs in the follicle. Rarely do adults of this species occur on the surface of the skin or under loose skin. Larvae and nymphs of *D. aurati* are usually found deeper in the follicle than the level of the sebaceous gland (Nutting and Rauch, 1958; Nutting, 1965).

In the life cycle of *Balaustium florale*, which feeds on flower pollen, three active instars occur: larva, deutonymph, and adult (Grandjean, 1959). Little is known of the habits of the larvae; they are either predatory or weakly parasitic without specific affinity for the host (Newell, 1963). The deutonymphs and adults are fairly aggressive predators of small insects (e.g., nymphal Hemiptera, Thysanoptera, mites, and others). Four inactive or quiescent calyptostases are also involved: egg, prelarva, protonymph, and tritonymph.

Astigmata. The general life cycle of the Astigmata differs from the cycle in the Oribatida in that usually only two nymphs are present (Hughes, 1959; Lukoschus et al., 1981) (Fig. 12-35). This is especially true of the skin parasites (Psoroptidia, Sarcoptidia). Modifications do occur, however. The egg hatches into a six-legged larva with a Claparede organ or urstigma. The larva changes into an octopod protonymph that exhibits one pair of genital setae and one pair of genital papillae. Neurosecretory cells and nutritional factors are evidently involved in the ecdysial change to the eight-

legged deutonymph (Hughes, 1964), which may or may not exist in the life cycle. This nymph may be facultative, and is usually heteromorphic with two pairs of genital setae and two pairs of genital papillae. The tritonymphal stage is octopod and has three pairs of genital setae and two pairs of genital papillae. The adults are octopod; females have a definitive genital area, a secondary genital opening (bursa copulatrix), and a spermatheca. All males are equipped with an aedeagus.

The existence of different male forms (andropolymorphism) occurs in Dermanyssidae and Macrochelidae (Gamasida), in Anystidae and Cheyletidae (Actinedida), and in Acaridae and Proctophyllodidae (Astigmata). The types include homeomorphic (distinguished from the female by sexual characteristics but with similar setae, legs, and body shape); bimorphic

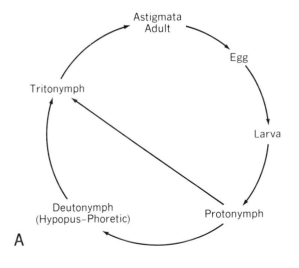

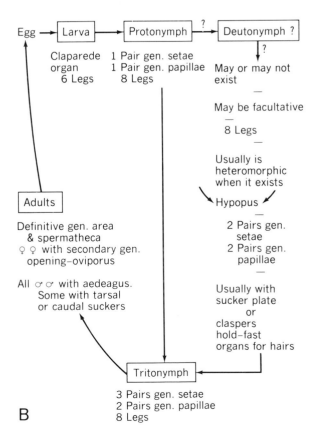

Figure 12-35. (A) General life cycle of Astigmata; (B) explanatory diagram of the life cycle with a phoretic deutonymph or hypopus.

(elongated body and much longer dorsal setae than the female); heteromorphic (one pair of legs much thicker than the female and with clawlike tarsus); and pleomorphic (one pair of legs thicker and with tarsus transformed into a claw-like structure, the dorsal setae much longer than in the female). In *Sancassania berlesei* (Michael) (Acaridae) pleomorphic males are produced when reared in isolation, independent of parental types (Timms et al., 1981). Retardation of pleomorphic males may occur because of pheromones (Timm et al., 1980).

Hypopi. The protonymph in many Astigmata may molt directly into the tritonymph or into a deutonymph with a subsequent change to the tritonymph. A number of species of the Astigmata pass through a heteromorphic stage that assures the survival of the species during unfavorable environmental conditions. These heteromorphic nymphs serve in the role of dispersal by means of phoresy, hitchhiking on a transport host. Such a heteromorphic deutonymph is called an *hypopus*.[2]

[2]Treat (1975) derives the word *hypopus* from the Greek *hypo* (under or less than ordinary) and *pous* (foot), but indicates that the word *hypopous* (having feet) is misleading. "Clearly, this was not what Duges had in mind in 1834, when he coined the generic name *Hypopus*, for his brief diagnosis refers to 'pedes brevissimi,' calling attention not to the mere possession of legs but to their much reduced size in the mites in question."

In 1873 and 1879 Megnin and Michael, respectively, showed that these nymphs were heteromorphic forms of familiar species. Subsequently, the word *Hypopus* became obsolete as a name for a taxon. "As a generic name, of course, the word required no plural, but when, as is in current usage, it is employed as a common noun, it is exactly equivalent in form to the word 'octopus' and its plural, if it is to reflect the meaning and derivation, should be 'hypopodes' (Greek, *feet*) and not 'hypopi'. The speciously scholarly forms 'hypopi' and 'hypopial' are absurd solecisms, incorrect even in 'New Latin'. They obscure not only the derivation of the terms, but their literal meaning as well" (Treat, 1975, p. 75; pp. 53–54).

Treat also describes the "reasonable though ugly" plural alternatives in English, . . . "hypopuses." He prefers the terms "hypopodes" (noun) and "hypopodial" (adjective). Without detracting in any way from his careful analysis and after communication with Fain, I have opted to use the established term.

The initiation and termination of the hypopial stage is marked by a molt (Wallace, 1960).

Hypopi are found in the majority of the families of the Astigmata, notably in the Acaridae, Saproglyphidae, Glycyphagidae, Chaetodactylidae, Histiostomatidae (=Anoetidae), Chortoglyphidae, and Hypoderidae. Because few of the species of these families have been studied beyond the hypopus stage, it is estimated that the 500 species repesented by this stage alone are not identified as adults. In the Hypoderidae, for example, life cycles are known for three species; of the other estimated 50 species, only hypopi are known. A similar situation exists for the Saproglyphidae and in the Glycyphagidae that are pilicolous (attached to hairs).

It was in 1735 that DeGeer observed a tiny red mite on a housefly. The mite was oval and enclosed in a distinctive carapace. Instead of having a mouth, it was armed with a closed, membranous tube with two terminal setae. One year earlier Duges coined the generic name *Hypopus* and referred to the reduced size of the legs. In 1873, Dujardin and Megnin and in 1883, Michael also used this name. In 1758 Linnaeus applied the name of *Acarus muscarum* to DeGeer's housefly mite.

As a heteromorphic form, the hypopus is structurally and behaviorally different from the protonymph, the normal homeomorphic deutonymph (when present), and the tritonymph. It has no mouth or functional mouthparts (Fig. 12-36A). The cheliceral anlagen are present internally (Woodring and Carter, 1974). If present, the gnathosoma (GN), recently called *palposoma*, is tactile in function (Fig. 12-36B). The palpal arms are attached to the subcapitulum and exhibit a long solenidion, a seta, and a lyriform pore on each (Krantz, 1978). The general appearance resembles that of the tritosternum of gamasids.

Hypopi generally are more active than other stages, but the extent of structural features and mobility of leg development vary with the species (Cutcher and Woodring, 1969). Depending on the type of hypopus, the form is flattened, discoidal (dorsally convex, ventrally

concave), or elongated. Both pliable and
sclerotized cuticle are characteristic, but the integument is frequently more sclerotized and
brown, compared to the soft cuticle of other instars (Wallace, 1960). Legs may or may not be
visible. Hypopi resemble tritonymphs in the
presence of two pairs of genital papillae and the
genital, anal, and trochanteral setae (Woodring
and Carter, 1974). Tarsal setae are often large
and ornate and may aid in dispersal by wind.
As phoretic nymphs (Frontispiece, hypopi on
head of termite), hypopi bear mechanisms for
transport which include a ventral sucker plate
(Fig. 12-36A), hair-clasping organs (Fig. 12-
36D), and distally spined tibia. Internally, the
alimentary canal is regressed and the instar
lives on food reserves accumulated during the
protonymphal instar (Cutcher and Woodring,
1969). An anus may or may not be present
(Oboussier, 1939). The central nervous system
is comparatively large. The distinguishable supraesophageal and subesophageal ganglia have
major nerves extending to the effector organs
(Woodring and Carter, 1974; Wallace, 1960),
and are similar to the adult system (Kuo and
Nesbitt, 1971).

Dipsersal is essential to the free-living
tyroglyphoid mites (Zakhvatkin, 1959). They
are normally wobbly, slow-gaited, clumsy, and
somewhat restricted in movement. *Glycyphagus* is probably the fastest hypopus of the
group (perhaps because of the attenuated
tarsi). The narrow ecological limits of these
mites reduce their active dispersal. The fluctuating humidities of their scattered habitats ne-
cessitate migrations for the preservation of the
populations. Zakhvatkin suggests that two
passive means of dispersal are employed: *anemohoria*, dispersal by air, and *zoohoria*, dispersal by animals. Dispersal by air is by means of
stages enclosed within protonymphal exuviae
(Fig. 12-36C) and by body hairs (Glycyphagidae, *Ctenoglyphus*) or leg hairs (macrosetae of
legs IV in some hypopi) that act as sails in
these wind-blown forms. Dispersal by animals
may include man (distribution in stored products, flour, grain, cheese, or in dust), or by

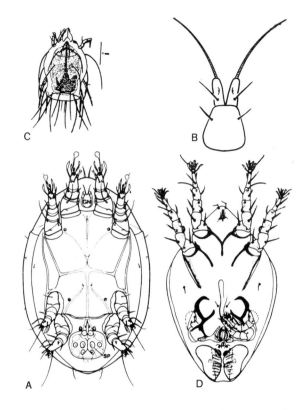

Figure 12-36. Hypopi: (A) hypopus of *Rhizoglyphus
echinopus*, ventral view showing sucker plate (GN,
gnathosoma) (after Hammen, 1978); (B) palposoma
(=gnathosoma) of hypopus showing subcapitulum,
setae, lyriform pores, and solenidia) (after Krantz,
1978); (C) *Glycyphagus domesticus* hypopus in protonymphal skin (after Hughes, 1961); (D) *Dermacarus hypudaei* showing hair-clasping organs (after
Fain, 1969).

means of invertebrate or vertebrate transport
hosts with which the mites have established an
epizoic method of dispersal.

Dr. Alex Fain, undoubtedly the world's authority on hypopi, asserts that *anemohoria* and
zoohoria are not completely accurate categories
(personal communication). He cites the instance of two hypopi in the same species
(Chaetodactylidae), one migratory and fixed to
a bee, the other inert, not dispersed by wind,
but remaining as a nidicole in the nest of the

bee and completing development after the bee reoccupies the nest.

Hypopial deutonymphs are differentiated into categories by Oboussier (1939) on the basis of external features and modifications of the gut: (1) active, motile hypopi with sucker plates (e.g., Carpoglyphidae); (2) motile, but less active types with shorter legs and larger sucker plates (*Carpoglyphus* sp.); and (3 and 4) immobile types that are usually enclosed within the protonymphal cuticle, with reduced or absent legs and sucker plate (cited by Woodring and Carter, 1974).

Zakhvatkin (1959) differentiates two major types of hypopi: (1) active, migratory forms; and (2) nonmigratory, passive, or inert types. IIe divides the active, migratory forms into two kinds: (1) entomophilic, with sucker plates for attachment to insects or terrestrial arthropods; and (2) terophilic, with muscular flaps and ribbed claspers for attachment to the hairs of mammals. The passive, inert, nonmigratory hypopi are stationary and some may be distributed by air currents. Free, follicular, and subdermal types of hypopi are usually grouped with these inert forms by other authors. Volgin (1973) reviews the types of hypopi and modifies the scheme of Zakhvatkin to include four basic morphological types (synonomous names in parentheses):

Acaroidal: (*entomochoral, entomophilic*)

> Most widely spread; characterized by sucker plate on opisthosoma; Acaridae, Saproglyphidae, Canestriniidae, Anoetidae

Labidophoridal: (*teriochoral, terophilic, pilicolous*)

> Characterized by a hair-clasping organ for attachment to mammals; Labidophorinae, Glycyphagidae; *Labidophorus, Dermacarus.*

Follicular: (*rodentiopal, echimyopidal*)

> Characterized by broad spines on tibiaetarsi of legs III, IV for grasping hair; *Melesodectes auricularis* Fain & Luk.; Ctenoglyphidae (Rodentopinae); mature forms nidicolous.

Resting: (*inert, nonmigratory*)

> Characterized by reduction of legs, sensory organs and attachment structures; enveloped in exuvium of protonymph; legs susceptible to unfavorable environmental conditions; *Glycyphagus, Lepidoglyphus, Chaetodactylus.*

Again, Alex Fain asserts that classification into active and inert types is artificial (personal communication). Follicular and subdermal hypopi (Hypoderidae), although considered inert, actively invade the tissues of the host. They also show two phases: an active–invasive form, and a passive, inert resting stage. Fain also indicates that the entomophilic–terophilic arrangement is artificial. In *Fibulanoetus* (Histiosomatidae) the hypopi have claspers that resemble the hair-claspers of phoretic nymphs on mammalian hosts, but are fixed to the hairs of beetles (Fain et al., 1980). They are thus pilicolous, but not on mammals.

The types of hypopi found on mammals include the following (Fain, 1969):

1. Pilicolous hypopi. The posterior–ventral aspect of the idiosoma is characterized by two external muscular valves that cover four clavate, ribbed, and sclerotized pegs, which enable the nymph to grasp hair. All of these are found in the subfamily Labidophorinae, and the adults of the species are habitually nidicolous in nests of rodents and insectivores.

2. Endofollicular hypopi. These are found exclusively in the depths of the hair follicles and are completely invisible externally. In certain of these (Metabidophorinae), there is a more or less well-developed hair-grasping or-

gan on the tibia of legs III and IV, rather than the idiosoma (*Rodentopus*) (Fig. 12-37B). Examples are found in many mammals and in all regions of the world. Such hypopi have been found in Europe, Africa, America, and Australia. Thirty-five species in 16 genera are described from 6 or more subfamilies. One species (*Pedetopus zumpti* (Fig. 12-37A), Fain, 1969; Glycyphagidae, Pedepotinae) has a suctorial plate posterior to the genital region.

3. Aural hypopi. This type of hypopus (*Melesodectes auricularis*) (Fig. 12-37C) is found in the earwax of badgers (Netherlands) and is known only from the hypopus and the tritonymph. The idiosoma of the hypopus is devoid of organs of attachment. The genital region is paramedial and the integument is smooth.

4. Subdermal hypopi. Representing the family Hypoderidae (=Hypodectidae), *Muridectes* (Fig. 12-37D) is known only in the hypopus stage. It is characterized by the ventral position of the legs and the large anus. Judged by the morphology and habitat, this type of hypopus is probably very primitive (Fain, 1969). It is also associated with primitive hosts, birds (Hypoderinae), and rodents (Muridectinae), where it is found in the cellular tissues and the skin. Such hypopi are also considered inert forms by some authors, but they actively invade the tissues of the host and are therefore active and migratory (A. Fain, personal communication; Pence, 1972).

Each hypopus of *Echimyopus dasypus* (Fig. 12-37E) resides in an epidermal gall (zoocaecidium) with a central opening and covered by hypertrophic corneal layers in the hairless venter of the nine-banded armadillo (Fain et al., 1973). Other species of this genus live within hair follicles of the host. This is different from the follicular types, where 11 hypopi have been found in a single follicle and over 2000 hypopi counted from a single chipmunk's tail (Tadkowski and Hyland, 1979).

The arrangement of hypopial categories indicated below is a combination of suggestions from A. Fain (personal communication, 1969, 1980) and other considerations. It is not final, however, because the diversity of hypopi is still incompletely known.

Active–Migratory

Pilicolous—hair-clasping organs (phoretic on mammals, insects, terrestrial arthropods).

Nonpilicolous—sucker plates fixing to bare surfaces (insects and terrestrial arthropods).

Endofollicular—ornate tibiotarsal spines.

Aural—without organs of attachment (ears of badgers).

Subdermal—includes active invasive and passive–inert forms

Epidermal galls (zoocaecidia)

Passive–Inert

Free—hypopi in protonymphal skins or with body/leg hairs typical of windblown forms.

Just what conditions or factors cause the formation of hypopi is not well understood. This heteromorphic stage may occur within a generation, or it may not, a condition called *facultative hypopody* by Krantz (1978). Endogenous and exogenous factors may be involved (Treat, 1975). Hypopi respond to and are resistant to environmental changes such as adverse temperatures and humidities (Woodring and Carter, 1974). They can withstand lower temperatures than adults (Cutcher and Woodring, 1969; Zakhvatkin, 1959). Hypopi of *Histiostoma polypori* survive in conditions of low humidity that are lethal to other stages. Elevation of the humidity to saturation causes deutonymphs to develop (Wallace, 1960). Fain (1977) identifies a method for provoking eclosion of hypopi.

Unfavorable conditions in the environment that affect the change of a protonymph into the hypopus include low humidity (drying of humus or substrate (Wallace, 1960)), deterioration of amount or quality of food, changes in temperature, alteration of the pH, and accumu-

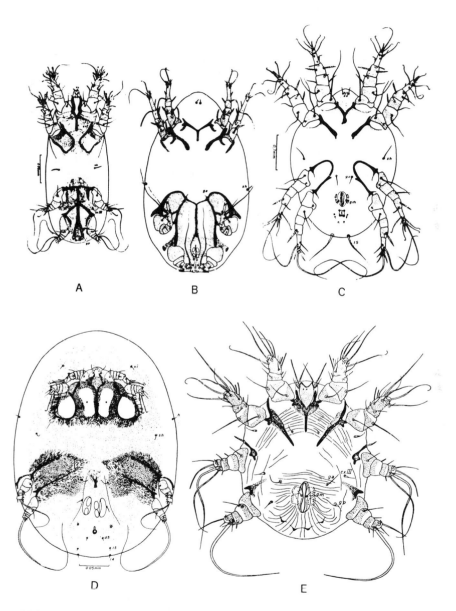

Figure 12-37. Variations of hypopi: (A) *Pedetopus zumpti*; (B) *Rodentopus (Scuiropsis) heterocephali*; (C) *Melesodectes auricularis* (after Fain, 1969). Variations of tissue hypopi. (D) *Muridectes heterocephali* (after Fain, 1969); (E) *Echinyopus dasypus* (found in epidermal galls = zoocaecidia).

lation of metabolic wastes (Hughes, 1964; Chmielewski, 1973).

The molt (apolysis) of hypopi is thought to be regulated by hormones as in other arthropods. Two interpretations of this influence are proposed: (1) adverse environmental conditions may inhibit production of the hormone so that as long as the adverse situation exists molting is prevented; or (2) in the absence of correct environmental stimulus, the ecdysonelike hormone is not produced or released. Temperature, relative humidity, and environmental salt concentration influence the retention of the hypopial stage in *Sancassania boharti* (Cross) (=*Caloglyphus boharti*). The quality or quantity of food, presence or absence of microorganisms, substrate pH, NH_4, CO_2, ultraviolet radiation, intra- or interspecific competition, or overcrowding do not directly cause termination of the hypopial stage. Some of the above factors, however, may indirectly alter the relative humidity or salt concentration and thus indirectly initiate or prevent hypopial molting (Cutcher and Woodring, 1969).

In *Glycyphagus domesticus*, a certain number of individuals within a generation of the species become hypopi, which suggests that genetic factors may be involved (Hughes, 1961, 1964; Zakhvatkin, 1959). Two genetically different strains are known in *Acarus siro*, one of which may be induced to form hypopi at the protonymphal molt. If this strain is deprived of food, large numbers of hypopi appear.

Examination of the central nervous system discloses that neurosecretory granules are absent. The fate of the protonymph is set at 24 h and starvation during this period produces hypopi, whether or not the protonymphs are fed subsequently (Griffiths, 1962, cited by Hughes, 1964). Wallace (1960) asserts that the integument, which is sensitive to changes in humidity, influences and controls the production of hypopi. A combination of extrinsic and intrinsic factors is most likely involved.

There are variations of factors and conditions that affect the formation of hypopi within populations. These depend upon the species of mite and the percentage of the populations with the hypopial stage (Zakhvatkin, 1959).

1. The hypopial stage is regularly produced in a percentage of the population in each generation as a result of a biological rhythmn independent of environmental conditions.

2. The hypopial stage is formed only under unfavorable environmental conditions. If the conditons of the habitat are good, the stage is not formed.

3. The hypopial stage is formed by compulsion for all individuals or is a stage associated with a definite rhythmn of the phoretic host, for example, nesting birds. In *Hericia*, hypopi are produced sporadically, and yet this production coincides with the period of oviposition of the phoretic hosts (flies or bees attracted to exuding sap of trees). Others (*Forcellinia*) are phoretic on ants as they leave the nest (Hughes, 1961), or bark beetles (*Dendroctonus ponderosae*) as they leave the gallery.

4. Hypopial stages are formed into alternate types: migratory or inert (some species have both types). The conditions for the change into one type or the other may be completely different.

The insect transport hosts for hypopi include the ground beetles (*Tenebrio molitor*), rhinoceros beetles, dung beetles, staphylinids, bark beetles (*Dendroctonus pondersosae*), ants, wasps, wood-boring and xylocopid bees, bumblebees, fleas, woodlice, crickets and cockroaches, flour moths, noctuid moths, and green and blue bottle flies (Zakhvatkin, 1959; Treat, 1975; Fain, 1965, 1969; Krantz, 1978).

Mammalian phoretic hosts for hypopi include rats, fieldmice, dormice, hamsters, squirrels, eastern chipmunks, Congolese rats, bats, badgers, nine-banded armadillos, opossums, Siberian marmots, shrews, and insectivores. Sea snakes also have hypopi (Zakhvatkin, 1959; Fain, 1969; and other sources).

The families in which hypopial stages are found include (Zakhvatkin, 1959; Fain, 1969):

Acaridae (Tyroglyphidae)

Saproglyphidae

Carpoglyphidae

Fusacaridae

Chaetodactylidae

Labidophorinae (Glycyphagidae)

Glycyphagidae (Labidophorninae; Metalabidophorinae; Alabidopinae; Ctenoglyphinae; Echimyopinae)

Hypoderidae

Hemisarcoptidae

Histiostomatidae (=Anoetidae)

Striking behavioral circumstances also are associated with the hypopial stage.

The saproglyphid mite *Vidia concellaria* Cooreman is a commensal in the nests of a sphecid wasp (*Cerceris arenaria* L.). Remains of the prey, curculionid larvae, are eaten by the feeding stages of the mite. Because the mite subsists on the food provided by the mother wasp, its stages of development are synchronized with the developmental cycle of the wasp and are completed within 15 days. The protonymphal mites transform into the hypopi to await the emergence of the overwintering wasp pupae. When the adult wasp emerges from the cocoon in the spring, the hypopi affix to the body and are transported to a new nest site (Cooreman and Crevecoeur, 1948).

The first acarinarium was described by Perkins (1899) in Xylocopine bees. This is a peculiar situation associated with hypopi in which a specific chamber in the body of the insect host serves as an assembly location for the hypopi as they wait for transportation. They are frequently packed together in rows within these cavities like soldiers closely arranged in a confined staging area.

Many saproglyphid hypopi are related in their mode of living with the insects with which they are associated, particularly the Hymenoptera. Some genera are known only from the hypopodial stadia (*Vidia*, *Macroharpa*, and *Zethacarus*) (Mostafa, 1970a). In the wasp genus

Zethus, the hypopi are not confined to specific areas, nor are mite chambers (acarinaria) present. In the wasp *Parancistrocerus fulvipes*, the hypopi become so tightly packed in an acarinarium between the abdominal tergites I and II that there appears to be no more room (Mostafa, 1971).

In a remarkable "venereal transmission" of hypopi by wasps, Cooper (1954) suggests that the phoretic deutonymphs of *Kennethiella trisetosa* are borne only on males of the wasp *Acistrocerus antilope*. Male wasps have loads of these hypopi on them whereas females emerge from the nests devoid of these "hitchhikers," but acquire the hypopi during mating. The hypopi forsake the female wasps during the construction or provisioning of the new nests.

Inasmuch as a full load of hypopi is confined to adult males and the females come out of the nest without hypopi, it appears that the hypopi are not released into a nest cell containing a diploid egg (female). The factors that determine fertilization and the production of a daughter also determine in which cell the hypopi are released. It is suggested that female wasp larvae eat all of the adult mites in the cell before spinning a cocoon, but male wasps do not harm the mites and thus emerge with their loads of hypopi (Krombein, 1961).

Krombein suggests that the rhythmic pulsations of the female wasp's abdomen during oviposition may stimulate the hypopi to leave the body and drop into the cell or nest. He observes that the transition to the tritonymph occurs between oviposition and the completion of the feeding by the wasp larva. The adult mites are on the wasp larva as it begins to spin the cocoon and thus become enclosed. Female mites lay eggs as soon as the pupa is formed and it is postulated that egg-laying by the mite is possibly related to the reduction or the lack of juvenile hormone, or to growth hormones in the wasp that stimulate ovulation and oviposition in the mites.

The protonymphs are identifiable as the wasp pupa prepares to molt. The mites cluster on the venter near the mouthparts and legs. The presence of the hypopi coincides with eclosion and the emergence of the adult wasps, perhaps stimulated by molting fluid of the insect. The hypopi of *Vespacarus* crawl around on the adult wasp and find the acarinarium, then turn around and back into the chamber. As more and more hypopi find the chamber, they line up in rows facing outward, like good soldiers ready and waiting for transport (Mostafa, 1970a,b).

Oribatida. Generalizations about the development of oribatids are limited because of the diversity of ecological, physiological, developmental, and physical requirements. Each species seems to be unique in its morphological patterns during this process. Some chaetotaxic changes may be restricted to species, but setal progress, setal regression, and anamorphic addition of segments may be similar for larger groups (suborders, superfamilies, families) (Arlian and Woolley, 1969; Woodring and Cook, 1962; Woodring, 1965; Grandjean, 1933).

Adult females usually have an ovipositor (Fig. 12-25F, G), some of which are extremely long and are telescoped within the body of the female when retracted. Adult males may have a short aedeagus. Oribatids pass through a prelarval, larval, and three nymphal stages (proto-, deuto-, and tritonymph) before the adult stasis is reached.

Eggs of oribatids are oval and cylindrical and surfaces range from glabrous to rough and sculptured. Eggs may develop (1) within the body of the living female, (2) following oviposition, or (3) after the death of the female.

A deutovial period and a prelarva are frequently present. The deutovum may occur in the body of the female parent (Michael, 1883). The prelarva is simple and relatively structureless. Appendages, if noticeable, are primarily raised bosses or ridges (Fig. 12–38); body setae are absent. In the prelarva the anlage of the chelicerae and palps are visible, a vestige of the pharynx, Claparede's organ, and the bases of three pairs of legs also may be observed (Fig. 12-38B–E).

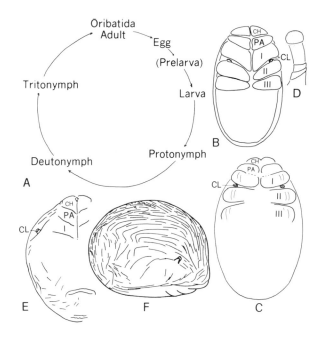

Figure 12-38. (A) Generalized life cycle of Oribatida; prelarva of (B) *Trhypochthonius fectorum* (after Krantz, 1978); (C) *Hermannia convexa* (after Hammen, 1978); (D) enlarged Claparede organ of *Phthiracarus anonymum* (after Grandjean, 1935); *Hoplophthiracarus pavidus*, (E) ventral and (F) lateral (after Hammen, 1963) (CH, chelicerae; CL, Claparede organ; PA, palp; I, II, III leg anlage).

The larval stage is hexapod and monodactylous with a distinct urstigmata at the base of coxa I. The anal opening is distinct with pseudanal segments visible. The genital anlage is not visible externally and genital papillae are absent. In some the integument is smooth and shiny; in others, the larval skin is plicated and wrinkled. Pigmentation varies from white to light amber and brown. Appendages are usually darker in color than the body. The distal tips of the chelicerae and the rutella may be almost black. Notogastral setation is in six transverse rows, reflective of possible segmentation (Grandjean, 1933). Gland openings and characteristic fissures are noticeable.

Octopod nymphal stages vary in their appearance among the major groups of oribatids. Many are homeomorphic and differ only slightly from each other. Others are distinct. Grandjean (1954) devises a system of differentiation that involves the method of molting, retention of scalps or exuviae, and the arrangement of dorsal setae. The system (with exemplary families) consists of the following categories:

Eupherederm: nymphs hold scalps of preceding instars close to dorsal cuticle; four shed skins may be in place above the body (Liodidae, Belbidae).

Apopherederm: exuviae are retained, but held aloft by centrodorsal setae (Oribatellidae)

Opsipherederm: the nymphs are naked, but the adult retains the tritonymphal exuvium (Hermaniellidae).

Apherederm: the nymphs and adults are naked and without exuviae on the dorsum. Pycnonotic forms may have smooth (Ceratoppiidae, Liacaridae) or plicate = wrinkled nymphs (Carabodidae). Poronotic mites may have (1) plicate nymphs (Achipteriidae, Scutoverticidae, Passalozetidae, Pelopsidae), (2) nymphs with large sclerites (Ceratozetidae, Galumnidae), or (3) nymphs with microsclerites (Oribatulidae, Scheloribatidae, Haplozetidae).

In some ptyctimous oribatids larvae and nymphs are unknown, and other groups are not identified in the system.

All of the oribatids have three active nymphal stages after the hexapod larvae (Figs. 12-38). Their morphology and size are usually different from the adults (Fig. 12-40), yet characteristic for each particular instar (Evans et al., 1961; Woodring and Cook, 1962; Hammen, 1978). Nymphal stages may be recognized by the presence of one or more pairs of genital papilla and genital setae, delineating the stage (Fig. 12-39C). Protonymphs have one pair of genital papillae; deutonymphs have two pairs; tritonymphs and adults have three pairs, thus affording a quick index to the nymphal stage. An exception is *Oppia neerlandica* where the female has two pairs; Fig. 12-40G). Corresponding changes in size and in the chaetotaxy of the notogaster and the venter are also distinguishable for each nymphal stage. During the nymphal stages the setae attain their greatest development and variety, hence chaetotaxy is extremely important.

Nymphal colors vary from clear or milky white to yellow, amber, brown, reddish brown, or crimson (Michael, 1883). The integument of nymphs may be wrinkled. In *Cepheus, Damaeus, Liodes, Pheroliodes,* and others the larval and nymphal scalps may be retained as camouflage through succeeding stages even to the adult (Liodidae; Woolley and Higgins, 1969). In Damaeidae the nymph may carry exuviae as well as detritus and eggs (Fig. 12-40I) (Lawrence, 1953). Where this occurs each shed skin is characteristically placed on the dorsum in sequence so that the tritonymph has all the scalps of the previous stages (Fig. 12-40H). These scalps are usually shed on transformation to the tritonymph and to the adult. Many adults are tridactylous; a few are monodactylous. Sclerotization and chaetotaxy are complete for the group or species involved.

An anomalous condition related to nutritional conditions is the phenomenon of albinism in the oribatid *Ceratozetes cisalpinus.* Lack of pigment and sclerotization characterize this condition. Even the legs and gnathosoma are without sclerotization, which gives the whole body a glassy appearance, except for black setae. Albinism may occur in newly hatched larvae, the immature stages, and adults. The adults, however, (1 to 2% of albino immatures reach this stage) never produce eggs or spermatophores and usually die within 2 days of molting. In albinos the legs are weak, which makes them susceptible to bending and makes movement difficult. Weakened legs also cause problems with molting, because in oribatids the strength of the legs is important to move the mite out of the shed skin. Maternal

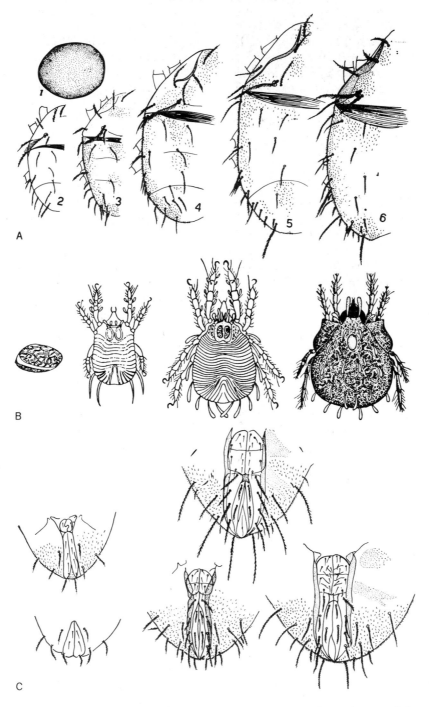

Figure 12-39. Developmental stages of (A) *Papillacarus aciculatus* (1, egg; 2, larva; 3, protonymph; 4, deutonymph; 5, tritonymph; 6, adult) (after Shereef, 1976); (B) *Pelops laevigatus* (after Michael, 1883); (C) genital and anal plates of developmental stages of *Papillacarus aciculatus* (after Shereef, 1976).

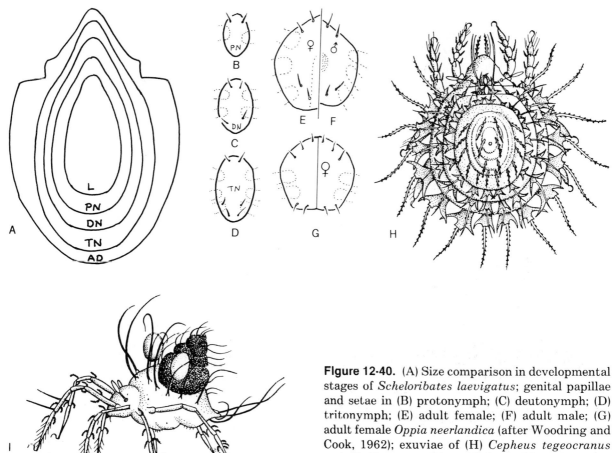

Figure 12-40. (A) Size comparison in developmental stages of *Scheloribates laevigatus*; genital papillae and setae in (B) protonymph; (C) deutonymph; (D) tritonymph; (E) adult female; (F) adult male; (G) adult female *Oppia neerlandica* (after Woodring and Cook, 1962); exuviae of (H) *Cepheus tegeocranus* and (I) *Damaeus* (after Michael, 1883, from Lawrence, 1953).

nutritional deficiency, overcrowding, and limited food source (lichen alone for one generation) are factors contributing to albinism (Woodring and Cook, 1962).

Numerous oribatid life cycles have been examined (e.g., Michael, 1883; Grandjean, 1954; Arlian and Woolley, 1969; Woodring and Cook, 1962). Peculiarities exist within families and among major groups. Biology is incompletely understood for many of these mites and much research needs to be accomplished before the developmental patterns and details of the life cycles will be understood. Grandjean's (1954) elaboration of the method of molting, nymphal stadia, their scalps, and chaetotaxy has done much to begin a systematization of this phase of oribatid biology.

The xylophagous species of the ptychoid oribatid, *Hoplophora magna*, tunnels in the heartwood of hemlock and yellow birch. These mites produce a complex system of burrows, usually filled with fecal pellets of finely shredded wood tissue. The side branches of the burrows contain nymphal stages and eggs are observed among the frass and fecal pellets within the tunnels.

The following example illustrates the specifics of the life cycle of an interesting oribatid.

The larvae and three stages of nymphs of *Orthogalumna terebrantis* Wallwork (=*Lepto-*

galumna of other authors) tunnel in the living tissue of the leaves of water hyacinths in South America. The adults overwinter on the outer surfaces of adjacent plants (e.g., water lettuce) and feed on the algae growing on the undersurface of the leaves. In the spring, the female excavates holes in the leaves of the water hyacinths and lays eggs in the parenchyma. The hatched larvae and three nymphal instars burrow and feed in the parenchyma of the leaves. After transformation from the tritonymph, the adults cut holes in about equal numbers in both the upper and lower surfaces of the leaf, emerge, and continue the life cycle. The adults do not normally feed on the leaf tissue, but are able to penetrate it for oviposition. Summer nymphal populations may peak at 10 000 galleries per plant, which causes severe damage to the host plant and results in a die-back of most leaves (Krantz and Lindquist, 1979).

REFERENCES

Aeschlimann, A. 1958. Développement embryonnaire d'*Ornithodoros moubata* (Murray) et transmission transovarienne de *Borrelia duttoni. Acta Trop.* 15:15–24.

Aeschlimann, A. and H. Hecker. 1969. Vitellogénèse et Formation Cuticulaire chez l'Oeuf d'*Ornithodoros moubata* Murray. (Ixodoidea: Argasidae). Etude au microscope électronique. *Acarologia* 11(2):180–192.

Aeschlimann, A. and E. Hess. 1984. What is our current knowledge of acarine embryology? *Acarol. [Proc. Int. Congr. Acarol.], 6th, 1982*, Vol. 1, pp. 90–99.

Alberti, G. 1973. Ernährungsbiologie und Spinnvermögen der Schnabelmilben (Bdellidae, Trombidiformes). *Zeitschr. Morphol. Tiere* 76(4):285–338.

Alberti, G. and V. Storch. 1976. Ultrastruktur untersuchungen am manliche Genitaltrakt und an Spermien von *Tetranychus urticae* (Tetranychidae: Acari). *Zoomorphologie* 83:283–296.

Amano, H. and D.A. Chant. 1978. Mating behavior and reproductive mechanisms of two species of predaceous mites, *Phytoseiulus persimilis*

Athias-Henriot and *Amblyseius andersoni* (Chant) (Acarina: Phytoseiidae). *Acarologia* 20(2):196–213.

Anderson, D.T. 1973. *Embryology and Phylogeny in Annelids and Arthropods*, Int. Ser. Monogr. Pure Appl. Biol., Zool. Div., Vol. 50. Pergamon, Oxford.

Andre, M. 1953. Observations sur la fécondation chez *Allothrombium fuliginosum* Herm. *Bull. Mus. Natl. Hist. Nat.* 4:383–386.

Arlian, L.G. and T.A. Woolley. 1969. Life stages of *Liacarus cidarus. J. Kans. Entomol. Soc.* 42(4):512–524.

Arthur, D.R. 1946. The feeding mechanism of *Ixodes ricinus. Parasitology* 37:154–167.

Arthur, D.R. 1951. The bionomics of *Ixodes hexagonus* Leach in Britain. *Parasitology* 41:82–90.

Arthur, D.R. 1962. *Ticks and Disease.* Pergamon, Oxford.

Athias, F. 1975. Observations morphologiques sur *Polyaspis patavinus* Berlese 1881 (Acariens: Uropodides) I. Morphologie de l'idiosoma au cours du développement postembryonnaire. *Acarologia* 7(3):410–435.

Athias-Henriot, C. 1970. Observations sur la morphologie externe des Gamasides mise au point terminologique. *Acarologia* 12:25–27.

Atyeo, W.T. 1974. *Doglielacarus uncitibia* Dubinin, 1949, redescribed and reassigned (Acarina: Analogidea). *J. Kansas Entomol. Soc.* 47(4):478–482.

Atyeo, W.T. and N.L. Braasch. 1966. The feather mite genus *Proctophyllodes* (Sarcoptiformes: Proctophyllodidae). *Bull. Univ. Nebr. State Mus.* 5:1–354.

Audy, J.R., M. Nadchatram and P.H. Vercammen-Grandjean. 1963. La "neosomie," un phénomène inédit de neoformation en acarologie, alliée à un cas remarquable de tachygénèse. *Bull. Cl. Sci., Acad. R. Belg.* (5) 49:1015–1027.

Audy, J.R., F.J. Radovsky and P.H. Vercammen-Grandjean. 1972. Neosomy: Radical intrastadial metamorphosis associated with Arthropod symbioses. *J. Med. Entomol.* 9(6):487–494.

Axtell, R.C. 1969. Macrochelidae (Acarina: Mesostigmata) as biological control agents for synanthropic flies. *Proc. Int. Congr. Acarol., 2nd, 1967*, pp. 401–416

Bacetti, B. 1979. Ultrastructure of sperm and its

bearing on arthropod phylogeny. *In*: Gupta, A.P., ed. *Arthropod Phylogeny*. Van Nostrand-Reinhold, New York, pp. 609-644.

Baker, E.W. 1979. A note on paedogenesis in *Brevipalpus* sp. (Acari: Tenuipalpidae), the first such record for a mite. *Int. J. Acarol.* 5(4):355-356.

Balashov, Y.S. 1972. *Bloodsucking Ticks (Ixodoidea)—Vectors of Diseases of Man and Animals*, Transl. 500 (T500). Med. Zool. Dep., USNAMRU No. 3, Cairo, Egypt, U.A.R. [published in *Misc. Publ. Entomol. Soc. Am.* 8(5):161-376 (1968)].

Beatty, R.A. 1967. Parthenogenesis in vertebrates. *In:* Metz, C.B. and A. Monroy, eds. *Fertilization: Comparative Morphology, Biochemistry and Immunology.* Academic Press, New York, Vol. 1, pp. 413-440.

Beer, R.E. 1960. A new species of *Pimeliaphilus* (Acarina: Pterygosomidae) parasitic on scorpions, with discussion of its postembryonic development. *J. Parasitol.* 46:443-440.

Berry, R.E. 1973. Biology of the predaceous mite, *Pergamasus quisquilarum* on the garden symphylan, *Scutigerella immaculata*, in the laboratory. *Ann. Entomol. Soc. Am.* 66(6):1354-1356.

Boczek, J. and D.A. Griffiths. 1979. Spermatophore production and mating behavior in the stored product mites *Acarus siro* and *Lardoglyphus konoi. In:* Rodriguez, J.G., ed. *Recent Advances in Acarology.* Academic Press, New York, Vol. 1, pp. 279-284.

Bolland, H.R. and W. Helle. 1981. A survey of chromosome complements in the Tenuipalpidae. *Int. J. Acarol.* 7(1-4):157-160.

Booth, T.F., D.J. Beadle and R.J. Hart. 1984. The ultrastructure of Gene's organ in the cattle tick *Boophilus microplus* Canestrini. *Acarol. [Proc. Int. Congr. Acarol.], 6th, 1982*, Vol. 1, pp. 261-267.

Boudreaux, H.B. 1963. Biological aspects of some phytophagous mites. *Annu. Rev. Entomol.* 8:137-154.

Boudreaux, H.B. 1979. *Arthropod Phylogeny with Special Reference to Insects.* Wiley, New York.

Brennan, J.M. and E.K. Jones. 1959. *Pseudoschongastia* and four new neotropical species of the genus (Acarina: Trombiculidae). *J. Parasit.* 45:421-429.

Brinton, L.P. and J.H. Oliver, Jr. 1971a. Gross ana-

tomical, histological and cytological aspects of ovarian development in *Dermacentor andersoni* Stiles (Acari: Ixodidae). *J. Parasitol.* 57(4):708-719.

Brinton, L.P. and J.H. Oliver, Jr. 1971b. Fine structure and development in *Dermacentor andersoni* Stiles (Acari: Ixodidae). *J. Parasitol.* 57(4):720-747.

Bruce, W.A. 1984. Temperature and humidity: Effects on survival and fecundity of *Pyemotes tritici* (Acari: Pyemotidae). *Int. J. Acarol.* 10(3):135-138.

Bruce, W.A. and G.L. LeCato. 1980. *Pyemotes tritici:* A potential new agent for biological control of the red imported fire ant, *Solenopsis invicta* (Acari: Pyemotidae). *Int. J. Acarol.* 6(4):271-274.

Camin, J.H., W.W. Moss, J.H. Oliver, Jr., and G. Singer. 1967. Cloacaridae, a new family of cheyletoid mites from the cloaca of aquatic turtles (Acari: Acariformes: Eleutherengona). *J. Med. Ent.* 4(3):261-272.

Campana-Rouget, Y. 1959. La teratologie des tiques. *Ann. Parasitol.* 34:209-260, 354-431 (cited by Sundman, 1965).

Chant, D.A. 1955. Notes on mites of the genus *Typhlodromus* Scheuten, 1857 (with descriptions of the males of some species and the female of a new species). *Canad. Entomol.* 87(11):296-503.

Chant, D.A. and Y.Y. Shaul. 1980. A world review of the *liliaceus* species group in the genus *Typhlodromus* Scheuten (Acarina: Phytoseiidae). *Can. J. Zool.* 58(6):1129-1138.

Chmielewski, W. 1973. A study on the influence of some ecological factors on the hypopus formation of stored products mites. *Proc. Int. Congr. Acarol. 3rd, 1971*, pp. 357-363.

Clark, K.U. 1973. *The Biology of the Arthropoda.* Arnold, London.

Coineau, Y. 1973. A propos de quelques caractères particulièrement primitifs de la prelarve et de la larve d'un Opilioacaridae du Gabon (Acariens). *C.R. Hebd. Seances Acad. Sci., Ser. D.* 276:1181-1184.

Coineau, Y. 1974a. Eléments pour une monographie morphologique, écologique et biologique des Caeculidae (Acariens). *Mem. Mus. Natl. Hist. Nat., Ser. A (Paris)* 81:1-299.

Coineau, Y. 1974b. Les Acariens. *In:* Introduction à l'etude des microarthropodes su sol et de ses annexes. Doc. Ecol., Doin, pp. 57-83.

Coineau, Y. 1976. La première prelarve connue du genre *Eupodes strandtmanni* n. sp. *Acarologia* 18(1):56–64.

Coineau, Y. 1979. Les Adamystidae, une étonnante familie d'Acarines prostigmates primitifs. *Proc. Int. Congr. Acarol., 4th, 1974*, pp. 431–435.

Coineau, Y. and L. van der Hammen. 1979. The postembryonic development of Opilioacarida, with notes on new taxa and on a general model for the evolution. *Proc. Int. Congr. Acarol., 4th, 1974*, pp. 437–441.

Collins, N.C. 1973. Misc. Publ. Entomol. Soc. Am. 9(5):250–254

Cook, D.R. 1974. Water mite genera and subgenera. *Mem. Amer. entomol. Inst.*, Ann Arbor, Michigan.

Cooper, K.W. 1937. Reproductive behavior and haploid parthenogenesis in the grass mite *Pediculopsis graminum* (Reut) (Acarina: Tarsonemidae). *Proc. Natl. Acad. Sci. U.S.A.* 23:41–44.

Cooper, K.W. 1939. The nuclear cytology of the grass mite *Pediculopsis graminum* (Reut) with special reference to karyokinesis. *Chromosoma* 1:51–103.

Cooper, K.W. 1940. Relations of *Pediculopsis graminum* and *Fusarium poae* to central bud rot of carnations. *Phytopathology* 39(10):853–859.

Cooper, K.W. 1954. Venereal transmission of mites by wasps and some evolutionary problems arising from the remarkable association of *Enliniella trisetosa* with the wasp *Ancistrocerus antilope*. *Trans. Am. Entomol. Soc.* 80:119–174.

Cooreman, J. 1948. Le cycle biologique de *Vidia concellaria* Cooreman (Acaridiae-Ensliniellidae) Acarien vivant dans les nids de *Cerceris arenaria* L. (Hymenoptera: Sphecidae). *Bull. Ann. Soc. Entomol. Belg.* 84:275–283.

Cross, E.A. 1965. The generic relationships of the family Pyemotidae (Acarina: Trombidiformes). *Univ. Kans. Sci. Bull.* 45(2):29–275.

Cunliffe, F. 1952. Biology of the cockroach parasite, *Pimeliaphilus podapolipophagus* Tragardh, with discussion of the genera *Pimeliaphilus* and *Hirstiella* (Acarina, Pterygosomidae). *Proc. Ent. Soc. Wash.* 54:153–162.

Cutcher, J.J. and J.P. Woodring. 1969. Environmental regulation of hypopial apolysis of the mite, *Caloglyphus boharti. J. Insect Phsiol.* 15:2045–2057.

Da Fonseca, J.P.C. 1969. Le Spermatophore de *Damaeus quadrihastatus* Markel et Meyer (Acarien,

Oribate). *Proc. Int. Congr. Acarol., 2nd, 1967*, pp. 277–232.

Dailey, M.D. and W.B. Nutting. 1979. *Demodex zalophi* sp. nov. (Acari: Demodicidae) from *Zalophus californianus*, the California sealion. *Acarologia* 49:421–428.

Davis, G.E. 1951. Parthenogenesis in the argasid tick *Ornithodoros moubata* (Murray, 1877). *J. Parasitol.* 37:99–101.

Delfinado-Baker, M. 1984. The nymphal stages and male of *Varroa jacobsoni* Oudemans, a parasite of honey bees. *Int. J. Acarol.* 19(2):75–80.

Desch, C.E. and W.B. Nutting. 1977. Morphology and functional anatomy of *Demodex folliculorum* (Simon) of man. *Acarologia* 19(3):422–462.

Dittrich, V. 1965. Embryonic development of tetranychids. *Boll. Zool. Agrar. Bachic* (2)7:101–104 (cited by Krantz, 1978).

Dosse, G. 1959. Uber den Kopulationsvorgang bei Raubmilben aus der Gattung *Typhlodromus* (Acari: Phytoseiidae). *Pflanzenschutzberichte* 22:125–133.

Douglas, J.R. 1943. The internal anatomy of *Dermacentor andersoni* Stiles. *Univ. Calif. Publ. Entomol.* 7(10):207–272.

Dubinina, E.V. 1964. Cycle of development of mites of the genus *Histiophorus* (Sarcoptiformes: Listrophoridae). *Zool. Zh.* 45:534–548 (in Russian, English summary).

Dusbabek, F. 1970. On Cuban species of the genus *Spelaeorhynchidae (Acarina). Acarologia* 12(2):258–261.

Dusbabek, F. and V. Cerny. 1970. The nasal mites of Cuban birds. *Acarologia* 12(2):269–281.

Efford, I.E. 1966. Observations on the life history of three stream-dwelling watermites. *Acarologia* 8(1):86–93.

Ehrenberger, R. 1974. Pralarval und Larvalentwickling bei Rhagidiien (Acarina: Prostigmata). *Osnabruck. Naturw. Mitt.* 3:85–117.

El Kammah, K.M., F.K. Adham, N.R. Tadros and M. Osman. 1982. Embryonic development of the camel tick *Hyalomma dromedarii* (Ixodoidea: Ixodidae). *Int. J. Acarol.* 8(1):47–54.

Evans, G.O. 1963. Observations on the classification of the family Otopheidomenidae (Acari: Mesostigmata) with description of two new species. *Ann. Mag. Nat. Hist.* (13)5:609–620.

Evans, G.O., J.G. Sheals and D. Macfarlane. 1961. *The Terrestrial Acari of the British Isles,* Vol. 1. Adlard & Son, Bartholomew Press, Dorking, England.

Evans, G.O. and W.M. Till. 1965. Studies on the British Dermanyssidae (Acari: Mesostigmata). Part I. External Morphology. *Bull. Brit. Mus. (Nat. Hist.)* 13:15–294.

Eveleigh, E.S. and W. Threlfall. 1975. The biology of *Ixodes (Ceratixodes) uriae* White, 1852 in Newfoundland. *Acarologia* 16(4):621–635.

Fain, A. 1963. La structure de la fente vulvaire chez les gamasides parasites. *Acarologia* 5(2):173–179.

Fain, A. 1964. Les Ophioptidae acariens parasites des ecailles des serpents (Trombidiformes). *Bull. Inst. Roy. Sci. Nat. Belg.* 40(15):1–57.

Fain, A. 1965. Les Acariens nidicoles et detricoles de la famille Pyroglyphidae Cunliffe (Sarcoptiformes). *Rev. Zool. Bot. Afr.* 72(3–4):257–288.

Fain, A. 1969a. Les deutonymphes hypopiales vivant en association phoretique sur les mammifères (Acarina: Sarcoptiformes). *Bull. Inst. R. Sci. Nat Belg.* 45:1–262.

Fain, A. 1969b. Adaptation to parasitism in mites. *Acarologia* 11(3):429–449.

Fain, A. 1970a. Un nouveau genre et une nouvelle espèce dans la sous-familie Rodhainyssinae (Sarcoptiformes: Gastronyssidae). *Acarologia* 12(1):160–163.

Fain, A. 1970b. Un nouvel acarien provenant des poussières d'une maison à Singapour (Sarcoptiformes: Pyroglyphidae). *Acarologia* 12(1):164–167.

Fain, A. 1977. Une méthode simple pour provoquer l'éclosion des Hypopes Pilicoles et Endofolliculaires et Mettre en évidence les Acariens parasites endocutanes. *Acarologia* 19(2):298–301.

Fain, A., A.M. Camerik, F.S. Lukoschus and F.M. Kniest. 1980. Notes on the hypopi of *Fibulanoetus* Mahunka, 1973, and anoetid genus with a pilicolous clasping organ. *Int. J. Acarol.* 6(1):39–44.

Fain, A. and G.O. Evans. 1966. The genus *Proctotydeus* Berl. (Acari: Iolinidae) with descriptions of two new species. *Ann. Mag. Nat. Hist.* 9(13):149–157.

Fain, A., F.S. Lukoschus, J.M.W. Louppen and E. Mendez. 1973. *Echimyopus dasypus,* n. sp., a hypopus from *Dasypus novemcinctus* in Panama (Glycyphagidae, Echimyopinae: Sarcoptiformes). *J. Med. Entomol.* 10(6):552–555.

Feldman-Muhsam, B. 1963. Function of the areae pososae of Ixodid ticks. *Nature (London)* 197:100.

Feldman-Muhsam, B. 1967. Spermatophore formation and sperm transfer in Ornithodoros ticks. *Science* 156:1252–1253.

Feldman-Muhsam, B. 1986. Observations on the mating behavior of ticks. *In:* Sauer, R.J. and J.A. Hair, eds. *Morphology, Physiology and Behavioral Biology of Ticks.* Ellis Horwood, Ltd., Chichester, England, pp. 217–232.

Feldman-Muhsam, B. and B.K. Filshie. 1976. Scanning and Transmission Electron Microscopy of the spermiophores of *Ornithodoros* ticks: An attempt to explain their mobility. *Tissue Cell* 8(3):411–419.

Feldman-Muhsam, B. and Y. Havivi. 1960. Accessory glands of gene's organ in ticks. *Nature (London)* 187:964.

Fernandez, N.A. 1981. Contribution à la connaissance de la faunc oribatologique d'Argentine. III. Spermatophore d'*Epilohmannia maurii* Fernandez, 1978. *Acarologia* 22(2):239–242.

Filipponi, A. 1955. Un nuova caso di arrenotochia nei Macrochelidi (Acarina: Mesostigmata). *Riv. Parassitol.* 16:145–168.

Filipponi, A. 1957. Arrentochia in *Macrocheles* subbadius (Acarina: Mesostigmata). *Boll. Zool. Agrar. Bachic.* (2)1:1–33.

Filipponi, A. 1964. Experimental taxonomy applied to the Macrochelidae (Acari: Mesostigmata). Acaralogia. *Proc. Int. Congr. Acarol., 1st, 1963,* pp. 91–100.

Gaud, J., W.T. Atyeo and H.F. Berla. 1972. Acariens sarcoptiformes plumicoles parasites des tinamou. *Acarologia* 14(3):393–453.

Gladney, W.J. and R.O. Drummond. 1971. Spermatophore transfer and fertilization of lone star ticks off the host. *Ann. Entomol. Soc. Am.* 64(2):378–381.

Grandjean, F. 1933. Etude sur le développement des Oribates. *Bull. Soc. Zool. Fr.* 58(1):30–61.

Grandjean, F. 1935. Observations sur les Acariens (7e série). *Bull. Mus. Natl. Hist. Nat.* 14(4):264–267.

Grandjean, F. 1936. Les Oribates de Jean Frederic Hermann et de son père (Arach. Acar.). *Ann. Soc. Entomol. Fr.* 105:27–110.

Grandjean, F. 1938. Description d'une nouvelle pre-larve et remarques sur la bouche des Acariens. *Bull. Soc. Zool. Fr.* 63:58–68.

Grandjean, F. 1939a. Quelques genres d'Acariens appartenant au group des endeostigmata. *Ann. Sci. Nat., Zool. Biol. Anim.* (11) (2):1–122.

Grandjean, F. 1939b. Observations sur les Acariens (5e série). *Bull. Mus. Nat. Hist. Natur.* (2)11:394–401.

Grandjean, F. 1941. Biologie—Statistique sexuelle et parthenogénèse chez les Oribates. *C.R. Hebd. Seances Acad. Sci.* 212:463–467.

Grandjean, F. 1943. Quelques genre d'Acariens appartenant au groupe des Endeostigmata (2e série) *Ann. Sci. Nat. Zool.* (II) 4:137–195.

Grandjean, F. 1946. Observations sur les Acariens (9e série). *Bull. Mus. Natl. Hist. Nat.* 18(4):337–344.

Grandjean, F. 1947. Observations sur les Acariens (10e série). *Bull. Mus. Natl. Hist. Nat.* 19(1):76–83.

Grandjean, F. 1948. Quelques caractères des Tetranyques. *Bull. Mus. Natl. Hist. Natur.* 20:517–524.

Grandjean, F. 1952. Observations sur les Oribates (24e série). *Bull. Mus. Natl. Hist. Natur.* 24:187–194.

Grandjean, F. 1954. Essai de classification des Oribates. *Bull. Soc. Zool. Fr.* 78(5/6):421–446.

Grandjean, F. 1955. Observations sur les Oribates. (32e série). *Bull. Mus. Natl. Hist. Natur.* 27:212–219.

Grandjean, F. 1956a. VIII. Caractères chitineux de l'ovipositeur, en structure normale, chez les Oribates (Acariens). *Arch. Zool. Exp. Gen.* 93(2):96–106.

Grandjean, F. 1956b. Observations sur les Oribates (34e série). *Bull. Mus. Natl. Hist. Natur.* 28(2):205–212.

Grandjean, F. 1957. L'evolution selon l'age. *Arch. Sci. Phys. Nat. Genève* 104(4):477–526.

Grandjean, F. 1959. Les stases du developpement ontogenetique chez *Balaustium florale* (Acarien: Erythroide). *Ann. Soc. Ent. Fr.* 128:159–181.

Grandjean, F. 1962. Prelarves d'Oribates. *Acarologia* 4:423–439.

Griffiths, D.A. and J. Boczek. 1977. Spermatophores of some Acaroid mites (Astigmata: Acarina). *Int. J. Insect Morphol. Embryol.* 6(5/6):231–238.

Guilfoy, F.M. and W.B. Nutting. 1987. In preparation.

Gupta, A.P., ed. 1979. *Arthropod Phylogeny.* Van Nostrand-Reinhold, New York.

Hammen, L. van der. 1963. The Oribatid family Phthiracaridae I. Introduction and resdescription of *Hoplophthiracarus pavidus* (Berlese). *Acarologia* 5(2):306–317.

Hammen, L. van der. 1976. *Glossaire de la terminologie acarologique,* Vols. 1 & 2. Junk, The Hague.

Hammen, L. van der. 1977. Studies on Opilioacaridae. IV. The genera *Panchaetes, Naudo, Salfacarus,* gen nov. *Zool. Meded.* 51(4):43–78.

Hammen, L. van der. 1978. The evolution of the chelicerate life cycle. *Acta Biotheor.* 27(1/2):44–60.

Hansell, R.E.C., M. Mollison and W.L. Putnam. 1964. A cytological demonstration of arrhenotoky in three mites of the family Phytoseiidae. *Chromosoma* 15:562–567.

Hartenstein, R. 1962. Soil Oribatei. V. Investigations on *Platynothrus peltifer* (Acarina: Camisiidae). *Ann. Entomol. Soc. Am.* 55(6):709–713.

Hazan, A., U. Gerson and A.S. Tahori. 1974. Spider mite webbing. I. The production of webbing under various environmental conditions. *Acarologia* 16(1):68–84.

Hazan, A., U. Gerson, and A.S. Tahori. 1975a. Spider mite webbing. II. The effect of webbing on egg hatchability. *Acarologia* 17:(1)270–273.

Hazan, A., A. Gertler, A.S. Tahori and U. Gerson. 1975b. Spider mite webbing. III. Solubilization and amino acid composition of the silk protein. *Comp. Biochem. Physiol. B* 51B:457–462.

Heineman, R.L. and R.D. Hughes. 1969. The cytological basis for reproductive variability in the Anoetidae (Sarcoptiformes: Acari). *Chromosoma* 28:357–369.

Helle, W. and J. Gutierrez. 1983. Karyotypes of tetranychid species from New Caledonia. *Int. J. Acarol.* 9(3):123–126.

Helles, W. and M.W. Sabelis, eds. 1985. *World Crop Pests. Spider Mites: Their Biology, Natural Enemies and Control.* Elsevier, Amsterdam.

Helle, W. and M. Wysoki. 1983. The chromosome and sex-determination of some actinotrichid taxa (Acari), with special reference to Eriophyidae. *Int. J. Acarol.* 9(2):67–71.

Helle, W., M.J. Tempelaar and L.J. Drenth-Diephuis. 1983. DNA-contents in spider mites (Tetranychidae) in relation to karyotype evolution. *Int. J. Acarol.* 9(3):127–129.

Hoogstraal, H. 1956. African Ixodoidea. I. Ticks of the Sudan. USNAMRU, Cairo, 1–1101.

Huffaker, C.B., M. van de Vrie and J.A. McMurtry. 1969. The ecology of tetranychid mites. *Annu. Rev. Entomol.* 14:125–173.

Hughes, A.M. 1961. *The Mites of Stored Food,* Tech. Bull. No. 9. H. M. Stationery Office, London.

Hughes, R.D. and C.G. Jackson. 1958. A review of the family Anoetidae. *Va. J. Sci.* 9(1):5–198.

Hughes, T.E. 1954. The internal anatomy of the mite *Listrophorus leuckarti. Proc. Zool. Soc. London* 124:239–257.

Hughes, T.E. 1959. *Mites or the Acari.* Athlone, London.

Hughes, T.E. 1964. Neurosecretion, ecdysis and hypopus formation in the Acaridei. Proc. Int'l. Congr. Acarology, 1st, 1963, Ft. Collins. *Acarologia* 6 (fasc. hors. ser.) pp. 338–342.

Hunter, W.D. and F.C. Bishopp. 1911. The Rocky mountain spotted fever tick. With special reference to the problems of its control in the Bitter Root valley in Montana. *USDA Bur. Ent. Bull.* 105:1–47. Washington, D.C.

Husband, R.W. 1979. The reproductive anatomy of male *Rhynchopolipus rhynchophori,* parasitic mite of the palm weevil. *Micron* 10:165–166.

Husband, R.W. and C.H. Flechtmann. 1972. A new genus of mite, *Rhynchopolipus,* associated with the palm weevil in Central and South America (Acarina: Podapolipidae). *Rev. Bras. Biol.* 32:519–522.

Husband, R.W. and R.N. Sinha. 1970. A revision of the genus *Locustacarus* with a key to genera of the family Podapolipidae (Acarina). *Ann. Entomol. Soc. Am.* 63(4):1152–1162.

Hussey, N.W., W.H. Read and J.J. Hesling. 1969. Order Acarina: Mites. *In: The Pests of Protected Cultivation.* Edward Arnold, London, pp. 190–228.

Jack, K.M. 1961. New species of Near Eastern agamid scale-mites (Acarina: Pterygosomidae) with notes on the developmental stages of *Geckobia hemidactyli* Law, 1936. *Parasitology* 51:241–256.

Jacot, A.P. 1933. Phthiracaroid mites of Florida. *J. Elisha Mitchell Sci. Soc.* 48(2):232–267.

Jakeman, L.A.R. 1961. The internal anatomy of the spiny rat mite, *Echinolaelaps echidninus* (Berlese). *J. Parasitol.* 47(2):329–349.

Jeppson, L.R., H.H. Keifer and E.W. Baker. 1975. *Mites Injurious to Economic Plants.* Univ. of California Press, Berkeley.

Khalil, G.M. 1969. Biochemical and physiological studies of certain ticks (Ixodoidea). Gonad development and gametogenesis in *Argas (Persicargas) arboreus* Kaiser, Hoogstraal and Kohls (Argasidae). *J. Parasit.* 55:1278–1297.

Khalil, G.M. 1972. Gonad development in the parthenogenetic *Hemaphysalis (Kaiseriana) longicornis* Neummann (Ixodoidea: Ixodidae). *J. Parasit.* 58(4):817–123.

Kinn, D.N. 1983. The life cycle of *Proctolaelaps dendroctoni* Lindquist and Hunter (Acari: Ascidae): A mite associated with pine bark beetles. *Int. J. Acarol.* 205–210.

Kirchner, W.P. 1967. Spermatophoreen bei Halacariden (Acarina). *Naturwissenschaften* 54(13): 345–346.

Krantz, G.W. 1978. *A Manual of Acarology,* 2nd ed. Oregon State Univ. Book Stores, Corvallis.

Krantz, G.W. and E.E. Lindquist. 1979. Evolution of phytophagous mites (Acari). *Annu. Rev. Entomol.* 24:121–158.

Krantz, G.W. and J.L. Mellott. 1968. Two new species of *Macrocheles* (Acarina: Macrochelidae) from Florida, with notes on their host-specific relationships with Geotrupine beetles (Scarabeidae: Geotrupinae). *J. Kans. Entomol. Soc.* 41(1):48–56.

Krantz, G.W. and J.G. Werntz. 1979. Sperm transfer in *Glyptholaspis americana. In:* Rodriguez, J.G., ed. *Recent Advances in Acarology.* Academic Press, New York, Vol. 2, pp. 441–446.

Krombein, K.V. 1961. Some symbiotic relations between saproglyphid mites and solitary vespid wasps. *J. Wash. Acad. Sci.* 51:89–93.

Kuo, J.S. and H.H. Nesbitt. 1971. Internal morphology of the hypopus *Caloglyphus mycophagus* (Megnin). *Acarologia* 13(1):156–169.

Lanciani, C.A. 1972. Mating behavior of water mites of the genus *Eylais. Acarologia* 14(4):631–637.

Lawrence, R.F. 1953. The biology of the cryptic fauna of forests. A.A. Balkema, Capetown/Amsterdam, pp. 1–408.

Lebrun, P. 1970. Ecologie et biologie de *Nothros pa-*

lustris (C.L. Koch, 1839). 3e Note Cycle de Vie. *Acarologia* 12(1):193–207.

Lees, A.D. and J.W.L. Beament. 1948. An egg-waxing organ in ticks. *Q. J. Microsc. Sci.* 89:291–332.

Legendre, R. 1968. La nomenclature anatomique chez les Acariens. *Acarologia* 10(3):411–417.

Lions, J.C. 1967. La prelarve de *Rhysotritia ardua* (C.L. Koch) (Acarien: Oribate). *Acarologia* 9(1):273–283.

Lions, J.C. 1973. Quelques prelarves nouvelles d'Oribates. *Acarologia* 15(2):356–370.

Lipa, J.J. and W. Chmielewski. 1966. Aparity observed in the development of *Caloglyphus* mite (Acarina: Acaridae). *Ekol. Pol., Ser. A* 14(37):741–748.

Lipovsky, L.J., G.W. Byers and E.H. Kardos. 1957. Spermatophores—the mode of inseminations of chiggers (Acarina: Trombiculidae). *J. Parasitol.* 43:256–262.

Lukoschus, F.S. and F.M. Driessen. 1970. *Myobia micromydis* spec. nov. (Myobiidae: Trombidiformes) from *Micromys minutus* Pallas. *Acarologia* 12(1):119–126.

Lukoschus, F.S., G. Scheperboer, A. Fain and M. Nadchatram. 1981. Life cycle of *Alabidopus asiaticus* sp. nov. (Acarina: Astigmata: Glycyphagidae) and hypopus of *Alabidopus malaysiensis* sp. nov. ex. *Rattus* spp. from Malaysia. *Int. J. Acarol.* 7(1–4):161–177.

Lukoschus, F.S., T. Woeltjes, E.A. Juckwer and A. Fain. 1979. Life cycle of *Orycteroxenus galemys* sp. nov. (Astigmata: Glycyphagidae). *Int. J. Acarol.* 5(1):29–38.

Mahunka, S. 1979. Some remarks on polymorphism (phoretomorphism) in Tarsonemid mites. *Folia Entomol. Hung.* 13(2):133–137.

Manual of Livestock Ticks, USDA, ARS. 1965, pp. 91–49.

Marchiando, A.A. 1981. Scanning electron microscopy of the egg of *Cheyletiella yasguri* Smiley (Acari: Cheyletiellidae) with a description of the egg burster. *Int. J. Insect Morphol. Ebryol.* 10(2):167–171.

Marshall, V. and M. Reeves. 1971. *Trichthonius majestus,* a new species of oribatid from North America (Cosmochthoniidae). *Acarologia* 12(3):623–632.

Michael, A.D. 1883. *British Oribatidae,* Vol. 1. Ray Society, London.

Michael, A.D. 1892. On the variation in the internal anatomy of the Gamasinae. *Trans. Linn. Soc. London, Zool.* 5:281–325.

Mitchell, R.D. 1958. Sperm transfer in the watermite *Hydryphantes ruber* Geer. *Am. Midl. Nat.* 60(1):156–158.

Mitchell, R.W. 1963. Comparative morphology of the life stages of the nasal mite *Rhinonyssus rhinolethrum* (Mesostigmata: Rhinonyssidae). *J. Parasitol.* 49:506–515.

Moser, J.C., E.A. Cross and L.M. Roton. 1971. Biology of *Pyemotes parviscolyti* (Acarina: Pyemotidae). *Entomophaga* 16:367–379.

Mostafa, A-R.I. 1970a. Saproglyphid hypopi (Acarina: Saproglyphidae) associated with wasps of the genus *Zethus* Fabricius. Part I. *Acarologia* 12(1):168–192.

Mostafa, A-R.I. 1970b. Saproglyphid hypopi (Acarina: Saproglyphidae) associated with wasps of the genus *Zethus* Fabricius. Part II. *Acarologia* 12(2):383–401.

Mostafa, A-R.I. 1971. Saproglyphid hypopi (Acarina: Saproglyphidae) associated with wasps of the genus *Zethus* Fabricius. Part III. *Acarologia* 12(3):546–582.

Nasr, A.K. and E.M. El-Banhawy. 1984. Effect of liquid nutrients on the development and reproduction of the predacious mite *Amblyseius gossipi* (Mesostigmata: Phytoseidae). *Int. J. Acarol.* 10(4):235–237.

Newell, I.M. 1963. Feeding habits in the genus *Balaustium* (Acarina: Erythraeidae) with special reference to attacks on man. *J. Parasit.* 49:498–502.

Newell, I.M. 1973. The protonymph of *Pimeliaphilus* (Pterygosomatidae) and its significance relative to the Calyptostases in the Parasitengona. *Proc. Int. Congr. Acarol., 3rd, 1971,* pp. 789–795.

Nuttall, G.H.F. 1913. Observations on ticks: (a) parthenogenesis (b) variation due to nutrition. *Cambridge Proc. Phil. Soc.* 17:210.

Nuttall, G.H.F. 1915. Observations on the biology of Ixodidae part 2. *Parasitol.* 7:408–455.

Nuttall, G.H.F. and G. Merriam. 1911. The process of copulation in *Ornithodoros moubata. Parasitology* 4(1):39–44.

Nutting, W.B. 1961. *Demodex aurati* sp. nov. and *D. criceti,* ectoparasites of the golden hamster *(Mesocricetus auratus). Parasitology* 51:515–522.

Nutting, W.B. 1964. Demodicidae—Status and prognostics. *Acarologia* 6(3):441–454.

Nutting, W.B. 1965. Host-parasite relations: Demodicidae. *Acarologia* 7(2):301–317.

Nutting, W.B. and F.M. Guilfoy. 1979. *Demodex cafferi* n. sp. from the African Buffalo, *Syncerus caffer*. *Int. J. Acarol.* 5(10):9–14.

Nutting, W.B. and H. Rauch. 1958. *Demodex criceti* n. sp. (Acarina: Demodicidae) with notes on its biology. *J. Parasitol.* 44(3):328–333.

Nutting, W.B. and H. Rauch. 1963. Distribution of *Demodex aurati* in the host *(Mesocricetus auratus)* skin complex. *J. Parasitol.* 49(2):323–329.

Oboussier, H. 1939. Beitrage zur Biologie und Anatomie der Wohnmilben. *Z. Angew. Entomol.* 26:253–296 (cited by Woodring and Carter, 1974).

Oldfield, G.N. and I.M. Newell. 1973. The role of the spermatophore in the reproductive biology of protogynes of *Aculus cornutus* (Acarina: Eriophyidae). *Ann. Entomol. Soc. Am.* 66(1):160–163.

Oldfield, G.N., R.F. Hobza and N.S. Wilson. 1970. Discovery and Characterization of Spermatophores in the Eriophyoidea (Acari). *Ann. Entomol. Soc. Am.* 63(2):520–526.

Oldfield, G.N., I.M. Newell and D.K. Reed. 1972. Insemination of Protogynes of *Aculus cornutus* from spermatophores and a description of the sperm cell. *Ann. Entomol. Soc. Am.* 65(5):1080–1084.

Oliver, J.H., Jr. 1962. A mite parasitic in the cocoons of earthworms. *J. Parasitol.* 48:12–123.

Oliver, J.H., Jr. 1964. Comments on karyotypes and sex determination in the Acari. *Proc. Int. Congr. Acarol., 1st, 1963*, pp. 288–293.

Oliver, J.H., Jr. 1966. Sex chromosomes of ticks (Acari: Ixodoidea). *Am. Zool.* 6(3):205.

Oliver, J.H., Jr. 1971a. Introduction to the Symposium on Parthenogenesis. *Am. Zool.* 11(2):241–243.

Oliver, J.H., Jr. 1971b. Parthenogenesis in mites and ticks (Arachnida: Acari). *Am. Zool.* 11(2):283–299.

Oliver, J.H., Jr. 1977. Cytogenetics of mites and ticks. *Annu. Rev. Entomol.* 22:407–429.

Oliver, J. H., Jr. 1986. Induction of oogenesis and oviposition in ticks. *In:* Sauer, R.J. and J.A. Hair, eds. *Morphology, Physiology and Behavioral Biology of Ticks.* Ellis Horwood, Ltd., Chichester, England, pp. 233–247.

Oliver, J.H., Jr. and E.D. Delfin. 1967. Gynandromorphism in *Dermacentor occidentalis* (Acari: Ixodidae). *Ann. Entomol. Soc. Am.* 60(5):1119–1121.

Oliver, J.H., Jr. and C.S. Herrin. 1974. Morphometrics of sexual dimorophism in an arrhenotokous mite, *Ornithonyssus bacoti* (Acari: Mesostigmata). *J. Exp. Zool.* 189(3):291–302.

Oliver, J.H., Jr. and B.C. Nelson. 1967. Mite chromosomes: An exceptionally small number. *Nature (London)* 214:809.

Oliver, J.H., Jr., J.H. Camin and R.C. Jackson. 1963. Sex determination in the snake mite Ophionyssus natricus (Gervais) (Acarina: Dermanyssidae). *Acarologia* 5(2):180–184.

Oliver, J.H., Jr., K. Tanaka and M. Sawada. 1973. Cytogenetics of ticks (Acari Ixodoidea). *Chromosoma* 42:269–288.

Oliver, J.H., Jr., Z. Al-Ahmadi and R.L. Osburn. 1974. Reproductions in ticks (Acari: Ixodoidea). 3. Copulation in *Dermacentor occidentalis* Marx and *Haemophysalis leporispalustris* (Packard) (Ixodidac). *J. Parasitol.* 60(3):499–506.

Pater de Kort, I., J.E.M.H. van Bronswijk and A. Fain. 1979. Redescription and biology of the mite *Acanthophthirius polonicus* Haitlinger, 1978, parasitic on the pond bat, *Myotis dasycneme* (Boie, 1825) (Prostigmata: Myobiidae). *Int. J. Acarol.* 5(4):291–298.

Pauly, F. 1952. Die "Copula" der Oribatiden (Moosmilben). *Naturwissenschaften* 39:572–573.

Pauly, F. 1956. Zur Biologies einiger Belbiden und zur Funktion ihrer pseudostigmatischen Organe. *Zool. Jarhb., Syst. (Oekol.), Geogr. Biol.* 84:275–328.

Pence, D. 1972. The hypopi (Acarina: Sarcoptiformes; Hypoderidae) from the subcutaneous tissues of birds in Louisiana. *J. Med. Entomol.* 9(5):435–438.

Perkins, C.L. 1899. A special acarid chamber formed within the basal abdominal segment of bees of the genus *Koptorthosoma* (Xylocopinae). *Entomol. Mon. Mag.* 35:37–39.

Pijnacker, L.P., M.A. Ferwerda and W. Helle. 1981. Cytological investigations on the female and male reproductive system of the parthenogenetic privet mite *Brevipalpus obovatus* Donnedieu (Phytoptipalpidae, Acari). *Acarologia* 22(2):157–163.

Pinichpongse, S. 1963a. A review of the Chirodiscinae with descriptions of new taxa (Acarina: Listrophoridae). (Part 1). *Acarologia* 5(1):81–91.

Pinichpongse, S. 1963b. A review of the Chirodiscinae with descriptions of new taxa (Acarina: Listrophoridae). (Part 2). *Acarologia* 5(2):266–278.

Pomerantzev, R.I. 1959. *Ixodid Ticks. Fauna of USSR. Arachnida,* Vol 4., No. 2 (translation from the Russian). Am. Inst. Biol. Sci., Washington, D.C.

Potter, D.A. 1979. Reproductive behavior and sexual selection in Tetranychine mites. *In:* Rodriguez, J.G., ed. *Recent Advances in Acarology.* Academic Press, New York, Vol. 1, pp. 137–146.

Pound, J.M. and J.H. Oliver, Jr. 1976. Reproductive morphology and spermatogenesis in *Dermanyssus gallinae* (De Geer) (Acari: Dermanyssidae). *J. Morphol.* 150:825–842.

Prasad, V. 1973. Description of life stages of the predatory mite *Phytoseiulus macropilis* (Banks) (Acarina: Phytoseiidae). *Acarologia* 15(3):391–399.

Pritchard, A.E. and E.W. Baker. 1955. *A Revision of the Spider Mite Family Tetranychidae,* Mem. Ser. Vol. 2. Pac. Coast Entomol. Soc., San Francisco, California.

Radinovsky, S. 1965. Biology and ecology of granary mites of the Pacific Northwest. II. Various aspects of the reproductive behavior of *Leiodinychus krameri. Ann. Entomol. Soc. Am.* 58:267–272.

Radovsky, F.J., J.K. Jones, Jr., and C.J. Phillips. 1971. Three new species of *Radfordiella* (Acarina: Macronyssidae) parasitic in the mouth of phyllostomatid bats. *J. Med. Entomol.* 8(6):737–746.

Rechav, Y. 1977. A case of gynandromorphism in *Amblyomma hebraeum* (Acarina: Ixodidae). *J. Med. Entomol.* 14(3):304.

Reger, J.F. 1963. Spermiogenesis in the tick *Amblyomma dissimilis* as revealed by electron microscopy. *J. Ultrastruct. Res.* 8:607–621.

Regev, S. 1974. Cytological and radioassay evidence of haploid parthenogenesis in *Cheyletus malaccensis* (Acarina: Cheyletidae). *Genetica* 45:125–132.

Robaux, P. 1970a. Thrombidiidae d'Amérique du Sud. III. *Valgothrombium paradoxus* n. sp. *Acarologia* 12(1):127–130.

Robaux, P. 1970b. La Prélarve de *Campylothrombium barbarum* Lucas (Acari: Thrombidiidae). *Acarologia* 12(1):131–135.

Robaux, P. 1971. Recherches sur le développement et la biologie des acariens Thrombiidae. Ph.D. thesis, CNRS Reg. No. 5616. Faculty of Science, University of Paris (cited by Krantz, 1978).

Robaux, P. 1974. Recherches sur le développement et la biologie des acariens "Thrombiidae." *Mem. Mus. Natl. Hist. Nat., Ser. A (Paris)* 85:1–186.

Robinson, L.E. and J. Davidson. 1914. The anatomy of *Argas persicus.* Part 3. *Parasitology* 6:382–424.

Rohde, C.J., Jr. and D.A. Oemick. 1967. Anatomy of the digestive and reproductive systems in an acarid mite (Sarcoptiformes). *Acarologia* 9(3):608–616.

Roshdy, M.A. 1961. Comparative internal morphology of subgenera of *Argas* ticks (Ixodoidea, Argasidae). I. Subgenus *Carios: Argas vespertilionis* Latreille, 1802. *J. Parasit.* 47:987–994.

Roshdy, M.A. 1962. Comparative morphology of subgenera of *Argas* ticks (Ixodoidea, Argasidae). 2. Subgenus *Chiropterargas: Argas boueti* Roubaud and Colas-Belcour, 1933. *J. Parasit.* 48(4):623–630.

Roshdy, M.A. 1963. Comparative morphology of subgenera of *Argas* ticks (Ixodoidea, Argasidae). 3. Subgenus *Secretargas: Argas transgariepinus* White, 1846. *J. Parasit.* 49(5):851–856.

Roshdy, M.A. 1966. Comparative internal morphology of subgenera of *Argas* ticks (Ixodoidea, Argasidae). 4. Subgenus *Ogadenus: Argas brumpti* Neumann, 1907. *J. Parasit.* 52(4):776–782.

Savory, T. 1977. *Arachnida,* 2nd ed. Academic Press, London.

Schroder, R.F.W. 1982. Effect of infestation with *Coccipolipus epilachnae* Smiley (Acarina: Podapolipidae) on fecundity and longevity of the Mexican bean beetle. *Int. J. Acarol.* 8(2):81–84.

Schuster, R. 1962. Nachweiss eines Paarungszeremoniells bei den Hornmilben. *Naturwissenschaften* 21:502–503.

Schuster, R. and I.J. Schuster. 1966. Uber das Fortpflanzungsverhalten von Anystiden-Mannehen (Acari: Trombidiformes). *Naturwissenschaften* 53(6):162–163.

Sharma, G.D., M.H. Farrier and A.T. Drooz. 1983. Food, life-history, and sexual differences of *Callidosoma metzi* Sharma, Drooz, and Treat (Acarina: Erythraeidae). *Int. J. Acarol.* 9(3):149–155.

Shereef, G.M. 1976. Biological studies and description of stages of two species: *Papillacarus aciculatus* Kunst and *Lohmannia egypticus* Elbadry and

Nasr (Oribatei—Lohmanniidae) in Egypt. *Acarologia* 18(2):351–359.

Skaife, S.H. 1952. The yellow-banded carpenter bee, *Mesotrichea caffra* and its symbiotic mite *Dinogamasus braunsi* Vitz. *J. Entomol. Soc. S. Afr.* 15:63–76.

Smiley, R.L. and J.C. Moser. 1984. Notes and a key to separate normal and heteromorphic males of *Pyemotes giganticus* Cross, Moser and Rack and *P. dimorphus* Cross and Moser (Acari: Pyemotidae). *Int. J. Acarol.* 19(1):11–14.

Smith, I.A. 1976. A study of the systematics of the water mite family Pionidae. *Mem. Entomol. Soc. Can.* 98:1–249.

Snodgrass, R.E. 1965. *Anatomy of Arthropoda.* Hafner, New York.

Solomon, M.E. 1945. Tyroglyphoid mites in stored products. Methods for the study of population density. *Ann. Appl. Biol.* 32:71–75.

Sonenshine, D.E. 1970. A contribution to the internal anatomy and histology of the bat tick *Ornithodoros kelleyi* Cooley and Kohls, 1941. *J. Med. Entomol.* 7(3):289–312.

Spicka, E.J and B.M. O'Connor. 1979. Description of life cycle of *Glycyphagus (Myacarus) microti*, Sp. n. (Acarina: Sarcoptiformes: Glycyphagidae) from *Microtus pinetorum* from New York. *Acarologia* 21:451–476.

Spickett, G.S. 1961. Studies on *Demodex folliculorum* Simon (1842). *Parasitology* 51:181–192.

Sternlicht, M. and D.A. Griffiths. 1974. The emission and form of spermatophores and the fine structure of adult *Eriophyes sheldoni* Ewing (Acarina, Eriophyoidea). *Bull. Entomol. Res.* 63(4):561–565.

Strandtmann, R.W. and O.E. Hunt. 1951. Two new species of Macronyssidae, with notes on some established genera (Acarina). *J. Parasitol.* 37(5):460–470.

Strandtmann, R.W. and G.W. Wharton. 1958. *A Manual of Mesostigmatid Mites Parasitic on Vertebrates,* Contrib. No. 4. Inst. Acarol., University of Maryland, College Park.

Sternlicht, M. and D.A. Griffiths. 1974. The emission and formation of spermatophores and the fine structure of adult *Eriophyes sheldoni* Ewing (Acarina, Eriophyidae). *Bull. Ent. Res.* 63(4):561–565.

Strenske, K. 1949. Zur Fortpflanzung der Moosmilben. *Mikrokosmos* 38(8):177–180.

Summers, F.M. and R.L. Witt. 1972. Nesting behavior of *Cheyletus eruditus* (Acarina: Cheyletidae). *Pan-Pac. Entomol.* 48(4):261–269.

Sundman, M.A. 1965. A case of gynandromorphism in *Amblyomma imitator* (Acarina: Ixodidae). *Ann. Entomol. Soc. Am.* 58(4):592–593.

Tadkowski, T.M. and K.E. Hyland. 1979. The developmental stages of *Aplodontopus sciuricola* (Astigmata) from *Tamius striatus* L. (Sciuridae) in North America. *Proc. Int. Congr. Acarol., 4th, 1974,* pp. 321–326.

Tatchell, R.J. 1962. Studies on the male accessory reproductive glands and the spermatophore of the tick, *Argas persicus* Oken. *Parasitology* 52:133–142.

Theis, G. and R. Schuster. 1974. Gestielte Tropfchenspermatophoren bei Calyptostomiden (Acari: Trombidiformes). *Zool. Inst. Mitt. Naturwiss. Ver Steiermark.* 104:183–185.

Timms, S., D.N. Ferro and J.B. Waller. 1980. Suppression of production of pleomorphic males in *Sancassania berlesei* (Michael) (Acari: Acaridae). *Int. J. Acarol.* 6(2):91–96.

Timms, S., D.N. Ferro and R.M. Emberson. 1981. Andropolymorphism and its heritability in *Sancassia berlesei* (Michael) (Acari: Acaridae). *Acarologia* 22(4):391–398.

Travé, J. 1967. *Phyllochthonius aoutti,* n. gen., n. sp. un Enarthronota nouveau de Côte d'Ivoire, avec la création d'une superfamille nouvelle (Phyllochthonoidea). *Zool. Meded.* 42:83–105.

Travé, J. 1976. Les prelarves d'Acariens. Mise au point et données recentes. *Rev. Ecol. Biol. Sol.* 13(1):161–171.

Treat, A. 1975. *Mites of Moths and Butterflies.* Comstock, Ithaca, New York.

Vercammen-Grandjean, P.H. and R.L. Langston. 1971. Two new species of *Leptotrombidium* (Acarina, Trombiculidae) from Malaysia. *J. Med. Ent.* 8:450–453.

Vitzthum, H.G. 1943. Acarina. *Bronns Kl.* 5:1–1011.

Volgin, V.I. 1973. The hypopus and its main types. *Proc. Int. Congr. Acarol., 3rd, 1971,* pp. 381–383.

Walker, N. 1964. *Euphthiracaroidea of California Sequoia Litter: With a Reclassification of the Families and Genera of the World (Acarina: Oribatei),* Fort Hays Stud., New Ser., Sci. Ser. No. 3. Kansas State College, Fort Hays.

Wallace, D.R.J. 1960. Observations on hypopus development in the Acarina. *J. Insect. Physiol.* 5(3/4):216–229.

Wallace, M.M.H. and J. A. Mahon. 1972. The taxonomy and biology of Australian Bdellidae (Acari). I. Subfamilies Bdellinae, Spinibdellinae and Cytinae. *Acarologia* 14(4):544–580.

Wallace, M.M.H. and J. A. Mahon. 1976. The taxonomy and biology of Australian Bdellidae (Acari). II. Subfamily Odontoscirinae. *Acarologia* 18(1):65–123.

Wallwork, J.A. 1962. Sexual dimorphism in the genus *Epilohmannia* Berlese, 1916. *Rev. Zool. Bot. Afr.* 65(1–2):90–96.

Whitmoyer, R.E., L.R. Nault and O.E. Bradfute. 1972. Fine structure of *Aceria tulipae*. *Ann. Entomol. Soc. Am.* 65(1):201–215.

Witalinski, W. 1975. Fine structure of vas deferens wall in the mite *Pergamasus viator* Halas (Mesostigmata: Parasitidae). *Acarologia* 17(2):197–207.

Witalinski, W. 1979. Fine structure of spermatozoa and some remarks on spermiogenesis in *Pergamasus viator* (Parasitidae). *Proc. Int. Congr. Acarol., 4th, 1974,* pp. 687–691.

Witte, H. 1975. Funktionsanatomie der Genital organe und Fortpflanzungsverhalt en bei dem Mannchen der Erythraeidae (Acari: Trombidiformes). *Z. Morphol. Tiere* 80:137–180.

Witte, H. and V. Storch. 1973. Licht- und Elektron-Mikroskopische Untersuchungen on Hodersekreten und Spermien der Trombidiformen Milbe *Abrolophus rubipes* (Trouessart, 1888). *Acarologia* 15(3):441–450.

Woodring, J.P. 1965. The biology of five new species of oribatids from Louisiana. *Acarologia* 7(3):564–576.

Woodring, J.P. 1970. Comparative morphology, homologies and functions of the male system in oribatid mites. *J. Morphol.* 139(4):407–430.

Woodring, J.P. and S.C. Carter. 1974. Internal and external morphology of the deutonymph of *Caloglyphus boharti* (Arachnida: Acari). *J. Morphol.* 144(3):275–295.

Woodring, J.P. and E.F. Cook. 1962. The biology of *Ceratozetes cisalpinus* Berlese, *Scheloribates laevigatus* Koch, and *Oppia neerlandica* Oudemans (Oribatei) with descriptions of all stages. *Acarologia* 4(1):101–137.

Woodring, J.P. and C.A. Gailbrath. 1976. The anatomy of the adult uropodid *Fuscouropoda agitans* (Arachnida: Acari), with comparative observations on other Acari. *J. Morphol.* 150(1):19–58.

Woolley, T.A. and H.G. Higgins. 1969. A new species of *Platyliodes* from the N.W. United States. *Proc. Entomol. Soc. Wash.* 71(2):143–146.

Wrensch, D.L. 1979. Components of reproductive success in spider mites. *In:* Rodriguez, J.G., ed. *Recent Advances in Acarology.* Academic Press, New York, Vol. 1, pp. 155–164.

Wysoki, M. and H.R. Bolland. 1983. Chromosome studies of phytoseid mites (Acari: Gamasida). *Int. J. Acarol.* 9(32):91–94.

Young, J.H. 1968. The morphology of *Haemogamasus ambulans*. II. Reproductive system. *J. Kans. Entomol. Soc.* 41(4):532–543.

Yunker, C.E. 1970. New genera and species of Ewingidae (Acari: Sarcoptiformes) from Pagurida (Crustacea) and notes on *Ewingia coenobitae* Pearse, 1929. *Rev. Zool. Bot. Afr.* 81(3–4):237–254.

Zakhvatkin, A.A. 1959. *Tyroglyphoidea (Acari)* (translation of *Fauna of USSR Arachnoidea,* Vol. 6. No. 1). Am. Inst. Biol. Sci., Washington, D.C. (translated by A. Ratcliffe and A.M. Hughes).

13

Pheromones

As humans all of us respond in various ways to chemical signals—the odor of fresh-baked bread, the scent of flowers or perfume, the repellant odor of a skunk or stink bug. Insects and other land arthropods produce and respond to chemical signals, usually between members of the same species. These substances are referred to as pheromones. Some animals produce only a few pheromones, while others produce many (Borror et al., 1976, 1979).

Semiochemicals are "information-bearing chemicals" produced by one animal that cause another animal to change behavior. These chemicals are found in both invertebrates and vertebrates and are categorized into three types. (1) Allomones are chemicals produced by members of one species that favorably modify behavior of another species toward the emitter species (e.g., defensive chemicals). (2) Kairomones are chemicals released by members of one species that excite members of another compatible species (e.g., attractants). Predatory mites such as *Phytoseiulus persimils* can detect prey species of spider mites (*Tetranychus urticae, Panonychus ulmi*) by the chemicals deposited in the feces of the prey species. Other attractants vary with the prey species involved. (3) Pheromone chemicals are produced and released by individuals of a species and modify the behavior of members of the same species in a way that is favorable to the survival of the species (e.g., sex pheromones) (Sonenshine, 1985, 1986).

Pheromones are chemicals of low volatility and long persistence that are secreted to the outside of the body and cause specific behavioral reactions by other members of the same species. They are sometimes called "social hormones" because they form a means of chemical communication between individuals or groups. By definition a pheromone is "a substance secreted by an animal that influences the behavior of other animals of the species" (Wilson, 1965); for example, invisible odor trails are well-known for ants, which are directed to food and nest locations by chemical trails left by their fellows. The word "pheromone" was coined for intraspecific chemical agents (sex, trail, alarm, swarming) (Hughes, 1959). These agents were originally referred to as "ectohormones," suggesting that they are elaborated externally and influence regulation of the external environment of the animals. The behaviors resulting from the influence of pheromones are conspecific or interspecific (Wilson, 1963). Pheromones constrast to hormones, which are internal organic regulators of physiology.

The synthesis of phenolic compounds within the bodies of land arthropods is an unusual phenomenon. Synthesis of pheromones, which are phenolic (= cresotic) compounds, is known to occur within the bodies of a number of ticks and mites. Wilson (1963) presaged some of the recent research on these agents by predicting that we would find chemical systems to be "the dominant means of communication in many animal species, perhaps even in most."

When the effect of the pheromone causes a more or less immediate yet reversible change in the behavior of the animals stimulated, this is called a "releaser" effect. It is assumed that such a response is due to the action of the pheromone directly on the nervous system. A second result, called "primer" effect, seems to elicit a chain of physiological events that influence behavior. Such substances account for the regulations of caste production in termites, or the acceleration of growth in the migratory locust. Other types of responses also are found in the acarines.

As is the case with hormones, a very small amount of pheromone elicits a response. The alluring perfume of the female may be in such dilution that in the molecular dispersal of 0.01 μg of gyplure from a single female, there is enough pheromone to lure one billion males to the presence of the female and to provoke sexual activity.

Much is known about pheromones and pheromonal communication in the insects, but relatively little has been determined for the arachnids (Leahy, 1979). Allomones are chemicals that affect other-species reactions (karo-

mone recipients). Pheromonal communication is documented in only a few orders of the Acari, but as research increases more information will become available. Some ordinal examples are the Gamasida (Dermanyssidae, *Dermanyssus gallinae, D. prognephilus*); Ixodida (Argasidae, *Ornithodoros, Argas*; Ixodidae, *Amblyomma, Dermacentor, Ixodes*); Actinedida (Tetranychidae, *Tetranychus urticae*).

Types of pheromones found in the Acari are (1) alarm pheromones, (2) assembly pheromones, (3) aggregation and attachment pheromones, and (4) sex pheromones.

Alarm pheromones are found in the Astigmata, specifically in the family Acaridae. Terpenoid chemicals from the body fluids of injured mites excite dispersal of conspecific mites. Some chemicals in this group may act as repellants of predators for other species of the family.

Assembly pheromones are known only in ticks where they cause aggregations of individuals in 14 species of Argasidae and 6 species of Ixodidae (Sonenshine, 1985). When ticks encounter this type of pheromone they cease movement and remain motionless in an aggregation. Such assembly behavior is thought to cause aggregation of tick populations in species-favorable environments (Leahy, 1979). Guanine is the assembly pheromone of several species of argasid and ixodid ticks (Sonenshine, 1986; Otieno et al., 1985).

Aggregation and attachment pheromones influence the attachment of ectoparasitic ticks to sites on the body of the host. These pheromones are produced by feeding males only, but the chemicals will attract unfed males, females, and nymphs in some species of *Amblyomma*. The pheromones are released after 5 days of feeding to a maximum of 7 or 8 days.

Sex pheromones change the behavior of one or both partners of a mating pair and thus regulate or modify various stages of the mating process. These compounds or mixtures of compounds are categorized as (1) arrestant, (2) attractant, and (3) contact or stimulant types of sex hormones.

Phytoseiidae and Tetranychidae are the only mite families that have arrestant sex pheromones. When males encounter deutonymphs or virgin females, they become quiescent, palpate the nymph, and assume a "guarding" or protective posture over the deutonymph or nymphochrysalis (Penman and Cone, 1972). Female deutonymphs emit sex hormones (e.g., farnesol, nerolidol, or geraniol) (Sonenshine, 1984). Silken threads secreted by spider mites aid in the distribution of the pheromones and guide the male in his search for a female. Arrestant behavior continues until emergence of the virgin female, at which time mating usually occurs. If another male approaches, the guarding male competes with the intruder by cheliceral thrusts at his body or attempts to entangle the competitor in silken threads (Potter, 1979).

Attractant sex pheromones occur exclusively in metastriate ticks. These volatile compounds are secreted by attached and feeding females and affect only feeding and fed males. They serve to bring mature partners together in a situation of maximum availability of food. A precise system of courtship and behavioral events follows the secretion of pheromones by females of *Dermacentor variabilis* and *D. andersoni*. The substance, 2,6-dichlorophenol (2,6-DCP) (and possibly other phenols) induces males to detach, identifies the location of sexually active partners, and facilitates mating (Sonenshine et al., 1976; Sonenshine, 1985; the latter reference details the chemistry of these agents).

Contact (mating stimulant) pheromones are known in both soft and hard ticks. These pheromones stimulate mating when males have identified and mounted a sexually active female. Mate-seeking males of *Ornithodoros* spp. (Argasidae) are attracted to such females and mate with them in response to sex pheromones released from coxal glands to the body surface. This appears to be a two-agent system of chemical and contact signals (Schlein and Gunders, 1981). Species of *Dermacentor* have specific sex hormones. The males are attracted to conspecific females by 2,6-DCP, which is detected by

sensilla on the cheliceral digits (Sonenshine et al., 1984a). Excitation of these sensilla results in probing of the gonopore, subsequent copulation, and placement of the spermatophore (Sonenshine et al., 1984a).

The first pheromones in ticks were reported by Berger (1972). Perhaps more work has been accomplished in pheromonal research on ticks than on all the other orders of mites; spider mites rank second.

The responses vary in this communication system, but fall mainly within the categories of orientation and recognition, excitation, identification of sexual partners, aggregation, and host location. The examples that follow will elucidate some of the present information about this important phase of mite biology.

Chemical and tactile stimuli elicit aggregation behavior in the martin mite (*Dermanyssus prognephilus* Ewing). Chemical stimuli initiate aggregation in this species and the response of the mites to previous aggregation sites. Chemoreceptors on the tarsi of the first legs are assumed to be involved in the response. An aggregation pheromone is produced by the mites that have become kinetic and the resulting behavior may serve to bring sexes together, reduce predation by sedentary predators, and increase success in host-finding. It may also reduce water loss and increase the retention of mites within the nest of the host (Davis and Camin, 1977). Sex hormones are produced by predatory phytoseids (Hoy and Smilanich, 1979).

In his studies of the methods by which the moth ear mite *Dicrocheles phalaenodectes* locates on a moth, Treat (1975) suggests that a female could leave a trail of pheromone to direct others to a chosen ear on the host. He tried to determine experimentally if a physical disturbance of the path traversed by the mites or a chemical trail was involved, but could not make that judgment. His assessment of the circumstances of travel suggests a pheromone, but is not confirmed.

The kinds of responses to pheromones that occur in ticks include (1) detachment, (2) detection and location of food, (3) aggregation, or assembly (Leahy et al., 1981; Petney and Bull, 1981; Rechav, 1978; Rechav and Whitehead, 1979; Rechav et al., 1976, 1977), (4) orientation to and recognition of females, (5) reproductive excitement and recruitment of sexual partners. It has been found (Sonenshine et al., 1979) that males detach earlier in the presence of females. Comparisons of solitary males and those in contact with females show that the solitary males remain without detachment much longer.

A second effect of pheromones in ticks is the excitation that occurs when males contact the chemical substances involved. When the pheromone is detected, the males wave their legs, search for and locate the females, mount them, and copulate. They will even attempt to copulate with other objects, such as dead or preserved females that have pheromone lingering on the body, or that have been tagged with pheromonal substance; however, copulation is rarely completed and the males depart without depositing a spermatophore. The pheromone is important in the recognition of and orientation to females. The male climbs on a female, palpates the foveolae dorsales, turns to the venter of the female so that the two partners are apposed, and copulates (Khalil et al., 1981, 1983a; Sonenshine et al., 1979).

Research has been done on the stimulus for the release of pheromones. There are evidently stretch receptors in the gut that affect the neurosecretory areas of the synganglion and trigger the foveal glands to produce the pheromones. In *D. variabilis* nonfeeding females injected with nonnutrient solution attracted males. The foveae dorsales are multilobed glands that have yielded as much as 17.9 ng of 2,6-dichlorophenol per gland. Autoradiography of foveal glands showed six times as much chlorine as elsewhere in the body.

Distance of pheromonal detection from a female is 2 to 3 cm. Concentration and dilution of the molecular substance evidently determines the intensity of response. Males are not affected until 4 days after the beginning of pheromone production in females. Spermatogenesis is influenced by early detachment. Females will continue to emit pheromones, but will not feed to repletion until mated.

Pheromones are known in both families of ticks. In soft ticks (Argasidae) the first demonstration of a pheromone-induced assembly was in *Argas persicus* (Leahy et al., 1973). Pheromones in other soft ticks are also known (Leahy et al., 1973, 1975). The assembly response of these ticks is also influenced by CO_2 and thigmotaxis. The production of the pheromones involved is related to feeding and to materials from conspecific females. Assembly response is interspecific.

The influence of pheromones in the argasid ticks shows an assembly response to materials from conspecific females and some interspecific assembly response. In *Argas persicus* pheromones from fed females elicit greater response than from unfed females (Leahy et al., 1973; Leahy, 1979).

Comparisons of pheromones from argasid ticks and ixodid ticks show that (1) feeding is required for release of the pheromones in ticks of both families (although assembly pheromones may be released by unfed argasid ticks); (2) sex attraction and host location and stimulation of attachment occurs in ixodids; no sex attraction occurs in argasids; (3) sources of the pheromones are unknown for argasids; in ixodids they are released from the foveolae dorsales and the genital aperture; (4) chemoreception by ticks is effected by the palps in soft ticks and by Haller's organ and sensilla of the cheliceral digits in hard ticks; (5) the chemistry of the pheromones differs in the soft ticks where low volatility, water-soluble, temperature-stable materials are found; in hard ticks the substances are phenolic in composition. Ixodid ticks often use a two-pheromone system: (1) a phenol, for example, 2,6-DCP, and (2) a genital sex pheromone or some other signal for specific recognition (Sonenshine, 1985).

An aggregation response of males of 10 of 11 species of soft ticks occurs within 1 h when exposed to pheromones by conspecific females. Female response to similar material from conspecific males occurs if the concentration of the material is at least twice that used for males, or if the period of assembly is extended for several hours.

Both male and female argasid ticks respond to pheromonal materials produced by the larvae and nymphs. The larvae and N1, N2 respond only to pheromones from adults and more to the male material than to that of the female.

Argasid ticks appear to detect pheromones more by means of palpal sensilla than by tarsal sensilla of the legs. When tarsi of legs I of *Ornithodoros moubata* are painted, the assembly response is reduced. Removal of the palps completely eliminates the assembly response in this species (Leahy, 1979).

When female Gulf Coast ticks (*Amblyomma maculatum*) were released on three separate areas of a bovine host (apart from a shoulder location) where extract from feeding males had been deposited, they migrated to the treated area and attached. The numbers of females that attached to the treated area increased proportionately as the distance from the treated area decreased.

When extracts from feeding males were mixed with a small amount of an insecticide (isoban) and applied on the bovine host, some of the females migrated to the treated area and were killed. Most females lured by the pheromones died without attaching. The Gulf Coast tick prefers the ear of the host as an attachment site. When one ear was treated with male extract and the other left untreated, over four times as many females had attached to the treated ear at 1 day after infestation (Gladney et al., 1974a).

Sexually active males of *D. andersoni* and *D. variabilis* respond to hexane or pentane extracts from partially fed females. Production of the sexual attractant is related to feeding and females of these species that have been attached less than 0.5 to 1.0 day fail to attract males. The production of the sex attractant in females diminishes after engorgement. However, even when the females were frozen

($-71\,^{\circ}$C) or chilled ($1\,^{\circ}$C) and then thawed, the attachment persisted. Males respond to feeding females even if separated from them by barriers and they exhibit orientation responses within 2 cm of the females. When females are killed the attractant diminishes rapidly at room temperature (Sonenshine et al., 1986).

In the Argasidae, sex pheromones evidently come from the coxal fluid (Schlein and Gunders, 1981). In ixodid ticks the foveal glands or sex pheromonal glands are identified as the source of pheromonal secretion. Phenolic compounds come from this set of multilobed glands beneath the foveae dorsales. These glands on the dorsum of the capitular base have as few as 15 and as many as 50 pores in each fovea. Pores, ducts, and subcuticular secretory lobes make up the structure of these glands. The production of 2,6-DCP is associated with this gland and the vapors of the pheromone are released through the tiny pore clusters (Chow, 1979; Soneshine et al., 1977). (Autoradiography indicates foveal glands have six times as much radioactive chlorine as normal.) When foveae of females are lacquered, no collection of pheromones is possible, but 5 h later pheromones can be detected around the margin of the glands (Sonenshine et al., 1977; Sonenshine, 1985).

The functional cycle of development, secretion, and degeneration of the foveae coincides with the metabolic activities in developing and reproducing stages. Primordia for these glands form in the nymphs and mature in adult females, where the pheromones occur during the maturation process (Sonenshine, 1985).

Evidently, the females begin synthesizing and storing the sex pheromone soon after they molt, but do not release the chemical until they have been stimulated by feeding. In *D. andersoni* the synthesis continues during the feeding (estimated approximately 22.2 ng/tick/day).

In *Amblyomma* the male ticks that gain attachment on a host seek to locate near other preattached males, evidently directed by male pheromones. Females, after gaining a host, move about at random and respond to pheromones secreted by these attached males by at-

taching to a site nearby. The assembly pheromones secreted by males of *Amblyomma* spp.
tend to regulate the behavior of the adult ticks
by attracting them to an attachment site and
thus increase the chances of contact with ticks
of the opposite sex.

Perception of the pheromones by species of
Ixodid ticks has been shown to be associated
with Haller's organ. *Amblyomma americanum*
has palpal reception for 2,6-DCP (Rechav and
Whitehead, 1978).

Stimulus for the release of pheromones is evidently related to feeding. Stretch receptors in
the gut elicit responses from neurosecretory
cells in the synganglion. These in turn stimulate the foveal glands to produce the pheromone. Females injected with nonnutrient solutions still produce pheromones that attract the
males (Sonenshine, 1985).

Pheromones are generally phenolic compounds. Berger et al. (1971) found 2,6-DCP in
Amblyomma americanum. Phenolic compounds have also been recovered from *Dermacentor andersoni*, *D. variabilis*, and *Amblyomma maculatum* (Gulf Coast tick). Paracresol
and salicylaldehyde are other compounds identified as pheromones in ticks. Both males and
females of *A. maculatum* have 2,6 dich-
lorophenol (Kellum and Berger, 1977). This is a
sex attractant pheromone in females; its function in males is unknown.

The quantity of the pheromone is detectable
in 2 days in females and the amount does not
change at feeding. In *Rhipicephalus appendiculatus* unfed females had no detectable amount
of pheromone, but after feeding, developed detectable amounts of phenol and paracresol, and
continued to produce pheromone after mating.
After molting, *D. variabilis* produced 9.2 ng per
female tick, with increases in amounts following feeding. The pattern of this production varies. In some ticks the unfed forms produced a
high level that is stabilized as to amount. In
other unfed ticks a high level was produced and
then declined. Feeding usually results in a high
level of production and then a stabilizing of the
amount produced.

Four genera of Ixodidae are known to have

pheromones (Sonenshine et al., 1979). Pheromones occur in *Amblyomma americanum*
(Berger et al., 1971). The pheromone 2,6-DCP
occurs in many species, e.g., *Dermacentor andersoni*, *D. variabilis*, *Amblyomma maculatum*
and *Rhipicephalus sanguineus* (Chow et al.,
1975; Sonenshine et al., 1981; Sonenshine,
1985). Paracresol is found in seven species of
ticks in three genera and salicylaldehyde is
identified from four species (Wood et al., 1975).

Up to 1979 all of the isolated phenolic compounds were from female ticks. The substances
appear to serve solely as sex pheromones. An
exception is found, however, in *Amblyomma
maculatum*, the Gulf Coast tick, where both
males and females contain substantial amounts
of 2,6-DCP (Kellum and Berger, 1977). This
substance has also been found to be distinctly
emitted by males of *A. maculatum* (W.J. Gladney, personal communication).

Spider mites exemplify a second group in
which pheromones are important, but they
have not been studied as extensively as the
ticks. Males of *Tetranychus urticae* and *T. cinnabarinus* are attracted strongly to quiescent
females destined to become sexually mature.
Males remain until the emergence of females,
then mating occurs. Thin-layer and gas–liquid
chromotography analyses show the sequiter-
pene alcohols farnesol and nerolidol in both species. Males are attracted by amounts of 10
ppm. There is an increase of population because
of the sex attractant; arrhenotokous females
that remain unmated produce haploid males
(Gerson, 1979). As a result of silk production in
Tetranychus cinnabarinus the web serves as a
sex pheromone carrier and is used by the males
to protect their claim on females from other
males.

Tetranychid behaviors resulting from pheromones (Cone et al., 1971a,b) include (1) hover-
ing over the pheric female, (2) guarding behavior against other males, and (3) attempted
mating with the telochrysalis or exuvium. Spider mites are located by their phytoseid predators through pheromones (Sabelis and Bann,
1984).

The normal life cycle of the two-spotted mite

(*Tetranychus urticae*) takes 5 to 7 days in warm temperature with high humidity. Since the mite has high resistance to acaricides, and fewer acaricides have been developed, it has been thought that pheromones could be used in the control of these mites (Cone, 1979).

In tetranychids the quiescent female, called a telochrysalis (?pheric female), is usually attended by males prior to emergence. Typically the male stands over the female or hovers in a guarding behavior with occasional attempts at mating with the quiescent form. Mating is usually accomplished upon emergence of the female. Males normally are not attracted to males. Occasionally a male is diverted, and the female emerges and mates with a nonguarding male. The original male returns and is occupied with the exuvium of the telochrysalis by which it originally stood guard. Evidently a sufficient amount of pheromone remains to attract such a guarding male.

The amount of pheromone perceived by tetranychid males is effective in phenomenally small amounts. It is perceived at 10–15 ng/deutonymph. Young males respond more readily than do older males. In deutonymphs a web aids the males in finding the females. Released pheromones act first as an attractant, then as a holdant for the males. The chemical can be detected for $2 \times$ the body length. A male makes a silk webbing over the female and rests with one appendage on the web, which enables him to detect the presence of another male or the eclosion of the female from the telochrysalis (Penman and Cone, 1974).

Production of pheromones in the tetranychids may occur from cuticular slits (lyriform fissures) in legs I, II of females (demonstrable with SEM and TEM). These fissures may be mechanoreceptors as well. Duplex setae on the legs of both males and females seem to negate the possibility of pheromone secretion from these setae.

Molting fluid has been suggested as an attractant because of the probing of the dorsum of a pheric female by the males. It is thought that the probing may be due to the fluid emitted from the telochrysalis (Penman and Cone, 1974).

Alarm pheromones are emitted by grain mites (Matsumoto et al., 1979). In 1906 Melander and Brues described defensive repellant substances as well as attractant, alluring substances in the two-spotted mite. Pheromones are used to control insect pests in various ways and have met with considerable success. The employment of pheromones as an adjunct in the control of ticks is documented (Ziv et al., 1981), but no commercial use has been developed, nor is it likely that such a strategy will be used in the near future. Use of assembly pheromones may be a means of tick control and some patents seem to be in process (Sonenshine, 1985).

REFERENCES

Axtell, R.C. and A. LeFurgey. 1979. Comparisons of the foveae dorsales in male and female ixodid ticks, *Amblyomma americanum, A. maculatum, Dermacentor andersoni* and *D. variabilis* (Acari: Ixodidae). *J. Med. Entomol.* 16:173–179.

Berger, R.S. 1972. 2,6-dichlorophenol, sex pheromone of the lone star tick. *Science* 177:704–705.

Berger, R.S., J.C. Dukes and Y.S. Chow. 1971. Demonstration of a sex pheromone in three species of hard ticks. *J. Med. Entomol.* 8:84–86.

Borror, D.J., D.M. Delong and C.A. Triplehorn. 1976. *An introduction to the Study of Insects.* Holt, Rinehart and Winston, New York.

Borror, D.J., D.M. Delong and C.A. Triplehorn. 1979. *An introduction to the Study of Insects.* Saunders, New York.

Chow, Y.S. 1979. Electro-olfactory potential of *Ixodes* ticks to 2,6-dichlorophenol. *Proc. Int. Congr. Acarol., 4th, 1974*, Vol. 1, pp. 501–505.

Chow, Y.S., C.B. Wang and L.C. Lin. 1975. Identification of a sex pheromone of the female brown dog tick, *Rhipicephalus sanguineus. Ann. Entomol. Soc. Amer.* 68:485–488.

Cone, W.W. 1979. Pheromones of Tetranychidae. *In:* Rodriguez, J.G., ed. *Recent Advances in Acarology.* Academic Press, New York, Vol. 2, pp. 309–317.

Cone, W.W., L.M. McDonough, J.C. Maitlen and Z. Burdajewic. 1971a. Pheromone studies of the two spotted spider mite. I. Evidence of a sex pheromone. *J. Econ. Entomol.* 64:355–358.

Cone, W.W., S. Predki and E. Klostermeyer. 1971b. Pheromone studies of the two spotted spider mite. II. Behavioral response of males to quiescent deutonymphs. *J. Econ. Entomol.* 64:379–382.

Davis, J.C. and J.H. Camin. 1977. Aggregation behavior in the martin mite, *Dermanyssus prognephilus* (Acari: Dermanyssidae). *J. Med. Entomol.* 14(3):373–378.

Dees, W.H., D.E. Sonenshine and E. Breidling. 1984a. Ecdysteroids in *Hyalomma dromedarii* and *Dermacentor variabilis* and their effects on sex pheromone activity. *Acarol., [Proc. Int. Congr. Acarol.], 6th, 1982,* Vol. 1, pp. 405–413.

Dees, W.H., D.E. Sonenshine and E. Breidling. 1984b. Ecdysteroids in the American dog tick, *Dermacentor variabilis* (Say), during different periods of tick development (Acari: Ixodidae). *J. Med. Entomol.* 21:514–523.

Dees, W.H., D.E. Sonenshine and E. Breidling. 1984c. Ecdysteroids in the camel tick, *Hyalomma dromedarii*, during different periods of adult tick development (Acari: Ixodidae). *J. Med. Entomol.* 21:514–523.

George, J. 1981. The influence of aggregation pheromones on the behavior of *Argas cooleyi* and *Ornithodoros concanensis* (Acari: Ixodoidea: Argasidae). *J. Med. Entomol.* 18:129–133.

Germond, J.E., P.A. Diehl and M. Orici. 1982. Correlations between integument structure and ecdysteroid titers in fifth-stage nymphs of the tick, *Ornithodoros moubata* (Murray, 1877: *sensu* Walton, 1962). *Gen. Comp. Endocrinol.* 46:255–266.

Gerson, U. 1979. Silk production in *Tetranychus* (Acari: Tetranychidae). *In:* Rodriguez, J.G., ed. *Recent Advances in Acarology.* Academic Press, New York, Vol. 2, pp. 177–187.

Gladney, W.J., R.R. Grabbe, S.E. Ernst, and D.D. Oehler. 1974a. The gulf coast tick: Evidence of a pheromone produced by males. *J. Med. Entomol.* 2(3):303–306.

Gladney, W.J., S.E. Ernst and R.R. Grabbe. 1974b. The aggregation response of the Gulf Coast Tick on cattle. *Ann. Entomol. Soc. Amer.* 67(5):750–752.

Gothe, R. and E. Burkhardt. 1981. Zur histologie der Fovealdrusen bei *Rhipicephalus evertsi evertsi* Neumann, 1897. *Ber. Muench. Tieraerztl. Wochenschr.* 94:492–496.

Haggart, D.A. and E.E. Davis. 1981. Neurons sensitive to 2,6-dichlorophenol on the tarsi of the tick, *Amblyomma american* (Acari: Ixodidae). *J. Med. Entomol.* 18:187–193.

Harris, K.F., ed. 1983. *Current Topics in Vector Research.* Vol. 1. Praeger, New York.

Harris, K.F., ed. 1984. *Current Topics in Vector Research.* Vol. 2. Praeger, New York.

Hess, E. and M. Vlimant. 1982b. The tarsal sensory system of *Amblyomma variegatum* Fabricus (Ixodidae: Metastriata). II. On pore sensilla. *Rev. Suisse Zool.* 90:157–161.

Hess, E. and M. Vlimant. 1982b. The tarsal sensory system of *Amblyomma variegatum* Fabricus (Ixodidae: Metastriata). II. No pore sensilla. *Rev. Suisse Zool.* 90:157–161.

Homsher, P.J. and D.E. Sonenshine. 1976. The effect of presence of females on spermatogenesis and mate-seeking behaviour in two species of *Dermacentor* ticks (Acaria: Ixodidae). *Acarologia* 18:226–233.

Hoy, M.A. and J.M. Smilanich. 1979. A sex pheromone produced by immature and adult females of the predatory mite, *Metaseilus occidentalus*, Acarina: Phytoseiidae. *Entomol. Exp. Appl.* 26:291–300.

Hughes, T.E. 1959. *Mites or the Acari.* Athlone, London.

Kellum, D. and R.S. Berger. 1977. Relationship of the occurrence and function of 2,6-dichlorophenol in two speices of *Amblyomma* (Acari: Ixodidae). *J. Med. Entomol.* 13(6):701–705.

Khalil, G.M., S.A. Nada and D.E. Sonenshine. 1981. Sex pheromone regulation of mating behavior in the camel tick, *Hyalomma dromedarii. J. Parasitol.* 67:70–76.

Khalil, G.M., D.E. Sonenshine, O. Sallam and P.J. Homsher. 1983a. Mating regulation and reprodductive isolation in the cammel ticks, *Hyalomma dromedarii* and *Hyalomma anatolicum excavatum* (Ixodoidea: Ixodidae). *J. Med. Entomol.* 20:136–145.

Khalil, G.M., D.E. Sonenshine, P.J. Homsher, W.H. Dees, K.A. Carson and V. Wang. 1983b. Development, ultra-structure and activity of the foveal glands and foveae dorsales of the camel tick, *Hyalomma dromedarii* (Acari: Ixodidae). *J. Med. Entomol.* 20:414–423.

Leahy, M.G. 1979. Pheromones of argasid ticks. *In:* Rodriguez, J.G., ed. *Recent Advances in Acarology.* Academic Press, New York, Vol. 2, pp. 299–308.

Leahy, M.G., R. Vandehey and R. Galun. 1973. Assembly pheromone(s) in the soft tick, *Argas persicus* (Oken). *Nature (London)* 246:515–516.

Leahy, M.G., G. Karuhize, C. Mango and R. Galun. 1975. An assembly pheromone and its perception in the tick *Ornithodoros moubata* Murray (Acari: Argasidae). *J. Med. Entomol.* 12:284–287.

Leahy, M.G., Z. Hajkova and J. Bouchalova. 1981. Two female pheromones in the metastriate tick, *Hyalomma dromedarii* (Acarina: Ixodidae). *Acta Entomol. Bohemosvlov.* 78:224–230.

Matsumoto, K., Y. Wada and M Okamoto. 1979. The alarm pheromone of grain mites and its antifungal effect. *In:* Rodriguez, J.G., ed. *Recent Advances in Acarology.* Academic Press, New York, Vol. 1, pp. 243–249.

Melander, A.L. and C. Brues. 1906. The chemical nature of some insect secretions. *Bull. Wis. Nat. Hist. Soc.* 4:22–36 (Milwaukee).

Otieno, D.A., Hassanali, A., Obenchain, F.D., Sternberg, A. and Galun, R. 1985. *Insect Sci. Appl.* 6:667–670.

Penman, D.R. and W.W. Cone. 1972. Behavior of male two-spotted spider mites in response to quiescent female deutonymphs and to web. *Ann. Entomol. Soc. Am.* 65:1289–1293.

Penman, D.R. and W.W. Cone. 1974. Structure of cuticular lyrifissures in *Tetranychus urticae. Ann. Entomol. Soc. Am.* 67(1):1–4.

Petney, T.N. and C.M. Bull. 1981. A non-specific aggregation pheromone in two Australian reptile ticks. *Anim. Behav.* 19:181–185.

Potter, D.A. 1979. Reproductive behavior and sexual selection in tetranychine mites. *In:* Rodriguez, J.G., ed. *Recent Advances in Acarology.* Academic Press, New York, Vol. 1, pp. 137–145.

Rechav, Y. 1978. Specificity in assembly pheromones of the tick *Amblyomma hebraeum* (Acarina: Ixodidae). *J. Med. Entomol.* 15:81–83.

Rechav, Y. and G.B. Whitehead. 1979. Male-produced pheromones of Ixodidae. *In:* Rodriguez,

J.G., ed. *Recent Advances in Acarology.* Academic Press, New York, Vol. 2, pp. 291–298.

Rechav, Y., G.B. Whitehead and M.M. Knight. 1976. Aggregation response of nymphs to pheromone(s) produced by males of the tick *Amblyomma hebraeum* Koch. *Nature (London)* 259:563–564.

Rechav, Y., H. Parolis, G.B. Whitehead and M.M. Knight. 1977. Evidence for an assembly pheromone(s) produced by males of the bont tick, *Amblyomma hebraeum* (Acarina: Ixodidae). *J. Med. Entomol.* 14:71–78.

Rechav, Y., R.A.I. Norval and J.H. Oliver. 1982. Interspecific mating of *Amblyomma hebraeum* and *Amblyomma variegatum* (Acari: Ixodidae). *J. Med. Entomol.* 19:139–142.

Rodriguez, J.G., M.F. Potts and C.G. Patterson. 1979. Allelochemic effects of some flavoring components on the acarid, *Tyroglyphus putrescentiae. In:* Rodriguez, J.G., ed. *Recent Advances in Acarology.* Academic Press, New York, Vol. 1, pp. 251–261.

Sabelis, M.W. and H.E. van de Bann. 1983. Location of distant spider mite colonies by phytoseiid predators: Demonstration of specific kairmones emitted by *Tetranychus urticae* and *Panonychus ulmi. Entomol. Exp. Appl.* 33:303–314.

Sabelis, M.W., B.P. Afman and P.J. Slim. 1984. Location of distant spider mite colonies by *Phytoseiulus persimilis:* Localization and extraction of a kairomone. *Acarol. [Proc. Int. Congr. Acarol.], 6th, 1982,* Vol. 1, pp. 431–440.

Schlein, Y. and A.E. Gunders. 1981. Pheromone of *Ornithodoros* spp. (Argasidae) in the coxal fluid of female ticks. *Parasitology* 83:467.

Silverstein, R.M. 1981. Pheromones: Background and potential for use in insect pest control. *Science* 213:1326–1332.

Silverstein, R.M., J.R. West, D.E. Sonenshine and G.M. Kahlil. 1984. Occurrence of 2,6-dichlorophenol in the hard ticks *Hyalomma dromedarii* and *Hyalomma anatolicum excavatum* and its role in mating. *J. Chem. Ecol.* 10(1):95–100.

Sonenshine, D.E. 1984a. Pheromones of Acari and their potential use in control strategies. *In:* Griffiths, D.A. and C.E. Bowman, eds., *Acarology VI.* Ellis Horwood, Chichester, England, pp. 100–108.

Sonenshine, D.E. 1984b. Tick Pheromones. *In:*

Harris, K.F., ed. *Current Topics in Vector Research.* Praeger, New York, pp. 225–263.

Sonenshine, D.E. 1985. Pheromones and semiochemicals of the Acari. *Annu. Rev. Entomol.* 30:1–28.

Sonenshine, D.E. 1986. Tick pheromones: an overview. *In:* Sauer, J.R. and J.A. Hair, eds. *Morphology, Physiology and Behavioral Biology of Ticks.* Ellis Horwood, Ltd., Chichester, England; Wiley (Halstead Press), New York, pp. 342–360.

Sonenshine, D.E., R.N. Silverstein, E.C. Plummer, J.R. West and T. McCullough. 1976. 2,6-dichlorophenol, the sex pheromone of the Rocky Mountain wood tick, *Dermacentor andersoni* Stiles, and the American dog tick, *Dermacentor variabilis* (Say). *J. Chem. Ecol.* 2:201–209.

Sonenshine, D.E., R.M. Silverstein, L.A. Collins, M. Saunders, C. Flynt and P.J. Homsher. 1977. The foveal gland: Source of sex pheromone in the ixodid tick, *Dermacentor andersoni* Stiles. *J. Chem. Ecol.* 3:697–706.

Sonenshine, D.E., R.N. Silverstein and P.J. Homsher. 1979. Female-produced pheromones of Ixodidae. *In:* Rodriguez, J.G., ed. *Recent Advances in Acarology.* Academic Press, New York, Vol. 2, pp. 281–290.

Sonenshine, D.E., D.M. Gainsburg, M.D. Rosenthal and R.M. Silverstein. 1981. The sex pheromone glands of *Dermacentor variabilis* (Say) and *Dermacentor andersoni* Stiles: Sex pheromone stored in neutral lipid. *J. Chem. Ecol.* 7:345–357.

Sonenshine, D.E., G.M. Kahlil, P.J. Homsher and S.N. Mason. 1982a. *Dermacentor variabilis* and *Dermacentor andersoni:* Genital sex pheromones. *Exp. Parasitol.* 54:317–330.

Sonenshine, D.E., R.M. Silverstein and Y.S. Rechav. 1982b. Tick pheromone mechanisms. *In:* Obenchain, F.D. and R. Galun, eds. *The Physiology of Ticks.* Pergamon, Oxford, pp. 439–468.

Sonenshine, D.E., G.M. Kahlil, P.J. Homsher, W.H. Dees, K.A. Carson and V. Wang. 1983. Development, ultra-structure and activity of the foveal glands and foveae dorsales of the camel tick, *Hyalomma dromedarii* (Acari: Ixodidae). *J. Med. Entomol.* 20:424–439.

Sonenshine, D.E., P.J. Homsher, K.A. Carson and V.G. Wang. 1984a. Evidence of the role of the cheliceral digits in the perception of genital sex pheromones during mating in the American dog tick, *Dermacentor variabilis* (Say) (Acari: Ixodidae). *J. Med. Entomol.* 21:296–306.

Sonenshine, D.E., R.M. Silverstein and J.R. West. 1984b. Occurrence of the sex attractant pheromone, 2,6-dichlorophenol, in relation to age and feeding in the American dog tick, *Dermacentor variabilis* (Say) (Acari: Ixodidae). *J. Chem. Ecol.* 10:95–100.

Treat, A. 1975. *Mites of Moths and Butterflies.* Comstock, Ithaca, New York.

Waladde, S.M. 1982. Tip-recording from ixodid tick olfactory sensilla: Responses to tick related odours. *J. Comp. Physiol.* 148:411–418.

Wood, W.F., M.G. Leahy, R. Galun, G.D. Prestwich, J. Meinwald, R. E. Purnell and R. C. Payne. 1975. Phenols as pheromones of ixodid ticks: A general phenomenon? *J. Chem. Ecol.* 1:501–509.

Ziv, M., D.E. Sonenshine, R.M. Silverstein, J.R. West and K.H. Gingher. 1981. Use of sex pheromone, 2,6-dichlorophenol, to disrupt mating by the American dog tick, *Dermacentor variabilis* (Say). *J. Chem. Ecol.* 7:829–840.

PART **III**

CLASSIFICATION

14

The Subclass Acari

Nomenclature being thought so difficult, its mastery has been the object of comparatively few.

—W. ARNOLD LEWIS, 1872

Nearly every phylum of animals has a class with numbers and diversity of species surpassing others in the phylum. For the Arthropoda, as now known, the class Insecta has this distinction. Similarly, within a class the members of a certain order will be more numerous and diverse than those of other orders. For mammals the Ungulata surpass all others in this respect. Within the Insecta the order Coleoptera exhibits the highest diversity: about one out of every five insects collected is a beetle. A similar condition exists in the class Arachnida where spiders and mites are the most diverse.

The classification of mites is an extensive subject of burgeoning growth. Because of the variety of mites and our limited knowledge, any study of this subject will be incomplete. Admittedly, the subject has not been completely mastered here, but from the basis of the classification included the reader may proceed through cited references to information applicable to different groups.

Regarding the classification of the mites, Krantz (1978) says:

Acarology . . . is in a state of systematic turmoil similar to that experienced in the field of entomology a century ago. A so-called "natural" classification for the Acari is impossible at our present level of understanding, nor may a phylogenetic system ever be realized in this fossil-poor group.

The study of acarology in its descriptive phases is expanding more rapidly than the descriptions can be assimilated or accommodated into the hierarchial classificatory system. Frequently the descriptions are not relegated to categories other than order or genus. In some instances familial relationships are not mentioned or identified, nor are comparisons made within the existing system; a monotypical new genus and new species may not have any other identified relationship than the subclass. The frequency with which new species are discovered and described, the relative infancy of acarology, and other factors make it difficult to keep abreast of the classificatory literature, but workers in the science persist. The broad taxonomic and phylogenetic relationships will be understood and the system of identifying and naming and categorizing the mites will be accomplished.

A simple incident may illustrate how changes have occurred. Some years ago when I became interested in classifying the oribatid mites I wrote to Dr. E.W. Baker and asked him who was the authority on oribatid taxonomy in the United States. I had hoped that there would be such a person of whom I could ask counsel and advice in my research. His reply was startling: "Having declared an interest in the group, you have become the authority in the United States." That situation has certainly changed!

The preface of the *International Code of Zoological Nomenclature* states:

Nomenclature does not determine the rank accorded to any group of animals, but it does provide the name that shall be applied to whatever rank any taxonomist may wish to assign to it.

The failure of the Code to deal with names of higher rank than superfamily or of lower rank than subspecies arises from no failure to recognize the necessity of such names. It exists because the practice of zoologists in regard to them is not sufficiently uniform to permit the formulation of rules covering them at this time.

The classification of mites is not free from the limitations and difficulties that arise from this situation.

Classification and taxonomy of mites are further complicated by the fact that the acarines appear to be at least diphyletic if not polyphyletic in origin; monophyly is also proposed (Lindquist, 1984). Classification is indicative of the diversity and differentiation of the animals and with its development come the valuable insights and understandings of the subclass.

Evolutionary derivation of the mites is often a matter of speculation and assumptions, which may affect taxonomic schemes.

Savory (1977) considers mites to be the real specialists among arachnids. He says of them:

They appear to be as rich in species as the spiders, but this inequality is annually disappearing as mites attract more and more attention; and like all specialists their successes are limited by the fact of specialization. Yet within these limits it must be admitted that they reign supreme.

In his earlier work Savory (1964) wrote that 6000 species of mites were known in 1939. Wharton (1964) estimated a total of 17 500 described species. Seven years later Krantz (1970) assumed 30 000 described species for the Acari. A more conservative estimate by Savory (1977) estimates 2000 genera of mites with 20 000 described species, but postulates 500 000 species in existence. It is affirmed by many acarologists that in the future mites will represent a greater number of species than any other subclass or order of Arachnida and possibly of the Arthropoda; they may even rival insects for numbers of species. With such diversity and extensive populations of species among mites, classification becomes a vital and challenging aspect of acarology.

There is always an almost inexpressible excitement and satisfaction in finding and describing a species new to science. The ornithologist or mammalogist would be ecstatic over such a find; the entomologist would be elated. But the occurrence is almost commonplace with acarologists, for new species are discovered almost daily, and new genera and species are described in multiples annually. In fact, at a summer institute a short time ago several new species were offered for description to participants willing to describe them, but there were no takers; specialists had more new species to describe than they could handle at the time.

Descriptive taxonomy of mites is a new frontier and a research specialty where any interested and determined pioneers may carve their way into a relatively little-known wilderness with resultant satisfaction and accomplishment.

HISTORICAL REVIEW

The brief notations regarding classification that follow are intended as a synopsis of names and categories from the past that may be found in the literature. Usage of these names in current systems will have changed, but their identity will give the reader some basic background in the classification system of mites.

In 1806 four families were included in Latreille's system, each with a mixture of different mites, and excluding ticks which are represented in his single family Acaridae, now a single family in the Astigmata (=Acaridida).

In 1905 Banks employed a relatively simple system for his classification of the order Acari, which was used for nearly two decades by zoologists and entomologists and included eight superfamilies.

Oudeman's unique system of 1906 stemmed from his use of the respiratory structures (spiracles, spiracular plates) to divide the class into eight categories, each ending in "-stigmata." In 1923 he changed the Acari from class

to order with six suborders.

Reuter's classification (1909) interpolates the higher categories of suborder and superfamily into a scheme with four major suborders: Gamasiformes, Eriophyiformes, Trombidiformes, and Sarcopiformes.

Count Vitzthum (1931) added the additional category, the supercohort, into the system below the level of suborder.

Suborder Notostigmata
Suborder Holothyroidea
Suborder Parasitiformes
    Supercohort Mesostigmata
    Supercohort Ixodides
Suborder Trombidiformes
    Supercohort Tarsonemini
    Supercohort Prostigmata
Suborder Sarcopiformes
    Supercohort Acaridiae
    Supercohort Oribatei

It should be noted, however, that the taxonomically conventional and generally accepted placement of the cohort is at a different level than the location made by Vitzthum. Mayr (1969) provides the conventional arrangement:

Conventional	Not	Nor
Class		
Subclass		
SUPERCOHORT		
COHORT		
SUBCOHORT		
Superorder		
Order		
Suborder	Suborder	Suborder
Superfamily (-oidea)	COHORT	SUPERCOHORT
Family (-idae)		COHORT
Subfamily (-inae)		SUBCOHORT
Tribe (-ini)		Superfamily
Genus		
Subgenus		
Species		
Subspecies		

Marc Andre's classification in the *Traité de Zoologie* (1949) includes six suborders with variations in name endings, including the employment of superfamily names at the suborder level.

Baker and Wharton (1952) include five suborders of the Acarina: Onychopalpida, Mesotigmata, Ixodides, Trombidiformes, and Sarcoptiformes. This system, along with intermediate names, genera, and families, makes *An Introduction to Acarology* a hallmark in mite classification.

Robertson (1959) recommends the elevation of the Acarina to subclass rank to overcome some of the difficulties resulting from the use of the cohort category at an unconventional level. Her work relates to *Tyrophagus* in the suborder Sarcoptiformes, and she describes the division of that suborder into supercohorts and cohorts by Zakhvatkin (1941, 1952), Yunker (1955), Hughes (1948), and Turk (1953), which preserves the basic concepts of Vitzthum. Robertson describes two alternatives for this dilemma: (1) retaining Acarina as a suborder and subdividing only at the superfamily level, or (2) elevating Acarina to a subclass or superorder and the Sarcoptiformes to an order.

Grandjean (1935) established two groups based on the types of refractive or nonrefractive setae when observed under polarized light. He proposed the names Anactinochitinosi for those without the polarizing property (actinochitin) in the setae and Actinochitinosi for those mites with this refractive property.

Evans et al. (1961) use the level of subclass for the Acari and couple this with the two major descriptive categories used by Grandjean (1935). They establish consistent endings for the categories using the "-stigmata" of Oudemans.

Johnston (1965) combines parts of other arrangements into three major orders: Parasitiformes, Opilioacariformes, and Acariformes. In a subsequent work (personal communication) he employs an arrangement used in the Institute of Acarology with the inclusion of the taxonomic synonyms of the subclass Acari.

Hammen (1961, 1968, 1973) follows the major dichotomy of Grandjean (1935) and arranges the subclass Acarida into superorders. He gives strict attention to the standardization of the name endings and to the phylogeny, but ignores the conventional placement of the cohort. In a later publication Hammen (1977) proposes a new classification for the Chelicerata based on leg articulation, cheliceral types, and other features. He also realigns arachnid fossil groups and their relationships. The resolution of such broad aspects of classification is beyond the purview of this writing.

Krantz (1970, 1978) employs a system of classification similar to Johnston's with slightly different arrangements in the later writing. Savory's (1977) most recent revision of Arachnida compares Krantz' (1970) system with the general system of Grandjean (1935) and references the system used by Evans et al. (1961). Savory's application of superorders for Grandjean's terms is the primary difference.

It appears that justification for the acceptance of the Acari as a subclass is particularly well-founded because of the morphological tagma—the gnathosoma. The Acari are the only arachnids with this body region. A second characteristic is shared with the Ricinulei: both mites and ricinuleids have hexapod larvae. These two groups are the only arachnids in which this condition exists; from the evidence of embryogeny all other arachnids have octopod juveniles. A third characteristic that supports the subclass status relates to feeding habits. There are some direct and simple similarities since many mites feed like other fluid-ingesting arachnids. The one type of feeding mechanism that separates the acarines from other arachnids, however, is the phytophagous habit of many trombidiform or actinedid mites.

Taken singly these features may not be considered by some as sufficient evidence for subclass status, except for the gnathosoma. Collectively, they reaffirm the basis for the designation of Subclass Acari.

While classification of the Acari remains somewhat problematical in many instances, the brief synoptic arrangement that follows is the one used in this text. It is a combination of the more modern terminology coupled with a traditional use of the taxonomic categories. The placement of cohort and its subordinates coincides with the conventional position within the taxonomic hierarchy and, I believe, simplifies the classification for greater utility. The descriptive terms Anactinochitinosi and Actinochitinosi initiated by Grandjean and modified by other authors (Evans et al., 1961; Hammen, 1973, 1979; Johnston, 1965, 1982; etc.) are changed to Anactinotrichida and Actinotrichida and are placed coincidentally with, but subordinate to the cohorts involved. They serve as terms of convenience and differentiation only, not as superorders (cf. Hammen, 1977).

A key to the orders of the subclass Acari will assist the reader in identifying and relating the groups of mites in the classificatory scheme that follows.

KEY TO THE ORDERS OF ACARI

After (Evans et al., 1961; Hammen, 1968, 1973, 1977; Krantz, 1978; Lindquist, 1984; OConnor, 1984).

1. Pedipalpal tarsus with a pair of terminal claws; (Fig. 5-1A) opisthosoma with four pairs of dorsolateral stigmata; peritremes absent; transverse sutures divide opisthosoma into 12 "segments"; lyrifissures present; propodosoma with two or three pairs of ocelli; one or two pairs of rutella on venter of gnathosoma and With's organ; more than seven pairs of hypostomal setae present; tritosternal base divided (Fig. 14-1A, B) . . . . Order Opilioacarida.
—Pedipalpal tarsus without terminal claws; apotele absent or in the form of a tined seta on distal edge of tarsus (Figs. 14-2 and 14-3) . . . . 2.

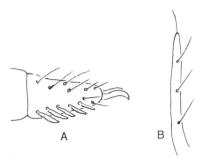

Figure 14-1. (A) Terminal claws on palps; (B) tritosternum (sternapophysis) of Holothyrida.

Figure 14-2. Tined seta on palp tarsus (Holothyrida).

Figure 14-3. Tined seta on palp tarsus (Gamasida).

2 (1). Hypostome armed with recurved (retrorse) teeth (Fig. 14-4); idiosoma with a pair of stigmata behind coxae IV or dorsolateral to coxae II - III, each stigmata usually in a stigmatal plate (Figs. 14-5, 14-6), peritremes never elongated; tarsus I with a dorsal, pitlike sensory organ (Haller's organ) (Fig. 14-7) . . . . (Metastigmata) . . . . Order Ixodida.

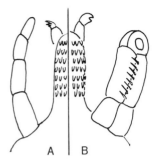

Figure 14-4. Hyptostome with retrose teeth: (A) Argasidae; (B) Ixodidae.

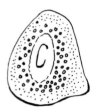

Figure 14-5. Stigmatal plate, hard ticks.

Figure 14-6. Stigmatal plate, soft ticks.

Figure 14-7. Haller's organ, tarsus I, Ixodida.

—Hypostome without recurved teeth; stigmata, if dorsolateral to coxae II - III, with elongated peritremes; sensory pit organs on tarsus I absent or simple (Fig. 14-8)........................3.

3 (2). Stigmata dorsolateral to coxae II–IV, usually with elongated peritremes (Figs. 14-9, 14-10, 14-11); a tined seta or apotele at inner basal margin of peidipalpal tarsus (Fig. 14-2) (may be absent in endoparasitic Gamasida); tritosternum usually present..........................4.

—Stigmata not dorsolateral to coxae, associated with gnathosoma, chelicerae, acetabula, or legs or absent; pedipalpal apotele, tritosternum absent.........5.

4 (3). Hypostome (venter of gnathsoma) with four pairs of setae (Fig. 14-12); tectum (epistome) and tritosternum usually present, the latter with lacinia or laci-

Figure 14-8. Haller's organ, tarsus I, Holothyrida.

Figure 14-9. Variation of stigmata and peritreme (Gamasida).

Figure 14-10. Variation of stigmata and peritreme (Gamasida).

Figure 14-11. Variation of stigmata and peritreme (Gamasida).

Figure 14-12. Hypostomal setae, Gamasida.

niae; anal covers with setae or naked....
....(Mesostigmata)....Order Gamasida.
—Hypostome with more than four pairs of setae (Fig. 14-13); tectum and tritosternum reduced or absent, laciniae weak if present; anal covers with many setae; tarsus I may have a dorsal sensory pit (Fig. 14-8); a pair of stigmata laterad of coxae I–III, with secondary stigmatal pores lateroposteriorly on dorsum; with a 2- or 3-tined subterminal pedipalpal claw; labrum radulalike; lyrifissures present on dorsum; ocelli absent................
.... Order Holothyrida.

5 (3). Pedipalps small, two-segmented and usually adpressed to sides of the infracapitulum (Figs. 14-14, 14-15); stigmata and tracheae absent; chelicerae invariably chelate or modified chelate; idiosoma never covered by overlapping sclerites and never vermiform; trichobothria ab-

Figure 14-13. Hypostomal setae, Holothyrida.

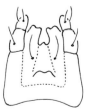

Figure 14-14. Infracapitulum, Astigmata, ventral.

Figure 14-15. Infracapitulum, Astigmata, lateral.

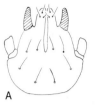 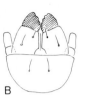 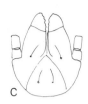

Figure 14-17. Infracapitulum in Oribatida showing rutellum (shaded): (A) anarthry; (B) diarthry; (C) stenarthry.

sent; ambulacrum of legs comprising a median claw, usually with a prominent pretarsus or with a membranous pad, or stalked suckerlike organ (Fig. 14-16A–D)......(Sarcoptiformes, Acaridei)......
.... Order Astigmata.
—Pedipalps usually conspicuous, 3- to 5-segmented, if minute, with fewer segments, then idiosoma is vermiform or has overlapping sclerites; a respiratory system usually present, stigmata associated with gnathosoma or with acetabula of legs; chelicerae chelate or variously modified into styliform or hooklike organs; idiosoma often with trichobothria; ambulacra of legs various, but not as above
..................................6.
6 (5). Gnathosoma with conspicuous rutella (Fig. 14-17A–C); chelicerae usually chelate–dentate; a pair of propodosomal trichobothria (Fig. 14-18) (pseudostigmata)

Figure 14-18. Trichobothrium (Oribatida).

nearly always present with a sensillum (pseudostigmatic organ) (piliform, setiform, clavate, barbed) arising from conical depression (bothridium) (Figs. 4-8; 21-1); pedipalps simple, 3- to 5-segmented, distal apotele absent, without a distal tibial spur; tracheal system when present, opening exteriorly in acetabulae of legs I and II, or as brachytracheae connecting to legs I and III or to bothridia; adults often heavily sclerotized, with ridgelike, platelike, or winglike expansions of propodosoma and hysterosoma (Fig. 14-19)..........(Sarcoptiformes, Oribatei; Cryptostigmata; Oribatoidea)
.... Order Oribatida.
—Gnathosoma rarely with rutella; chelicerae rarely chelate–dentate; propodosomal trichobothria, when present, usually without conspicous bothridia (pseudostigmata); pedipalps varied, simple to

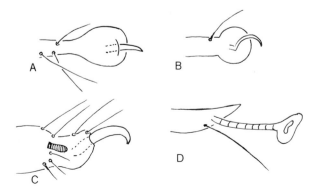

Figure 14-16. Ambulacra of Astigmata: (A–C) with membranous pad and median claw; (D) stalked claw.

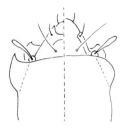

Figure 14-19. Pteromorphs, winglike expansions in Oribatida.

clawlike, raptorial, some with tibial spur and simple tarsus (thumb–claw arrangement) (Fig. 14-20); tracheae, when present, opening into paired stigmata associated with base of gnathosoma, chelicerae, or propodosoma, peritremes varied; immatures and adults usually weakly sclerotized, integument varies from soft, wrinkled ("fingerprint") surface to sclerotized and ornate structures; ridgelike, or winglike expansions of idiosoma absent; (Trombidiformes; Prostigmata)................Order Actinedida.

Regardless of what appears in print, there will be those who may not accept a particular system of classification or part of the arrangement as taxonomically presented. The system that follows is perhaps a compromise between several that are available. It is presented with the desire that it may form a basis for uniform-

Figure 14-20. Palpal thumb–claw complex (Actinedida).

ity. That ideal may not be possible given the complex parameters of this topic and the state of classification in acarology. If nothing else, it is hoped that we can revise the system to purge the erroneous location of the category "cohort" as it has been used in the past and employ this and some other categories in their conventional taxonomic locations in the system.

It is not the intention of this book to duplicate information that is available in such excellent references as Krantz (1978), *A Manual of Acarology;* McDaniel (1979), *How to Know the Mites and the Ticks;* or the acarological part of Parker (1982), *Synopsis and Classification of Living Organisms.* This text is envisioned as a companion volume that will elaborate on information on the anatomy, physiology, and biology of the Acari found in these other volumes. References located at the end of this chapter other than those cited will extend information on classification for the reader. In many instances a family name will follow the citation (e.g., SEJIDAE).

Because acarology is a young science, the classification of mites is in a state of change. The major orders have a somewhat stable classification and many genera are well-established; however, opinions differ as to the higher categories and some of the subordinate names. The material in the chapters that follow is not intended to be complete, or to detail the familial and generic names as in Baker and Wharton (1952). The arrangement is more of a telegraphic approach, to identify the categories within an order, summarize some of the distinguishing characteristics, and exemplify selected families or other categories.

What follows is a comparison of two cohorts and a listing of seven orders of the Acari. Table 14-1, which contrasts the cohorts, is included initially (after Woodring and Galbraith, 1976). The orders are then characterized in sequence and subordinal categories are identified. Superfamilies and families are listed and some illustrative examples are given. It should be noted, however, that the names of superfamilies are not static, especially for the water mites and

TABLE 14-1 Morphological Comparison of Cohorts[a]

Characteristic	Cohort Parasitiformes	Cohort Acariformes
1. Setae	Not birefringent	Birefringent
2. Tagmata	Gnathosoma, idiosoma tritosternum present	Gnathosoma, propodosoma–hysterosoma (and variations) tritosternum absent
3. Chelicerae and palps	Variable reduction of segments; modifications slight; chelicerae with lyrifissures; palps with an apotele	Variable reduction of segments; modifications extensive; chelicerae without lyrifissures; palps without apotele
4. Legs	Coxae free, trochanters rotatable	Coxae fused to venter; no rotatable segments
5. Muscles	Striated	Striated (exception in Eriophyidae)
6. Digestive system	Ventriculus small, caeca elongated, narrow	Ventriculus large, caeca short, wide
7. Respiratory system	Peritremes variable; stigmata usually lateral, tracheae idiosomal	Peritremes in Actinedida only; stigmata variable or absent; tracheae varied in location
8. Circulatory system	Heart present in some; hemocytes present; periganglionic blood sinus present in Ixodida	?Heart absent in most; hemocytes present; periganglionic blood sinus absent
9. Excretory system	Excretory tubules well-developed	Excretory tubules rudimentary or lacking
10. Nervous system	Synganglion; sensory structures varied; trichobothria absent	Synganglion; sensory structures varied; trichobothria present in many
11. Reproductive system		
Gonopore	Midbody or anterior	Midbody or posterior
Male	Synchronous meiosis and indirect sperm transfer; aedeagus reduced or absent	Synchronous meiosis in Oribatids only; sperm transfer direct or indirect (spermatophores); aedeagus present in some
Female	Stalked ova common; ovipositor in few	Stalked ova rare; ovipositor in many
12. Glands		
Coxal glands	Form variable; reduced or absent	Well-formed with sacculus and utriculus usually
Podocephalic glands	Absent; sternal taenidia present	Complex; sternal taenidia absent

[a]Modified after Woodring and Galbraith, 1976).

TABLE 14-2 The Cohorts and Orders of the Subclass Acari
Subclass ACARI Leach, 1817
(= Monomerosomata Leach, 1815; Acarina Nitsch, 1818; Acaromorpha Dubinin, 1956;
Acarida Hammen, 1972, 1973)

I. COHORT PARASITIFORMES Reuter, 1909

Anactinotrichida Zakhvatkin, 1952

(Gamasiformes Reuter, 1909; Peritremata Ewing, 1909; Anactinochitinosi Grandjean, 1935; Anactino-chaeta Evans et al., 1961; Anactinotrichida Zakhvatkin, 1952; Hammen, 1960)

ORDER Opilioacarida

(Opilioacarina With, 1902; Eucarina With, 1903; Notostigmata With, 1903–04; Onychopalpida Wharton, 1947; Opilioacariformes Johnston, 1968)

ORDER Holothyrida Thon, 1909

(Holothyrina Thon, 1905; Holothyroidea Reuter, 1910; Tetrastigmata Evans et al., 1961)

ORDER Gamasida Hammen, 1968

(Gamasides, Leach, 1814; Mesostigmata G. Canestrini, 1891; Distigmata Oudemans, 1906)

ORDER Ixodida Leach, 1815

(Ixodides Leach, 1815; Metastigmata G. Canestrini, 1891; Evans et al., 1961; Ixodoidea Reuter, 1909)

II. COHORT ACARIFORMES Zakhvatkin, 1952

Actinotrichida Hammen, 1972

(Trombidi–Sarcoptiformes Oudemans, 1931; Actinochitinosi Grandjean, 1935; Actinotrichida Zachvatkin, 1952; Hammen, 1960; Actinochaeta Evans et al., 1961)

ORDER Actinedida Hammen, 1968

(Prostigmata Kramer, 1877; Evans et al., 1961; Trombidiformes Reuter, 1909)

ORDER Astigmata G. Canestrini, 1891

(Acaridei Latreille, 1802; Sarcoptides C.L. Koch, 1842; Atracheata Kramer, 1877; Astigmata G. Canestrini, 1891; Sarcoptoidea Reuter, 1909; Sarcoptiformes Reuter, 1909 (part))

ORDER Oribatida Hammen, 1968

(Oribatides Duges, 1839 (and Auct); Carabodides C.L. Koch, 1842; Cryptostigmata G. Canestrini, 1891; Octostigmata Oudemans, 1906; Oribatoidea Reuter, 1909 (and Auct); Sarcoptiformes Reuter, 1909 (part); Stegasima Grandjean, 1932; and including Palaeacariformes Tragardh, 1932; Astegasima Grandjean, 1932; Palaeosomata Grandjean, 1969)

oribatids. Because of the enormous volume of taxonomic literature in any one or all of the orders, this section of the book cannot be as complete as one would like. Selected references on the various families will lead the reader to accounts in the literature of genera and species, morphological details, biology, and bibliogra-phies that can provide additional information.

The summary of the major categories of the Acari is shown in Table 14-2. Descriptive characteristics and subordinal categories will be discussed in detail in each of the chapters that follow.

REFERENCES[1]

Andre, M. 1949. Acari. *In:* Grassé, P.-P., ed. *Traité de Zoologie.* Masson, Paris.

Athias-Henriot, C. 1973. Idiosomal sigilotaxy of gamasids. Preliminary observations towards the knowledge of morphology of these Arachnida. *Proc. Int. Congr. Acarol., 3rd, 1971,* pp. 257–261.

Baker, E.W. and G.W. Wharton. 1952. *An Introduction to Acarology.* Macmillan, New York.

Baker, E.W., J.H. Camin, F. Cunliffe, T.A. Woolley and C.E. Yunker. 1958. *Guide to the Families of Mites.* Inst. Acarol., University of Maryland, College Park.

Balogh, J. 1972. *The Oribatid Genera of the World.* Akadémiai Kiadó, Budapest.

Bock, W.J. 1982. Biological classification. *In:* Parker, S., ed. *Synopsis and Classification of Living Things.* McGraw-Hill, New York, pp. 1067–1071.

Bregetova, N.G. 1973. Some considerations on the system and phylogeny of gamasid mites. *Proc. Int. Congr. Acarol., 3rd, 1971,* pp. 263–267.

Coineau, Y. 1973. Ontophylogenetic considerations on the chaetotaxy of Caeculidae (Acari–Prostigmata). *Proc. Int. Congr. Acarol., 3rd, 1971,* pp. 269–273.

Corliss, J.O. 1982. The history and role of nomenclature in the taxonomy and classification of organisms. *In:* Parker, S., ed. *Synopsis and Classification of Living Organisms.* McGraw-Hill, New York, pp. 1065–1066.

Evans, G.O., J.G. Sheals and D. Macfarlane. 1961. *Terrestrial Acari of the British Isles.* Adlard & Son, Bartholomew Press, Dorking.

Gilyarov, M.S. and N.G. Bregetova, eds. 1977. *A Key to the Soil-inhabiting Mites* (in Russian). Nauka, Leningrad.

Grandjean, F. 1935. Observations sur les Acariens (1e série). *Bull. Mus. Hist. Nat.* (2) 7:119–126.

Hammen, L. van der. 1961. Descriptions of *Holothyrus grandjeani* nov. spec., and notes on the classification of the mites. *Nova Guinea* 10(9):173–194.

Hammen, L. van der. 1968. Introduction générale à la classification, la terminologie, morphologique, l'ontogénèse et l'évolution des Acariens. *Acarologia* 10(3):401–412.

Hammen, L. van der. 1972. A revised classification of the mites (Arachnidea, Acarida) with diagnoses, a key and notes on phylogeny. *Zool. Meded.* 47(22):273–292.

Hammen, L. van der. 1973. Classification and phylogeny of mites. *Proc. Int. Congr. Acarol., 3rd, 1971,* pp. 275–282.

Hammen, L. van der. 1977. A new classification of Chelicerata. *Zool. Meded.* 51(20):307–319.

Hennig, W. 1950. *Grundzuge einer Theorie der Phylogenetischen Systematik.* Deutscher Zentralverlag, Berlin.

Hennig, W. 1965. Phylogenetic systematics. *Annu. Rev. Entomol.* 10:97–116.

Hennig, W. 1966. *Phylogenetic Systematics.* Univ. of Illinois Press, Urbana.

Hennig, W. 1975. "Cladistic analysis or Cladistic classification?" Reply to Ernst Mayr. *Syst. Zool.* 24:244–256.

Hughes, A.M. 1948. The mites associated with stored food products. Minist. Agric. Fish. London, pp. 1–168.

Johnston, D.E. 1965. *An Atlas of Acari. I. The Families of Parasitiformes and Opilioacariformes.* Acarol. Lab., Ohio State University, Acarol. Publ., Columbus.

Johnston, D.E. 1982. ACARI: Opilioacariformes, Parasitiformes, Mesostigmata, Ixodida. *In:* Parker, S., ed. *Synopsis and Classification of Living Things.* McGraw-Hill, New York, pp. 111–117.

Krantz, G.W. 1970. *A Manual of Acarology.* Oregon State Univ. Book Stores, Corvallis.

Krantz, G.W. 1978. *A Manual of Acarology.* Oregon State Univ. Book Stores, Corvallis.

Lindquist, E.E. 1984. Current theories on the evolution of major groups of acari and on their relationships with other groups of Arachnida, with consequent implications for their classification. *Acarol.*

[1]The reference listed at the ends of the chapters on the orders in Part III include citations found in the text as well as general references that will lead the reader to more specific information about the families, genera and species, biology, etc. Families are identified parenthetically at the ends of references (e.g., Varroidae) where the title may not include the name. Extensive literature on each of the orders and the limited space in this volume precluded extensive referencing of families.

[Proc. Int. Congr. Acarol.], 6th, 1982, Vol. 1, pp. 18–62.

Mayr, E. 1969. *Principles of Systematic Zoology.* McGraw-Hill, New York, pp. 1–428.

McDaniel, B. 1979. *How to Know the Mites and Ticks.* Brown, Dubuque, Iowa.

OConnor, B.M. 1984. Phylogenetic relationships among taxa in the Acariformes with particular reference to the Astigmata. *Acarol. [Proc. Int. Congr. Acarol.],* 6th, 1982, Vol. 1, pp. 19–27.

Parker, S.P. ed. 1982. *Synopsis and Classification of Living Organisms.* McGraw-Hill, New York.

Reuter, E. 1909. Zur Morphologie und Ontogenie der Acariden mit besonderer Berücksichtung von *Pediculopsis graminum. Helsingfors Acta Soc. Sci. Fenn.* 36(4):1–288.

Robertson, P.L. 1959. A revision of the genus *Tyrophagus* with a discussion of its taxonomic position in the Acarina. *Aus. J. Zool.* 7(2):146–181.

Savory, T.H. 1964. *Arachnida.* Academic Press, London, pp. 1–291.

Savory, T. 1977. *Arachnida,* 2nd ed. Academic Press, New York.

Stoll, N.R. ed. 1961. *International Code of Zoological Nomenclature.* Int. Trust Zool. Nomenclature, London.

Treat, A. 1975. *Mites of Moths and Butterflies.* Cornell Univ. Press, Ithaca, New York.

Turk, F.A. 1953. A synonymic catalogue of British Acari. Parts I & II. *Ann. Mag. Nat. Hist.* (12)6:1–26; 81–99.

Vitzthum, H.G. 1931. Acari. *In:* Kükenthal u. Krumbach *Handbuch der Zoologie,* Vol. 3(2). Berlin & Leipzig, pp. 1–112.

Vitzthum, H.G. 1940–43. Acarina. *In:* Bronn's *Klassen und Ornungen des Tierreiches,* Band 5, Abt. 4, Buch 5. Leipzig, pp. 1–1011.

Wharton, G.W. 1964. (Keynote address, Proc. Intl. Congr. Acarology 1st, 1963) *Acarologia* 6(fasc. hors série):37–43.

Wiley, E.O. 1981. *Phylogenetics: The Theory and Practice of Phylogenetic Systematics.* Wiley, New York.

Yunker, C.E. 1955. A proposed classification of the Acaridiae (Acarina: Sarcoptiformes). *Proc. Helminth. Soc. Wash.* 22:98–105.

Zakhvatkin, A.A. 1941. Arachnoidea. Acariens, Tyroglyphoides. Faune de l'U.R.S.S. *Inst. Zool. Acad. Sci.* 6(1):1–475 (in Russian with keys in French).

Zakhvatkin, A.A. 1952. The division of the Acarina into orders and their position in the system of the Chelicerata. *Mag. Parazit. (Moscow)* 14:5–46 (in Russian).

COHORT PARASITIFORMES

15

Order Opilioacarida

(Anactinotrichida Zakhvatkin 1952) (Notostigmata With 1903–04; Eucarina With 1903; Onychopalpida Wharton 1947; Opilioacariformes Johnston 1968) Family Opilioacaridae With 1902 (Neoacaridae Chamberlin and Mulaik 1942).

These purplish, violet-blue or greenish-blue mites represent the most primitive of the acarines and have the greatest number of primitive characters among the anactinotrichid mites. The podocephalic canal, so characteristic of actinotrichid forms, is absent. Coxal glands are present, however, and empty into sternal taenidia associated with the subcapitular groove (Hammen, 1968a,b, 1970). Opilioacarida are called "synthetic acarines" because of the combinations of characters that bridge both anactinotrichid and actinotrichid forms (Grandjean, 1936).

Opilioacarids are found in the warmer, semi-arid parts of the world—the southwestern United States, Puerto Rico, South America, the Mediterannean area of southern Europe, West Africa, South Africa, Madagascar, and central Asia. Hammen (1970) indicates they are gonwandian in location.

Purplish opilioacarids were collected on the island of Viegas, south of Puerto Rico. These mites clung to the under surface of rocks, as many as eight beneath each boulder, protected from the south wind and the sun most of the day (T. Tibbetts, personal communication, 1952). In Madagascar groups of 6 to 12 mites were collected by Legendre under rocks in eucalyptus forests near rice fields (Hammen, 1977).

These are large mites (1.5–2.5 mm) that superficially resemble unspecialized arachnids of the opilionid group Cyphophthalmi. Generally regarded as predaceous in habit, the opilioacarids may also be palenophagous, as pollen grains have been found in the gut of some. Chitinous remains in the gut of others also suggests they are carnivorous (Baker and Wharton, 1952). Some are found in association with isopods, spiders, and Japygida (Hammen, 1977).

Opiliocarids are elongated or oval, relatively unsclerotized, with a soft, slightly leathery and yellowish cuticle. The pigment granules of the cuticle are frequently in bands (Hammen, 1977). A distinct rostral lobe and a disjugal furrow are present. No forward-projecting epistome is present. Two (or three) pairs of dorso-lateral ocelli adorn the propodosoma, along with broad setae. Three pairs of ocelli are considered primitive (Fig. 15-2F) (Hammen, 1970). No tergites, sternites, or pleurites are present. On the hysterosoma weak, indistinct transverse sutures exhibit elliptical or rounded scars of muscle attachments and delimit 12 segments posterior to legs IV and the genital opening. Hammen (1969) identifies 19 segments. Between the sutures are rows of numerous

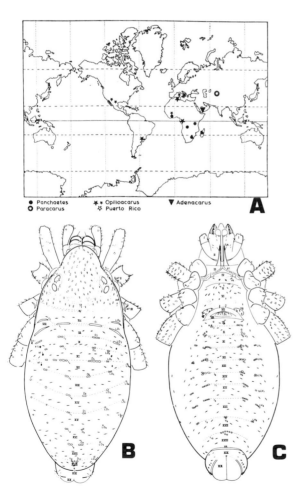

Figure 15-1. (A) Distribution of genera in the world (after Hammen, 1970); (B) dorsal view of *Opilioacarus texanus;* (C) ventral view (after Hammen, 1966); with permission.

lyriform pores (lyrifissures) (Fig. 15-1B). Dorsal setae are short and somewhat spatulate or flattened. Primitively these setae are found from segments VII to XVII; in some species the setae are absent—a derived character according to Hammen (1970). Peritremes are absent, but four pairs of stigmata placed in a crescentic arrangement are visible dorsolaterally. Specific placement of the stigmata varies with genera and species. In *Panchetes* the larva is without stigmata; two pairs are present in the protonymph, three pairs in the deutonymph, and four pairs in the tritonymph and adult (Hammen, 1969).

The gnathosoma has several pairs of ventral setae and shows a pair of hypertrophied paralabial setae (With's organs) laterad of the setate hypostome (Fig. 15-2A–D). These are postulated to have been forked in the primitive condition (Hammen, 1970) and are furcate in *Paracarus* (Hammen, 1968b). A pair of sclerotized rutella is also present, which may be used as food handlers or perhaps in cleaning the chelicerae. A single pair of rutellar setae is considered primitive (Hammen, 1970). The labrum is a radulalike organ with tiny retrorse teeth on its surface. The chelicerae are three-segmented and chelate. The fixed digit has three to four setae and two lyrifissures on segment II. The pedipalps are five-segmented with free coxae. Distally, the palpi exhibit a clawed apotele. These claws perhaps enable the palps to be used as grasping organs. (Such apoteles in the more advanced Parasitiformes mites are subterminal in position (Krantz, 1978).) Posterior to the palpal coxae on the venter is a tritosternum (sternapophysis) with paired and separate bases (Fig. 15-1C) (compared to the fused, single tritosternal base in the Gamasida).

Coxae of the legs are movable. Legs I and II show six segments. The trochanters of legs III and IV are divided into basi- and telotrochanters, resulting in seven segments. False divisions of all femora, some tibiae, and tarsi may be observed in some species. All of the tarsi have ambulacral claws. Tarsi of legs I exhibit a small sensory organ possibly a homolog of Hal-

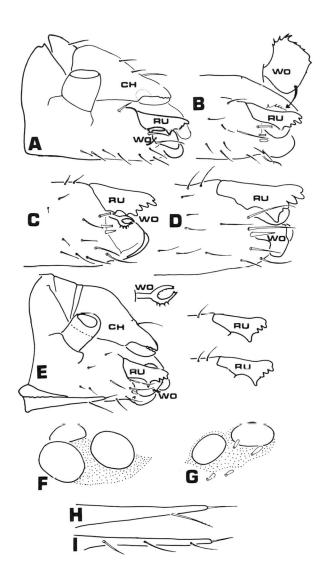

ler's organ (Evans et al., 1961). Legs of Opilioacarida exhibit a distinct neotrichy commensurate with the increased length. Legs break off easily for escape when the mites are disturbed.

Uncovered transverse genital openings in both males and females are located between coxae III and IV (Fig. 12-2). Males have specialized capsules laterad of the genital pore. Females exhibit a simple, protrusible ovipositor. Eggs laid by the females develop into a quiescent prelarva, a hexapod larva (with limb buds at the locations of legs IV), three octopod nymphal stages, and the adult. Adults evidently may continue to molt. (This contrasts with the mites of the Acariformes where there is an egg, prelarva, larva, protonymph, deutonymph, tritonymph, and nonmolting adult. Both the prelarva and larva are hexapod; both have a Claparede organ, but the genital field is undifferentiated) (Coineau, 1973).

Genera include *Opilioacarus* from the southwestern United States, South America (Argentina, Uruguay), Western Africa, and the Mediterranean; *Panchaetes* from Western Africa and Madagascar; *Phalangiacarus* from Gabon in French Equatorial Africa; *Salfacarus* from East Africa and Madagascar; *Adenacarus* from the Arabian peninsula, and *Paracarus* from Central Asia, in Kazakhstan, the most northern location of opilioacarids (42° 30′ to 45° north latitude) (Hammen, 1968b). See the map in Figure 15-1A.

Figure 15-2. Gnathosoma of Opilioacarida from the lateral aspect: (A) *Opilioacarus segmentatus;* (B) *Salfacarus;* (C) *Paracarus;* (D) *Adenacarus;* (E) *O. texanus* (CH, chelicerae; RU, rutellum; WO, With's organ) ((A) after Evans et al., 1961; (B) after Hammen, 1977; (C) after Hammen, 1968b; (D) after Hammen, 1969; (E) after Hammen, 1966)). Eyes of (F) *Paracarus* (after Hammen, 1968b); (G) *Opilioacarus texanus* (after Hammen, 1966). Tritosternum or sternapophysis of (H) *Adenacarus* (after Hammen, 1969), (I) *O. texanus* (after Hammen, 1966).

REFERENCES

Chamberlin, R.V. and S. Mulaik. 1942. On a new family in the Notostigmata. *Proc. Biol. Soc. Wash.* 55:125–132.

Coineau, Y. 1973. A-propos de quelques caractères particulièrement primitifs de la prelarve et de la larva d'un Opilioacarides du Gabon (Acariens). *C. R. Hebd. Seances Acad. Sci., Ser. D* 276:1181–1184.

Coineau, Y. and L. van der Hammen. 1979. The postembryonic development of Opilioacarida, with notes on new taxa and on a general model for the evolution. *Proc. Int. Congr. Acarol., 4th, 1974,* pp. 437–441.

Evans, G.O., J.G. Sheals and D. MacFarlane. 1961. *The Terrestrial Acari of the British Isles.* Adlard & Son, Bartholomew Press, Dorking.

Grandjean, F. 1936. Un acariens synthetique: *Opilioacarus segmentatus* With. *Bull. Soc. Hist. Nat. Afr. Nord* 27:413–444.

Hammen, L. van der. 1966. Studies on Opilioacarida (Arachnida). I. Description of *Opilioacarus texanus* (Chamberlin and Mulaik) and revised classification of the genera. *Zool. Verh.* 86:1–80.

Hammen, L. van der. 1968a. Stray notes on Acarida (Arachnida) I. *Zool. Meded.* 42(25):261–280.

Hammen, L. van der. 1968b. Studies on Opilioacarida (Arachnida). II. Redescription of *Paracarus hexophthalmus* (Redikorzev). *Zool. Meded.* 43(5):57–76.

Hammen, L. van der. 1969. Studies on Opiliocarida (Arachnida). III. *Opilioacarus platensis* Silvestri and *Adenacarus arabicus* (With). *Zool. Meded.* 44(8):113–131.

Hammen, L. van der. 1970. La phylogénèse des Opilioacarides, et leur affinités avec les autres Acariens. *Acarologia* 12(3):465–473.

Hammen, L. van der. 1972. A revised classification of the mites (Arachnida, Acarida) with diagnoses, a key and notes on phylogeny. *Zool. Meded.* 47(22):273–292.

Hammen, L. van der. 1973. Classification and phylogeny of mites. *Proc. Int. Congr. Acarol., 3rd, 1971,* pp. 275–282.

Hammen, L. van der. 1977. Studies on Opilioacarida (Arachnida) IV. The genera *Panchetes, Naudo,* and *Salfacarus,* gen. nov. *Zool. Meded.* 51(4):43–78.

Hirst, S. 1923. On some Arachnid remains from the Old Red Sandstone (Rhynie Chert Bed, Aberdeenshire). *Ann. Mag. Nat. Hist.* (9) 12:455–474.

Krantz, G.W. 1978. *A Manual of Acarology,* 2nd ed. Oregon State Univ. Book Store, Corvallis.

Lehtinen, P.T. 1980. A new species of Opilioacarida (Arachnida) from Venezuela. *Acta Biol. Venez.* 10:205–214.

Naudo, M.H. 1963. Acariens Notostigmata d'Angola. *Publ. Cult. Co. Diam. Angola., Lisboa* 63:13–24.

Redikorzev, V. 1937. Eine neue Opilioacarus. *Zool. Anz.* 118:10–12.

With, C. 1903. A new Acarid, *Opilioacarus segmentatus. C. R. Congr. Natl. Med. Nord., Helsingfors,* 1902, *Sect.* 6:4.

With C. 1904. The Notostigmata, a new suborder of Acari. *Vidensk. Medd. Dan. Naturh. Foren. Kjobenhavn,* pp. 137–192.

16

Order Holothyrida

(Anactinotrichida Zakhvatkin 1952) (Holothyrina Thon 1909; Holothyroidea Reuter 1909; Tetrastigmata Oudemans 1910; Evans et al. 1961). Families: Holothyridae; Allothyridae; Neothyridae.

These epigaeic (free, earth-crawling), large mites (2–7 mm) are strongly sclerotized. Oval in outline and somewhat tortoiselike in shape, the body consists of two regions, the gnathosoma and an unsegmented idiosoma (Fig. 16-1A, B). Anteriorly, a slight dorsal epistome is present. The surface of the idiosoma is hairless and glabrous in the Holothyridae, but hairy in the Allothyridae and Neothyridae. Coloration ranges from bright red to dark brown with two-toned legs and blackish-brown with reddish-brown legs. These mites differ in several respects from the Opilioacarida (lack of ocelli and tritosternum, fewer lyrifissures, two pairs of stigmata) and yet are similar in other ways (radulalike labrum, cheliceral setae, and lyrifissures). The similarities are considered plesiomorphic. The diagnostic characters of the three families represent the last remnants of three divergent phyletic lines (Lehtinen, 1981). Lehtinen compares the features in a comprehensive table.

Holothyrids are reported to be lethal to poultry and toxic to humans. Handling of these mites results in inflammation of the mucous membranes of the mouth and pharynx if accidentally ingested, or if the secretions of the mites come in contact with these tissues. The mites are thought to have a vesicating saliva, for it blisters the fingers of collectors who try to pick them up (Fredrickson, 1968). Southcott (1976) reports that an entomologist who accidentally put his fingers in his mouth five hours after touching one of these red mites experienced an "extraordinarily pungent, galvanic sensation or taste" which occurred immediately and spread rapidly to his mouth and throat. This lasted several hours, and was accompanied by excessive salivation. The entomologist, Green, and a medical colleague both tried out these mites, confirming that they were the source of the symptoms. This species of mite (*Holothyrus coccinella* Gervais, 1842) "was stated to be 'sometimes swallowed by ducks and geese, causing their death'. . . . Children who put these mites in their mouths, or

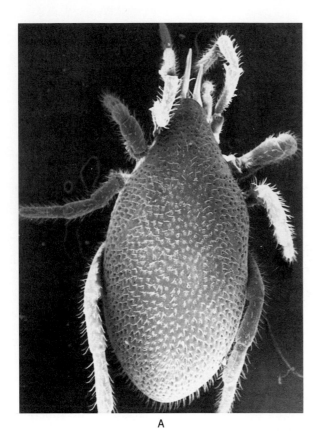

A

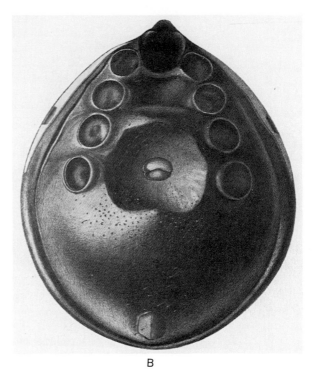

B

Figure 16-1. Scanning electron micrograph of (A) *Holothyrus australasiae* Womersley, 1935, dorsal aspect (by permission of Dr. R.V. Southcott, 1976) (Dr. J. Kethley indicates this is *Allothyrus*); (B) male *H. grandjeani*, ventral view (after Hammen, 1961).

touched them and put their fingers in their mouths were stated to have suffered ill effects.''

Depending on the family, two pairs of stigmata are present laterodorsally, marginally, or submarginally (Hammen, 1961). This is a distinctive characteristic and is reflected in the former name Tetrastigmata. The stigmata are connected to external peritremes and to internal air sacs (Lehtinen, 1981; Krantz, 1978) (Fig. 16-2G, H). An Australian species (*Allothyrus australasiae* Womersley) has six stigmata (Hammen, 1972).

The gnathosoma is enclosed in a ventral camerostome not visible from the dorsal aspect

(Fig. 16-1B). Characteristically from 6 to more than 10 pairs of infracapitular setae are present. A distinct corniculus is located at the dorsolateral edges of hypostome (Fig. 16-3A–C). With's organs and rutellae, both present in the Opilioacarida, are absent. The labrum is strongly developed and radulalike. It is thought to be protruded by hydrostatic pressure of the body (Fredrickson, 1968). The chelicerae are chelate–dentate with variable arrangements of the teeth, depending on the family. One or two cheliceral setae are present; a cheliceral lyrifissure may be present (Fig. 16-2I). The five-segmented palps have two or more clawlike, tripartite, and subterminal apoteles.

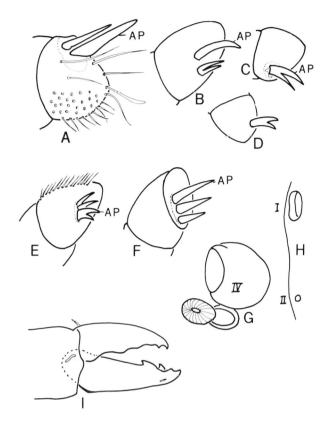

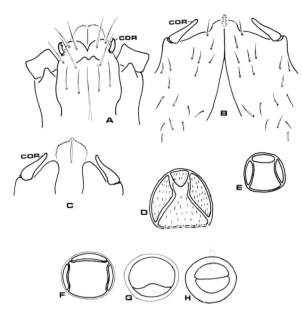

Figure 16-3. Hyptostome of (A) *Holothyrus sp.* (after Evans et al., 1961); (B, C) *Neothyrus* (COR, corniculus) (after Lehtinen, 1981); genital openings of (D) female *Allonothrus* (after Domrow, 1955); males: (E) *Neothyrus;* (F, G) *Hammenius;* (H) *Neothyrus* (after Lehtinen, 1981).

Figure 16-2. Palpal apoteles (AP), stigmata, and chelicerae of Holothyrida: (A) *Holothyrus sp.* (after Evans et al., 1961); (B) *Holothyrus sp.* (after Krantz, 1978); (C) *H. grandjeani* (after Hammen, 1961); (D) *H. sp.*; (E) *Allothyrus* (after Domrow, 1955); (F) *Neothyrus* (after Lehtinen, 1981); (G) peridium of *Allothyrus* (after Domrow, 1955); (H) stigmata I, II of *Holothyrus* (after Hammen, 1961); (I) chelicera of *H. sp.* (after Evans et al., 1961).

A picket-fence-like row of setae is present on the palpal tibia. (Fig. 16-2A–F). This palpal comb is present in the Holothyridae and Neothyridae. The arrangement of the comb row consists of a spoonlike seta separated from 10 to 16 other setae in the row (Fredrickson, 1968; Lehtinen, 1981). The tarsal apotele and palpal comb may be used for grasping or holding food of these predators.

A tritosternum represented by two simple

setae in the Holothyridae is absent in Neothyridae and Allothyridae.

The seven-segmented legs are usually shorter than the idiosoma and vary in presence of apophyses, especially in males where these processes are characteristic. Tarsal segmentation is usually absent, but most tarsi exhibit an ambulacrum of pulvillus and claws. Tarsi of legs I have a distal sensory structure, Haller's organ, whose solenidia and sensory hairs differ in numbers and location between families. This organ resembles the Haller's organ of ticks in several respects, especially in the family Holothyridae. A curious pit, the peridium, lies behind the legs IV in the Allothyridae, but is usually absent in the other two families (Domrow, 1955; Hammen, 1968; Lehtinen, 1981). Womersley regards this organ as a very large scent gland (Hammen, 1961).

Sexes are dimorphic. The female genital plate is generally quadrangular with four setate genital shields: an anterior shield, a median epigynial shield, and two lateral shields (Fig. 16-3D). Males have divided, setate genital valves (Fig. 16-3E–H). The anus is subterminal; anal covers have more than two setae (in contrast to those of the Gamasida).

Little is known of the biology of these forms. The feeding habits are similar in the Opilioacarida, Holothyrida, and some of the unspecialized Gamasida, which is attributed to the similarities in the gnathosomas of these groups (Lehtinen, 1981). R.W. Fredrickson (personal communication) assumes Holothyrids eat oribatid mites because larger parts of these beetle mites have been found in their stomachs.

Development is unknown.

The Holothyrida occupy humus in the forest floors of upland and forested regions of tropical areas of the world (Domrow, 1955; Fredrickson, 1968). They seem to be confined to mesic and xeric habitats, particularly on islands, which implies specific adaptation to insular situations. Species in New Guinea are found in ferns and moss (Hammen, 1961). These mites have not been reported from the mainlands of Africa or Asia, which reinforces the assumption that they are principally insular in habitat. Records from Louisiana place the mites in the United States (R.W. Fredrickson, personal communication).

Neothyridae are neotropical (Peru), Allothyridae are Australian, and Holothyridae are from Australia, New Guinea, New Caledonia, and islands of the Indian Ocean. One species has been reported from Formosa. Some representative genera are:

Neothyrus	Peru
Holothyrus	Mauritius Island
Thonius	Ceylon, Indonesia, New Guinea
Thonius	Seychelle Islands
Holothyrus	Australia, New Zealand, New Caledonia; Australia, Woodlark Island

Allothyrus	Australia: Queensland
Hammenius	New Guinea

REFERENCES

Domrow, R. 1955. A second species of Holothyrus (Acarina: Holothyroidea) from Australia. *Proc. Linn. Soc. N.S.W.* 79:159–162.

Evans, G.O., J.G. Sheals and D. MacFarlane. 1961. *The Terrestrial Acari of the British Isles,* Vol. 1. Adlard & Sons, Bartholomew Press, Dorking, England.

Fredrickson, R.W. 1968. Personal communication. Symposium on Acarology, Waltham, Massachusetts.

Gervais, P. 1842. Une quinzaine d'espèces d'insectes aptères. *Ann. Soc. Entomol. Fr.* 11:45–48.

Hammen, L. van der. 1961. Descriptions of Holothyrus grandjeani nov. sp., and notes on the classification of mites. *Nova Guinea Zool.* 9:173–194.

Hammen, L. van der. 1965. Further notes on the Holothyrina (Acarina). I. Supplementary description of Holothyrus coccinella Gervais. *Zool. Meded.* 40:253–276.

Hammen, L. van der. 1966. Studies on Opiliocarida (Arachnida). I. Description of *Opilioacarus texansu* (Chamberlin and Mulaik) and revised classification of the genera. *Zool. Verh.* 86:1–80.

Hammen, L. van der. 1968. Stray notes on Acarina (Arachnida). I. *Zool. Meded.* 43:261–280.

Hammen, L. van der. 1969. Studies on Opilioacarida (Arachnida). III. *Opilioacarus platensis* Silvestri, and *Adenacarus arabicus* (With). *Zool. Meded.* 44(8):113–131.

Hammen, L. van der. 1972. A revised classification of the mites (Arachnida, Acarina) with diagnoses, a key and notes on phylogeny. *Zool. Meded.* 47:273–292.

Hirst, S. 1917. Species of Arachnida and Myriapoda (Scorpions, spiders, mites, ticks and centipedes) injurious to man. *Bull. Br. Mus. (Nat. Hist.) Econ. Ser.* 6:1–60.

Krantz, G.W. 1978. *A Manual of Acarology.* Oregon State Univ. Book Stores, Corvallis.

Legendre, R. 1968. La nomenclature anatomique chez les Acariens. *Acarologia* 10:413–417.

Lehtinen, P.T. 1980. A new species of Opilioacarida (Arachnida) from Venezuela. *Acta Biol. Venez.* 10:205–214.

Lehtinen, P.T. 1981. New Holothyrina (Arachnida, Anactinotrichida) from New Guinea and South America. *Acarologia* 22(1):3–13.

Southcott, R.V. 1976. Arachnidism and allied syndromes in the Austrialian region. *Rec. Adelaide Children's Hosp.* 1(1):97–186.

Thon, K. 1906. Die aussere Morphologie und die Systematik der Holothyriden. *Zool. Jahrb., Syst.* 23:677–724.

Womersley, H. 1935. A species of Acarina of the genus Holothyrus from Austrialia and New Zealand. *Ann. Mag. Nat. Hist.* (10) 16:154–157.

17

Order Gamasida

(Anactinotrichida, Zakhvatkin, 1952) (Gamasides Leach 1814;
Mesostigmata G. Canestrini 1891, Baker and Wharton 1952, Evans
et al. 1961; Distigmata Oudemans 1906).

The order Gamasida (Mesostigmata of some authors) is highly diverse. The majority are free-living, but others are ecto- and endoparasites of reptiles, birds, and mammals, as well as a few invertebrates. In size they range from 200–2000 μm. The characteristics of the order are quite distinctive.

The principal feature of the body relates to the lateral stigmata. These are usually located opposite coxae II, IV or between coxae III and IV. The level of location of the stigmata is implied in the older name Mesostigmata (Figs. 9-2, 17-1). Usually the stigmata are accompanied by an elongated peritreme, a groove or tube that extends anteriorly from the stigmata. No tracheal connections are associated with the peritreme; the tracheae extend inward from the stigmata. The peritreme varies in length, shape, and linear configuration. It may extend anteriorly to the gnathosoma and have posterior peritremalia (peritremal plate) (Fig. 9-2). SEM observations show setalike papillae within this groove (Fig. 9-3) (Witalinski, 1979).

The peritreme is reduced in some parasitic forms. A sensory, hygroreceptive function is suggested.

Dorsally, Gamasida exhibit a cover over the chelicerae called an epistome (or tectum capituli). This dorsal extension of the skeleton is simple in some, ornate in others (Figs. 5-6; 17-2A). The gnathosoma is usually distinctly separate from the idiosoma so that the palps and chelicerae, the hypostome and gnathosomal base are easily seen. The chelicerae are mainly chelate–dentate, except for some parasitic forms. Each has a pilus dentilus (=cheliseta of Hammen (1964)) on the fixed digit, together with antiaxial and paraxial lyrifissures. Males exhibit special processes, spermadactyls (=spermatodactyls, spermatophoral processes), for the transfer of sperm. Palps are leg-like with a palpal apotele of two or three tines on the tarsus. This apotele is thought to be a remnant of a claw (Figs. 17-1; 5-19).

Ventrally the gnathosoma exhibits a basal hypostome with three pairs of setae set in a tri-

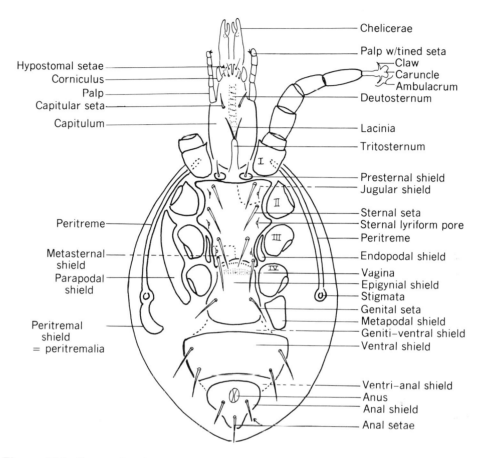

Figure 17-1. Composite drawing of venter of a gamasid mite showing shields, setae, and other features. Dotted lines indicate separation and/or fusion of shields.

angular arrangement or in a straight line. External malae (=corniculi) are anterolateral hornlike extensions of the hypostome. They may be serrate or bifid. Posterior to the gnathosoma in the synarthrodial membrane between the gnathosoma and the idiosoma is the tritosternum (=sternapophysis of Hammen, 1964, 1969). The shape of the tritosternum varies. The lacinia (flagellated anterior extensions) vary in numbers from zero to three in the Gamasina and are free; in the Uropodoidea the lacinia are covered by coxae I. No function is ascribed to the tritosternum or to the tectum. Gnathosomal structures vary between the subgroups of Gamasida (Fig. 17-1).

The gamasids are usually heavily sclerotized with colors that tend toward brown or reddish-brown shades. Parasitic forms may be colorless. Patterns of sclerotization vary. In larvae this sclerotization begins in the propodosomal region and proceeds posteriorly; then fusion occurs (Fig. 4-16). Adults with an entire dorsal plate are assumed to have a fused condition of anterior and posterior plates. The pattern of ventral plates may vary some with fused parts; some plates are absent, other plates approximate each other. The plates are not homologous between orders of mites, but seem to have taxonomic significance within this order. The dorsal plate may be single and entire, paired, or

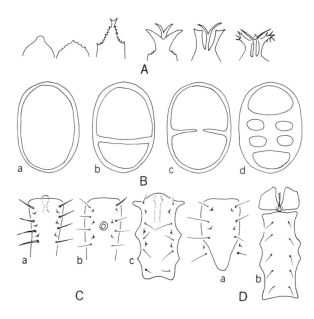

Figure 17-2. Characteristic structures of gamasid mites: (A) examples of epistome (=tectum); (B) types of arrangements of dorsal shields: a, entire; b, divided podonotal and opisthonotal (Ascidae: Zerconiidae); c, incompletely divided (Veigaiidae); d, podonotal, mesonotal scutellae, pygidial (Sejidae); (C) male apertures: a, gamasine; b, uropodine; c, Ascidae; (D) nymphal shields: a, gamasine; b, uropodine.

modified into platelets (Fig. 17-2B). Ocelli are absent.

The sclerotization of the venter is diagnostic for sexual dimorphism and for group differentiation within the order (Figs. 4-19; 17-2C–D). Among the ventral plates or shields, the sternal plate is most anterior. It typically has four pairs of sternal setae and three pairs of sternal pores (lyrifissures). In those families where the anterior corner of the sternal plate is separated and bears the first sternal seta, it is called a jugular plate. When the posterior part of the sternal plate is separate and has a pair of sternal setae and a pair of sternal pores, it is called a metasternal plate (e.g., Parasitidae, Fig. 17-1).

Genital apertures are usually ventral and sclerotized. In females the gonopore is trans-verse and intercoxal with one to four shields present. The single female epigynial (genital) plate is posterior to the sternal shield and usually has two genital setae in the monogynaspids (Fig. 17-1). Three genital plates are characteristic of the trigynaspids (Fig. 4-20). However, J.B. Kethley (personal communication, 1986) maintains that these females are technically quadrigynaspid in character. Male apertures are located anteriorly or medially in a holoventral or sternoventral shield complex (Fig. 17-2C); the aperture has two small plates.

Sexes are determined by diplodiploidy, haplodiploidy, or thelytoky. Tocospermy and podospermy occur in the Gamasida. Tocospermy implies direct transfer of sperm into the vagina of the female. Podospermy involves the transfer of sperm by means of spermadactyls on the movable digit of the male chelicerae to sperm induction pores in the coxae, trochanters or shields of the venter of the females (see Chapter 12).

Development in the Gamasida proceeds through an orderly arrangement. The larva is almost always a nonfeeding instar without stigmata. The dorsum exhibits a podonotal shield. A sternal shield is present with three pairs of setae. The protonymph is larger and exhibits a podonotal shield and either a pygidial or a larger opisthosomal shield. Stigmata are present and the peritreme may be elongated in this and other postlarval stages, except where neoteny occurs (an increase in the numbers of setae and pores on the idiosoma and on the appendages (Krantz, 1978)). Palpal and leg chaetotaxy are important in determining specific instars. The deutonymph shows dorsal shield arrangments that approximate the adults, but metasternal setae are added ventrally (Fig. 17-2D). In adults the general pattern of shield development from the larva results in increased sclerotization in most instances. In some parasitic forms (Entonyssidae, Halarachnidae) the shield development is reduced in the nymphal stages and incomplete or weak in the adults.

The hypothetical development of the venter of female gamasids is supposed to have occurred with a differentiation of sclerotization. This process does not, however, follow segmentation. The forward movement of the pores and the sternal setae is correlated with the fusion of the sternal shield and other accompanying changes. Variations occur in the stigmata, dorsal and ventral plates, tectum, and gnathosoma of different subgroups (J.H. Camin, personal communication) (Fig. 4-19).

As indicated above, Monogynaspida and Trigynaspida comprise the two main divisions of Gamasida. The single epigynial shield of the former (reduced or wanting in parasitic forms) and a single pair of genital setae (or none) are accompanied by a cheliceral brush and sperm induction pores; males thus have spermadactyls for podospermy. Eighty to ninety percent of the gamasids belong to this group (Fig. 4-20A). The suborders of the division are indicated below.

About 10–20% of the Gamasida belong in the Trigynaspida where the genital shields of the females are composed of two latigynials and a mesogynial shield, although these may be variously placed, coalesced, and fused (Fig. 4-20B). Chelicerae in this group typically have excrescences (filamentous, dendritic, or brushlike) either medially or terminally. Male trigynaspids lack a spermadactyl, so tocospermy, direct transfer of sperm into the genital tract of the females, is typical.

The classification below follows Johnston (1982) with some modifications.

SUBORDER SEJINA

These are assumed to be the most primitive of the Gamasida. They are medium-size mites with short, triangular or bidentate corniculi. Females have two to six dorsal shields and the plates of the sternal region are also fragmented. The genital shields of the females are large with three or more pairs of setae. Males have a sternal aperture.

Sejidae (Fig. 17-3A, B)

Uropodellidae?
Ichthyostomatogasteridae (Fig. 17-3C, D)

SUBORDER ARCTACARINA

Arctacaridae—boreal regions of eastern Russia and North America (Alaska to Oregon)

SUBORDER MICROGYNINA

A single family comprises this suborder of small, lightly sclerotized mites of boreal regions. Dorsally a podonotal shield, two mesonotal shields, and a small pygidial shield characterize adults and deutonymphs. The sternal shield of females is fragmented with sternal setae on separate platelets. The female epigynial shield is reduced and bears a single pair of setae anterior to a ventrianal shield. Chelicerae are small and chelate–dentate. Peritremes are short and do not extend beyond legs II. These mites are found in woody debris.

Microgyniidae (Fig. 17-3E, F)

SUBORDER EPICRIINA

These soil-dwelling mites are restricted to the Northern Hemisphere where they reside in decaying wood, duff, litter, and moss. The two families in this suborder are characterized by a dorsum of two subequal shields or by a single shield. Surface ornamentation consists of reticulations or polygonal shapes. The posterior dorsal surface of Zerconidae exhibits two pairs of large muscle scars, a distinctive character that is easily observed. Leg setation is normal, except that legs I of Epicriidae lack claws. The corniculi are short and the tectum is usually denticulate. Peritremes are short, sometimes convoluted. The epigynial shield of females is rounded or truncate with one or two pairs of setae.

Epicriidae (Fig. 17-3G–I)
Zerconidae (Fig. 17-3J, K)

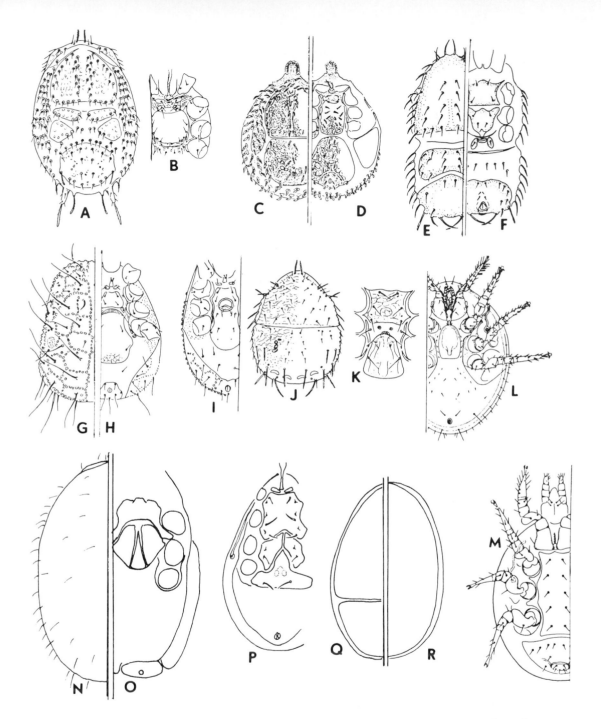

Figure 17-3. Representative families of Gamasida. Sejidae: (A) dorsal view; (B) podosoma of female. Ichthyostomatogasteridae (=Uropodellidae); (C) dorsal view; (D) ventral view of female. Microgyniidae: (E, F) dorsal, ventral view of female. Epicriidae: (G, H) dorsal, ventral view of female; (I) ventral view of male. Zerconidae: (J) dorsal view; (K) ventral podosoma of female. Uropodidae: (L) ventral view of female; (M) venter of nymph. Triplogyniidae: (N, O) dorsal, ventral view of female (after Funk). Parasitidae: (P) ventral view of female; (Q, R) divided and entire dorsal plates. (A–M after Baker et al., 1958).

SUBORDER UROPODINA

Like their insect counterparts, the tortoise-shelled beetles, these mites are medium size and resemble miniature turtles because of their complete, convex dorsal shield. Marginal shields may be present, either as fused shields or separate platelets. Ventrally the sternal area is fused with a trapdoor-like epigynial plate without setae in females. Excavations of the venter (foveae pedales or pedofossae) enable the mites to retract the legs into these concavities for protection. Corniculi are usually short, but the hypostomal processes may be elaborate. Chelicerae are chelate–dentate and in some may be very elongated, extending from a broadened base in the idiosoma to an attenuated shaft with small chelae. Some heteromorphic deutonymphs may be phoretic and attach to transport hosts by means of the styloproct.

Protodynichidae
Thinozerconidae
Polyaspididae
Dithinozerconidae
Uropodidae (Fig. 17-3L, M)
Trachyuropodidae

SUBORDER DIARTHROPHALLINA

The single family of this suborder is ectoparasitic on passalid beetles. They are lightly sclerotized mites with a single dorsal shield covering the entire idiosoma. Peritremes are absent. Chelicerae are chelate–dentate with a hyaline process or weak spurs. Palps have reduced setation. Legs I are elongated and an amulacrum is lacking. All stages are ovoid in shape with greatly enlarged, elongate lateral setae and reduced dorsal setae. The sternal and ventral shields are fused surrounding the ovoid epigynial shield of the females; males have a smaller genital opening in the same region.

Diarthrophallidae

SUBORDER CERCOMEGISTINA

These mites are characterized by chelicerae with numerous small digits and small excrescences on the digitus mobilus. Dorsal shields may be complete and entire, subequally divided, or paired mesonotal and pygidial shields. The ventral shield arrangement in females varies in size and sclerotization with sternal, mesogynial, and latigynial elements differently arranged. There are many undescribed taxa in this group of free-living mites associated with bark beetles, spiders, and other terrestrial forms.

Cercomegistidae
Asternoseiidae
Davacaridae
Seiodidae

SUBORDER ANTENNOPHORINA

Characteristically these mites are trigynaspids with entire dorsal shields and pronounced filamentous or dendritic excrescences on the movable digit of the chelicerae. The tritosternal laciniae are usually separate and fimbriated. An ambulacrum is missing on legs I. Five superfamilies and 15 families comprise this suborder, each with definitive characteristics varying with regard to the arrangements of the typical genital plates. Some are myrmecophilous; others are associated with beetles and millipedes of both the New World and Old World.

Aenictoquoidea—associated with ants
    Aenictequidae
    Messorcaridae
    Physalozerconidae
    Ptochacaridae
Antennophoroidea—associated with ground-dwelling ants
    Antennophoridae

Caelenopsoidea
  Caelenopsidae—associates of wood beetles
  Schizogyniidae—beetles and burrowing snakes
  Megacaelenopsidae—forest litter
  Triplogyniidae—forest litter (Fig. 17-3N, O)
  Meinertulidae
  Diplogyniidae—litter; beetle associates
  Euzerconidae—litter, beetles, worm snakes, millipedes
  Neotenogyniidae—neotropical millipedes
Fedrizzioidea—cosmopolitan predators in compost, litter, dung; nidicoles in nests of bumblebees, silphids, small mammals
  Fedrizziidae
  Klinckowstroemiidae
  Promegistidae
  Paramegistidae

Megisthanoidea are among the largest of the Gamasida. They may be 3 mm in length with long legs. Chelicerae are well-developed with short, branched excrescences on the movable digit. A massive subcapitulum with forked corniculi is distinctive. Latigynial or sternogynial shields characterize the females; the male aperture is central in the sternal area. Again, these mites are associated with passalid beetles. The adults are phoretic and the immatures are found in the galleries of the hosts (Johnston, 1982).

Megisthanoidea
  Megisthanidae
  Hoplomegistidae
Parantennuloidea
  Parantennulidae
  Philodanidae

SUBORDER PARASITINA

Parasitoidea
  Parasitidae (Fig. 17-3P, Q)
    Parasitinae
  Pergamasidae
    Pergamasinae

These terrestrial mites are medium-size with fused or separate podonotal and opisthonotal dorsal shields in females; dorsal shields in males are entire. Peritremes are usually long, but may be widened in a few species. The chelicerae are strongly chelate–dentate with a spermadactyl (=spermatophoral process) in males. Corniculi are usually short and strong. The sternum in females bears three sternal setae; metasternal shields are large and usually flank the triangular epigynial shield midventrally. The genital aperture of males is presternal in position. Legs usually have ambulacra on tarsi I; legs II of males may be crassate and hypertrophied for copulation and sperm transfer.

SUBORDER DERMANYSSINA

This is the largest, most diversified suborder of the Gamasida. Distinctively, the males exhibit a spermatodactyl on the movable digit of the chelicerae; females have auxiliary spermatheca connected to a sacculus foemineus. The male aperture is presternal in location with an eversible ejaculatory organ for sperm transfer to the chelicerae.

Rhodacaroidea—predators in soil, litter
  Rhodacaridae (Fig. 17-4A–C)
  Ologamasidae
  Digamasellidae (Fig. 17-4D, E)
  Laelaptonyssidae (?)
Veigaioidea—predators in soil
  Veigaiidae (Fig. 17-4F, G)
Eviphidioidea—predators, or associated with beetles

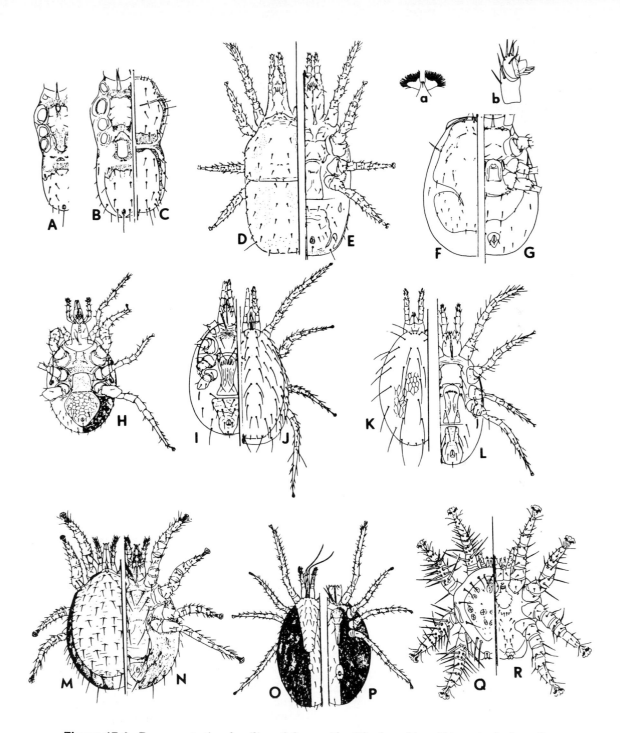

Figure 17-4. Representative families of Gamasida. Rhodacaridae: (A) ventral view of male; (B, C) ventral, dorsal views of female. Digamasellidae: (D, E) dorsal, ventral views of female. Veigaiidae (F, G) dorsal, ventral views of female: a, hypostomal process and b, palp tibia–tarsus. Macrochelidae: (H) ventral view of female. Ascidae: (I, J) ventral, dorsal views of female. Phytoseiidae: (K, L) dorsal, ventral views of female. Laelapidae: (M, N) dorsal, ventral views of female. Dermanyssidae: (O, P) dorsal, ventral views of female. Spinturnicidae: (Q, R) dorsal, ventral views of male. (A–R after Baker et al., 1958).

Macrochelidae (Fig. 17-4H)

Parholaspidae

Pachylaelapidae

Megalolaelapidae

Eviphididae

Ascoidea—predators

    Ascidae (Fig. 17-4I, J)

    Antennoseiidae

Phytoseiidae—predators of spider mites (Fig. 17-4K, L)

Otopheidomenidae—moth ear mites

Ameroseiidae—soil, litter

Dermanyssoidea—nidicoles, paraphages, parasites of terrestrial arthropods, vertebrates

Laelapidae—soil; nidicoles; small mammals, insects (Fig. 17-4M, N)

Haemogamasidae—ectoparasitic, small mammals

Dermanyssidae—ectoparasitic, birds, rodents (Fig. 17-4O, P)

Hystrichonyssidae—parasites of Malayan porcupine; tree snake

Macronyssidae—ectoparasites of birds, mammals

Rhinonyssidae—nasal parasites of birds (Fig. 17-5A, B)

Spinturnicidae—ectoparasites of bats (Fig. 17-4Q, R)

Spelaeorhynchidae—ectoparasites of bats (Central, South America) (Fig. 2-4)

Halarachnidae—parasites of seals and monkeys (Fig. 17-5C, D)

Raillietidae—aural parasite of cattle, antelope

Entonyssidae—entoparasites of snakes (Fig. 17-5E, F)

Ixodorhynchidae—ectoparasites of snakes (*Crotalus*)

Omentolaelapidae—snake parasites

Dasyponyssidae—parasites of armadillos (Fig. 17-5G, H)

Manitherionyssidae—parasites of edentates

Varroidae—paraphages of honeybees

The mites of the Dermanyssoidea fall into some interesting habit and host categories. Obligate hematophages include Dermanyssidae, Spinturnicidae, Hystrichonyssidae, Macronyssidae, and Ixodorhynchidae. Ectoparasites of bats comprise the families Spinturnicidae, Spelaeorhynchidae, and many of the Macronyssidae. Reptiles are hosts for Entonyssidae, Omentolaelapidae, Macronyssidae, and Histriconyssidae. The families Dermanyssidae, Macronyssidae, and Rhinonyssidae parasitize birds. Mammals are hosts for Halarachnidae, Rhinonyssidae, Dasyponyssidae, Manitherionyssidae, and Histrichonyssidae. Varroidae are arthropod paraphages, but are of greatest concern because of the damage they do to brood forms, particularly honeybees.

SUBORDER HETEROZERCONINA

Not much is known about the suborder Heterozerconina, but these mites have been taken from centipedes and millipedes (and one from a snake) in subtropical and tropical areas. Few species are adequately described. Female chelicerae have slender digits with or without denticles. Male spermadactyls are elaborate on the fixed digit (Heterozerconidae) and more slender and on the movable digit (Discozerconidae). The dorsal shield in both sexes is entire and somewhat sclerotized. The sternal shields are incompletely developed, but an endopodal shield is present. Most adults of both families exhibit enlarged, suckerlike organs.

Heterozerconoidea

    Heterozerconidae—snakes, scolopendrine centipedes

    Discozerconidae—myriapodophilous, snakes

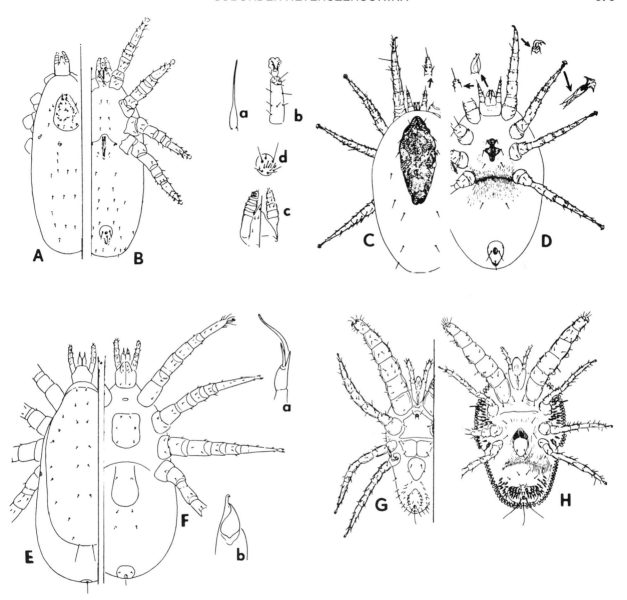

Figure 17-5. Representative families of Gamasida. Rhynonyssidae: (A, B) dorsal, ventral views of female: a, chelicera; b, tarsus I; c, palps; d, sensory area of palp. Halarachnidae: (C, D) dorsal, ventral views of female. Entonyssidae: (E, F) dorsal, ventral views of female: a, male chelicera; b, female chelicera. Dasyponyssidae: (G) ventral view of male; (H) ventral view of female. (A–H after Baker et al., 1958).

REFERENCES

Akratanatul, P. and D.M. Burgett. 1975. *Varroa jacobsoni:* A prospective pest of honeybees in many parts of the world. *Bee World* 56(3):119–121. (Varroidae)

Athias-Henriot, C. 1969. Notes sur la morphologie externe des gamasides (Acariens Anactidotriches). *Acarologia* 11(4):609–629.

Athias-Henriot, C. 1972. Gamasides Chiliens (Arachnides). II. Revision de la famille Ichthyostomatogasteridae Sellnick, 1953 (=Uropodellidae Camin, 1955). *Arq. Zool.* 22(3):113–191.

Athias-Henriot, C. 1978. Sur le genre *Eugamasus* Berlese 1892 (Parasitiformes, Parasitidae). I. Redefinition. Description des petites espèces. *Acarologia* 20(1):3–18.

Athias-Henriot, C. 1980. Sur le genre *Eugamasus* Berl. (Parasitiformes, Parastidae) II. Les grandes espèces. *Acarologia* 21(3):313–329.

Baker, E.W. and G.W. Wharton. 1952. *An Introduction to Acarology.* Macmillan, New York.

Baker, E.W. and C.E. Yunker. 1964. New blattisociid mite (Acarina, Mesostigmata) recovered from neotropical flowers and hummingbirds' nares. *Ann. Entomol. Soc. Am.* 57(1):103–126. (Ascidae)

Baker, E.W., T.M. Evans, D.J. Gould, W.B. Hull and H.L. Keegan. 1956. *A Manual of Parasitic Mites of Medical or Economic Importance.* Natl. Pest Control Assoc., New York.

Baker, E.W., J.H. Camin, F. Cunliffe, T.A. Woolley and C.E. Yunker. 1958. *Guide to the Families of Mites,* Contrib. No. 3. Institute of Acarology.

Baker, E.W., M. Delfinado-Baker and F. Reyes Ordaz. 1983. Some laelaptid mites (Laelaptidae: Mesostigmata) found in nests of wasps and stingless bees. *Int. J. Acarol.* 9(1):3–10.

Bregetova, N.G. 1961. The veigaiid mites (Gamasoidea, Veigaiidae) in the USSR. *Parazitol. Sb.* 20:10–107.

Bregetova, N.G. and E.V. Koroleva. 1960. The Macrochelidae Vitzthum, 1930 in the USSR. *Parazitol. Sb.* 19:32–154.

Butler, L. and P.E. Hunter. 1968. Redescription of *Megisthanus floridanus* with observations on its biology. (Acarina: Megisthanidae). *Fla. Entomol.* 51:187–197.

Camin, J.H. 1953a. A revision of the cohort Trachytina Tragardh, 1938, with the description of *Dyscritaspis whartoni,* a new genus and species of polyaspid mites from treeholes. *Bull. Chicago Acad. Sci.* 9(17):335–385.

Camin, J.H. 1953b. Metagynellidae, a new family of uropodine mite with the description of *Metagynella parvula,* a new species from tree holes. *Bull. Chicago Acad. Sci.* 9(18):391–409.

Camin, J.H. 1955. Uropodellidae, a new family of mesostigmatid mites based on *Uropodella laciniata* Berlese, 1888 (Acarina: Liroaspina). *Bull. Chicago Acad. Sci.* 10(5):65–81. (Ichthyostomatogasteridae)

Chant, D.A. 1959. Phytoseiid mites (Acarina: Phytoseiidae). Part I: Bionomics of seven species in southeastern England; Part 2: A taxonomic review of the family Phytoseiidae, with descriptions of 38 new species. *Can. Entomol. Suppl.* 12.

Chant, D.A. 1963. The subfamily Blattisocinae Garman (=Aceosejinae Evans) (Acarina: Blattisociidae Garman) (=Aceosejidae Baker and Wharton) in North America, with descriptions of new species. *Can. J. Zool.* 41:243–305. (Ascidae)

Chant, D.A. 1965. Generic concepts in the family Phytoseiidae (Acarina: Mesostigmata). *Can. Entomol.* 97(4):351–374. (Otopheidomenidae)

Delfinado, M.D. and E.W. Baker. 1974. A new record for the bee mite *Mellitiphis* in New Zealand. *Bee World* 55(4):148–149.

Dusbabek, F. and V. Cerny. 1970. The nasal mites of Cuban birds. I. ASCIDAE, Ereynetidae, Trombiculidae (Acarina). *Acarologia* 12(1):169–281.

Elzinga, R.J. and C.W. Rettenmeyer. 1974. Some new species of *Circocylliba* (Acarina: Uropodina) found on army ants. *Acarologia* 16(4):595–611.

Evans, G.O. 1957. An introduction to the British Mesostigmata (Acarina) with keys to the families and genera. *J. Linn. Soc. London, Zool.* 43:203–259.

Evans, G.O. 1958. A revision of the British Aceosejinae (Acarina: Mesostigmata). *Proc. Zool. Soc. London* 131(1):177–229. (Ascidae)

Evans, G.O. 1963a. Observations on the chaetotaxy of the legs in the free-living Gamasina (Acari: Mesostigmata). *Bull. Br. Mus. (Nat. Hist.) Zool.* 10(5):277–303.

Evans, G.O. 1963b. Observations on the classification of the family Otopheidomenidae (Acari: Mesostigmata) with descriptions of two new species. *Ann. Mag. Nat. Hist.* (5) 13:609–620.

Evans, G.O. 1963c. The genus *Neocypholaelaps* Vitzthum (Acari: Mesostigmata). *Ann. Mag. Nat. Hist.* (6) 13:209–230. (Ameroseiidae)

Evans, G.O. 1964. Some observations on the chaetotaxy of the pedipalps in the Mesostigmata (Acari). *Ann. Mag. Nat. Hist.* (6) 13:513–527.

Evans, G.O. 1965. The ontogenetic development of the chaetotaxy of the tarsi of legs II–IV in the Antennophorina (Acari: Mesostigmata). *Ann. Mag. Nat. Hist.* (6) 13:513–527.

Evans, G.O. 1969. A new mite of the genus *Thinoseius* Halbt. (Gamasina: Eviphididae) from the Chatham Islands, New Zealand. *Acarologia* 11(3):505–514.

Evans, G.O. 1972. Leg chaetotaxy and the classification of the Uropodina (Acari: Mesostigmata). *J. Zool.* 167:193–206.

Evans, G.O. and W.M. Till. 1965. Studies on the British Dermanyssidae (Acari: Mesostigmata). I. External morphology. *Bull. Br. Mus. (Nat. Hist.) Zool.* 13(8):249–294. (Dermanyssoidea)

Evans, G.O. and W.M. Till. 1966. Studies on the British Dermanyssidae (Acari: Mesostigmata). II. Classification. *Bull. Br. Mus. (Nat. Hist.) Zool.* 14:107–370. (Dermanyssoidea)

Evans, G.O. and W.M. Till. 1979. Mesostigmatic mites of Britain and Ireland (Chelicerata: Acari-Parasitiformes). *Trans. Zool. Soc. London* 35:139–270.

Evans, G.O., J.G. Sheals and D. MacFarlane. 1961. *The Terrestrial Acari of the British Isles*, Vol. 1. Adlard & Son, Bartholomew Press, Dorking, England.

Funk, R.C. 1968. Revision of the family Euzerconidae and its relationship within the superfamily Celaenopsoidea (Acarina: Mesostigmata), aided by the techniques of numerical taxonomy. Ph.D. dissertation, University of Kansas, Lawrence.

Gilyarov, M.S. and N.G. Bregetova, eds. 1977. *A Key to the Soil-Inhabiting Mites, Mesostigmata* (in Russian). Nauka, Leningrad.

Hammen, L. van der. 1964. The morphology of *Glyptholaspis confusa* (Foa, 1900) (Acarida, Gamasina). *Zool. Verh.* 71:3–56.

Hammen, L. van der. 1969. Studies on Opilioacarida (Arachnida) III. *Opilioacarus platensis* Silvestr

and *Adenacarus arabicus* (With). *Zool. Meded.* 44(8):113–131.

Hirschmann, W. 1957–1966. *Gangsystematik der Parasitiformes,* Ser. 1–22, Parts 1–232. Acarol., Schriftenr. Vergl. Milbenkd. Hirschmann-Verlag Inh., Furth/Bayern.

Hirschmann, W. 1973. "Gangsystematik" of the Parasitiformes and the family Uropodidae Berlese. *Proc. Int. Congr. Acarol., 3rd, 1971,* pp. 287–292.

Hirschmann, W. 1975. Larvalsystematische, Gleiderung des Suborder Mesostigmata (Teilgang: Larve, Protonymphe, Deutonymphe). Novae Supercohortes Trichopygidiina Hirschmann 1975, Atrichopygidiina Hirschmann 1975, Nova Cohors Trachyuropodina Hirschmann 1975. *Teilgangsyst. Parasitiformes, Schrftenr. Vergl. Milbenkd.* 21(1):93–100.

Hirschmann, W. 1979. Stadienfamilien und Stadiengattungen der Atrichopygidiina, erstellt im Vergleich zum Gangsystem Hirschman, 1979. *Acarol., Schriftenr. Vergl. Milbenkd.* 26:57–68.

Hirschmann, W. and M. Hutu. 1974. Uropodiden-Forschung und die Uropodiden der Erde, geordnet nach dem Gangsystem und nach den Landern in zoogeographischen Reichen und Unterreichen. *Teilgangsyst. Parasitiformes, Schriftenr. Vergl. Milbenhd.* 20(187):6–36.

Hoffmann, A. and I.B. de Barrera. 1970. Acaros de la familia Spelaeorhynchidae. *Rev. Latino. Am. Microbiol.* 12:145–149.

Hoy, M.A. 1977. Inbreeding in the arrhenotokous predator *Metaseiulus occidentalis* (Nesbitt) (Acari: Phytoseiidae). *Int. J. Acarol.* 3(2):117–121.

Hoyt, S.C. and E.C. Burts. 1974. Integrated control of fruit pests. *Annu. Rev. Entomol.* 19:231–252. (Phytoseiidae)

Hunter, P.E. and M. Costa. 1970. Two new African species of *Megisthanus* (Thorell) (Mesostigmata: Megisthaniidae). *Fla. Entomol.* 53(4):233–240.

Hunter, P.E. and S. Glover. 1968. The genus *Diarthrophallus* Tragardh, 1946 (Acarina: Diarthrophallidae). *Proc. Entomol. Soc. Wash.* 70(3):193–197.

Hurlbutt, H. 1967. Digamasellid mites associated with bark beetles in North America. *Acarologia* 9(3):497–534.

Ishikawa, K. 1970. Studies on the mesostigmatid

mites in Japan. III. Family Podocinidae Berlese. *Annot. Zool. Jpn.* 43(2):112–122.

Jeppson, L.R., H.H. Keifer and E.W. Baker. 1975. *Mites Injurious to Economic Plants.* Univ. of California Press, Berkeley, Biological enemies of mites. pp. 75–90.

Johnston, D.E. 1961. A review of the lower uropodoid mites (formerly Thinozerconidae, Protodinychoidea, and Trachytoidea) with notes on the classification of the Uropodina. *Acarologia* 3(4):522–545.

Johnston, D.E. 1968. *An Atlas of Acari. I. The Families of Parasitiformes and Opilioacariformes,* Acarol. Lab. Publ. 172. Acarol. Publ., Columbus, Ohio.

Johnston, D.E. 1982. Parasitiformes. *In:* Parker, S. ed. *Synopsis and Classification of Living Organisms.* McGraw-Hill, New York, pp. 111–116.

Johnston, D.E. and A. Fain. 1964. *Ophiocelaeno sellnicki,* a new genus and species of Diplogyniidae associated with snakes (Acari: Mesostigmata). *Bull. Soc. R. Entomol. Belg.* 100(6):79–91.

Karg, W. 1965a. Larvalsystematische und phylogenetische Untersuchung sowie Revision des Systems der Gamasina Leach, 1815 (Acarina, Parasitiformes). *Mitt. Zool. Mus. Berlin* 41(2):193–340.

Karg, W. 1965b. Neue Erkenntnisse zum System der Gamasina (Acarina, Parasitiformes) durch Larvalsystematische Untersuchungen. *Zesz. Problm. Postepow Nauk Roln.* 65:89–114.

Karg, W. 1965c. Die Anwendung Systematisch-Phylogenetischer Arbeitsmethoden bei iner Bearbeitung der Gamasina (Acarina, Parasitiformes). *Zesz. Problm. Postepow Nauk Roln.* 65:115–138.

Karg, W. 1965d. Entwicklungsgeschichtliche Betrachtung zur Okologie der Gamasina (Acarina, Parasitiformes). *Zesz. Problm. Postepow Nauk Roln.* 65:139–155.

Kethley, J.B. 1974. Developmental chaetotaxy of a paedomorphic celaenopsoid, *Neotenogynium malkini* n. g., n. sp. (Acari: Parasitiformes: Neotenogyniidae, n. fam.) associated with millipedes. *Ann. Entomol. Soc. Am.* 67(4):571–579.

Kethley, J.B. 1977. A review of the higher categories of Trigynaspida (Acari: Parasitiformes). *Int. J. Acarol.* 3:129–149.

Kinn, D.N. 1966. A new genus and species of Schizo-gyniidae from North America with a key to the genera. *Acarologia* 8(4):576–586.

Krantz, G.W. 1978. *A Manual of Acarology.* University of Oregon Book Stores, Corvallis.

Lee, D.C. 1970. The Rhodacaridae (Acari: Mesostigmata); classification, external morphology and distribution of genera. *Rec. S. Aust. Mus.* 16(3):1–219. (Rhodacaridae, Ologamasidae)

Lee, D.C. 1974. Rhodacaridae (Acari: Mesostigmata) from near Adelaide, Australia. III. Behaviour and development. *Acarologia* 16(1):21–44. (Ologamasidae)

Lee, D.C. and P.E. Hunter. 1974. Arthropoda of the subantarctic islands of New Zealand. 6. Rhodacaridae (Acari: Mesostigmata). *N. Z. J. Zool.* 1(3):295–328. (Ologamasidae)

Lindquist, E.E. 1962. *Mucroseius monochami,* a new genus and species of mite (Acarina: Blattisociidae) symbiotic with sawyer beetles. *Can. Entomol.* 94:972–980. (Ascidae)

Lindquist, E.E. 1975. *Digamasellus* Berlese, 1905, and *Dendrolaelaps* Halbert, 1915, with descriptions of new taxa of Digamasellidae (Acarina: Mesostigmata). *Can. Entomol.* 107(1):1–43. (Digamasellidae)

Lindquist, E.E. and G.O. Evans. 1965. Taxonomic concepts in the Ascidae, with a modified setal nomenclature for the idiosoma of the Gamasina (Acarina: Mesostigmata). *Mem. Entomol. Soc. Can.* 47:1–64.

Merwe, G. van der. 1968. A taxonomic study of the family Phytoseiidae (Acari) in South Africa with contributions to the biology of two species. *S. Afr. Dep. Agric. Tech. Serv. Entomol. Mem.* 18:1–198.

Metz, L.J. and M.H. Farrier. 1969. Acarina associated with decomposing forest litter in the North Carolina piedmont. *Proc. Int. Congr. Acarol., 2nd,* pp. 43–52.

Micherdzinski, W. 1969. Die Familie Parasitidae Oudemans 1901 (Acarina, Mesostigmata). *Zak. Zool. Syst. Pol. Akad.* Nauk, Krakow.

Newell, I.M. 1947. Studies on the morphology and systematics of the family Halarachnidae Oudemans 1906 (Acari: Parasitoidea). *Bull. Bingham Oceanogr. Collect.* 10(4):235–266.

Pence, D.B. 1975. Keys, species and host list, and bibliography for nasal mites of North American birds (Acarina: Rhinonyssinae, Turbinoptinae,

Speleognathinae and Cytoditidae). *Spec. Publ. Mus. Tex. Tech. Univ.* 8:1–148.

Radovsky, F.J. 1967. The Macronyssidae and Laelapidae (Acarina: Mesostigmata) parasitic on bats. *Univ. Calif. Publ. Entomol.* 46:1–288.

Radovsky, F.J. 1969. Adapative radiation in the parasitic Mesostigmata. *Acarologia* 11(13):450–483.

Schuster, R.D. and F.M. Summers. 1978. Mites of the family Diarthrophallidae (Acarina: Mesostigmata). *Int. J. Acarol.* 4:279–385.

Strandtmann, R.W. and G.W. Wharton. 1958. *Manual of Mesotigmatid Mites Parasitic on Vertebrates*, Contrib. No. 4. Inst. Acarol., University of Maryland, College Park.

Witalinski, W. 1979. Fine structure of the respiratory system in mites from the family Parasitidae. *Acarologia* 21(13–14):330–339.

Yunker, C.E. 1973. Mites. *In*: Flynn, R.J., ed. *Parasites of Laboratory Animals.* Iowa State Univ. Press, Ames, pp. 425–492.

18

Order Ixodida

(Reuter 1909) (Anactinotrichida Zakhvatkin 1952; Hammen 1960)
(Ixodides Leach 1815; Ixodei Duges 1834; Metastigmata G.
Canestrini 1891, Evans et al. 1961; Ixodoidea Reuter 1909; Ixodida
Sundevall 1833; Hammen 1968).

All ticks are mites, but not all mites are ticks. The long-standing semantical separation in the description of "mites and ticks," is not necessarily true. Ticks are not any more distinctive as a group than water mites or velvet mites, the eriophyids, or others. However, the common usage of the verbal description has established a slight dichotomy that persists.

Ticks are obligate, hematophagous parasites of vertebrates (mammals, birds, reptiles, and amphibians) that are also among the largest of the Acari. Some velvet mites may be larger than unengorged ticks, but are still smaller than the larger engorged female ticks (e.g., *Ornithodoros acinus*, which exceeds 3 cm in length). Their striking morphology and large size distinguish ticks from most other Acari (Hoogstraal, 1970, Vol. 1). They range in size from 2000 to over 30 000 μm (Krantz, 1978).

From a historical sense, ticks have been studied the longest and perhaps are the best known of the Acari. These ectoparasites have been associated with humans and domestic animals for over three millennia. The oldest record and figure of ticks is one from a hyenalike animal in the tomb of Ora Abnel-Nago (1060–1061 B.C.) and from Egypt in 1500 B.C. (Arthur, 1965b). Homer described the annoyance caused by these mites and the use of the crushed bodies of engorged ticks as a remedy for disease and for love potions. In the Old World up to the eighteenth century, the Miana bug (*Argas persicus*) and the castor bean ticks (*Ixodes*) were well-known forms. In 1746 Linnaeus included the ticks in the genus *Acarus,* and in 1795 Latreille described 11 genera, including *Argas* and *Ixodes.* Then followed a long list of specialists whose efforts in Europe, Africa, Russia, and the United States resulted in the extensive literature currently available and summarized so well in Hoogstraal's seven-volume work (1970–1982), *Bibliography of Ticks and Tickbone Diseases, from Homer (about 800 B.C.) to 31 December 1981.* Additional references are available in Smith (1973).

Porcius Cato (200 B.C.) indicated that following treatment for ticks on sheep no sores would be present and wool would be more plentiful. Columella (60 B.C.) recommended that passing the hands under the belly of a cow would cause ticks to be removed. In 1730 Haller perpetuated Varro's recommendation that "ears and toes of dogs should be smeared because flies, ticks (ricini) and fleas produce sores in these situations" (Arthur, 1965a).

All ticks are stationary or transitional ectoparasitic acarines. It is little wonder that they have received so much attention. They are the most important biological vectors of diseases affecting people and animals in the temperate and tropical regions of the earth and surpass all other arthropods in the number and variety of diseases transmitted. Although humans are not the normal hosts for ticks, as transmitters of human disease alone, ticks are surpassed only by the lowly mosquitoes. In addition to vectoring diseases, the bites of some soft ticks may be painful. Secondary infections and injuries add to the discomforts and problems of hosts affected by ticks.

Ticks are cosmopolitan distributed through many habitats and different ecological situations. Some species are very specific in their adaptations to selected habitats (e.g., Boutonneuse fever is limited to the moist environment of the coastal Mediterannean). Forest, marsh, deserts, steppes, mountains, and high meadows are among the various locations of ticks. They have few natural enemies and an extremely wide host range.

MORPHOLOGY

Broadly characterized, ticks (formerly Metastigmata; Ixodides) have stigmata without sinuous peritremes (compared to Gamasida). The stigmata are located posterior to coxae IV (hard ticks) or anterodorsal to coxae IV (soft ticks) (Fig. 18-1). All ticks lack an epistome (tectum), corniculi, tritosternum, and an apotele on the palp tarsus. The denticulate hypostome (with recurved or retrorse teeth) is formed from

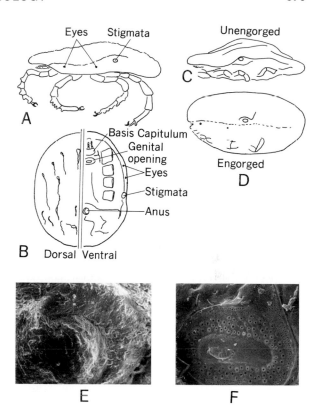

Figure 18-1. Diagram of a soft tick, *Ornithodoros savignyi:* (A) lateral view; (B) dorso lateral view; (C) unengorged; (D) engorged (after Hoogstraal, 1956); (E) SEM of stigmatal plate of *Argas persicus* (1000×); (F) SEM of stigmatal plate of female *Dermacentor andersoni* (170×).

the hypostomal endites and is diagnostic in most ticks. This toothed tongue (Fig. 18-2) anchors the tick to its host. The chelicerae are two-segmented with a membranous mantle around the toothed digit except for the cutting edge. The denticulate movable digit is modified for horizontal incisiton of the host integument; cheliceral muscles are extensive. The chelicerae lack setae and lyrifissures. Palps are three- or four-segmented, without claws, leglike in Argasidae, and with knifelike edges (cultriform) and a fused tibiotarsus in Ixodidae (Figs. 18-2B, 18-3). Soft ticks have no pulvillus on the

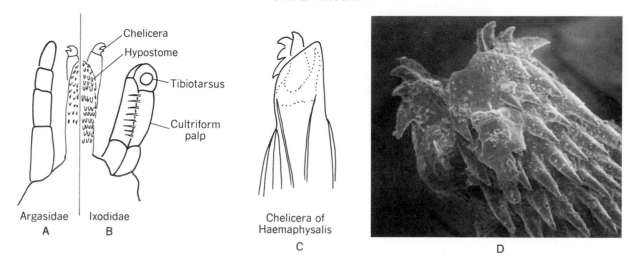

Figure 18-2. Chelicerae and hypostome of ticks: (A) Argasidae; (B) Ixodidae. (C) Enlarged chelicera of *Haemaphysalis punctata* (after Nuttall et al., 1916); (D) SEM of chelicera and hypostome of *Otobius megnini* (1000×).

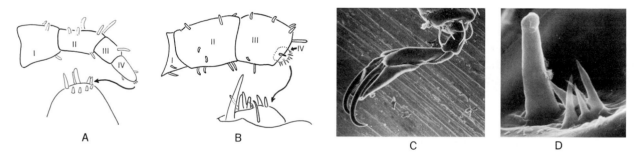

Figure 18-3. Schematic drawing of palps of ticks: (A) *Ornithodoros savignyi*; (B) *Hyalomma dromedarii* (after Hammen, 1968). (C) SEM of tarsal claws and pulvillus of *Dermacentor andersoni* (100×); (D) sensilla of Haller's organ of *Boophilus annulatus* (SEM, 1000×).

tarsi; hard ticks exhibit a pulvillus (Fig. 18-3C). The first leg of ticks exhibits a tarsal complex of sensory organs and sensilla, Haller's organ, with humidity receptors, chemoreceptors, olfactory receptors, and mechanoreceptors (Figs. 11-8; 18-3D).

The integument varies between the families of ticks from a dull-hued, leathery surface that may be mammilliform, tuberculate, papillated, or wrinkled, to a smooth, slightly sculptured,

striated, sclerotized, and colorful structure. The genital aperture is intercoxal in both sexes, usually between coxae I and II. The anal valves are subterminal in position and exhibit setae on their surfaces.

With respect to the stigmata of ticks, in Argasidae, where these respiratory openings are anterodorsad of coxae IV, the spiracular plate is crescentic in shape with a distinct macula or cone shaped. Aeropyles or goblets are absent

(Fig. 18-1E, SEM) in Argasidae. Ixodidae have a single pair of stigmatal plates posterior to coxae IV in both nymphs and adults. The hexapod larvae are without stigmatal openings. Octopod nymphs have stigmatal plates. The shape of the plate varies in the adults. A macula (or ecdysial scar) is present in the center of the plate or eccentric in placement. Goblets or goblet cells (aeropyles) are present in the surface of the plate and may be taxonomically diagnostic (Fig. 18-1F, SEM).

CLASSIFICATION

The classification of ticks involves three families, Argasidae, Nuttalliellidae, and Ixodidae, characterized in the following comparative summary. The reader is referred to Camicas and Morel (1977), Clifford et al. (1964, 1973), Hoogstraal et al. (1973), Morel (1969, 1976), and Pospelova-Shtrom (1969) for details. (The late Dr. Harry Hoogstraal (1986) identifies the phylogeny of families and subfamilies of ticks together with genera and numbers of species described in each genus.)

Family Argasidae	Family Ixodidae
1. Scutum absent.	1. Scutum present (all stages).
2. Slight sexual dimorphism.	2. Marked sexual dimorphism male scutum almost completely covering dorsum; female scutum limited, anterior; alloscutum allows engorgement.
3. Capitulum subterminal, ventral, inferior; hood, camerostome may be present.	3. Capitulum anterior and terminal; hood absent.
4. Stigmata small, anterior to coxae IV; aeropyles absent.	4. Stigmata large, distinctly platelike, posterior to coxae IV; aeropyles present.
5. Porose areas absent in both females and males.	5. Porose areas present in dorsum of female capitulum.
6. Palps free, leglike; palp tarsus not inserted or depressed into penultimate segment.	6. Palps cultriform, modified, ridged, adpressed to hypostome; palp tarsus inserted into penultimate segment, with prominent chemosensory setae.
7. Integument leathery, wrinkled, mammillated, tuberculated; if present, eyes located in supracoxal folds.	7. Integument smooth, striated; eyes in lateral edges of scutum in both sexes.
8. Pulvilli absent or rudimentary (in some larvae); claws present; dorsal humps present on tibia, tarsus, and metatarsus.	8. Pulvilli and claws present; dorsal humps absent on leg segments; coxae sometimes spined (males).
9. Blood meal before each molt (larva, nymph(s), adult); feedings intermittent, rapid, multiple blood meals.	9. Blood meal before each molt (larva, nymph, adult); feeding slow, sustained, large blood meal.
10. Coxal glands present.	10. Coxal glands absent.
11. Under 1000 eggs laid in small batches; oviposit several times; life cycle extended.	11. Over 1000 eggs laid in one massive batch; life cycle concentrated.
12. Resistant to starvation; low dispersal.	12. Not resistant to starvation; high dispersal.

Family Nuttalliellidae

1. Weakly scutate, scutum weakly sclerotized, "pseudoscutum," mainly like remaining integument.

2. Dimorphic.

3. Capitulum anterior, visible from above; hood absent; hypostome usually concealed by palp, teeth rudimentary (scattered hooklets).

4. Stigmata posterolateral to coxae IV, spiracular plates absent.

5. Porose areas absent.

6. Palps 3-segmented, freely articulated, joints short, not ridged; palpal tarsus terminal; segment I massive, enfolding hypostome; inner side of segment II grooved (suggestion of reduction in Ixodidae).

7. Integument leathery, extremely convoluted with closely spaced pits and elevated rosettes; peripheral integument undifferentiated.

8. Organs of unknown function posterior to coxae III; ball and socket leg joints; Haller's organ modified.

The following brief synopsis identifies the superfamilies, families and subfamilies of the order Ixodida.

Superfamily Argasoidea, Family Argasidae; Subfamilies Anticolinae, Argasinae, Ornithodorinae, Otobinae

Superfamily Nuttallielloidea, Family Nuttalliellidae

Superfamily Ixodoidea, Family Ixodidae; Subfamilies (Prostriata); Ixodinae; (Metastriata) Amblyomminae, Haemaphysalinae, Hyalomminae, Rhipicephalinae

Superfamily Argasoidea, Family Argasidae

The Argasidae are the soft or nonscutate ticks. Relatively scant in numbers in temperate and arctic regions, they occur in more arid and semiarid habitats, are able to tolerate long periods of dessication, and reach their greatest abundance in dry regions. If they move into more humid areas, they tend to select the drier niches (Smith, 1973).

Argasids are multihost ticks. They are nidicolous (nest loving) or are found in caves, burrows, cracks, and crevices of buildings in semitropical and tropical areas (Cooley and Kohls, 1944). Argasids with flattened bodies live in cracks and crevices; those with "raisinlike" bodies burrow in the soil. These ticks parasitize birds (e.g., poultry are infected by a spirochaetosis transmitted by *Argas persicus*), snakes, turtles, bats, and eight other orders of mammals, including humans. Although the argasids are blood feeders, they feed intermittently on hosts that return to the nest, burrow, or shelter and remain on the host only long enough to obtain a blood meal (2 min to 2 h) (Krantz, 1978). They do not remain for replete engorgement as do hard ticks. The rapid and frequent feeding of soft ticks necessitates several nymphal instars (Lees, 1952). Their rapid feeding on animals that are usually at rest reduces the danger of their transportation during bad weather and to unfavorable locations. They normally remain close to the burrow, den, or house of the host animals. Lees (1952) suggests that the several blood meals a female argasid takes to nourish the several egg batches is an adaptation to maintain her fecundity. Survival of an argasid population does not depend on large numbers of eggs, as it does in the Ixodidae. Some of these soft ticks may live for years without food; for example, *Ornithodoros papillipes* Birula may survive for 11 years (Pavlovsky and Skrynnik, 1960). Balashov (1972) indicates other argasids that lived up to 10 years without food.

Argasids are considered to be less important than hard ticks as disease transmitters, but do act as vectors for arboviruses and bacterial and rickettsial infections (Krantz, 1978). Three of the genera of the family are medically important (Smith, 1973).

Between 140 to 167 species of soft ticks are

known in five genera (Krantz, 1978; Hoogstraal, 1986). Examples of these ticks include *Argas persicus*, the fowl tick; *Ornithodoros hermsi*, Herms' relapsing fever tick; and *Otobius megnini*, the spinose ear tick. A neotropical species, *Nothoaspis redelli* Keirans and Clifford (1975) bears a pseudoscutum also, but most of its other characteristics conform to the features defined for the family Argasidae (Fig. 18-4).

Superfamily Nuttallielloidea, Family Nuttalliellidae

The family Nuttalliellidae is considered intermediate in taxonomic position between the Ix-

odidae and the Argasidae. Nuttalliellids have an apical capitulum, a "pseudoscutum" (?true scutum like the Ixodidae), lack a ventral paired organ and coxal and supracoxal folds as in Argasidae, and have structural similarity of the dorsal and ventral integument. The principal characters relating this family to the Argasidae are the unarmed coxae, hypostomal features, the integumental structure, and the lack of areae porosae (Keirans et al., 1976). *Nuttalliella namaqua* Bedford is the monotypical species of the family and is found in Tanzania and South and Southwest Africa in bird and mammal habitats. Apparently it is also a parasite of hyrax (*Procavia*) (Krantz, 1978). The Nuttalliellidae are of no medical importance (Smith, 1973).

Superfamily Ixodoidea, Family Ixodidae

The family Ixodidae is the largest family of ticks. The scutate or hard ticks are cosmopolitan on terrestrial vertebrates such as snakes, lizards, land tortoises, some amphibians, mammals, and birds. They are found in habitats where moisture is more abundant. The ixodids comprise 650 species in 19 genera (Krantz, 1978; Camicas and Morel, 1977; Hoogstraal, 1986) e.g., *Aponoma*, *Amblyomma*, *Boophilus*, *Dermacentor*, *Haemaphysalis*, *Hyalomma*, *Ixodes*, and *Rhipicephalus*. *Aponoma* species parasitize snakes and varanid lizards. Mammals are parasitized by *Dermacentor*, *Haemaphysalis*, *Boophilus*, *Rhipicephalus*, and *Amblyomma*. Species of *Amblyomma* also are parasites of land tortoises, sea snakes, lizards, and some birds (Krantz, 1978; Hoogstraal, 1986).

The hard ticks are called scutate ticks because of the sclerotized prodorsal, podonotal scutum in immatures and adults. The opisthosoma is covered with a soft cuticle, the alloscutum, in immatures and adult females, which permits engorgement. In males the scutum or dorsal sheild covers the entire idiosoma and thus limits engorgement. The venter of these ticks is variously sclerotized. Hard ticks are flattened dorsoventrally and are discoidal in outline. The capitulum is terminal and directed

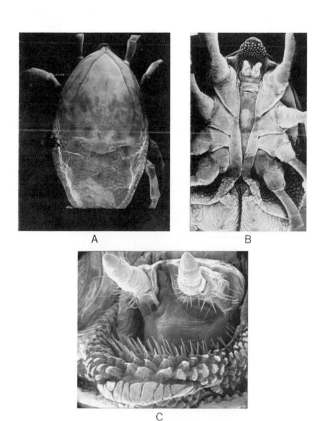

Figure 18-4. *Nothoaspis reddelli*: (A) dorsal view; (B) ventral view; (C) gnathosoma. (SEMs by permission of Keirans and Clifford (1975).)

TABLE 18-1[a]

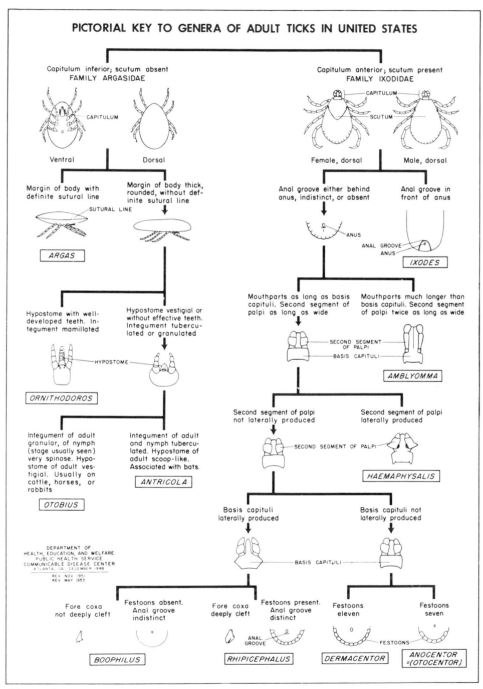

[a]From H.D. Pratt and K.S. Littig, USDHEW, PHS, *Ticks of Public Health Importance and Their Control*, Atlanta, Georgia, 1954.

anteriorly. Coxae may be spurred; coxae II and III are immovable.

Hard ticks represent three categories related to the number of hosts required to complete the life cycle. In one-host ticks (e.g., *Boophilus*) the larval, nymphal, and adult stages engorge on the same animal and molts occur on the hosts. Two-host ticks (e.g., *Rhipicephalus*) require two hosts to complete the life cycle. The larva feeds and molts on the first host and the engorged nymph drops to the ground. The adult seeks another host for completion of the cycle. Three-host ticks require a different host for each instar and each molt takes place on the ground before the new host is sought (Smith, 1973). The feeding of ixodids is slower than that of argasids and is extended over a much longer period of time, sometimes several months in the case of males; other males do not feed (*Ixodes*). Because of the flexible alloscutum, females have the capability of feeding to repletion in 2 or 3 days. The blood meal is requisite for oviposition and females may lay as many as 18 000 eggs, with an average of 2000 to 8000 (Krantz, 1978).

See Table 18-1 for a key to adult ticks in the United States.

TICKS AND DISEASE

People are not normal hosts for these vectors, but are accidentally affected when a tick is available in the location, the disease of the native fauna is transmissible to humans, and the tick uses a person as a host. Humans must be regarded as incidental hosts and play no part in the enzootic reservoirs of wild animals and ticks (Smith, 1973).

Ticks affect all vertebrates except fishes and are distributed throughout the world. Some species are adapted to specific environments (e.g., Boutonneuse fever, coastal Mediterranean), but suitable habitats include forest, marsh, steppes, and desert. In addition to vectoring, ticks may inflict painful bites (some Argasidae), and many serve as reservoirs for etio-

logical agents of disease (e.g., *Rickettsiae* (RMSF), tick-borne typhus, Boutonneuse fever) and promote transphasic and transovarial involvement with the disease.

Some of the features that make ticks effective vectors are:

1. They attach firmly with the use of the shearing chelicerae and the anchoring effect of the toothed hypostome. The saliva has a narcotizing effect and deadens the tissue being penetrated. Some hard ticks attach the gnathosoma or capitulum to the host by means of a cement (Fig. 7-1).

2. Ticks suck blood and must have a blood meal for development. The method of feeding is to alternatively secrete the saliva and inject it into the wound, which is followed by sucking of the blood. J. D. Gregson (personal communication) indicates that there is a flow of clear fluid into the wound, followed by a returning surge of blood. This feeding facilitates the passage of etiological agents into the wound of the host.

3. Ixodid ticks feed slowly and it may take days for the female to engorge, which greatly assures the passage of pathogens into the host. Argasids feed more rapidly, but the blood meal enables transmission of the disease agents.

4. The narcotizing effects of the saliva enable the tick to go unnoticed by the host, although some argasida cause a painful bite.

5. Ticks are resistant to environmental stresses and may live extended periods of time without food. Argasid ticks may live for months without a blood meal.

6. Ticks accept and use a wide range of hosts.

7. Some species are preyed upon by birds, but ticks have few natural enemies.

8. Ticks may transmit disease agents transovarially (to the egg) and transtadially (larva to nymph, to nymph to adult).

Six principal types of diseases are transmitted by ticks:

1. Rickettsial—Rickettsiosis: RMSF; Japanese river fever, Tsutsugamushi disease.
2. Viral: Colorado tick fever.
3. Bacterial, spirochaetal: relapsing fever.
4. Protozoan: (e.g., *Babesia bigemina*).
5. ?Helminthic infections.
6. Toxicosis—tick paralysis—descending motor paralysis due to neurotoxic salivary secretions.

Other effects resulting from the parasitism of ticks include the irritation from the bites, especially from seed ticks and argasids; dermatitis; exsanguination of baby chicks (*Argas persicus*); secondary infections after the bites from irritation and scratching; effects of the pests on people and animals.

Two major categories of human viral diseases are vectored by ticks: (1) the large, varied Russian spring–summer complexes involving antigenic group B (e.g., Powassan encephalitis, Negishi encephalitis, Langat encephalitis, Kyasuanur forest disease (KFD), Omsk hemorrhagic fever (OHF), Russian spring–summer encephalitis (RSSE), central European tick-borne encephalitis (TE), and louping ill); (2) the arbovirus infections with separate subgroups of viruses in antigenic group B as well as antigenically independent viruses (Colorado tick fever (CTF), Quaranfil fever, Crimean–Congo hemorrhagic fever (CHF–Congo), Kemerovo

TABLE 18-2 Some Important Tick Species and Their Hosts[a]

Genus	Species	Common Name	Common Hosts
		Ixodidae (Hard Ticks)	
Amblyomma	*americanum*	Lone star tick	Livestock, dog, deer, birds, man
Amblyomma	*cajennense*	Cayenne tick	Livestock, deer, dog, man
Amblyomma	*maculatum*	Gulf Coast tick	Livestock, deer, birds, dog man
Boophilus	*annulatus*	Cattle tick	Livestock, deer
Dermacentor	*albipictus*	Winter tick	Large domestic and wild animals, man
Dermacentor	*andersoni*	Rocky Mountain wood tick	Livestock, wild mammals, man
Dermacentor	*occidentalis*	Pacific Coast tick	Wild mammals, livestock
Dermacentor	*variabilis*	American dog tick	Dogs, small mammals, large mammals, man
Haemaphysalis	*leporispalustris*	Rabbit tick	Rabbits, small mammals, birds
Ixodes	*scapularis*	Black-legged tick	Mammals, birds, man
Rhipicephalus	*sanguineus*	Brown dog tick	Dogs, other mammals
		Argasidae (Soft Ticks)	
Argas	*persicus*	Fowl tick	Fowl
Ornithodoros	*coriaceus*	Pajaroello tick	Mammals, man
Ornithodoros	*hermsi*	Herms' relapsing fever tick	Chipmunk, man
Ornithodoros	*parkeri*	Parker's relapsing fever tick	Man, rodents
Ornithodoros	*turicata*	Relapsing fever tick	Man, rodents, hogs
Otobius	*megnini*	Spinose ear tick	Livestock, large wild mammals, man

[a]From H.D. Pratt and Littig, K.S., USDHEW, PHS, *Ticks of Public Health Importance and Their Control*, Atlanta, Georgia, 1954.

tick fever, and Tribee) (Smith, 1973). Ticks also may be involved as vectors of eastern and western equine encephalitis and lymphocytic choriomeningitis.

Some important tick species and their hosts are shown in Table 18-2.

Colorado Tick Fever and Rocky Mountain Spotted Fever

On the eastern slope of the Rocky Mountains the most common infection from wood ticks is Colorado tick fever. This is a viral disease that infects about 25% of the people exposed to the bite of *Dermacentor andersoni*. The disease cannot be treated with antibiotics, but Rocky Mountain spotted fever, a tick-borne typhus disease, does respond to such treatment. The latter disease affects 2 to 3% of tick-associated fevers in Colorado; it is more prevalent in the eastern United States.

Ticks acquire the etiological agent of CTF by feeding on infected rodents and in turn reinfect other rodents in a cyclic way. Ticks may become infected as larvae, nymphs, or adults since each stage requires a blood meal for maturation. In RMSF transmission is not as dependent upon rodents and the ticks may transfer the causative agent from the egg to progeny (transovarial and transtadial transmission).

The incubation period for CTF is about 5 days following the bite of an infected tick. The victim is usually ill for several days, appears to recover and then may suffer a relapse. The relapse occurs in about half of the cases and the victims may feel sicker than during the initial phase and may have temperatures of 102–104°F.

Symptoms of these two diseases are similar in the early stages: headaches, fever, sore muscles, and general malaise. A rash may occur in each of these diseases; it lasts only a few days in CTF. In RMSF a severe rash develops on the palms, trunk, arms, and legs within 3 to 7 days after the onset of fever. The rash spots (petechae) are reddened to purplish in color. If untreated, RMSF may be fatal.

Following are some of the tick-borne diseases that affect humans and the causative agent of each.

Disease	Etiological Agent
Boutonneuse fever	*Rickettsia conori*
Bullis fever	*Rickettsia sp.*
Colorado tick fever	Filtrable virus
Louping ill (trembling disease of sheep; accidental in humans)	Virus
Encephalitis, Russian	*Erro sylvestris* virus
Maculatum disease	*Rickettsia sp.*
Plague	*Yersenia pestis*
Q fever (9 mile fever)	*Coxiella burnetti*
Relapsing fever	*Borrelia duttoni* (spirochaete)
Texas cattle fever	*Babesia bigemina* (piroplasm)
Spotted fever (*Haemaphysalis, Dermacentor, Amblyomma*)	*Rickettsia rickettsi*
Tularemia	*Francisella tularensis*
Tick paralysis	tick saliva protein
Sweating sickness	??

REFERENCES

Anastos, G. 1950. The scutate ticks, or Ixodidae, of Indonesia. *Entomol. Am.* 30:1–144.

Anastos, G., S.T. Kaufman and S. Kadarson. 1973. An unusual reproductive process in *Ixodes kopsteini* (Acarina; Ixodidae). *Ann. Entomol. Soc. Am.* 66(2):483–484.

Anderson, J.F., I.A. Magnarelli and J.E. Keirans. 1984. Ixodid and argasid ticks in Connecticut, U.S.A.: *Aponomma latum, Amblyomma dissimile, Haemaphysalis leachi* group, and *Ornithodoros kelleyi. Int. J. Acarol.* 19(3):149–151.

Arthur, D.R. 1960. *Ticks. A Monograph of the Ixodoidea. Part 5. On the genera* Dermacentor, Anocentor, Cosminomma, Boophilus *and* Margaropus. Cambridge Univ. Press, London and New York.

Arthur, D.R. 1962. *Ticks and Disease.* Pergamon, Oxford.

Arthur, D.R. 1963. *British Ticks.* Cambridge University Press, London.

Arthur, D.R. 1965a. *Ticks of the Genus Ixodes in Africa.* University of London, London.

Arthur, D.R. 1965b. Ticks in Egypt in 1500 B.C. *Nature (London)* 206:1060–1061.

Arthur, D.R. 1966. The ecology of ticks with reference to the transmission of Protozoa. *In:* Soulsby, E.J.L., ed. *Biology of Parasites.* Academic Press, New York, pp. 61–84.

Balashov, Y.S. 1972. Bloodsucking ticks (Ixodoidea)—vectors of disease in man and animals. *Misc. Publ. Entomol. Soc. Am.* 8(5):161–376.

Bedford, G.A.H. 1931. *Nuttalliella namaqua,* a new genus and species of tick. *Parasitology* 23(2):230–232.

Beklemishev, V.N. 1948. On interrelationships between the systematic position of the agent and the vector of transmissible diseases of terrestrial vertebrates and man. *Med. Parazito. Parazit. Bolezni* 17:385–400.

Berland, L. 1932. *Les Arachnides (Biologie, Systematique).* Encyl. Ent., Vol. 16, Paris, pp. 1–485.

Booth, T.F., D.J. Beadle and R.J. Hart. 1984. The ultrastructure of Gene's organ in the cattle tick *Boophilus microplus* Canestrini. *Acarol. [Proc. Int. Congr. Acarol.], 6th, 1982,* Vol. 1, pp. 261–267.

Burgdorfer, W. and M.G.R. Varma. 1967. Trans-stadial and transovarial development of disease agents in arthropods. *Annu. Rev. Entomol.* 12:347–376.

Camicas, J.L. and P.C. Morel. 1977. Position systematique et classification des tiques (Acarida: Ixodida). *Acarologia* 18:410–420.

Casals, J., H. Hoogstraal, K.M. Johnson, A. Shelokov, N.H. Wiebenga and T.H. Work. 1966. A current appraisal of hemorrhagic fevers in the U.S.S.R. *Am. J. Trop. Med. Hyg.* 15:751–764.

Casals, J., B.E. Henderson, H. Hoogstraal, K.M. Johnson and A. Shelokov. 1970. A review of Soviet viral hemorrhagic fevers, 1969. *J. Infect. Dis.* 122:437–453.

Clifford, C.M., G.M. Kohls and D.E. Sonenshine. 1964. The systematics of the subfamily *Ornithodorinae* (Acarina: Argasidae). I. The genera and subgenera. *Ann. Entomol. Soc. Am.* 57(4):429–437.

Clifford, C.M., D.E. Sonenshine, J.E. Keirans and G.M. Kohls. 1973. Systematics of the subfamily Ixodinae (Acarina: Ixodidae). I. The subgenera of *Ixodes. Ann. Entomol. Soc. Am.* 66(3):489–500.

Cooley, R.A. 1938. The genera *Dermacentor* and *Otocentor* (Ixodidae) in the United States with studies in variation. *Natl. Inst. Health Bull.* 171:1–89.

Cooley, R.A. and G.M. Kohls. 1944. The genus *Amblyomma* (Ixodidae) in the United States. *J. Parasitol.* 30:77–111.

Cooley, R.A. and G.M. Kohls. 1945. *The Genus Ixodes in North America.* U.S. Govt. Printing Office, Washington, D.C.

Elbl, A. and G. Anastos. 1966. *Ixodid ticks (Acarina, Ixodidae) of Central Africa.* 4 volumes. *Mus. R. Afr. Cent., Tervuren, Belg., Anna., Ser. 8,* Nos. 145–148.

Feldman-Muhsam, B. 1973. New evidence on the function of porose areas of Ixodid ticks. *Experientia* 29:799–800.

Hammen, L. van der. 1968. Introduction générale à la classification, la terminologie morphologique, l'ontogénèse et l'évolution des Acariens. *Acarologia* 10(3):401–412.

Herms, W.B. 1917. Contribution to the life history and habits of *Ornithodorus megnini. J. Econ. Entomol.* 10:407–411.

Homsher, P.J. and D.E. Sonenshine. 1979. Scanning electron microscopy of ticks for systematic studies: 3. Structure of Haller's organ in five species of the subgenus *Multidentatus* of the genus Ixodes. In: Rodriguez, J.G., ed. *Recent Advances in Acarology.* Academic Press, New York, Vol. 2, pp. 485–490.

Hoogstraal, H. 1956. *African Ixodoidea. I. Ticks of the Sudan.* USNAMRU, Cairo, Egypt U.A.R.

Hoogstraal, H. 1966. Ticks in relation to human diseases caused by viruses. *Annu. Rev. Entomol.* 11:261–308.

Hoogstraal, H. 1967a. Tickborne hemorrhagic fevers, encephalitis and typhus in U.S.S.R. and souther Asia. (Theobald Smith Memorial Lecture.) *Exp. Parasitol.* 21:98–111.

Hoogstraal, H. 1967b. Ticks in relation to human diseases caused by *Rickettsia* species. *Annu. Rev. Entomol.* 12:377–420.

Hoogstraal, H. 1970–1982. *Bibliography of Ticks and Tickborne Diseases*, Vols. 1–7, Spec. Publ. USNAMRU No. 3, Cairo, Egypt, U.A.R. (Vol. 1, Oct. 1970; Vol. 2, Dec. 1970; Vol. 3, June 1971; Vol. 4, June 1972; Vol. 5, Pt. I, Aug. 1974; Vol. 5, Pt. II, April 1978; Vol. 6, July 1981; Vol. 7, May 1982).

Hoogstraal, H. 1986. Theobald Smith: His scientific work and impact. *Bull. Entomol. Soc. Am.* 32(1):23–34.

Hoogstraal, H., M.N. Kaiser, M.A. Taylor, S. Gaber and E. Guindy. 1961. Ticks (Ixodoidea) on birds migrating from Africa to Europe and Asia. *Bull. W.H.O.* 24:197–212.

Hoogstraal, H., C.M. Clifford, Y. Saito and J.E. Keirans. 1973. *Ixodes (Partipalpiger) ovatus* Neumann, subgen. nov.: identity, hosts, ecology, and distribution (Ixodoidea; Ixodidae). *J. Med. Entomol.* 10(2):157–164.

Kaiser, M.N. 1966a. Viruses in ticks. I. Natural infections of *Argas (Persicargas) arboreus* by Quaranfil and Nyamanini viruses and absence of infections in *A. (P.) persicus* in Egypt. *Am. J. Trop. Med. Hyg.* 15:964–975.

Kaiser, M.N. 1966b. Viruses in ticks. II. Experimental transmission of Quaranfil virus by *Argas (Persicargus) arboreus* and *A. (P.) persicus. Am. J. Trop. Med. Hyg.* 15:976–985.

Keirans, J.E. and C.M. Clifford. 1975. *Nothoaspis reddelli*, new genus and new species (Ixoidoidea: Argasidae), from a bat cave in Mexico. *Ann. Entomol. Soc. Am.* 68(1):81–85.

Keirans, J.E., C.M. Clifford, H. Hoogstraal and E.R. Easton. 1976. Discovery of *Nuttalliella namaqua* Bedford (Acarina: Ixodoidea Nuttalliellidae) in Tanzania and redescription of the female based on scanning electron microscopy. *Ann. Entomol. Soc. Am.* 69(5):926–932.

Lees, A.D. 1952. The role of cuticle growth in the feeding of ticks. *Proc. Zool. Soc. London* 121:759–772.

Morel, P.C. 1969. Contribution à la connaissance de la distribution des tiques (Acariens, Ixodidae et Amblyommidae) en Afrique ethiopienne continentale. Thèse D.Sc., Orsay, Ser. A, No. 575.

Morel, P.C. 1976. Étude sur les tiques d'Ethiopie (Acariens, Ixodides). *Inst. d'Élevage et de Med. Veterinaire des Pays Tropicaux.* Maisons-Alfort, France, 326 pp.

Naumov, R.L., E.N. Levkovich and O.E. Rzhakhova. 1963. The part played by birds in the circulation of tick-borne encephalitis virus. *Med. Parazitol. Parazit. Bolezni* 1:18–29. (In Russian, English translation, United States Medical Research Unit No. 3, Cairo, translation 141.)

Nuttall, G.H.F. 1911. On the adaptation of ticks to the habits of their host. *Parasitology* 4:46–67.

Nuttall, G.H.F. and C. Warburton. 1911. *Ticks: A Monograph of the Ixodoidea. Part II. Ixodidae,* pp. 105–348. Cambridge Univ. Press, London and New York.

Nuttall, G.H.F. and C. Warburton. 1915. *Ticks: A Monograph of the Ixodoidea. Part III. The Genus* Haemaphysalis, pp. 349–550. Cambridge Univ. Press, London and New York.

Nuttall, G.H.F., C. Warburton, W.F. Cooper and L.E. Robinson. 1908. *Ticks: A Monograph of the Ixodoidea. Part I. Argasidae.* Cambridge Univ. Press, London and New York.

Parker, R.R. 1934. Recent studies of tick-borne diseases made at the U.S. Public Health Service Laboratory at Hamilton, Montana. *Proc. Pac. Sci. Congr., 5th, 1933,* Vol. 5, pp. 3367–3374.

Pavlovsky, E.N. and A.N. Skrynnik. 1960. Laboratoriumbeobachtungen an der Zecke. *O. hermsi* Wheeler, 1935. *Acarologia* 2:62–65.

Phillip, C.B. 1963. Ticks as purveyors of animal ailments. *Adv. Acarol.* 1:285–325.

Phillip, C.B. and G.M. Kohls. 1952. Elk, winter ticks and Rocky Mountain spotted fever; a query. *Public Health Rep.* 66:1672–1675.

Pomerantzev, B.I. 1950. (Ixodid ticks (Ixodidae).) *Fauna SSR* 4(2):1–224. (In Russian, translated by Alena Elbl, Am. Inst. Biol. Sci., Washington, D.C., 1959.

Pospelova-Shtrom, M.V. 1969. On the system of classification of ticks of the family Argasidae Can. 1890. *Acarologia* 11(1):1–22.

Rehacek, J. 1965. Development of animal viruses and rickettsiae in ticks and mites. *Annu. Rev. Entomol.* 10:1–24.

Rich, G.B. 1971. Disease transmission by the Rocky Mountain wood tick, *Dermacentor andersoni* Stiles, with particular reference to tick paralysis in Canada. *Vet. Med. Rev.,* pp. 1–27.

Robinson, L.E. 1926. *Ticks: A Monograph of the Ixodoidea. Part IV. The Genus* Amblyomma. Cambridge Univ. Press, London and New York.

Santos Dias, J.A.T. 1963. Contribuicao para o estudio da sistematica dos acaros da subordem *Ixodoidea* Banks, 1894. I. Familia *Ixodidae* Murray, 1877. *Mem. Est. Mus. Zool. Univ. Coimbra* (285):1–34.

Sauer, J.R. and J.A. Hair, eds. 1986. *Morphology, Physiology and Behavioral Biology of Ticks.* Ellis Horwood, Ltd., Chichester, England; Wiley (Halstead), New York.

Schulze, P. 1935. Zu vergleichenden Anatomie der Zecken. (Das Sternale, die Mundwerkzeuge, Analfurchen und Analbeschilderung, ihre Bedeutung Ursprunglichkeit und Luxureiren). *Z. Morphol. Oekel. Tiere* 30(I):1–40.

Smith, C.E.G. 1962. Ticks and viruses. *Symp. Zool. Soc. London* 6:199–211.

Smith, K.G.V., ed. 1973. *Insects and Other Arthropods of Medical Importance.* Trustees of the British Museum (Nat. Hist.), London.

Sonenshine, D.E., C.M. Clifford and G.M. Kohls. 1962. The identification of larvae of the genus *Argas* (Acarina: Argasidae). *Acarologia* 4(2):193–214.

Southcott, R.V. 1976. Arachnidism and allied syndromes in the Australian region. *Rec. S. Aust. Children's Hosp.* 1(1):97–186.

Taylor, R.M., H.S. Hurlbut, T.H. Work, J.R. Kingston and H. Hoogstraal. 1966. Arboviruses isolated from *Argas* ticks in Egypt: Quaranfil, Chenuda, and Nyamanini. *Am. J. Trop. Med. Hyg.* 15:76–86.

Theiler, G. 1962. *The Ixoidea Parasites of Vertebrates in Africa South of the Sahara (Ethiopian Region),* Vol. 1, Proj. S. 9958. Report to the Director of Veterinary Services, Onderstepoort.

Thomas, L.A., R.C. Kennedy and C.M. Ecklund. 1960. Isolation of a virus closely related to Powassan virus from *Dermacentor andersoni* collected along the North Central Cache la Poudre River, Colo. *Proc. Soc. Exp. Biol. Med.* 104:355–359.

Trapido, H., M.G.R. Varma, P.K. Rajagopalan, K.P.R. Singh and M.J.A. Rebello. 1964. A guide to the identification of all stages of *Haemaphysalis* ticks of South India. *Bull. Entomol. Res.* 55:249–270.

Usakov, U.Y. 1961. Intraspecific parasitism (homovampirism) in ixodid ticks. *Zool. Zh.* 40:608–609.

Vachon, M. 1970. L'évolution du concept d'Arachnide. (Compte Rendus du 4th Congr. Int. d'Arachnologie, Paris, 8–13 Avril 1968). *Bull. Mus. Natl. Hist.* (2) 41 (Suppl. No. I):184–187.

Varma, M.G.R. 1962. Transmission of relapsing fever spirochaetes by ticks. *Symp. Zool. Soc. London* 6:61–82.

Varma, M.G.R. 1964. The acarology of louping ill. *Acarologia* 6:241–254.

Walton, G.A. 1962. The *Ornithodoros moubata* superspecies problem in relation to human relapsing fever epidemiology. *Symp. Zool. Soc. London* 6:83–156.

Walton, G.A. 1979. A taxonomic review of the *Ornithodoros moubata* (Murray) 1877 (*sensu* Walton, 1962) species group in Africa. In: Rodriguez, J.G., ed. *Recent Advances in Acarology.* Academic Press, New York, Vol. 2, pp. 491–500.

Warburton, C. 1907. Notes on ticks. *J. Econ. Biol.* 2(3):89–95.

Wilkinson, P.R. and M.B. Garvie. 1975. Notes on the role of ticks feeding on lagomorphs and ingestion of ticks by vertebrates in the epidemiology of Rocky Mountain spotted fever. *J. Med. Entomol.* 12(4):480.

Woolley, T.A. 1972. Scanning electron microscopy of the respiratory apparatus of ticks. *Trans. Am. Microsc. Soc.* 91(3):348–363.

Yeoman, G.H. and J.B. Walker. 1967. *The Ixodid Ticks of Tanzania.* Commonwealth Institute of Entomology, London.

Zdrodovskij, P.F. 1964. Les rickettsioses en U.R.S.S. *Bull. W.H.O.* 31:33–43.

Zumpt, F. 1950. Preliminary study to a revision of the genus *Rhipicephalus* Koch. *Doc. Mocambique* 60(1949):57–123.

COHORT ACARIFORMES

Actinotrichida (=Trombidi–Sarcoptiformes Oudemans 1931;
Actinochitinosi Grandjean 1935; Actinotrichida Zachvatkin 1952;
Hammen 1960; Actinochaeta Evans et al. 1961).

Diversified morphology is characteristic of this cohort. The tactile setae of
both the body and appendages actively refract in polarized light because of
the physical aspects of the chitin and thus place these mites in the
Actinotrichida.

The body is generally divided into two regions by the dorsosejugal suture
(Actinedida is an exception). The anterior division, the propodosoma,
includes insertions of legs I and II. The prodorsum of this region typically
bears six pairs of setae and two pairs of trichobothria. The hysterosoma is
the region posterior to the dorsosejugal suture and has legs III and IV
attached. Dorsally, the hysterosoma has nine transverse rows of setae that
seem to represent the primitive arachnid segmentation and six pairs of
cupules. The ventral genital aperture and anus are located in this region.
Stigmata are associated with the gnathosoma or the cheliceral bases.
Peritremes may be present with the stigmata found on the gnathosoma or
chelicerae (Fig. 19-1A).

The chelicerae are basically chelate–dentate, but modifications occur. The
cheliceral bases are capable of independent action and lie above the
subcapitulum. Coxal glands release their secretions into a podocephalic canal
that leads to the cheliceral bases. The palps usually are linear and
primitively have five segments; some are modified.

Leg coxae are fused with the ventral body wall and show as coxal fields
with variously modified apodemes. Legs I–IV are generally homeomorphic
with six articulating segments (trochanter, basifemur, telofemur, genu, tibia,

and tarsus). A pretarsus bears paired claws and an empodium; in the absence of true claws, an empodial claw may be present. The legs bear two types of sensory structures: tactile setae and chemosensory sensilla called solenidia. The solenidia may be present on the tarsus, tibia, and genu of the four legs.

Sexual dimorphism varies, but sperm transfer is either indirect by means of spermatophores or direct by use of an aedeagus. The life cycle includes the egg, an inactive hexapod prelarva, an active hexapod larva, protonymphal, deutonymphal, and tritonymphal stages, and adult males and females. Anamorphic development is the rule with segments added to the body after eclosion. Opisthosomatal segments C, D, E, F, H, and PS (pseudanal) are designated in the hexapod larva. Segment AD (adanal) is added in the protonymph along with a fourth pair of legs. Segment AN (anal) is added in the deutonymph and segment PA (postanal) is found in the tritonymph. Genital and aggenital setae are added beginning with the protonymph; the number is diagnostic in the adults, when internal genitalia and eugenital setae are formed. Genital acetabula (papillae; suckers) develop in pairs through the stages as well: one pair—protonymph; two pairs—deutonymph; three pairs—tritonymph and adult.

The biology of Acariformes is diverse. They are found in many different terrestrial niches, in varied freshwater and marine habitats. Symbiotic relationships include predation and parasitism of both vertebrate and invertebrate hosts. Phytophagous, fungivorous, and algophagous forms occur.

Disagreement exists regarding the classification of the higher categories of Acariformes because of the biological and taxonomic diversity. The system used in this writing includes the orders Actinedida, Astigmata, and Oribatida. (The reader is referred to Hammen (1972), Kethley (1982), Krantz (1978), Lindquist (1976), and OConnor (1982, 1984) for additional details about the classification.)

19

Order Actinedida

(=Prostigmata Kramer 1877; Hughes 1959, Evans et al. 1961; Trombidiformes Kramer 1909; Hammen 1960).

The greatest biological diversity within the Acariformes probably occurs within this order. Morphological modifications and ecological locations reflect these variations. Feeding capabilities range from phytophagous to fungivorous, algivorous, and saprophagous, and varied types of parasitic habitats involving both vertebrate and invertebrate hosts (Kethley, 1982).

The heterogeneity and complexity of this order make categorization and classification difficult. The elusive characterization results from the comparing features that are *not present,* rather than positive distinguishing structures. The extreme diversity of morphological features and body form lead one to conjecture that the order may be a complex composite of several subordinal groups (Krantz, 1978). Clear trends in the diversity of populations and biology appear to exist.

The evolution from fungivory to predation to plant or animal feeding reflects major patterns in speciation. Fewer than 1% of the total species are fungivores, and only 7% are free-living predators. Twenty-four percent of the species are plant feeders, 20% are vertebrate parasites and nearly 47% are associates of other arthropods as parasitoids or parasites. (Kethley, 1982)

The body is incompletely sclerotized or weakly sclerotized (an exception would be Labidostommatidae=Nicolettiellidae). The presence of shields and sclerotized plates is variable, but no sternal shield is present. The body surface may be wrinkled ("fingerprinted"). Integument ranges from colorless and pastel shades to brilliant hues (e.g., bright red of the Trombidiidae).

The body is divided by a dorsosejugal suture and the body regions in this order represent the most complex arrangement among mites. The gnathosoma is usually distinct, but may be noticeable only from the ventral aspect. The regions of the body vary among the groups within the order, namely, gnathosoma and idiosoma; propodosoma and hysterosoma; propodosoma, metapodosoma, and opisthosoma; podosoma and opisthosoma (Figs. 4-14; 19-1). Some structural reorganization may occur so that elements of the hysterosoma may migrate into the propodosomal region. Genital and anal apertures have migrated to the dorsum in some groups. The opisthosoma may show clear evidence of segmentation with distinct patterns. Anamorphosis may be completely or partially regressed (Kethley, 1982). Genital acetabula (=papillae) and eugenital setae may or may not be present.

Body setation is variable, either in rows or at random (Fig. 19-1D) and the chaetotaxy is in a

chaotic state. The setae of the anal valves may belong to the PA, AN, AD, or PS series. The hypertrichy of the hysterosoma may extend beyond the normal developmental patterns in different groups, apparently independently. The setae are tactile or chemosensory and complex setae are present in some families. Special sensory plates (shields) may be present, for example, trichobothria and crista metopica; the sensilla arising from the bothridia vary in form. Trichobothria may also be found on the legs. Ocelli are present in some and are propodosomal and lateral in position; one or two pairs may be present.

As the older name, Prostigmata, implies, the one or two pairs of stigmata are usually in the region of the gnathosoma associated at or near its base or with the chelicerae (Fig. 19-1A). Peritremes may or may not be present. If present they may be sessile or emergent, chambered or unchambered. A tracheal system is usually present, although the Eriophyidae have no respiratory system (Krantz, 1978).

Chelicerae may be exposed or hidden. Predators exhibit chelate chelicerae, but others of these mites exhibit hooklike, sickle-shaped, styletlike or long, whiplike styletiform chelicerae. In some the cheliceral bases are fused medially to form a stylophore, or when fused to the dorsal face of the subcapitulum to form a stylophore–capsule. In others the chelicerae may move laterally like scissors (Fig. 19-1B).

Palps vary from simple free or adpressed forms to fanglike, raptorial types and in many of the actinedids a thumb–claw complex is present with its typical tibial spur and palpal "thumb."

A number of modifications occur in the legs: false segmentation may be present, leg segments may be fused, pretarsi and one or more pairs of legs may be missing. Apodous parasites of vertebrates represent the ultimate reduction of legs (Kethley, 1982). True claws with or without empodia are present. In many the claws and empodia are rayed (Fig. 6-5F); tenent hairs may be present (Fig. 19-1C).

Sexual dimorphism may be pronounced,

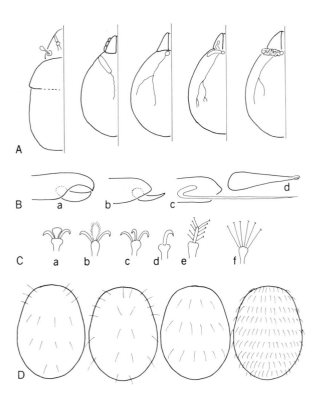

Figure 19-1. Schematic characteristics of Actinedida: (A) idiosomal types and respiratory structures; (B) types of chelicerae: a, chelate; b, B-type, hooklike digit; c, styletiform with stylophore; d, scissors type; (C) tarsi: a, claws and empodium; b, rayed empodium; c, d, empodial claws; e, f, tenent hairs; (D) variations of body setation.

with a definitive aedeagus in the male for direct insemination; indirect insemination by spermatophore also occurs. The development follows the acariform pattern, but within different groups suppression or elimination of immature stages may occur, which results in parthenogenetic species where adult females give birth to adult females. In some groups heteromorphic development occurs in which parasitic larvae occur or in which nymphal stages are alternatively actively predaceous and inactive (Parasitengona) (Kethley, 1982).

Because disagreement exists about higher categories the following arrangement is not completely satisfactory. It is a modification of the classification from Kethley (1982) and Krantz (1978). The order at present contains 8 suborders, 31 superfamilies and 135 families, with 1100+ genera, 14 000+ named species.

SUBORDER ENDEOSTIGMATA

Many of the characters of this suborder are primitive and derived. Chelicerae may be chelate–dentate, or edentate with one or two cheliceral setae. Palps are usually linear, nonraptorial, and five-segmented (rarely four). A rutellum may be present along with a well-developed naso. Stigmata and peritremes are usually absent. One or two pairs of trichobothria are present on the prodorsum and a dorsosejugal suture is present. Eugenital setae and two or three pairs of genital papillae characterize adults. Fertilization is accomplished by spermatophores. Eight families comprise this heterogeneous assemblage within a single superfamily. (The Pediculochelidae were placed here by earlier authors, but that position has been changed (Norton et. al., 1983).)

Superfamily Bimichaeloidea
  Alicorhagiidae—mosses, forest leaf litter (Fig. 19-2A)
  Bimichaelidae—mosses, forest leaf litter

Grandjeanicidae—dry soil, forest leaf litter
Lordalychidae—mosses, *Acacia* litter (Fig. 19-2B)
Micropsammidae—Mediterranean coastal dunes
Nanorchestidae—mosses, leaf litter, seashore algae
Nematalycidae—grasslands (Fig. 19-3A)
Oehserchestidae—dry leaf litter
Sphaerolichidae—deciduous leaf litter, pasture soils
Terpnacaridae—dry leaf litter

SUBORDER EUPODINA

The biology of these mites varies from free-living and parasitic to fungivores, phytophages, and predaceous habits. Cheliceral bases are separate or fused medially. Chelate chelicerae occur in predaceous forms, in others piercing stylets are present. Phytophagous and parasitic mites in this group have styletiform chelicerae, with the fixed digit reduced or absent. Palps have one to five articles and may be simple, linear, or raptorial. Stigmata, if present, open at the base of the chelicerae. A single pair of trichobothria is present on the prodorsum. Body surface varies from weakly sclerotized to armored integument. Genital setae and papillae are usually present. The aedeagus in males appears to have arisen independently twice. Three nymphal stages occur, except in parasitic forms where one or more may be suppressed.

The chelicerae of the Bdelloidea are movable and scissorslike with tiny chelae at the ends. Palps may be antenniform or raptorial. These mites are predaceous on other mites, small arthropods, and their eggs.

Superfamily Bdelloidea
  Bdellidae—predators of arthropods and their eggs

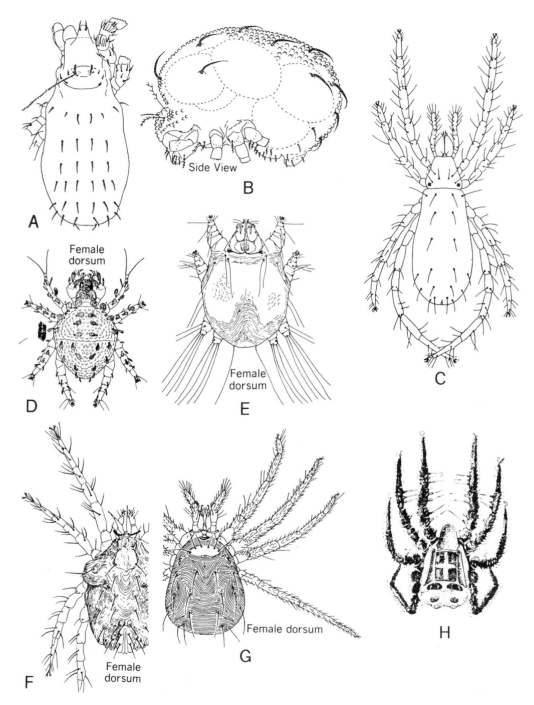

Figure 19-2. Representative examples of Actinedida: (A) dorsal view of female Alicorhagidae; (B) lateral view of Lordalychida; (C) dorsal view of female Rhagidiidae; (D)
dorsal view of female Cheyletidae; (E) dorsal view of female Harpyrhynchidae; (F) partial dorsal view of Pterygosomidae; (G) partial dorsal view of Anystidae; (H) dorsal view
Caeculidae (after Baker et al., 1958).

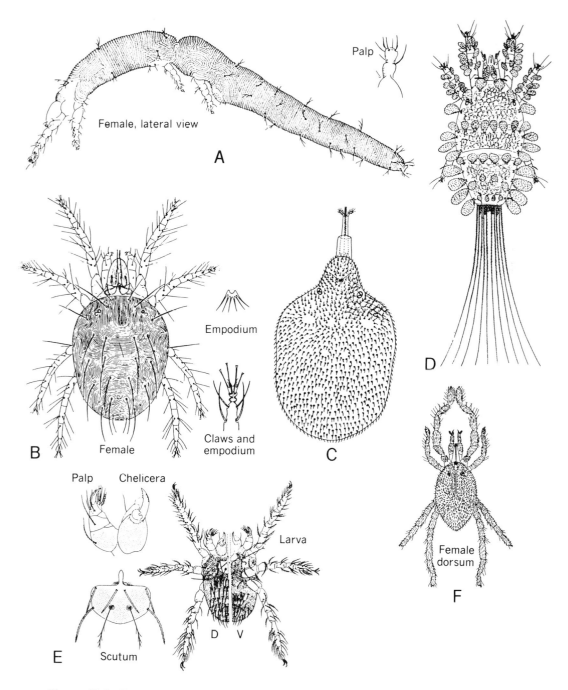

Figure 19-3. Representative examples of Actinedida (continued): (A) lateral view of Nematalycidae a, palp; (B) dorsal view of Tetranychidae, tenent hairs of empodium; (C) dorsal view of Calyptostomidae; (D) dorsal view of Tuckerellidae; (E) dorsal, ventral views of chigger larva (Trombiculidae), chelicera, palp and scutum; (F) dorsal view of Erythraeidae (after Baker et al., 1958).

Cunaxidae—predators of phytophagous insects, eggs (Fig. 2-3D)

Superfamily Eupodoidea

The feeding habits of the Eupodoidea are not well known. Some are assumed to be predators, others fungivores, and many inhabit mosses and leaf litter without an identified food habit.

Eupodidae—soil, humus, mosses, mushrooms

Penthalodidae—mosses, leaf litter, algae?, lichens?

Penthaleidae—grain, cotton, legumes, ornamentals, peanuts

Rhagidiidae—soil, humus, mosses, bark, caves (Fig. 19-2C)

Strandtmanniidae—oak litter, Czechoslovakia; Hawaii

Halacaroids comprise an assemblage of over 300 species of marine mites of interstitial to abyssal locations. In the vernacular they are referred to as "sea-going Cunaxids," because of resemblances to the Bdelloidea. Some are intertidal in habit and feed on algae. Others appear to be parasitic on a variety of marine organisms from decapods to chitons to sea urchins

Superfamily Halacaroidea
Halacaridae
10 subfamilies

Not all authors agree on the classification of Tydeoidea, but these mites are characterized by short, curved, sicklelike chelicerae. They are predaceous, fungivorous, facultatively phytophagous, and some are parasitic.

Superfamily Tydeoidea
Ereynetidae—mosses, leaf litter, lichens parasites in nasal passages of frogs, toads, birds, mammals (Speleognathinae)

Iolinidae—ectoparasites of *Blaberus*, grasshoppers

Tydeidae—omnivores, fungivores, phytophages, predators

SUBORDER LABIDOSTOMMATINA

Labidostommids are cosmopolitan, medium to large, heavily sclerotized predators that occur in soil, lichens, mosses humus, and litter. They feed on other mites and arthropods by means of their large, chelate–dentate chelicerae. The integument is highly sculptured (Fig. 4-4A).

Superfamily Labidostommatoidea
Labidostommatidae (=Nicoletiellidae)

SUBORDER ANYSTINA

This assemblage of mites comprises five superfamilies about which there is little consensus as to taxonomic relationship. They share primitive characteristics with the Endeostigmata and Eupodina, and at the same time resemble the Parasitengona and Eleutherengona. Higher and lower categories are not treated by same by taxonomists. Most of these mites are known predators or parasites. Whirlygig mites (Anystidae) are bright reddish-orange with a crablike gait (Fig. 19-2G) (referred to in the Introduction as possibly the object of interest in Robert Frost's poem). The rake-legged mites (Caeculidae) may be found under rocks in mountain areas, in xeric desert locations, or in seashore sands. They are large mites with characteristic setae on the first legs and are predatory in habit, although fungus has been used for food in culture (Crossley and Merchant, 1971). Another family of questionable taxonomic placement, Pterygosomatidae, parasitize Old World and New World lizards and cockroaches. (These mites are sometimes placed with Parasitengona.)

Superfamily Anystoidea

Anystidae—predators of phytophagous mites; predators in soil, on vegetations (Fig. 19-2G)

Adamystidae—litter, soil

Pseudocheylidae—leaf litter, under bark, cosmopolitan

Tenereffiidae—predators, xeric conditions, seashore

Superfamily Caeculoidea

Caeculidae—xeric habitats, seashore, deserts, mountains (Fig. 19-2H)

Superfamily Paratydeoidea

Paratydeidae—soil, humus, grasslands

Superfamily Pomerantzioidea

Pomerantziidae—deep soil inhabitants (15–20 cm)

Superfamily Pteryogosomatoidea

Pterygosomatidae—scale parasites of lizards; parasites of cockroaches, reduviids in culture

SUBORDER PARASITENGONA

One of the striking features of these mites is the complexity of their life cycles. The larvae are heteromorphic with respect to the nymphs and adults (with the exception of Calyptostomatidae) and parasitize some arthropods and vertebrates. Inactive calyptostases occur in the protonymph and tritonymph, but the deutonymphs and adults are active predators. This assemblage comprises highly specialized terrestrial and aquatic mites. The chelicerae have separate bases, the fixed digit is lacking, and the movable digit is usually sickle-shaped.

The aquatic parasitengones (water mites or Hydrachnellae) are usually strongly sclerotized and lack dorsal trichobothria in both adults and larvae. Active swimmers frequently show hypertrichy on legs. Terrestrial parasitengones are more weakly sclerotized with one or two pairs of trichobothria. Sexual dimorphism is limited to the genital region. Both aquatic and terrestrial species in this suborder are characterized by a distinctive "thumb–claw" palp, with a strong tibial spur, or claw, and a prominent tarsal thumb. Many of the adult Parasitengona are strongly hypertrichous with simple, uniramous, pectinate, or plumose setae. Coloration tends toward the brighter hues of red, orange, and reddish-brown.

Of all the superfamilies the Trombidioidea is one of the most difficult to classify because of its heterogeneity (Wellbourne, 1984). Most of the Leeuwenhoekiidae and the Trombiculidae are known from immatures, while the Trombidiidae are identified by the adult stage. The Trombidiidae parasitize invertebrates as larvae; the Trombiculidae and Leeuwenhoekiidae are larval parasites of vertebrates. Chiggers (the larvae of Trombiculidae and Leeuwenhoekiidae), for example, are medically important because they transmit some of the most serious human diseases.

The classification and intrasubordinal relationships are not understood.

Superfamily Calyptosomatoidea

Calyptosomatidae—predators, sphagnum, subaquatic sites (Fig. 19-3C)

Superfamily Erythraeoidea

Erythraeidae—parasitic as larvae on terrestrial arthropods; predators; parasitic in lizards (Indio-Malaya) (Fig. 19-3F)

Smaridiidae - mosses, leaf litter

Superfamily Trombidioidea

Chyzeriidae—leaf litter of mesic forests

Johnstonianidae—subaquatic, litter at streamsides

Leuwenhoekiidae—skin parasites of birds, mammals

Podothrombiidae—soil, leaf litter; larvae parasitic on homopterans

Trombellidae—mesic forest leaf litter

Neotrombidiidae—leaf litter, under bark; larvae parasitize streblid bat flies

Trombiculidae—larvae parasitize all

classes of vertebrates except fishes
(Figs. 2-6E; 19-3E)

Trombidiidae—semiarid, sandy soils; larvae parasitic on grasshoppers

Superfamily Stygothrombioidea

Stygothrombiidae—larvae parasitic on
stone flies; aquatic habitats

The water mites (Hydrachnellae, Hydra-
charina, Hydrachnida, Hydrachnidia of other
authors) comprise a large number of highly diverse species found in aquatic or semiaquaitic
habitats. Specializations for life in restricted
niches (i.e., surface of submerged rocks, stream
bottoms, surface film of wet substrates, in hot
springs, etc.) and under more or less constant
environmental conditions has resulted in distinctive, and occasionaly rather startling, modifications in both external and internal morphology. Water mites have cuticular glands,
stigmata, and tracheae peculiar to themselves
(Fig. 9-10). Two pairs of eyes are present and
may be separate or coalesced; an imperfectly
developed median eye is also present at times.
The tarsi usually possess two claws but are
without an empodium or pretarsus. The palpi
usually have five movable segments. Chelicerae are only the movable digits, and are strong
and sicklelike. The chelicerae work with the
raptorial palpi for capturing prey. The prodorsum lacks trichobothria in these mites, but
there are distinctive glands associated with
some dorsal setae. Both males and females exhibit a profusion of small genital acetabula adjacent to the genital aperture (gonopore). Larvae of water mites are heteromorphic and
possess an urstigma between coxae I and II.

These brightly colored mites are found in
lakes and ponds, along shores, and in streams.
A few are parasitic in the gills of freshwater
mussels and the larvae of other forms are parasitic on aquatic insects, much like chiggers on
vertebrates. A few water mites are parasitic in
the gill chambers of crabs, where they grasp
the gill filaments. Water mites are found at all
times of the year, but most species appear as
adults in the late summer and fall. With the exception of the Antarctic continent, water mites
are found throughout the world (Krantz, 1978).

Familial classification of the water mites,
based principally on external morphology, is
quite robust and extensive for larvae and genera (Cook, 1974; Prasad and Cook, 1972). Early
use of specific morphological modifications as
familial characteristics resulted in the naming
of innumerable families, many of which were
based on a single species. More recently, specialists of this group have recognized the diffi-
culty of erecting a familial classification based
on morphology alone, yet this system continues to be workable. Rodger Mitchell, an emi-
nent authority in this group, suggests that a
valid classification must be based on comparative biology, coupled with cataloging the diversity of forms. There remains, however, the necessity to develop a satisfactory key to the
families and to couple that key with an extensive compilation that would enable the investi-
gator to determine genera and species as well.

Water mites are monophyletic (except for
the Hydrovolzioidea) (Cook, 1974). They perhaps evolved from and are most closely related
to the terrestrial Trombidia (Krantz, 1978). The
many families of water mites are summarized
by Kethley (1982), Krantz (1978), Cook (1974),
as well as the numerous works of K. O. Viets
(e.g., 1936, 1956, 1982).

Water mites form an ecological rather than a
morphological group and are currently placed
within the Parasitengona. Some are swimming
forms; others are nonswimmers. Except for the
marine Pontarachnidae, water mites are inhabitants of fresh water.

Superfamily Hydryphantoidea—red water
mites; aerial larvae

Ctenothyadidae

Hydrodromidae

Hydryphantidae

Rhnychohydracaridae

Teratothyadidae

Thermacaridae

Superfamily Eylaioidea—red water mites, aerial larvae

Eylaidae

Limnocharidae

Piersigiidae

Superfamily Hydrovolzioidea—predators in streams, seepage water, springs; parasites of hemipterans, dipterans

Hydrovolziidae

Superfamily Hydrachnoidea—predators in slow streams, sluggish water; larvae aquatic

Hydrachnidae

Superfamily Lebertoidea—fast running streams, some thermal waters; larvae parasitic

Anistiellidae

Lebertiidae

Oxidae

Rutripalpidae

Spherchontidae

Teutonidae

Torrenticolidae

Superfamily Hygrobatoidea—aquatic, streams, lakes, ponds, mosses; decapod, dipteran, stone fly parasites

Astacocrotidae

Aturidae

Axnopsidae

Feltriidae

Ferradasiidae

Hygrobatidae

Lymnesiidae

Omartacaridae

Pionidae

Pontarachnidae

Unionicolidae

Superfamily Arrenuroidea—aquatic, interstitial, streams, springs; dipteran parasites

Acalyptonotidae

Arenohydracaridae

Arrenuridae

Athienemanniidae

Bogatiidae

Chappuisididae

Harpagonpalpidae

Hungarohydracharidae

Kantacaridae

Krendowskiidae

Laversiidae

Mideidae

Mideopsidae

Momoniidae

Neoacarida

Nipponacaridae

Uchidastygacaridae

SUBORDER ELEUTHERENGONA

This suborder is a vast assemblage of superfamilies of different aggregations that exhibit a palpal thumb-claw complex not unlike the Parasitengona, but with reduced elements of both the tibial spur and the tarsal thumb. In some groups this characteristic is lacking. Two divisions, Raphignathae and Heterostigmae, comprise superfamilies and families with a variety of biological habits; some are predaceous, others phytophages, and some parasitic. The stigmata and peritremes vary together with chelicerae, body setation, legs, and ambulacra. Males usually have an aedeagus. Developmental stages include the egg, prelarva, larva, two homeomorphic nymphs, and adults.

Raphignathae

In the Raphignathae the stigmata and peritremes are found at the base of the chelicerae or the cheliceral bases, which are fused medially or with the subcapitulum. Trichobothria are absent on the prodorsum. Tarsi may be rayed or with tenent hairs. The Raphignathoidea are phytophagous and predaceous mites. Females exhibit a longitudinal genital opening usually contiguous with the anal aperture. Males have a subterminal or ventral aedeagus (rarely dorsal).

Cheyletoidea are characterized by styletiform chelicerae and the presence or absence of peritremes on the dorsum of the stylophore (fused cheliceral bases). Female genital openings are generally coalesced.Male genital openings are usually dorsoterminal or dorsal with an aedeagus. Peritremes may be present on the dorsum of the stylophore (fused cheliceral bases) or absent. Solenidia are usually not present on tarsi III and IV.

Cheyletidae are predators of other arthropods or may be parasitic on arthropods and vertebrates (Summers and Price, 1970; Volgin, 1969). Many are associated with stored products insects and mites and with bark beetles, and prey in leaf litter. Species of *Cheyletiella*, especially *parasitivorax*, may cause mange on dogs and cats and eczema in humans and have been placed in a separate family, Cheyletiellidae (Fain, 1980). Cloacaridae are skin parasites in the cloacal region of freshwater and terrestrial turtles. These were mentioned earlier as being transmitted venereally from host to host. Demodicidae are vermiform parasites found in hair follicles and sebaceous glands and ducts of mammals. None is known to vector human disease, but several cause demodectic mange in dogs and goats and lead to secondary infection by *Staphylococcus* (Desch and Nutting, 1972).

Harpyrhynchidae are skin parasites of birds and may cause tumorlike growths (Fain, 1972). Myobiids are skin parasites of birds, marsupials, bats, and other mammals (Yunker, 1973). Psorergatids are skin parasites of ungulates (Fain, 1960). Syringophilidae are skin parasites of birds, where they live in the quill cavities of feathers. The genera appear to be host-specific at the ordinal level of birds and four or five species may inhabit different feather tracts on the same bird (Kethley, 1970, 1971).

Superfamily Raphignathoidea
  Barbutiidae—coniferous, deciduous litter

Caligonellidae—leaf, grass litter

Camerobiidae—foliage, bark inhabitants; predator on scale insects

Cryptognathidae—?algivores, ?phytophages; leaf litter, bark, lichens (Fig. 2-3B)

Eupalopsellidae—soil, leaf litter, citrus foliage

Homocaligidae—mosses, semiaquatic habitats

Raphignathidae—predators in leaf litter, under tree bark, in grain storage

Stigmaeidae—arboreal predators, phytophagous insects; leaf litter; mosses, ?phytophages

Xenocaligonellidae—foliage; phoretic on elaterid beetles

Superfamily Cheyletoidea

Cheyletidae—predators in leaf litter, foliage, bark; parasitic, phoretic (Fig. 19-2D)

Cloacaridae—skin parasites of turtles

Demodicidae—skin parasites, hair follicles of mammals (Fig. 1-2A, B)

Harpyrhynchidae—skin parasites of birds (Fig. 19-2E)

Myobiidae—skin parasites of mammals (Fig. 1-1A)

Ophioptidae—ectoparasites of snakes

Psorergatidae—skin parasites of mammals

Syringophilidae—skin parasites of bird; quill mites

The superfamily Tetranychoidea comprises a large group of important obligate phytophagous mites. Tenuipalpidae and Tetranychidae include the largest majority of species and are of considerable economic importance. The spider mites (Tetranychidae) are found on grasses, low plants, grains, or legumes (Bryobinae) and on higher flowering plants, ornamentals, conifers, and fruit trees (Tetranychinae). They are called spider mites because of their ability to spin silk and form webbing for anchoring of eggs, protection, or pheromonal transfer. The literature on this family is extensive (e.g., Baker, 1979; Jeppson et al., 1975; Helle and Sabelis, 1985) and studies of the group disclose much in ecological, genetic–cytogenetic, behavioral, and pesticide-control data.

Linotetranidae, Allochaetophoridae and the bizaare Tuckerellidae are monogeneric families.

Superfamily Tetranychoidea

Allochaetophoridae—Bermuda grass

Linotetranidae—grass litter

Tenuipalpidae—false spider mites; phytophages of grasses, tea, orchids, pomegranates, fruit trees

Tetranychidae—spider mites; grasses, ornamentals, conifers, deciduous trees (Fig. 19-3B)

Tuckerellidae—cosmopolitan, bizaare, flagelliform setae

The bizzare, vermiform, four-legged eriophyoids (Fig. 2-6D) comprise an array of mites-in-miniature that are worldwide in distribution. These are obligate phytophages on perennial deciduous hosts in temperate and subarctic regions where their feeding causes galls, blisters, erinea, witches broom, rusts, and other hypertrophications of plant growth. The host reactions to the insertion of the styletlike chelicerae into plant cells demonstrate a high degree of specificity. Edge-rolling of leaves, leaf spotting, leaf puckering, and chlorosis are other forms of injury that result from their feeding. Eriophyidae is the only family of mites known to transmit plant viruses (Jeppson et al., 1975; Krantz, 1978).

Superfamily Eriophyoidea

Dipstilomiopidae (=Rhyncaphytoptidae)

Eriophyidae

Phytoptidae (=Sierraphytoptidae)

SUBORDER HETEROSTIGMATA

Many of the mites in this suborder have stigmata at the anterodorsal margins of the body that lack peritremes but have tracheae. The idiosoma may also exhibit a longitudinal series of dorsal shields suggestive of segmentation. Most of these mites are parasitic or phoretic on insects. Many of them (Pyemotidae, Pygmephoridae) have a stylophore-capsule from which the tiny styletlike chelicerae extend. The prodorsum exhibits trichobothria?=capitate sensilla that are characteristic. The life cycle is reduced, but with modifications that facilitate phoretic dispersal. The Pyemotidae parasitize the immature stages of beetles, flies, hymenopterans, and lepidopterans. One pyemotid species *Pyemotes tritici* is a potential biological control agent for the fire ant (Fig. 12-1). This species is also interesting because the female is physogastric and all development occurs within the body of the female. This straw itch mite is known to cause skin lesions, asthma, and nausea in humans (Kethley, 1982). In the Podapolipidae, parasites of cockroaches, grasshoppers, and beetles, the legs are reduced to one pair in some species. Scutacarids have distinctive dorsal sclerites and are found in leaf litter and humus; some are nidicolous in bird and mammal nests. The family Tarsonemidae includes plant pests of economic importance, parasites of bark beetles, predators of tetranychid eggs and parasites in the stink glands of true bugs. The species *Acarapis woodi* is a tracheal parasite of honeybees and a causative agent of Isle of Wight disease, recently reported of serious concern in Texas and a number of other states.

Superfamily Tarsocheyloidea

Tarsocheylidae—rotting wood, treeholes, leaf litter

Superfamily Heterocheyloidea

Heterocheylidae—subelytral parasites of passalid beetles

Superfamily Pyemotoidea

Acarophenacidae—parasitic on grain beetles (*Tribolium*)

Caraboacaridae—subelytral parasites of carabid beetles

Dolichocybidae—under bark of deciduous trees, associated with beetles

Pyemotidae—parasites of immature insects

Superfamily Pygmephoroidea

Microdispidae—soil, leaf litter

Pygmephoridae - leaf litter, soil, humus, nidicoles in insect, mammal nests; vectors of fungal diseases of plants

Scutacaridae—leaf litter, humus, nests of birds, mammals, associated with insects

Superfamily Tarsonemoidea

Podapolipidae—tracheal parasites of grasshoppers, hymenopterans; external, subelytral parasites of beetles

Tarsonemidae—phytophages; parasites of bark beetle eggs; predators of plant mite eggs; tracheal parasite of honeybees (Fig. 1-2C)

REFERENCES

Andre, H.M. 1979. A generic revision of the family Tydeidae (Acari: Actinedida). IV. Generic descriptions, keys and conclusions. *Ann. Soc. R. Belg.* 116:103–168.

Andre, H.M. 1984. Redefinition of the Iolinidae (Acari: Actinedida) with a discussion of their familial and superfamilial status. *Acarol. [Proc. Int. Congr. Acarol.], 6th, 1982,* Vol. 1, pp. 180–185

Athias-Henriot, C. 1961. Nouveaux acariens phytophages d'Algérie (Actinotrichida, Tetranychoidea: Tetranychidae, Linotetranidae). Ann. Ec. Natl. Agric. Alger 3(3):1–10.

Atyeo, W.T. 1960. A revision of the mite family Bdellidae in North and Central America. *Univ. Kans. Sci. Bull.* 40(8):345–499.

Atyeo, W.T. 1963a. The Bdellidae (Acarina) of the Australian Realm. *Bull. Univ. Nebr. State Mus.* 4(8):113–210.

Atyeo, W.T. 1963b. New and redescribed species of Raphignathidae (Acarina) and a discussion of the chaetotaxy of Raphignathoidea. *J. Kans. Entomol. Soc.* 36:172–186.

Atyeo, W.T., E.W. Baker and D.A. Crossley, Jr. 1961. The genus *Raphignathus* Duges (Acarina, Raphignathidae) in the United States with notes on the Old World species. *Acarologia* 3(1):14–20.

Baker, E.W. 1965. A review of the genera of the family Tydeidae (Acarina). *Adv. Acarol.* 2:95–133.

Baker, E.W. 1979. Spider mites revisited—a review. *In:* Rodriguez, J.G., ed. *Recent Advances in Acarology.* Academic Press, New York, pp. 387–394.

Baker, E.W. and A. Hoffman. 1948. Acaros de la familia Cunaxidae. *An. Esc. Nac. Cienc. Biol. (Mexico City)* 5(3–4):229–273.

Baker, E.W. and A.E. Pritchard. 1953. The family categories of tetranychoid mites, with a review of the new families Linotetranidae and Tuckerellidae. *Ann. Entomol. Soc. Am.* 46(2):243–258.

Baker, E.W. and G.W. Wharton. 1952. *An Introduction to Acarology.* Macmillan, New York.

Baker, E.W., J.H. Camin, F. Cunliffe, T.A. Woolley, and C.E. Yunker. 1958. *Contrib. No. 3, Inst. Acarology,* Univeristy of Maryland, College Park.

Barr, D. 1972. The ejaculatory complex in water mites (Acari: Parasitengona) morphology and potential value for systematics. *Life Sci. Contrib. R. Ont. Mus.* 81:1–87.

Barr, D. 1982. Comparative morphology of the genital acetabula of aquatic mites (Acari, Prostigmata): Hydrachnoidea, Elayeoidea, Hydryphantoidea and Lebertoidea. *J. Natl. Hist. Agric. Bot. (G.B.)* 16:147–160.

Barr, D. and D.T. Smith. 1979. Contribution of setal blades to effective swimming in the aquatic mite *Limnochaeres americanum* (Acari: Prostigmata): Lymnochaeridae. *J. Linn. Soc. London, Zool.* 65:55–69.

Barr, D. and D.T. Smith. 1980. Stable swimming by diagonal phase synchrony in Arthropods. *Can. J. Zool.* 58:782–795.

Beer, R.E. 1954. A revision of the Tarsonemidae of the western hemisphere. *Univ. Kans. Sci. Bull.* 36, Pt. 2 (16):1091–1387.

Blauvelt, W.E. 1945. The internal morphology of the common red spider mite (*Tetranychus telarius* Linn.). *Mem.-N.Y. Agric. Exp. Stn. (Ithaca)* 270:3–46.

Brennan, J.M. and M.L. Goff. 1977. Keys to the genera of chiggers of the western hemisphere (Acarina: Trombiculidae). *J. Parasitol.* 63(3):554–566.

Brennan, J.M. and E.K. Jones. 1959. Keys to the chiggers of North America with synonymic notes and descriptions of two new genera (Acarina: Trombiculidae). *Ann. Entomol. Soc. Am.* 52(1):7–16.

Bronswijk, J.E.M.H. van and E.J. de Kareek. 1976. Cheyletiella (Acari: Cheyletiellidae) of dog, cat and domesticated rabbit, a review. *J. Med. Entomol.* 13:315–327.

Camin, J.H., W.W. Moss, J.H. Oliver, Jr. and G. Singer. 1967. Cloacaridae, a new family of cheyletoid mites from the cloaca of aquatic turtles (Acari: Acariformes: Eleutherengona). *J. Med. Entomol.* 4(3):261–272.

Coineau, Y. 1974. Eléments pur une monographie morphologique et biologique des Caeculidae (Acariens). *Mem. Mus. Natl. Hist. Nat. Ser. A (N.S.)* 81:1–229.

Coineau, Y. and P. Theron. 1983. Les Micropsammidae, N. Fam. d'Acariens Endeostigmata des sables fins. *Acarologia* 24(3):275–280.

Cook, D.R. 1974. Water mite genera and subgenera. *Mem. Am. Entomol. Inst.* 21:1–860.

Cooper, K.W. 1940. Relations of *Pediculopsis graminum* and *Fusarium poae* to bud rot of carnation. *Phytopathology* 30(10):853–859. (Pygmephoridae)

Cooreman, J. 1959. Notes sur quelques acariens de la faune cavernicole (2me série). *Bull. Inst. R. Sci. Nat. Belg.* 35(34):1–40. (Johnstonianidae)

Cross, E.A. 1965. The generic relationships of the family Pyemotidae (Acarina: Trombidiformes). *Univ. Kans. Sci. Bull.* 45(2):29–275.

Cross, E.A. and J.C. Moser. 1975. A new dimorphic species of *Pyemotes* and a key to previously described forms (Acarina: Tarsonemioidea). *Ann. Entomol. Soc. Am.* 68(4):723–732.

Crossley, D.A. and V. Merchant. 1971. Feeding by caeculid mites on fungus demonstrated with radioactive tracers. *Ann. Entomol. Soc. Am.* 64(4):760–762.

Cunliffe, F. 1955. A proposed classification of the trombidiforme mites. *Proc. Entomol. Soc. Wash.* 57(5):209–218.

Cunliffe, F. 1956. A new species of *Nematalycus* Strenske with notes on the family (Acarina, Nematalycidae). *Proc. Entomol. Soc. Wash.* 58(6):335–355.

Danks, H.Z., ed. 1979. Canada and its insect fauna. *Mem. Entomol. Soc. Can.* 108:252–290.

Delfinado, M.D. 1963. Mites of the honeybee in southeast Asia. *J. Apic. Res.* 2(2):113–114. (Tarsonemidae)

Delfinado, M.D. and E.W. Baker. 1976. New species of Scutacaridae (Acarina) associated with insects. *Acarologia* 18(2):264–301.

Delfinado-Baker, M. and E.W. Baker. 1982. Notes on honey bee mites of the genus *Acarapis* Hirst (Acari: Tarsonemidae). *Int. J. Acarol.* 8(4):211–209.

Delfinado, M.D., E.W. Baker and M.J. Abbatiello. 1976. Terrestrial mites of New York. III. The family Scutacaridae. *J. N. Y. Entomol. Soc.* 84(2):106–145.

Desch, C. and W.B. Nutting. 1972. *Demodex folliculorum* (Simon) and *D. brevis* Akbulatova of man: Redescription and reevaluation. *J. Parasitol.* 58(1):169–177.

Dusbabek, F. 1969. To the phylogeny of the genera of the family Myobiidae (Acarina). *Acarologia* 11(3):537–584.

Ehara, S. 1975. A guide to the tetranychid mites of agricultural importance in Japan. *IIBP Synth* 7:15–23.

Fain, A. 1960. Les acariens psoriques parasites des chauves-souris. XIII. La famille Demodicidae Nicolet. *Acarologia* 2(1):80–87.

Fain, A. 1963. Chaetotaxie and classification des Speleognathinae. *Bull. Inst. R. Sci. Nat. Belg.* 39:1–80.

Fain, A. 1964. Ophioptidae, Acariens parasites des écailles des serpents (Trombidiformes). *Bull. Inst. R. Sci. Nat. Belg.* 40(15):1–57.

Fain, A. 1968. Notes sur les Acariens de la famille Cloacaridae Camin et al., parasites du cloaque et des tissus profondes des Tortues (Cheyletoidea: Trombidiformes). *Bull. Inst. R. Sci. Nat. Belg.* 44:1–33.

Fain, A. 1971. Note sur la répartition géographique des Coreitarsoneminae parasites des Hemiptères coreides avec description de taxa nouveaux (Acarina: Trombidiformes). *Bull. Ann. Soc. R. Entomol. Belg.* 107:81–88. (Tarsonemidae)

Fain, A. 1972. Notes sur les Acariens des familles Cheyletidae et Harpyrhynchidae producteurs de gale chez les oiseaux ou les mammifères. *Acta Zool. Pathol. Antverp.* 56:37-60.

Fain, A. 1973. Notes sur la nomenclature des poils idiosomaux chez les Myobiidae, avec description de taxa nouveaux (Acarina: Trombidiformes). *Acarologia* 15(2):279-309.

Fain, A. 1975. Observations sur les Myobiidae parasites des rongeurs. Evolution parallele hôtes-parasites (Acariens: Trombidiformes). *Acarologia* 16(3):441-475.

Fain, A. 1979. Idiosomal and leg chaetotaxy in the Cheyletidae. *Int. J. Acarol.* 5:305-310.

Fain, A. 1980. Observations on Cheyletid mites parasitic on mammals (Acari: Cheyletidae and Cheyletiellidae). *Acarologia* 21(3):408-422.

Fain, A., F. A. Lukoschus and M. Nadchatram. 1980. Two new species of *Cheletophyes* Oudemans, 1914 (Prostigmata: Cheyletidae) from the nest of a carpenter bee in Malaysia. *Int. J. Acarol.* 6(4):309-312.

Feider, Z. and N. Vasiliu. 1969. Revision critique de la famille Nicoletiellidae. *Proc. Int. Congr. Acarol., 2nd, 1967,* pp. 202-207.

Feider, Z. and N. Vasiliu. 1970. Six especes de Nicoletiellides d'Americque du Sud. *Acarologia* 12(2):282-309.

Feldman-Muhsam, B. and Y. Havivi. 1972. Two new species of the genus *Podapolipus* (Podapolipodidae, Acarina), redescription of *P. aharonii* Hirst, 1921 and some notes on the genus. *Acarologia* 14(4):657-674.

Gould, H., J. Hussey and W.J. Parr. 1969. Large scale commercial control of *Tetranychus urticae* Koch on cucumbers by the predator *Phytoseiulus persimilis* A.H. *In:* Evans, G.O., ed. *Proc. Int. Congr. Acarol., 2nd, 1967,* (Akad Kaido) Budapest, pp. 383-388.

Gerson, U. 1972. A new species of *Camerobia* Southcott, with a redefinition of the family Camerobiidae (Acari: Prostigmata). *Acarologia* 13(3):502-508.

Grandjean, F. 1939. Quelques genres d'acariens appartenant au groupe des Endeostigmata. *Ann. Sci. Nat., Zool. Biol. Anim.* (11) 2:1-122.

Grandjean, F. 1942a. Observations sur les Labidostommidae. *Bull. Mus. Natl. Hist. Nat.* (2) 14(2):118-125; (3):185-192; (5):319-326; (6):414-418.

Grandjean, F. 1942b. Quelques genres d'acariens appartenant au groupe des Endeostigmata (2 Ser.), Première partie. *Ann. Sci. Nat., Zool. Biol. Anim.* (11) 4:85-135.

Grandjean, F. 1943. Le développement postlarvaire d'*Anystis* (Acariens). *Mem. Mus. Natl. Hist. Nat. (N.S.)* 18:33-77.

Grandjean, F. 1944. Observations sur les Acariens de la famille des Stigmaeidae. *Arch. Sci. Phys. Nat.* (5) 26:103-131.

Guilyarov, M.S. and N.G. Bregetova, eds. 1977. *A Key to the Soil-Inhabiting Mites* (in Russian). Nauka, Leningrad.

Hammen, L. van der. 1972. A revised classification of the mites (Arachnidea, Acarida) with diagnoses, a key and notes on phylogeny. *Zool. Meded.* 47(2):273-292.

Hammen, L. van der. 1981. Numerical changes and evolution in Actinotrichid mites (Chelicerata). *Zool. Verh.* 182:1-47.

Helle, W. and M.W. Sabelis, eds. 1985. *World Crop Pests. Spider Mites: Their Biology, Natural Enemies and Control,* Vols. 1A and 1B. Elsevier, Amsterdam.

Heyer, J. den. 1981. Systematics of the family Cunaxidae Thor, 1902 (Actinedida: Acarida). *Publ. Univ. North. Ser. A* 24:1-19.

Hoy, M.A. and D.C. Herzog. 1985. *Biological Control in Agricultural IPM Systems.* Academic Press, Orlando, Florida.

Hoy, M.A., G.L. Cunningham and L. Knutson, eds. 1983. Biological control of pests by mites. *Agric. Exp. Stn. Univ. Calif., Berkeley, Spec. Publ.* 3304.

Husband, R.W. 1980. Review of the genus *Podapolipus* (Acarina: Podapolipidae) with emphasis on species associated with Tenebrionid beetles. *Int. J. Acarol.* 6(4):257-270.

Husband, R.W. 1984. *Dilopolipus, Panesthipolipus, Peripolipus* and *Stenopolipus,* new genera of Podapolipidae (Acarina) from the Indo-Australian region. *Int. J. Acarol.* 10(4):251-269.

Husband, R.W. and R.N. Sinha. 1970. A revision of the genus *Locustacarus* with a key to genera of the family Podapolipidae (Acarina). *Ann. Entomol. Soc. Am.* 63(4):1152-1162.

Hussey, N.W., W.H. Read and J.J. Hesling, eds. 1969. Order Acarina. *In: The Pests of Protected Cultivation: The Biology and Control of Glass-*

house and Mushroom Pests. Edward Arnold, London pp. 190–228.

Immamura, T. and R.D. Mitchell. 1967. The ecology and life cyce of the water mite, *Persigia limophila* Protz. *Annot. Zool. Jpn.* 40(1):37–44.

Jalil, M. and R.D. Mitchell. 1972. Parasitism of mosquitoes by water mites. *J. Med. Entomol.* 9(4):305–311.

Jeppson, L.R., H.H. Kiefer and E.W. Baker. 1975. *Mites Injurious to Economic Plants*. Univ. of California Press, Berkeley.

Johnston, D.E. 1964. *Psorergates bos*, a new mite parasite of domestic cattle (Acari: Psorergatidae). *Ohio Agric. Exp. Stn. Res. Circ.* 129:1–7.

Keifer, H.H. 1952. The eriophyid mites of California (Acarina: Eriophyidae). *Bull. Culif. Insect Surv.* 2(1):1–123.

Kethley, J.B. 1970. A revision of the family Syringophilidae (Prostigmata: Acarina). *Contrib. Am. Entomol. Inst.* 5(6):1–76.

Kethley, J.B. 1971. Population regulation in quill mites (Acarina: Syringophilidae). *Ecology* 52(6):1113–1118.

Kethley, J.B. 1982. Acariformes. *In*: Parker, S., ed. *Synopsis and Classification of Living Organisms*. McGraw-Hill, New York, pp. 117–145.

Krantz, G.W. 1978. *A Manual of Acarology*, 2nd ed. Oregon State Univ. Book Stores, Corvallis.

Lindquist, E.E. 1974. Nomenclatural status and authorship of some family-group names in the Eriophyidae (Acarina: Prostigmata). *Can. Entomol.* 106:209–212.

Lindquist, E.E. 1976. Transfer of the Tarsocheylidae to the Heterostigmata, and reassignment of the Tarsonemina and Heterostigmata to lower hierarchy status in the Prostigmata (Acari). *Can. Entomol.* 108:23–48.

Lindquist, E.E. 1984. Current theories on the evolution of major groups of Acari and on their relationships with other groups of Arachnida, with consequent implications for their classification. Acarol. VI [Proc. Int. Congr. Acarol.], 6th, 1982, Vol. 1, pp. 28–62.

Lindquist, E.E. 1985a. The Tetranychidae. Anatomy, phylogeny and systematics. *In:* Helle, W. and M.W. Sabelis, eds. *World Crop Pests. Spider Mites: Their Biology, Natural Enemies and Control.* Elsevier, Amsterdam, Vol. 1A, pp. 3–28.

Lindquist, E.E. 1985b. The Tetranychidae. Diagnosis and phylogenetic relationships. *In:* Helle, W. and M.W. Sabelis, eds. *World Crop Pests. Spider Mites: Their Biology, Natural Enemies and Control.* Elsevier, Amsterdam, Vol. 1A, pp. 63–74.

Lindquist, E.E. and J.B. Kethley. 1975. The systematic position of the Heterocheylidae Tragardh (Acari: Acariformes: Prostigmata). *Can. Entomol.* 107:887–898.

Lukoschus, F.A., F. Dusbabek and E.W. Jameson, Jr. 1972. Parasitic mites of Surinam. IV. *Archemyobia philander* spec. nov. (Myobiidae: Trombidiformes) from *Philander opposum. Acarologia* 14(2):179–189.

Lundblad, O. 1949. Hydrachnellae. *Expl. Parc Natl. Albert Mission Damas* 18:1–87.

Mahunka, B. 1970. Considerations on the systematics of the Tarsonemina and the descriptions of new European taxa (Acari: Trombidiformes). *Acta Zool. Acad. Sci. Hung.* 16(1–2):137–174.

Mahunka, S. and G. Rack. 1982. Bibliographica Tarsonemidologica VII. (1980–81). *Folia Entomol. Hung.* 43(1):77–86.

McDaniel, B., D.K. Morihara and J.K. Lewis. 1975. A new species of *Tuckerella* from South Dakota and a key with illustrations of all known described species. *Acarologia* 17(1):274–283.

McDaniel, B., D. Morihara and J.K. Lewis. 1976. The family Teneriffiidae Thor, with a new species from Mexico. *Ann. Entomol. Soc. Am.* 69(3):517–537.

McMurtry, J.A., C.B. Huffaker and M. van de Vrie. 1970. Ecology of tetranychid mites and their natural enemies: A review. I. Tetranychid enemies: Their biological characters and the impact of spray practicies. *Hilgardia* 40(11):331–390.

Mitchell, R.D. 1964. An approach to the classification of water mites. *Proc. Intr. Congr. Acarol., 1st, 1963,* pp. 75–79.

Moser, J.C. and E.A. Cross. 1975. Phoretomorph: A new phoretic phase unique to the Pyemotidae (Acarina: Tarsonemoidea). *Ann. Entomol. Soc. Am.* 68(5):820–822.

Nadchatram, M. 1970. Correlation of habitat, environment and color of chiggers, and their potential significance in the epidemiology of scrub typhus in Malaya (Prostigmata: Trombiculidae). *J. Med. Entomol.* 7(2):131–144.

Newell, I.M. 1957. Studies on the Johnstonianidae (Acari: Parasitengona). *Pac. Sci.* 11:396–466.

Norton, R.A., B.M. OConnor and D.E. Johnston. 1983. Systematic relationships of the Pediculochelidae (Acari: Acariformes). *Proc. Entomol. Soc. Wash.* 85:493–512.

Nutting, W.B. 1975. Pathogenesis associated with hair follicle mites (Acari: Demodicidae). *Acarologia* 17(3):493–507.

Nuzzaci, G. 1979. Studies on the structure and function of mouthparts of Eriophyid mites. *In*: Rodriguez, J.G., ed. *Recent Advances in Acarology*. Academic Press, New York, Vol. 2, pp. 411–415.

OConnor, B.M. 1982. Acari: Astigmata. *In:* Parker, S.P., ed. *Synopsis and Classification of Living Organisms*. McGraw-Hill, New York, pp. 146–169.

OConnor, B.M. 1984. Phylogenetic relationships among higher taxa in the Acariformes, with particular reference to the Astigmata. *In:* Griffiths, D.A. and C.E. Bowman, eds., *Acarology VI*, Vol. 1. Ellis Horwood, Chichester, England, pp. 19–27.

Oldfield, G N. and I.M. Newell. 1973. The role of the spermatophore in the reproductive biology of protogynes of *Aculus cornutus* (Acarina: Eriophyidae). *Ann. Entomol. Soc. Am.* 66(1):160–163.

Prasad, V. and D.R. Cook. 1972. The taxonomy of water mite larvae. *Mem. Am. Entomol. Inst.* 18:1–326.

Price, D.W. and G.S. Beham, Jr. 1976. Vertical distribution of pomerantziid mites (Acarina: Pomerantziidae). *Proc. Entomol. Soc. Wash.* 78(3):309–313.

Rack, G. 1974. Neue und bekannte Milbenarten der Uberfamilie Pygemoprhoidea aus dem Saalkreis bei Halle (Acarina: Tarsonemidae). *Entomol. Mitt. Zool. Mus. Hamburg* 4:499–521.

Robaux, P. 1967. Contribution à l'étude des acariens Thrombidiidae d'Europe. *Mem. Mus. Nat. Hist. Nat.* 46:1–412.

Robaux, P. 1973. Importance de l'étude des caractères morphologiques, de la biologie et de l'écologie à toutes les stases, pour établir la phylogénèse des Acariens voisins des Thrombidions. *Acarologia* 15(1):121–128.

Robaux, P. 1976. Observations sur quelques Actinedida (=Prostigmates) du sol d'Amérique du Nord. VII. Sur deux espèces nouvelle de Raphignathidae. *Rev. Ecol. Biol. Sol* 13:505–516.

Schevtchenko, V.G. 1971. The phylogenetic relationships and basic trends in the evolution of the four-legged mites (Acariformes, Tetrapodili). *Proc. Int. Congr. Entomol., 13th, 1968*, Vol. 1, p. 295 (in Russian).

Slykhuis, J.T. 1969. Mites as vectors of plant viruses. *In*: K. Maramorosch, ed. *Viruses, Vectors and Vegetation*, Interscience Contrib. 613, pp. 121–141.

Slykhuis, J.T. 1972. Transmission of plant viruses by eriophyid mites. *In*: Kado, C.I. and H.O. Agarwal, eds. *Principles and Techniques in Plant Virology*. Van Nostrand-Reinhold, New York, pp. 204–225.

Smiley, R.L. 1970. A review of the family Cheyletiellidae (Acarina). *Ann. Entomol. Soc. Am.* 63(4):1056–1078.

Smiley, R.L. 1977. Further studies on the family Cheyletiellidae (Acarina). *Acarologia* 19:225–241.

Smiley, R.L. and J.C. Moser. 1968. New species of mites from pine (Acarina: Tarsocheylidae, Eupalopsellidae, Caligonellidae, Cryptognathidae, Raphignathidae and Neophyllobiidae). *Proc. Entomol. Soc. Wash.* 70:307–317.

Smith, I.M. and D.R. Oliver. 1976. The parasitic associations of water mites with imaginal aquatic insects, especially Chironomidae. *Can. Entomol.* 108:127–144.

Southcott, R.V. 1961. Studies on the systematics of the Erythraeoidea (Acari), with a critical review of the genera and subfamilies. *Aust. J. Zool.* 9(3):367–610.

Southcott, R.V. 1963. The Smaridiidae (Acarina) of North and Central America and some other countries. *Trans. R. Soc. S. Aust.* 14:687–819.

Southcott, R.V. 1976. Arachnidism and allied syndromes in the Australian region. *Rec. Adelaide Children's Hosp.* 1(1):97–186.

Strandtmann, R.W. 1964. Insects of Campbell Island. Prostigmata: Eupodidae, Penthalodidae, Rhagidiidae, Nanorchestidae, Tydeidae, Ereynetidae. *Pac. Insects Monogr.* 7:148–156.

Strandtmann, R.W. 1967. Terrestrial Prostigmata (Trombidiform mites). Antarct. Res. Ser. 10:51–80.

Summers, F.M. 1960. *Eupalopsis* and eupalopsellid mites (Acarina: Stigmaeidae, Eupalopsellidae). *Fla. Entomol.* 43(3):119–138.

Summers, F.M. 1966a. Key to families of the Raphignathoidea (Acarina). *Acarologia* 8(2):226–229.

Summers, F.M. 1966b. Genera of the mite family Stigmaeidae Oudemans (Acarina). *Acarologia* 8(2):230–250.

Summers, F.M. and D.W. Price. 1970. Review of the mite family Cheyletidae. *Univ. Calif. Publ. Entomol.* 61:1–153.

Summers, F.M., R.H. Gonzales-R and R.L. Witt. 1973. Mouthparts of *Bryobia rubrioculus* (Sch.). (Acarina: Tetranychidae). *Proc. Entomol. Soc. Wash.* 75(1):96–111.

Theron, P.D. 1974. Hybalicidae, a new family of endeostigmatic mites (Acari: Trombidiformes). *Acarologia* 16(3):397–412.

Travé, J. and M. Vachon. 1975. Francois Granjean, 1882–1975 (Notice Biographique et Bibliographique). *Acarologia* 17(1):1–19.

Treat, A.E. 1975. *Mites of Moths and Butterflies.* Cornell Univ. Press, Ithaca, New York. (pp. 199–238, Eythraeidae; pp. 239–270, Tydeidae).

Tuttle, D.M. and E.W. Baker. 1968. *Spider Mites of the Southwestern United States and a Revision of the Family Tetranychidae.* Univ. of Arizona Press, Tucson.

Vercammen-Grandjean, P.H. 1973. Sur les statuts de la familie des Trombidiidae Leach, 1815 (Acarina: Prostigmata). *Acarologia* 15(1):102–114.

Viets, K.O. 1936. Wassermilben oder Hydracarina (Hydrachnellae und Halacaridae). *Tierwelt Dtsch.* 31–32:1–574.

Viets, K.O. 1956. *Die Milben der Susswassers und des Meeres,* Parts 2 and 3. Fischer, Jena.

Viets, K.O. 1982. Die milben des Susswassears (Hydrachnellae and Halacaridae (part) (Acari). l: Bibliographie. *Naturwiss. Ver. Hambourg* 6:1–116.

Volgin, V.I. 1969. Acarina of the family Cheyletidae. World Fauna. *Akad. Nauk SSSR, Zool. Inst. Opredel. Fauna SSSR* 101:1–432 (in Russian).

Wallace, M.M.H. and J.A. Mahon. 1972a. The taxonomy and biology of Australian Bdellidae (Acari). I. Subfamilies Bdellinae, Spinibdellinae and Cytodinae. *Acarologia* 14(4):544–580.

Wallace, M.M.H. and J.A. Mahon. 1972b. The taxonomy and biology of Australian Bdellidae (Acari). II. Subfamily Odontoscirinae. *Acarologia* 18(1):65–123.

Wallace, M.M.H. and M.C. Walters. 1974. The introduction of *Bdellodes lapidaria* (Acari: Bdellidae) into South Africa for the biological control of *Sminthurus viridis* (Collembola). *Aust. J. Zool.* 22:505–517.

Wellbourn, W.C. 1983. Potential use of trombidioids and erythraeoids as biological control agents of insect pests. *Agric. Exp. Stn. Univ. Calif., Berkeley, Spec. Publ.* 3304:103–140.

Welbourn, W.C. 1984. Phylogenetic studies on Trombidioidea. 1984. *In:* Griffiths, D.A. and C.E. Bowman, eds., *Acarology VI,* Vol. 1. Ellis Horwood, Chichester, England, pp. 135–142.

Wharton, G.W. and H.S. Fuller. 1952. A manual of the chiggers; The biology, classification, distribution, and importance to man of the larvae of the family Trombiculidae (Acarina). *Mem. Entomol. Soc. Wash.* 4:1–185.

Whitaker, J.O., Jr. and N. Wilson. 1974. Host and distribution lists of mites (Acari), parasitic and phoretic, in the hair of wild mammals of North America, north of Mexico. *Am. Midl. Nat.* 91(1):1–67.

Yunker, C.E. 1973. Mites. *In:* Flynn, R.J., ed. *Parasites of Laboratory Animals.* Iowa State Univ. Press, Ames, pp. 425–492.

20

Order Astigmata

(G. Canestrini 1891) (=Acaridei Latreille 1802; Sarcoptidae C.L. Koch 1842; Atracheata Kramer 1877; Astigmata Evans et al. 1961; Sarcoptiformes Reuter 1909 (part); Acaridida Hammen 1968).

Few of the Astigmata are exclusively free-living in all their stages because most of the taxa are associated with arthropods, vertebrates, or other animals. Their feeding habits range from filter feeding on microorganisms and absorption of food materials to the maceration and ingestion of solid materials (gramnivorous, etc.) These mites are common in where decaying organic materials (rotting vegetation and fungus, carrion, dung) occur, in nidicolous locations (nests of insects, birds and mammals), and in subcortical habitats. Many are direct pests of stored products (grains, cereals, cheese, vegetables), for which they are most noted. A large number of species in this order are parasites of birds and mammals, including humans; fewer species are parasites of insects and crustaceans. A minor number are aquatic or plant feeders (OConnor, 1982b).

The order contains a group of mites with a most abundant and diverse complex of species associated with stored products. Thought to have been fungivorous ancestrally and dispersed through the highly specialized deutonymph (=hypopus) stage, 34 genera of 10 families have invaded the habitats of stored products and/or house dust. The phoretic deutonymphs associated ancestrally with insects effected dispersal that has allowed radiation into many temporary habitats. Origins of these mites are said to be from specific sources, for example, dung, rotting vegetation, hay; grassland soil and litter; nests of mammals where the nesting materials serve as food or the mammals retrieve and store plant parts in burrows and nests; nests of birds; some may invade the dust of houses and cause severe allegies (Arlian, 1976; OConnor, 1979a). One group of the Astigmata is generally referred to as "cheese" mites; another major group is parasitic and falls into the vernacular categories of "feather," "fur," and "itch" mites.

These are soft skinned mites without trichobothria on the prodorsum.The integument is usually pale in color, ranging from whitish to yellowish, light tan, or lavender. The surface of the body is smooth or with fine striae, wrinkled or rough. Cuticular microtrichae occur in Glycyphagidae, Echimyopodidae, and Glycacaridae. Sometimes distinctive plates or shields are present (propodosomal and/or opisthosomal (e.g., Rosensteiniidae, Sarcoptidae); some (Fusacarinae; Chortoglyphidae) are even as sclerotized as the Oribatida; many exhibit slight ventral epimera and insunk coxal apodemes. Some parasitic taxa, for example, Hypoderatidae, appear to be sclerotized, but have not been described (B.M. OConnor, personal communication, 1985).

The gnathosoma is fairly movable, usually visible from above, and may be retractable beneath the prodorsum (or stegasime as in Oribatida), and may be covered by a dorsal prolongation of the propodosoma. A dorsal tectum is lacking. The infracapitulum or subcapitulum bears two pairs of setae, and in the primitive *Schizoglyphus biroi* may have adoral setae (as in Oribatida). True pantelebasic rutella are present or are reduced with apical rutellar teeth, or ventrally may have cuticular flaps (pseudorutella, Rosensteiniidae, Heterocoptidae) (B.M. OConnor, personal communication, 1985).

Chelicerae are usually exposed, basally depressed at an angle, and compressed laterally. They are usually chelate–dentate with a swollen fixed digit and have relatively small, slightly dentate chelae; fungal spore feeders have heavier teeth. The chelicerae work in a vertical plane and are held at an angle to the food (Fig. 20-2). The shears work independently of the vertical action. The fixed digit has a conical spur and a mandibular spine (paraxial). Where there is a reduction of the chela the movable digit remains (Family Linobiidae, Lemurnyssidae). The chelicerae of Histiostomatidae (=Anoetidae) have the movable and fixed digits fused, with many divided teeth used for filtering food in diverse habitats (Fig. 20-1B). Chelicerae are nearly absent in the Cytoditidae.

Generally, the palps are simple and one-segmented, but may be secondarily divided into two podomeres (false segments of Krantz,

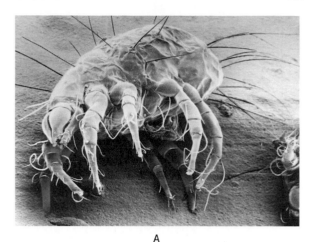

A

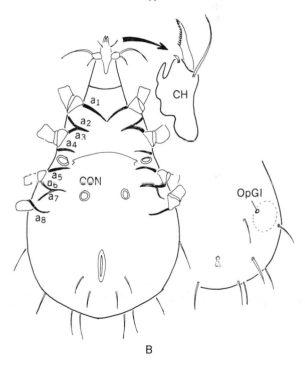

B

Figure 20-1. (A) SEM of *Tyrophagus putrescentiae* (by permission of Cambridge Instruments, Inc., New York); (B) diagram of Histiostomatidae (=Anoetidae) female from the ventral aspect showing apodemes (a1–8) (CON, conoids, and enlarged view of chelicera, CH); (C) dorsal aspect of hysterosoma showing opening of opisthosomal gland (OpGl).

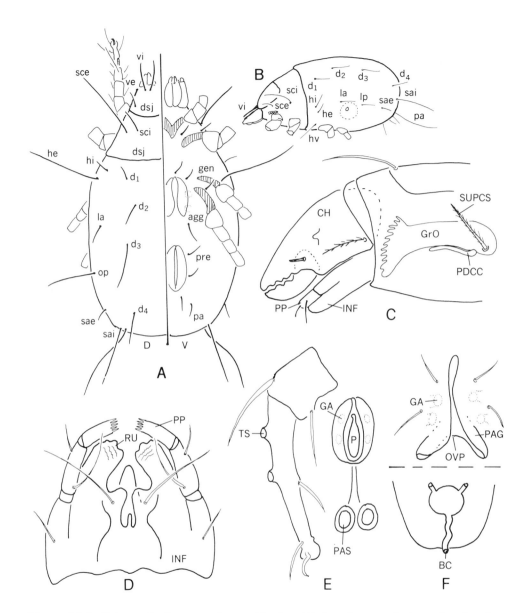

Figure 20-2. Schematic diagram of acaroid mite: (A) dorsal–ventral view, showing dorsosejugal suture (dsj), propodosomal shield, setae; ventral view showing epimera (shaded), genital opening with genital acetabula, anal aperture, and ventral setae; (B) lateral view showing setae, scapulars: *sci,* interior; *sce,* exterior; verticals: *vi,* interior; *ve,* exterior; humerals: *hi,* interior; *he,* exterior; dorsals: *d1-4*; laterals: *la,* anterior; *lp,* posterior; sacrals: *sae,* external; *sai,* internal; genital: *gen*; aggential: *agg*; preanal: *pra*; postanal: *pa*; (C) lateral view of gnathosoma showing chelicerae with seta and conical process; infracapitulum, INF; Grandjean's organ, GrO; pedipalp, PP; podocephalic canal, PDCC; supracoxal seta, SUPCS; (D) ventral view of gnathosoma showing rutella, RU; palps, PP; and infracapitulum, INF; dimorphic sexual characters of astigmatid mites, (E) male, showing copulatory suckers on tarsus of leg IV; penis, P; paranal suckers, PAS; and genital acetabula, GA; (F) female mite showing bursa copulatrix, BC; oviporus, OVP; genital acetabula, GA; paragynial folds, PAG.

1978). The segments are free but are closely apressed to the subcapitulum (Fig. 20-2). Three-segmented palps occur in Schizoglyphidae. Palpal coxae are fused ventrally to form the base of the gnathosoma or subcapitulum and the anterior malae. The epistome=labrum above the mouth forms the attachment for the dilator muscles of the pharynx. In Histiostomatidae (Anoetidae) the palpal solenidia are long eupathidia with flagellated tips for skimming the food from the fluid medium in which they live (Fig. 20-1B).

The idiosoma is oval in outline without evidence of external segmentation and without overlapping sclerites. None is vermiform (although this may be a relative term when considering the families Winterschmitiidae and Hemisarcoptidae, mites that live in spore tubes of polysporic fungi). The idiosoma may be divided into propodosoma and hysterosoma. The propodosoma (with legs I and II) is with or without a podonotal shield, lacks trichobothria (cf. Oribatida), and is indistinctly separated from the hysterosoma by a dorsosejugal suture. A maximum of four pairs of propodosomal setae occurs and these are useful in taxonomic differentiation (Vi, vertical internal setae; Ve, vertical external setae; Sci, scapular internal setae; Sce, scapular external setae). External verticals are absent in many lineages and internal scapulars as well. Vertical setae are on the podonotal shield in Winterschmidtiidae (=Saproglyphidae) and off the shield in parasitic taxa and Acaridae (Fig. 20-2). Both pairs of vertical setae arise from the podonotal shield in the Histiostomatidae. A maximum of 12 pairs of dorsal setae occur on the opisthosoma on segments C–PS. Setae of segment F are absent as are setae $h3$ and $ps3$. The paraproctal segments AD and AN each bear a maximum of three pairs of setae, although reduction is common. Segment PA is lacking. Lyrifissures ia, im ip, ih are usually present (OConnor, 1982a, also personal communication, 1985).

The sides of the propodosoma exhibit Grandjean's organ over coxa I and the podocephalic canal. This canal is an extension of the supra-coxal canal, and possibly conducts supracoxal gland fluids to the mouth (as in Pyroglyphidae). The canal may be hollow (Canestriniidae; Rhizoglyphinae, Acaridae). A supracoxal seta is present in some lineages (Glycyphagidae), but in most Astigmata is inconspicuous or absent. (At one time it was considered equivalent to the trichobothria of Oribatida, but is not now (OConnor, 1982a) (Fig. 20-C).

The hysterosoma (metapodosoma and opisthosoma) exhibits a typical reduced chaetotaxy with paired humeral dorsal, lateral, and sacral setae (Fig. 20-2A, B). This body region may have an opisthonotal shield (insect and mammal parasites, feather mites). Paired opisthosomal glands are typical and occur between setae $d2$ and $e2$ in almost all Astigmata, both free-living and parasitic forms. In free-living forms the hysterosoma is generally rounded, globular, swollen, and glabrous. In parasitic astigmatids the body is frequently flattened, discoidal, or rectangular. Body setae are highly variable (filiform, plumose, comblike or fanlike). The setae are generally short and simple, not extraordinary in appearance. Finely striated integument occurs in many taxa (free-living, insect and mammal parasites). Scales occur in some (Sarcoptidae). Commonly males of parasitic taxa have a bilobed or an oddly shaped posterior end. This modification is used by the males to attach to immature females (OConnor, 1982a).

On the venter the coxal fields are less sclerotized and each bears a maximum of one pair of setae. Genital valves are weakly sclerotized and sclerotized epimera and genital valves accompany sclerotized cuticle and a stegasime prodorsum. Each genital valve bears a true seta; a single pair of aggenital setae is present. Immatures and males have the aggenital setae off the genital valves. Sexual dimorphism is pronounced to moderate and may involve secondary sex characteristics, but differs from group to group. Males are generally smaller than females and exhibit a sclerotized aedeagus together with tarsal (legs IV) and caudal (paranal) copulatory suckers (an ancestral character) (Fig. 20-2E). Females have a longitudinal or transverse aperture, the oviporus, frequently with genital valves (paragynial flaps of Krantz, 1978), and many have an epigynial apodeme or sclerite. Sperm transfer occurs through a secondary copulatory bursa or sperm duct rather than through the genital oviporus; the sperm duct may be extended into a long, projecting tube (Fig. 20-2F). A maximum of two pairs of genital papillae occur in adults regardless of habitat, but are reduced in parasitic taxa, hygrophilous, and halophilous forms. Vestiges of these papillae may be observed in other taxa, however (B.M. OConnor, personal communication, 1985). Various modifications of legs accompany these sexual characteristics.

Leg coxae are fused into the ventral surface and the coxal fields are identified by the subepidermal epimera (Figs. 20-1A; 20-2A). Leg chaetotaxy in the Astigmata is highly variable and neotenous; most ancestral conditions are lost. Chaetotaxy may be reduced, but the larval setal complex obtains through ontogeny without many additions. Solenidia are added to leg IV as postprotonymphal features. The tarsal morphology varies among the suborders, but usually exhibits a pretarsus. An empodial claw may be present, highly modified, or absent, with an expanded membranous pad or sucker (sessile or stalked) involved in the ambulacrum. True claws are absent in this order (Fig. 6-5A).

The ancestral life cycle appears to be similar to that of other Acariformes mites. The paurometabolous life cycle passes through egg, larva, protonymph, (sometimes a deutonymph=hypopus), and tritonymph to adult without much morphological change except in size, modification of setation, and addition of genital papillae (acetabulae). Compression of life stages to egg, deutonymph, and adult occurs in some (*Hypodectes*, Hypoderatidae). The larva is hexapod and usually exhibits an urstigma (Claparede organ) on coxal field I. The modification of the deutonymph (or hypopus) is a distinctive character of the order. This heteromorphic stage (compare hypopus, Chapter 12) is highly modified for phoresy (transport on dif-

ferent hosts) or for resisting adverse environmental conditions. Morphologically, the stage is different from the other life stages because it lacks chelicerae and a mouth, hence does not feed in the normal manner (it may absorb food through the cuticle) (B.M. OConnor, personal communication, 1985). The palps and capitulum are greatly reduced or absent. Generally, the heavily sclerotized body is flattened dorsoventrally; the body setae are very short. An attachment organ (either a sucker plate or claspers for grasping hair) is formed in the paraproctal area of the deutonymphs that are carried by insects or mammals. Cutaneous hypopi have the attachment organs reduced or absent (parasitic deutonymphs of the Metalabidophorinae, Glycyphagidae). The setae of segment AD usually are large, bulbous, and suckerlike; setae of segment AN are lacking, but the alveoli form into large suckers. Legs of the hypopi are short and stout and have modified leg setae. Development of the hypopial deutonymph appears to be facultative. The protonymph develops directly into the hypopus or into the tritonymph. Hypopi are not found in the Astigmata that parasitize vertebrates (OConnor, 1982a, also personal communication, 1985).

Compared with other orders of mites, the Astigmata exhibit a relatively simple hierarchial classification. The existing homogeneity of the higher categories may preclude further division at present, but with future cladistic analysis of data the higher classification may change. Good hypotheses about the true evolutionary history of the group will eventually lead to stable classification (B.M. OConnor, personal communication, 1985; Krantz, 1978).

The Astigmata seem to be a natural group characterized by a relatively large array of "derived character states." Their affinities are not completely defined, but proposed affiliations with the Actinedida and the Oribatida differ. They are considered mostly closely allied to the higher Oribatida. Cladistic data strongly support the proposition that Astigmata are derived from oribatids that are neotenous and not related to the Prostigmata or "Endeostigmatal" groups (OConnor, 1984a, also personal communication, 1985). The comparison in Table 20-1 should help with some of those affinities.

As is true in many of the acarine orders, the classification of the higher categories and phyletic relationships of the Astigmata are being clarified by cladistic analysis. Family groups are undergoing taxonomic revisions. The phyletic system below is adapted from OConnor (1982b) (nonvertebrate parasites) and Krantz (1978) (the vertebrate parasites) with modifications in the superfamilial arrangements. Families identified with question marks are considered to be of uncertain placement, to represent ecological specialties, or to be without sufficient comparative morphological basis at present.

TABLE 20-1 Acariformes: Comparison of Astigmata and Oribatida

Astigmata	Oribatida
Cheese, itch, feather, and fur mites.	Beetle, moss mites, oribatids (Hornmilben).
Tracheal system ancestrally absent; adults without tracheal system.	Tracheal system ancestrally absent; tracheal system hidden with acetabular stigmata in adults.
Weakly sclerotized, neotenous.	Strongly sclerotized in adults.
Trichobothria absent; present ancestrally.	Trichobothria present.
Stegasimy absent ancestrally.	Stegasimy absent ancestrally.
Gnathosoma not in camerostome.	Gnathosoma in a camerostome (stegasime) adults; some without a camerostome (astegasime).
Infracapitulum with 2 pairs of setae; adoral setae may be present.	Infracapitulum with several pairs of setae; with 2 or 3 pairs of adoral setae (primitive).
Rutella present; gena and mentum obscure.	A pair of rutellae with paired gena and mentum and paired genal, mental setae.
Chelicerae chelate-dentate; most with conical spine and mandibular process=seta; palps 1-,2-segmented, adpressed to subcapitulum.	Chelicerae chelate-dentate, some highly modified (Gustaviidae); usually with 2 setae; palps usually 5-segmented (trochanter, femur, genu, tibia, tarsus; with acanthions).
Grandjean's organ at coxa I, not ancestral; podocephalic canal present; tarsi with ambulacrum of caruncles, empodial claws, membranous pad, stalked suckers; without true claws (true claws lost ancestrally, neotenous).	Grandjean's organ absent; podocephalic canal sometimes obscure; tarsi with caruncles, some with empodial claws; with true claws.
Tarsal solenidia bacilliform (ancestral).	Tarsal solenidia filiform or setiform (primitive).
Sexual dimorphism distinctive; males with copulatory suckers on legs IV, anal suckers; penis or aedeagus present; females with or without genital apodemes; ovipositor absent; with 2 pairs of genital acetabulae (=papillae=discs=sense organs); some females with copulatory bursa; genital opening (oviporus), intercoxal, unsclerotized, longitudinal, with paragynial folds; one pair of setae in adults; genital setae not added in immatures.	Sexual dimorphism indistinct; males without copulatory or anal suckers; aedeagus, spermatophores present, females with or without genital apodemes; prominent ovipositor present; 3 pairs of genital acetabulae/papillae in adults; females without copulatory bursa; genital opening intercoxal; heavily sclerotized genital covers (trapdoor like) in adults; covers with varying setae, setal pairs added in nymphal stages.

ACARIDIA

The superfamily Schizoglyphoidea is considered the earliest lineage of the Astigmata and is known only from the deutonymphal stage.

Superfamily Schizoglyphoidea
Schizoglyphidae—tenebrionid beetle in New Zealand

Filter-feeding mites in the Histiosomatoidea have flattened, nonchelate chelicerae with many fine teeth. Palps are reduced in some, enlarged in others. Legs have spinelike setae. The adults are usually soft-bodied.

Superfamily Histiosomatoidea
Histiosomatidae (=Anoetidae)—rotting, decaying organic matter (Fig. 20-1B)
Guanolichidae—burrowing mites from bat guano; South Africa

The Canestrinioidea are relatively large, flattened external parasites or ectocommensals. Canestriniid mites are diverse in morphology and live in the subelytral spaces of several families of beetles (Carabidae, Scarabeidae, Lucanida, Passalidae, Tenebrionidae, Chrysomelidae). The mites apparently feed on exudates under the hard wing covers, but some are known to feed from the mouthparts of passalids. No Nearctic species are known.

Superfamily Canestrinioidea
Canestriniidae—various beetles
Heterocoptidae—chrysomelid beetles

Mites in the Hemisarcoptoidea are usually weakly sclerotized and frequently have ocelli on the prodorsum. The pretarsi have well-developed ambulacra with empodial claws. The mites are found in association with various insects in wood-related habitats.

Superfamily Hemisarcoptoidea
Chaetodactylidae—exclusive in nests of solitary bees

Hyadesiidae—algophagous, intertidal mites
Carpoglyphidae—pests of stored products and beehives
Algophagidae—algivores in wet habitats, treeholes, algae
Hemisarcoptidae—predaceous on scale insect eggs, nidicoles
Winterschmidtiidae—(=Saproglyphidae) fungivores, arboreal, nidicoles
?Heterocoptidae?

Glycyphagoids are sclerotized with small microtrichae on the body surface. Setae are usually long and strongly barbed, but may be short and simple. Males usually lack paranal suckers and copulatory suckers on legs IV. Legs are elongated with attenuated tarsi and a membranous ambulacrum. The deutonymphs are variously adapted for attachment to mammals and insects; some are parasitic in the tissues of their hosts.

Superfamily Glycyphagoidea
Euglycyphagidae—nidicoles in raptorial bird nest
Chortoglyphidae—nidicoles in mammal nests
Pedetopodidae—phoretic deutonymphs on African rodent
Echimyopodidae—nidicoles of marsupials, edentates, and rodents
Aeroglyphidae—bat roosts and guano; nidicoles
Rosensteiniidae—bat roosts and guano; blaberid cockroaches
Glycyphagidae—stored products and house dust

The superfamily Acaroidea includes a wide diversity of mites found in associations with some vertebrates and insects. They occur in mealworm cultures, termite cultures, different types of stored products and innumerable locations in which decaying organic matter is present. The body is usually weakly sclerotized, rounded, and seldom flattened dorsoventrally. Paired longitudinal shields may be present dorsally or lacking. The female ovipore is an inverted-Y shape. Males have paranal suckers and copulatory suckers on tarsi IV; sometimes legs III are enlarged. The legs exhibit a short ambulacra, each with an empodial claw; condylophores are short and stout.

Superfamily Acaroidea
    Suidasiidae—nidicoles of vertebrates and insects
    Lardoglyphidae—nidicoles of vertebrates and insects; stored products, meat, fish meal
    Glycacaridae—nidicoles of Antarctic petrels
    Acaridae—diverse habitats; nidicoles of vertebrates and insects (Hymenoptera); organic matter, dung, carrion, rotting materials; (Figs. 20-1A; 20-3A) (Ewingidae, gill parasites of land hermit crab)

The Hypoderatoidea (=Hypoderidae) are known mainly from the deutonymphs and share characteristics with other families parasitic on birds and mammals listed below. Chelicerae and palps may be vestigial in some (*Hypodectes*). The deutonymph is highly modified for a subcutaneous parasitic existence and is without attachment organs. Neosomy is common as this stage grows in the host. The genera appear to be host-specific and the life cycles correlate closely with those of the hosts.

Superfamily Hypoderatoidea
    Hypoderatidae—specific nidicoles of many bird orders, some mammals

PSOROPTIDIA

With very few exceptions the categories in this group are parasites of birds and mammals (OConnor (1982b) uses the term "Psoroptidia"

without categorical rank). The absence of setae from the opisthosomal segment AN and AD is characteristic. Genital valves of both sexes are reduced and fused to the ventral body surface. Empodial claws of the pretarsi may be reduced or fused and expanded into an ambulacral disc. Females usually show genital apodemata and are different morphologically from males. The deutonymph is missing in the life cycle (OConnor, 1982b).

Few characters are consistently present in the Pterolichoidea. The pretarsus has large, secondarily developed sclerites. This superfamily represents the oldest lineage of the Psoroptidia. The majority of the species in this superfamily comprise ectocommensal or parasites on most of the orders of birds, where they are parasites on the wing and tail feathers. The Pterolichidae constitutes the stem group for the superfamily and the genera in this family (with few exceptions) are restricted to single orders of their bird hosts.

Superfamily Pterolichoidea
    Pterolichidae—numerous orders of birds
    Gabuciniidae—wing parasites
    Falculiferidae—wing parasites of Columbiformes
    Eustathiidae—Apodiformes (New World or Old World)
    Crypturoptidae—Tinaniformes, Neotropical
    Thoracosathesidae—Galliformes, New Guinea
    Rectijanuidae—wing and neck feathers of ducks
    Syringobiidae—endoparasites in quills of flight feathers
    Kramerellidae—parasites of ibises
    ?Freyaniidae?—large mites on wing feathers of aquatic birds
    Vexillariidae—parasites of hornbills, passeriforms
    Caudiferidae—wing parasites of ibises

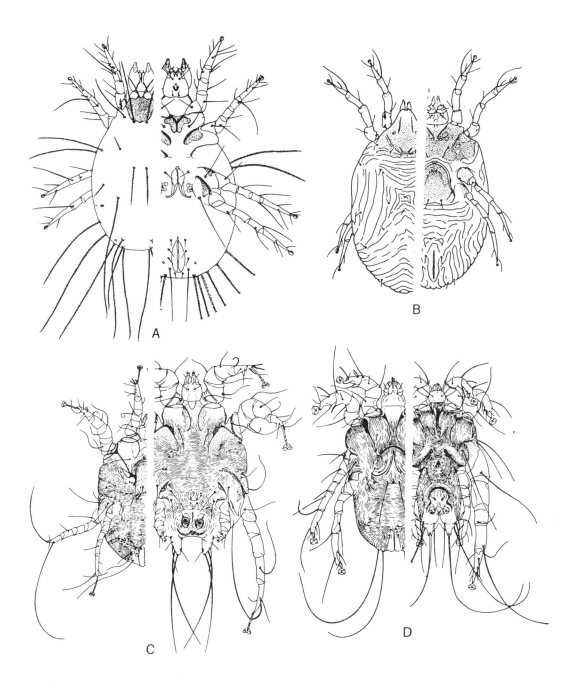

Figure 20-3. Representative examples of Astigmata: (A) dorsal and ventral views of Acaridae; (B) dorsal and ventral views of Pyroglyphidae; (C) ventral view of Psoroptidae female (left), male (right); (D) ventral view of Analgidae female (left), male (right) (after Baker et al., 1958).

Most of the Pyrolglyphoidea are parasitic on birds, but the family Pyroglyphidae includes species of *Dermatophagoides*, the house dust mites, principal allergens in house dust allergy (Arlian, 1976) (Fig. 1-1B). The mites of Pyroglyphidae range in habitat from free-living to parasitic status. One species is the only nonparasitic form in the Psoroptidia. Pyroglyphids are associated with many avian groups, but most are nidicoles of Passeriformes. Morphology and biology of these mites seems to indicate that the free-living forms are descended from permanently parasitic ancestors (OConnor, 1982b).

Superfamily Pyroglyphoidea

Pyroglyphidae—external parasites on feathers, nidicoles; (Fig. 20-3B)

Ptyssalgidae—quills of flight feathers of hummingbirds

Turbinoptidae—upper respiratory tract of birds

Analgoids are feather mites in which the pretarsi are retracted into the end of the tarsus. These are mites associated with birds, either externally, or in the feather follicles, in quills or in space around the quills.

Superfamily Analgoidea

Analgidae—nonaquatic bird parasites; on body feathers, wing coverts (Fig. 20-3D)

Xolalgidae—down feathers of many bird orders

Avenzoariidae—parasites on flight feathers of aquatic birds

Proctophyllodidae—wing mites of many birds (Fig. 20-4A–C)

Trouessartidae—wing mites of passeriform birds

Alloptidae—wing parasites of aquatic birds

Epidermoptidae—skin parasites of birds

Apionacaridae—quill inhabitants of birds

Dermoglyphidae—quill inhabitants of birds

Laminosioptidae—developing feather follicles or subcutaneous tissues of birds (Fig. 20-4G)

Knemidokoptidae—parasites of feather follicles, cause "scaly-leg" in wild and domestic fowl

?Cytoditidae?—parasites of lower respiratory tract of birds (Fig. 20-4F)

The superfamily Psoroptoidea comprises mites that are parasitic on skin, hair, in hair follicles, respiratory tract, and subcutaneous tissues of mammals. They differ from Analgoidea in the lack of retractable pretarsi. Propodosomal and opisthosomal sclerites are present. Sexual dimorphism is marked: in males legs III and/or IV are greatly hypertrophied and copulatory suckers are present at the posterior end of the venter.

Superfamily Psoroptoidea

Psoroptidae—skin parasites of mammals (Fig. 20-3C) (=Galagalgidae—skin parasites of lorisid primates)

Lobalgidae—skin parasites of sloths, edentates, rodents

Myocoptidae—skin, hair mites of marsupials and rodents

Rhynchoptidae—hair follicle mites of monkeys and porcupines

Audycoptidae—hair follicle parasites of carnivores, primates

Listrophoridae—skin, hair parasites of mammals (Fig. 20-4D, E)

Chirodiscidae—mammal hair-clasping mites

Atopomelidae—hair-clasping mites

Chirorhynchobiidae—wing ectoparasites of bats

Gastronyssidae—endoparasites of digestive and respiratory tracts of bats and rodents

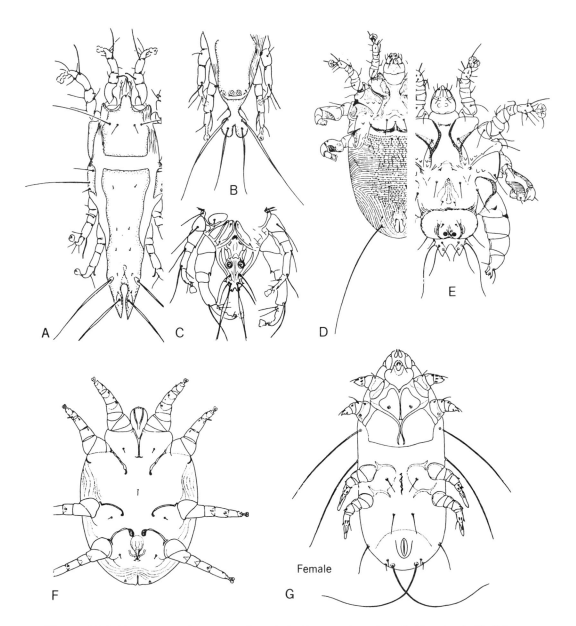

Figure 20-4. Representative examples of Astigmata (continued): Proctophyllodidae, (A) dorsal view of female; (B) dorsal view of posterior hysterosoma of male; (C) ventral view of posterior hysterosoma of male; Listrophoridae, (D) ventral view of female; (E) ventral view of male; (F) ventral view of Cytoditidae; (G) ventral view of Laminosioptidae (after Baker et al., 1958).

Lemurnyssidae—endoparasites of respiratory tracts of Primates

Pneumocoptidae—lungs of rodents

?Sarcoptidae?—follicle and skin parasites of mammals (Fig. 1-1C)

REFERENCES

Arlian, L.G. 1976. Mites and house dust allergy. *J. Asthma Res.* 13(4):165-172.

Arlian, L.B. and D.L. Vyszenski-Moher. 1986. *In:* Slansky, F., Jr. and J.C. Rodriguez, eds. *Nutritional Ecology of Insects, Mites and Spiders.* Wiley, New York, pp. 765-790.

Arlian, L.G., R.A. Runyan and S.A. Estes. 1984a. Cross infestivity of *Sarcoptes scabei. J. Am. Acad. Dermatol.* 10(6):979-986.

Arlian, L.G., D.P. Geis, D.L. Vyszenski-Moher, I.L. Bernstein and J.S. Gallagher. 1984b. Cross antigenic and allergenic properties of the house dust mite *Dermatophagoidea farinae* and the storage mite *Tyrophagus putriscentiae. J. Allergy Clin. Immunol.* 74(2):172-179.

Arlian, L.G., R.A. Runyan, S. Achar and S.A. Estes. 1984c. Survival and Infestivity of *Sarcoptes scabei* var. *canis* and var. *hominis. J. Am. Acad. Dermatol.* 11(2):210-215.

Atyeo, W.T. and J. Gaud. 1971. Comments on nomenclatural systems for idiosomal chaetotaxy of sarcoptiform mites. *J. Kans. Entomol. Soc.* 44(3):414-419.

Atyeo, W.T. and P.C. Peterson. 1972. The feather mite family Alloptidae Gaud, new status. I. The subfamilies Trouessartinae Gaud and Thysanocercinae, new subfamily (Analgoidea). *Zool. Anz.* 188(1/2):56-60.

Atyeo, W.T. and P.C. Peterson. 1976. The species of the feather mite family Rectijanuidae (Acarina: Analgoidea). *J. Ga. Entomol. Soc.* 11(4):349-366.

Atyeo, W.T., E.W. Baker and M.D. Delfinado. 1974. *Gaudiella minuta,* a new genus and species of mite (Acarina: Acaridia) belonging to the new family Gaudiellidae. *Wash. Acad. Sci.* 64(49):295-298.

Baker, E.W. 1951. *Pneumocoptes,* a new genus of lung-inhabiting mite from rodents (Acarina: Epidermoptidae). *J. Parasitol.* 37(6):583-586. (Pneumocoptidae)

Baker, E.W. and D.A. Crossley, Jr. 1964. A new species of mite, *Fusoherecia lawrenci,* from an artificial tree-hole (Acarina: Glycyphagidae). *Ann. Natal Mus.* 16:1-6

Baker, E.W. and G.W. Wharton. 1952. The suborder Sarcoptiformes Reuter, 1909. *In: An Introduction to Acarology.* Macmillan, New York, pp. 320-386.

Baker, E.W., T.M. Evans, D.J. Gould, W.B. Hull and H.L. Keegan. 1956. *A Manual of Parasitic Mites of Medical or Economic Importance.* Natl. Pest Control Assoc. Tech. Publ.

Baker, E.W., M.D. Delfinado and M.J. Abbatiello. 1976. Terrestrial mites of New York. II. Mites in birds' nests. *J. N. Y. Entomol. Soc.* 84(1):48-66.

Baker, E.W., J.H. Camin, F. Cunliffe, T.A. Woolley, and C.E. Yunker. 1958. *Contrib. No. 3, Inst. Acarol.,* University of Maryland, College Park.

Boczek, J. 1956. Einfluss mancher Faktoe der Umwelt auf die Bildung der Hypopen bei *Tyroglyphus farinae* L. *Ekol. Pol.* 4A:213-218.

Boczek, J. 1979. Spermatophore production and mating behavior in the stored products mites *Acarus siro* and *Lardoglyphus konoi. In:* Rodriguez, J.G., ed. *Recent Advances in Acarology.* Academic Press, New York, Vol. 1, pp. 279-284.

Bronswijk, J.E.M.H. van. 1973. *Dermatophagoides pteronyssus* (Trousessart, 1897) in mattresses and floor dust in a temperate climate (Acari: Pyroglyphidae). *J. Med. Entomol.* 10(1):63-70.

Bronswijk, J.E.M.H. van and R.N. Sinha. 1971. Pyroglyphid mites (Acari) and house dust allergy. *J. Allergy* 47(1):31:52.

Chmielewski, W. 1971. Morfoligia, biologia i ekologia *Caropoglyphus lactis* (L., 1758) (Glycyphagidae, Acarina). *Pr. Nauk. Inst. Ochr. Rosl.* 1382:63-166.

Cooper, K.W. 1955. Venereal transmission of mites by wasps, and some evolutionary problesm arising from the remarkable association of *Ensliniella trisetosum* with the wasp *Ancistrocerus antilope. Trans. Am. Entomol. Soc.* 80:119-174. (Saproglyphidae)

Cooreman, J. 1954. Acariens Canestriniidae de la collection A. C. Oudemanns, Leiden. *Zool. Meded.* 33(13):83-90.

Dodd, K. 1972. The identity of *Knemidocoptes laevis* (Raillet, 1885) (Acari: Knemidocoptidae). *Acarologia* 14(4):675-680.

Domrow, R. 1970. *Chirodiscus amplexans* Trouessart and Neumann redescribed (Acari: Listrophoridae). *Acarologia* 12(2):415–420. (Chirodiscidae)

Dubinin, V.B. 1953. Arachnoidea—Feather Mites (Analgesoidea). Part 2. Families Epidermoptidae and Freyanidae. *Fauna USSR* 6.6:1–412.

Fain, A. 1956. Une nouvelle famille d'acariens endoparasites des chauves-souris: Gastronyssidae fam. nov. *Ann. Soc. Belge Med. Trop.* 36(1):87–98.

Fain, A. 1960. Revision du genre *Cytodites* (Megnin) et description de deux espèces et un genre nouveaux dans la famille Cytoditidae. *Acarologia* 2(2):38–249.

Fain, A. 1965a. Les acariens producteurs de gale chez les edentes et les marsupiaux (Psoroptidae et Lobalgidae, Sarcoptiformes). *Bull. Inst. R. Sci. Nat. Belg.* 41(17):1–42.

Fain, A. 1965b. A review of the family Epidermoptidae Trouessart parasitic on the skin of birds. Parts I–II. *Proc. K. Ned. Akad. Wet.* 27(84):1–176 + ix:5–144.

Fain, A. 1965c. A review of the family Rhyncoptidae Lawrence parasitic on procupines and monkeys. *Adv. Acarol.* 2:135–159.

Fain, A. 1965d. Les Acariens nidicoles et detricoles de la famille Pyroglyphidae lCunliffe (Sarcoptiformes). *Rev. Zool. Bot. Afr.* 72(3–4):257–288.

Fain, A. 1967. Les hypopes parasite des tissu cellulaires des oiseaux (Hypodectidae: Sarcoptiformes). *Bull. Inst. R. Sci. Nat. Belg.* 43(4):1–139. (Glycyphagidae)

Fain, A. 1968. Notes sur trois acariens remarquables (Sarcoptiformes). *Acarologia* 10(2):276–291. (Audycoptidae, Chirorhynchobiidae)

Fain, A. 1969. Adaptation to parasitism in mites. *Acarologia* 11(3):429–449.

Fain, A. 1973. Les listrophorides d'Amerique Neotropicale (Acarina: Sarcopitformes). I. Familles Listrophoridae et Chirodiscidae. *Bull. Inst. R. Sci. Nat. Belg. Entomol.* 49:1–149.

Fain, A. and R. Domrow. 1974. The subgenus *Metacytostethum* Fain (Acari: Atopomelidae): Parasites of macropodid marsupials. *Acarologia* 16(4):719–738.

Fain, A. and K. Hyland. 1974. The listorphorid mites of North America. II. The family Listrophoridae. *Bull. Inst. R. Sci. Nat. Belg. Entomol.* 50(1):1–69.

Fain, A. and D.E. Johnston. 1975. A new algophagin mite. *Algophagopsis pneumatica*, gen. n., sp. n., living in a river. (Astigmata: Hyadesiidae). *Bull. Inst. roy. Sci. Belg. Entomol.* 111:66–70.

Fain, A. and F. Lukoschus. 1970. Parasitic mites of Surinam. II. Skin and fur mites of the families Pstoroptidae and Lobalgidae. *Acta Zool. Pathol. Antverp.* 51:49–60.

Fashing, N.J. and B.M. OConnor. 1984. *Sarraceniopus* A new genus of histiostomatid mites inhabiting the pitchers of the Sarraceniaceae (Astigmata: Histiostomatidae). *Int. J. Acarol.* 10(4):217–228.

Gaud, J. and W.T. Atyeo. 1975. Ovacaridae, une famille nouvelle de sarcoptiformes plumicoles. *Acarologia* 17(1):169–176.

Gaud, J. and W.T. Ateyo. 1977. A new name for *Ovacarus* and Ovacaridae (Acarina: Analgoidea). *Acarologia* 18(3):568–569. (Apionacaridae)

Hughes, A.M. 1976. The mites of stored food and houses. *Tech. Bull.-Minis. Agric., Fish. Food (G.B.)* No. 9.

Hughes, R.D. and C.G. Jackson. 1958. A review of the Anoetidae (Acari). *Va. J. Sci.* 9:1–198.

Hughes, T.E. 1959. *Mites, or the Acari.* Athlone Press, London.

Jeppson, L.R., H.H. Keifer and E.W. Baker. 1975. Astigmata (Acaridae) and Cryptostigmata. *In: Mites Injurious to Economic Plants.* Univ. of California Press, Berkeley, pp. 307–326.

Kanungo, K. 1969. Acarine moulting—the migration of hemocytes through the epidermis of *Caloglyphus berlesei. Ann. Entomol. Soc. Amer.* 62:155–157.

Krantz, G.W. 1978. *A Manual of Acarology,* 2nd ed. Oregon State Univ. Book Stores, Corvallis.

Lindquist, E.E. and R.C. Belding. 1949. A report on the subcutaneous or flesh mite of chickens. *Mich. State Coll. Vet.* 10:20–21. (Laminosioptidae)

Lindquist, E.E., B.M. OConnor, F. Clulow and H.H.J. Nesbitt. 1979. Acari: Acaridiae. *Mem. Entomol. Soc. Can.* 108:277–284.

Lombert, H.A.P.M, F.S. Lukoschus and B.M. OConnor. 1982. The life-cycle of *Cosmoglyphus inaequalis* Fain and Caceres, 1973, with comments on the systematic position of the genus. Results of the Namaqual and Namibia Expedition of the King Leopold III Foundation for the exploration and protection of nature (1980). *Bull. Inst. R. Sci. Nat. Belg. Entomol.* 54(10):1–17.

Lukoschus, F.S., A. Fain and F.M. Driessen. 1972. Life cycle of *Apodemopus apodemi* (Fain, 1965) (Glycyphagidae: Sarcoptiformes). *Tijdsch. Entomol.* 115(8):325–339.

Mahunka, S. 1974. Beitrage zur Kenntnis der an Hymenopteraen lebenden Milben (Acari). II. *Folia Entomol. Hung.* 27(1):99–108.

McDaniel, B. 1968a. The superfamily Listrophoroidea and the establishment of some new families (Listrophoridae: Acarina). *Acarologia* 10(3):477–482. (Listrophoridae, Myocoptidae, Rhyncoptidae)

McDaniel, B. and E.W. Baker. 1962. A new genus of Rosensteiniidae (Acarina) from Mexico. *Fieldiana, Zool.* 44(16):127–131.

Miyamoto, T. 1972. House dust and house dust mites. *In*: Patterson, R., ed. *Allergic Disease: Diagnosis and Treatment.* Lippincott, Philadelphia, Pennsylvania, pp. 114–123. (Pyroglyphidae)

Mostafa, A.-R.I. 1970a. Saproglyphid hypopi (Acarina: Saproglyphidae) associated with wasps of the genus *Zethus* Fabricius. *Acarologia* 12(1):168–192.

Mostafa, A.-R.I. 1970b. Saproglyphid hypopi (Acarina: Saproglyphidae) associated with wasps of the genus *Zethus* Fabricius. *Acarologia* 12(2):383–401.

Norton, R.A., B.M. OConnor and D.E. Johnston. 1983. Systematic relationships of the Pediculochelidae (Acari: Acariformes). *Proc. Entomol. Soc. Wash.* 85:493–512.

OConnor, B.M. 1979a. Evolutionary origins of astigmatid mites inhabiting stored products. *In*: Rodriguez, J.G., ed. *Recent Advances in Acarology.* Academic Press, New York, Vol. 1, pp. 273–278.

OConnor, B.M. 1979b. A review of the family Heterocoptidae (Acari: Astigmata). *In*: Rodgriguez, J.G., ed. *Recent Advances in Acarology.* Academic Press, New York, Vol. 2, pp. 429–433.

OConnor, B.M. 1981a. A new genus and species of Hypoderidae (Acari: Astigmata) from the nest of an owl (Aves: Strigiformes). *Acarologia* 22:299–304.

OConnor, B.M. 1981b. A systematic revision of the family group taxa in the non-psoroptoid Astigmata. Ph. D. thesis. Cornell University, Ithaca, New York.

OConnor, B.M. 1982a. Evolutionary ecology of astigmatid mites. *Annu. Rev. Entomol.* 27:385–409.

OConnor, B.M. 1982b. Acari: Astigmata. *In*: Parker, S.P., ed. *Synopsis and Classification of Living Organisms.* McGraw-Hill, New York, pp. 146–169.

OConnor, B.M. 1984a. Co-evolutionary patterns between astigmatid mites and Primates. *Acarol. [Proc. Int. Congr. Acarol.], 6th, 1982,* Vol. I, pp. 186–195.

OConnor, B.M. 1984b. Nomenclatural status of some family-group names in the non-Psoroptidid Astigmata (Acari: Acariformes). *Int. J. Acarol.* 10(4):203–208.

Pence, D.B. 1972. The hypopi (Acarina: Sarcoptiformes: Hypoderidae) from the subcutaneous tissues of birds in Louisiana. *J. Med. Entomol.* 9(5):435–438. (Glycyphagidae)

Pence, D.B. 1975. Keys, species and host list, and bibliography for nasal mites of North American birds (Acarina: Rhynonyssinae, Turbinoptinae, Speleognathinae and Cytoditidae). *Spec. Publ.-Mus. Tex. Tech. Univ.* 8:1–148.

Pinichpongse, S. 1963. A review of the Chirodiscinae with descriptions of new taxa (Acarina: Listrophoridae). *Acarologia* 5(1):81–91, 5(2):266–278, 5(3):397–404, 5(4):620–627. (Chirodiscidae)

Rupes, V., C.E. Yunker and N. Wilson. 1971. *Zibethacarus* n. gen., and three new species of *Dermacarus* (Acari: Labidophoridae). *J. Med. Entomol.* 8(1):17–22. (Glycyphagidae)

Samsinak, K. 1971. Die auf *Carabus*-Arten (Coleoptera, Adephaga) der palaearktischen Region lebenden Milben der Unterordnung Acariformes (Acari): ihre Taxonomie und Bedeutung fur die Losung zoogeographischer, entwicklungsgechichtlicher und parasitophyletischer Fragen. *Entomol. Abh.* 38(6):145–234. (Canestriniidae)

Santana, F.J. 1976. A review of the genus *Trouessartia* (Analgoidea-Alloptidae). *J. Med. Entomol. Suppl.* 1:1–128.

Smiley, R.L. 1965. A new mite, *Prosarcoptes scanloni*, from monkey. *Proc. Entomol. Soc. Wash.* 67:166–167. (Sarcoptidae)

Sorokin, S.V. 1951. The causes of the formation of the hypopi in cereal mites. *Zool. Zh.* 30:523–589.

Southcott, R.V. 1976. Arachnidism and allied syndroms in the Australian region. *Rec. Adelaide Children's Hosp.* 1(1):97–186.

Spicka, E.J. and B.M. OConnor. 1980. Description and life cycle of *Glycyphagus (Myacarus) microti* n. sp. (Acarina: Sarcoptiformes: Glycyphagidae) from *Microtus ppinetorum* in New York, U. S. A. *Acarologia* 21:451–476.

Sweatman, G.K. 1957. Life history, non-specificity and revision of the genus *Chorioptes,* a parasitic mite of herbivores. *Can. J. Zool.* 35:641–689.

Sweatman, G.K. 1958. On the life history and validity of the species in *Psoroptes,* a genus of mange mites. *Can. J. Zool.* 36:905–929.

Unger, L. and M.C. Harris. 1974. Stepping stones in Allergy. *Ann. Allergy* 33(4):228–240.

Volgin, V.I. 1973. The hypopus and its main types. *Proc. Int. Congr. Acarol., 3rd, 1971,* pp. 381–383.

Wharton, G.W. 1970. Mites and commercial extracts of house dust. *Science* 167(3923):1382–1383.

Wharton, G.W. 1976. House dust mites. *J. Med. Entomol.* 12(6):577–621.

Whitaker, J.O., Jr. and N. Wilson. 1974. Host and distribution lists of mites (Acari), parasitic and phoretic, in the hair of wild mammals of North America, north of Mexico. *Am. Midl. Nat.* 91(1):1–67.

Yunker, C.E. 1970. A second species of the unique family Chirorhynchobiidae Fain, 1967 (Acarina: Sarcoptiformes). *J. Parasitol.* 56(1):151–153.

Yunker, C.E. 1973. Mites. *In:* Flynn, R.J., ed. *Parasites of Laboratory Animals.* Iowa State Univ. Press, Ames, pp. 425–492.

Zakhvatkin, A.A. 1941. *Fauna of USSR Arachnoidea 6(1), Tyroglyphoidea (Acari),* N. S. 28. Zool. Inst. Acad. Sci., USSR (translated by Am. Inst. Biol. Sci., Washington, D.C.).

21

Order Oribatida

(=Oribatides Duges 1839 (and Auct); Carabodides C.L. Koch 1842; Cryptostigmata G. Canestrini 1891, Evans et al. 1961; Octostigmata Oudemans 1906; Oribatoidea Reuter 1909 (and Auct); Sarcoptiformes Reuter 1909 (part); Stegasima Grandjean 1932).

In a vernacular description these mites are referred to as "cheese mites with plates," but the common name, "beetle mites," comes from the fact that the majority of the oribatids have a hard or coriaceous integument, glabrous, moderately or heavily sculptured, and in miniature resemble their insect counterparts. The name "moss mites" results from their common location in mosses. They are called oribatids because originally they were considered to be "mountain dwellers, mountain ranging" by Duges. The German name "hornmilben" identifies the variable lamellae and the trichobothria and sensilla so characteristic of the order. Oribatids are much less aggressive in behavior than many of their relatives. When disturbed, they characteristically "play possum" and remain motionless for a few moments, depending on the intensity of the stimulus, then resume activity when the stimulus is lessened or removed.

These mites inhabit the upper layers of the soil, but have been found as deep as 18 in. in the Pawnee Grasslands of Colorado (J. Leetham, personal communication).

They are likely the most numerous of the soil mites, and, like beetles, one of which is found in every five insects collected, the oribatids are common in collections and in about the same proportion. High levels of oribatid diversity occur in forest ecosystems and that diversity is high when compared to other groups of mites in the soil, for example, Gamasida, Actinedida, Astigmata, and the ever-present Collembola (Wallwork, 1983). They are called soil mites but not all are found in soil. Some are also arboreal (Aoki, 1973). Oribatids are important members of the soil fauna where they are also evident in mosses, lichens, duff, litter, punk, fungi, leaf mats, turf, and humus. Those species that are phoretic occur on beetles (Johnston, 1982; Woolley, 1969) and other insects (Norton, 1980). Most are fungivorous and make up large part of the faunal secondary decomposers that contribute to the formation of humus and facilitate soil tilth (Jacot, 1936).

Six main feeding types are recognized for forest soil oribatids (Luxton, 1972a; Wallwork, 1983). These include macrophytophages, which feed mainly on fragments of higher plants and rarely on fungi; microphytophages that consume a wide range of microflora (fungi, bacteria, yeast, algae); panphytophages, which select from a wide range of higher plant materials and microflora (mainly fungi); coprophages, which have nonobligatory feeding habits with a wide enough range of food materials to be place near or in panphytophagous feeding; zoophages and necrophages, which occur sporadically in the literature and are not considered significant in the overall biology of these mites. The prevalence of panphytophagous oribatids suggests that this type of feeding behavior has important benefits for forest soil oribatids (Wallwork, 1983).

The interaction between oribatids and the microflora is not completely understood; much is yet to be discovered and interpreted. We do not know whether grazing by oribatids on microflora is inhibitory or stimulatory. The reciprocal question is also unanswered: Are certain strains of fungi inhibitory to oribatids? Feeding specificity in oribatids is yet to be completely understood. When that becomes known, we may gain an appreciation for the factors that determine succession of species in the sequence of decomposition (Wallwork, 1983).

Oribatids are not true "decomposers" in the sense that fungi and bacteria are in a forest ecosystem; they are much less efficient in degradation of chemicals in lignified plant systems. It appears that their direct effects on the decomposition of plant materials are minimal, but they produce indirect effects that stimulate microfloral growth and activity, and they free nutrient pools in the microflora in a substantial way (Wallwork, 1983).

Another important relationship of the oribatids in nature is their function as vectors of anoplocephaline tapeworms. The moss mites are intermediate hosts for the transmission of cysticercoids of these tapeworms. Twelve families, 25 genera, and 32 species are confirmed vectors (Rajski, 1959, cited by Sengbusch, 1977).

The principal, rather singular characteristic that distinguishes the oribatid mites is the peculiar trichobothrium (=pseudostigmata) and its sensillum (pseudostigmatic organ). These structures are located at the posterolateral corners of the prodorsum (Fig. 21-1A–F). It was originally thought to be a spiracle and associated with the respiratory system, but the sensory sensillum made it difficult to assess as a respiratory organ. Its function is not thoroughly known, but it is assumed to detect air currents and is possibly a humidity receptor.

The body of the oribatids is usually divided by a dorsosejugal suture into the propodosoma and the hysterosoma (Fig. 21-1A), although exceptions to this condition occur. The setae of the prodorsum are distinctive and include the rostral, lamellar, interlamellar, and exobothridial hairs (one or two pairs) in addition to the distinctive trichobothria of this region (Fig. 21-1B, M). The trichobothria and sensilla not only characterize the order, but exemplify one of the many morphological variations that occur. Primitively five pairs of normal setae and one pair of trichobothria distinguish the prodorsum. Aquatic species lack trichobothria (Johnston, 1982). Other diagnostic integumental structures of the prodorsum include costulae, lamellae (with or without cusps), and in some higher forms a translamella (Fig. 21-1A, E). A tutorium (=second lamella or tectopedium of some authors) and pedoctectum I (=tectopedium I) are associated with the lateral aspects of the propodosoma; prolamellae and sublamellae may also occur (Fig. 21-1F).

Because of the projection of a prodorsal sclerite over the chelicerae, the gnathosoma may be found in a camerostome (stegasime) (Fig. 21-1P) or dorsally visible and exposed (astegasime). The gnathosoma is peculiarly constructed with simple palps of three to five segments. The chelate–dentate chelicerae are sclerotized to meet the needs and feeding of the mites (with bizarre exceptions in some, e.g., Gustaviidae, Pelopidae, Figs. 5-18; 21-5E). The

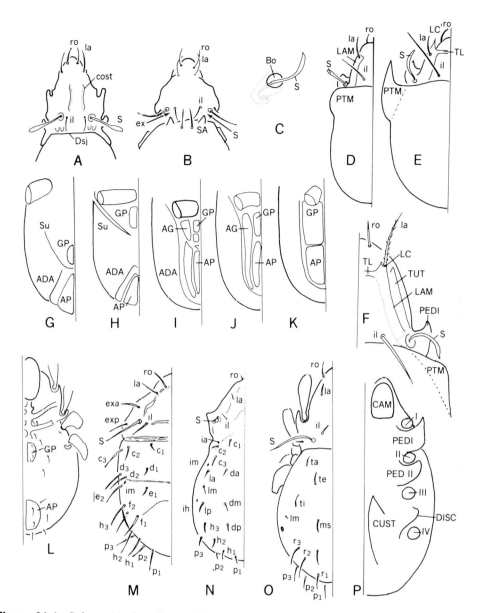

Figure 21-1. Schematic drawings of characteristics: (A) prodorsum of eremaeid showing rostral (ro), lamellar (la), interlamerllar hairs (il), sensillum (S), costula (COST) and dorsosejugal suture (Dsj); (B) prodorsum of belboid showing similar features and spinae adnatae (SA) of hysterosoma; (C) enlarged view of bothridium (Bo) and sensillum (S); (D) prodorsum and hysterosoma of oribatulid showing prodorsal setae, lamella (LAM), and pteromorph (PTM); (E) same of achipteriid showing forward-extended cusp of pteromorph (PTM); (F) prodorsum of ceratozetid showing lamella with cusp (LC), translamella (TL), tutorium (TUT), and pedotectum I (PED I). Ventral views showing genital and anal plates in different families (after G.O. Evans, personal communication, 1968): (G) Epilohmanniidae; (H) Nanhermanniida; (I) Lohmanniidae; (J) Nothroidea; (K) Phthiracaridae; (AP, anal plate; ADA, adanal plate; AG, aggenital plate; GP, genital plate; Su, suture); (L) ventral view of Circumdehiscentiae; varied dorsal setation, (M) Lohmannoidea (16 pairs); (N) Ameronothridae (15 pairs) (after Krantz); (O) Oppiidae (10 pairs) (after Wallwork); (P) composite of ventral structures in oribatids, (CAM), camerostome, custodium (CUST), discidium (DISC), pedotecta (I, II), coxae (I,II, III,IV).

chelicerae bear two setae in most species. The infracapitulum (=subcapitulum) is specifically structured, comprising strongly developed rutella (transversely moving auxiliary "jaws"), one to three pairs of adoral setae, and variously separated or coalesced plates, the gena and mentum. The progressive fusion of the gena and mentum results in the conditions Grandjean described as anarthry, diarthry and stenarthry (Fig. 5-10) (Grandjean, 1957; Woolley, 1969).

The hysterosoma (=notogaster, notaspis) shows varying degrees of sclerotization. The opisthosomal segmentation includes C, D, E, F, H, PS, AD, and AN (Figs. 4-15E, 4-23). In some primitive forms the peranal segment AN occurs. Immatures exhibit 16 pairs of notogastral setae, but this is variable and modified in the adults. Dorsal chaetotaxy is important in the identification and taxonomy. Specific pairs of distinctive lyrifissures occur in addition to the setae (*ia, im, ip, ih, ips, iad)* (Figs. 4-15E, 4-23; 21-1N–O). In addition to the setae and lyrifissures the dorsum of the hysterosoma may have lateral pteromorphs that are fixed or movable (to protect the legs), anterior spinae adnatae (Damaeidae), and humeral processes (anteriorly directed processes) (Fig. 21-1B). Areae porosae, sacculi and pori are also found in the integument (Figs. 9-13; 21-2D–G).

The venter and pleura of the hysterosoma are variously sclerotized, coalesced, or sculptured with the epimera and apodemata. Setae of the epimeral region are diagnostic and distinctive. The genital and anal openings have trapdoor-like covers, each with a diagnostic number of setae (Fig. 21-3).

Sexual dimorphism is relatively rare (Wallwork, 1962b), but the presence of an ovipositor in the females and a short aedeagus in the males is characteristic. Both sexes exhibit three pairs of genital acetabula.

The life cycle consists of prelarva, larva, protonymph, deutonymph, tritonymph, and adult. "The evolution of immature–adult heteromorphy is a pervasive trend throughout the oribatids. Sex determination is diplodiploid, al-

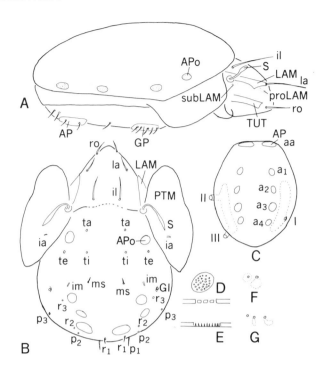

Figure 21-2. (A) Lateral view of higher oribatid showing areae porosae (APo), prolamella (proLAM), sublamella (subLAM), tutorium (TUT) (after Krantz); (B) dorsal setation and arrangement of areae porosae and fissures in galumnoid mite; (C) schematic of location of areae porosae; different locations of opisthosomatic gland opening: I, usual, II, Hermanielloidea, III, Parhypochthoniidae; schematic of (D) area porosa and section; (E) section of area spinosa; (F) saculae; (G) pori.

though thelytoky is common" (Johnston, 1982).

Leg segmentation exhibits primitive elements (Johnston, 1982). The legs consist of six joints or articles: coxa, trochanter, femur, genu (patella), tibia, and tarsus. The coxae articulate in acetabula that receive the openings of the tracheae (some brachytracheae) and hence connotate the older name Cryptostigmata (Fig. 9-12). No external stigmata are visible; peritremes are absent.

The chaetotaxy of the legs is distinctive and important to identification and classification

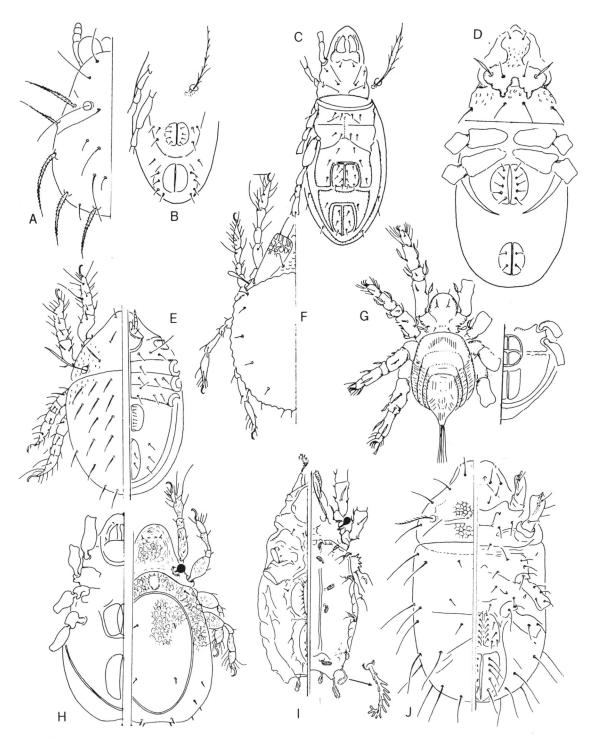

Figure 21-3. Representative oribatids: (A) Acaronychidae; (B) Eulohmanniidae; (C) Epilohmanniidae; dorsal–ventral view of (D) Nanhermanniidae; (E) Hermanniidae; (F) dorsal view of Cepheidae; dorsal–ventral view of (G) Liodidae; (H) Cymbaeremiidae; (I) Camisiidae; (J) Lohmanniidae.

(Grandjean, 1935; Norton, 1977). Much of the new literature employs the established system, which lends exactness to the description of species. Solenidia are numerous on legs of oribatids (four to seven on tarsus I; three to seven on genu I; seven on genu and tibia III) (Johnston, 1982). The tibial solenidia are always long and flagelliform, as is true in the Astigmata. The tarsus has a complex of specialized setae, including the dorsal set of solenidia (hollow sensory setae) and a famulus (specialized microsolenidion). Immatures usually exhibit a single claw on each tarsus; adults may have one, two, or three claws. If one claw is present (monodactylous), it is generally considered an empodial claw. If, in company with two lateral claws, (tridactylous) the central claw is empodial, the lateral claws are true claws. Two true claws (bidactylous) may be present without an empodial claw. The larger size of the lateral claws in comparison to the middle claw results in a heterodactylous condition.

According to Johnston (1982): "Although species diversity is greatest in the tropics, the group is well represented in the temperate and extreme northern and southern zones. Oribatids are the only mite group in which diversity of species richness has been achieved in the absence of parasitism." The order Oribatida has been well-studied, but probably is also the order with the greatest need for taxonomic clarification and revision.

An estimated 8000–9000 species belonging to over 500 genera comprise this enormous order. The morphology has been elucidated brilliantly by such authors as the late François Grandjean. Details from his numerous efforts have been used effectively in the classification of these mites. Balogh (1972) and many others have contributed to the classification.

There has been a tendency to elevate or upgrade the classification in the order and many works have moved the classification into an arrangement of suborders and superfamilies that is still in a state of flux. The reader is referred to Balogh's (1972) *The Oribatid Genera of the World*, to the compendium of Grandjean's

(1972–1976) *Oeuvres Acarologique Completes,* and to Johnston's (1982) part of *Synopsis and Classification of Living Organisms.* An estimated 145 families exist in the order (Johnston, 1982).

The reader is also referred to Lee (1984) whose classification divides the Oribatida into two suborders each with three subgroups ("cohorts").

The classification suggested below contains six suborders:

Suborder Palaeosomata

Superfamily Archeonothroidea
    Acaronychidae (Fig. 21-3A)
    Archeonothridae
    Palaeacaridae
Superfamily Palaeacaroidea
    Palaeacaridae
Superfamily Ctenacaroidea
    Adelphacaridae
    Aphelacaridae
    Ctenacaridae

Suborder Enarthronota

Superfamily Parhypochthonoidea
    Parhypochthoniidae
    Elliptochthoniidae
    Gehypochthoniidae
Superfamily Hypochthonoideaa
    Hypochthoniidae (Fig. 21-4D)
    Eniochthoniidae
Superfamily Brachychthonoidea
    Brachychthoniidae
Superfamily Cosmochthonoidea
    Cosmochthoniidae
    Haplochthoniidae
    Sphaerochthoniidae
    ?Pediculochelidae

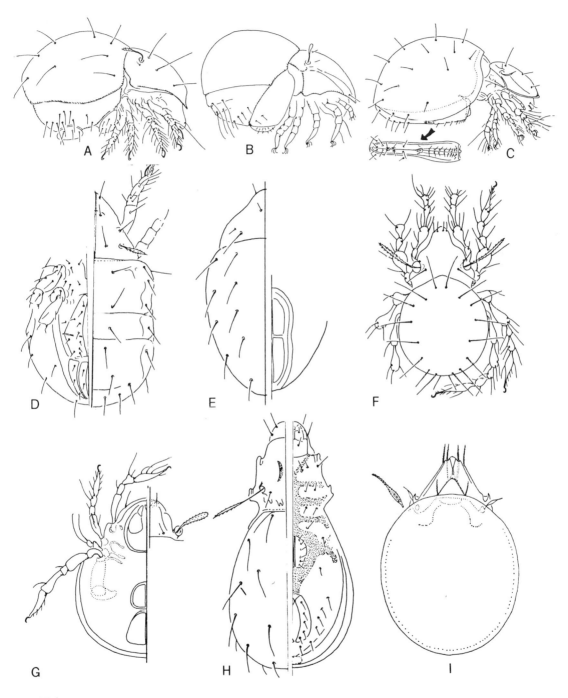

Figure 21-4. Representative oribatids: Lateral view of (A) Protoplophoridae; (B) Meso-plophoridae; (C) Rhysotritiidae; dorsoventral views of (D) Hypochthoniidae; (E) Mala-conothridae; (F) dorsal view of Belbidae; dorsoventral views of (G) Zetorchestidae, (H) Eremaeidae; (I) dorsal view of *Liacarus nitens,* Liacaridae.

Superfamily Heterochthonoidea
  Heterochthoniidae
Superfamily Phyllochthonoidea
  Phyllochthoniidae
  Atopochthoniidae (Fig. 21-6)
  Pterochthoniidae
Superfamily Protoplophoroidea
  Protoplophoridae (Fig. 21-4A)
Superfamily Mesoplophoroidea
  Mesoplophoridae (Fig. 21-4B)
  Archoplophoridae

Suborder Parhypochthonata

Superamily Parhypochthonoidea
  Parhypochthoniidae
  Gehypochthoniidae
  Elliptochthoniidae

Suborder Mixonomata

Superfamily Eulohmannoidea
  Eulohmanniidae (Fig. 21-3B)
  Lohmanniidae (Fig. 21-3J)
  Epilohmanniidae (Fig. 21-3C)
  Neuhypochthoniidae

Suborder Euptyctima

Superfamily Collohmannoidea
  Collohmanniidae
  Euphthiracaridae
  Phthiracaridae (Fig. 21-1K; 21-4C)

Suborder Nothronata

Superfamily Nothroidea
  Nothridae
  Crontoniidae
  Camisiidae (Fig. 21-3I)
  Trhypochthoniidae
  Malaconothridae (Fig. 21-4E)
Superfamily Nanhermannioidea
  Nanhermanniidae (Fig. 21-3D)
Superfamily Hermannioidea
  Hermanniidae (Fig. 21-4D)
  Perhohmanniidae
Superfamily Liodoidea
  Liodidae (Fig. 21-3G)
Superfamily Gymnodamaeoidea
  Gymnodamaeidae
Superfamily Belbidoidea
  Damaeidae
  Belbidae (Fig. 21-4F)
  Belbodamaeidae

Circumdehiscentiae
Pycnonoticae (79 families)
Superfamily Eremaeoidea
  Eremaeidae (Fig. 21-4H)
  Megeremaeidae
Superfamily Zetorchestoidea
  Zetorchestidae (Fig. 21-4G)
Superfamily Eremuloidea
  Amerobelbidae
  Eremulidae
  Staurobatidae
  Damaeolidae
  Basilobelbidae
  Heterobelbidae
  Ameridae
  Ctenobelbidae
Superfamily Liacaroidea
  Gustaviidae (Fig. 2-6B)

Metrioppiidae
Multoribulidae
Liacaridae (Fig. 21-4I)
Xenillidae
Astegistidae
Tenuialidae (Fig. 21-5A)
Superfamily Polypterozetoidea
  Charassobatidae
  Eutegaeidae
  Tumerozetidae
  Polypterozetidae
  Eremaeozetidae
  Nodocepheidae
  Podopterotegaeidae
Superfamily Cepheoidea
  Cepheidae (Fig. 21-3F)
Superfamily Carabodoidea
  Niphocepheidae
  Carabodidae
  Nipobodidae
  Tectocepheidae
Superfamily Oppioidea
  Tuparezetidae
  Sternoppiidae
  Oxyameridae
  Oppiidae
  Autognetidae
  Thyrosomidae
  Suctobelbidae
  Rhynchoribatidae
  Caleremaeidae
  Eremellidae
  Aceremaeidae
  Machadbelbidae
  Spinozetidae
  Anderemaeidae
Superfamily Otocepheoidea
  Dampfiellidae
  Otocepheidae
  Tokuncepheidae

Superfamily Hydrozetoidea
  Hydrozetidae
  Limnozetidae
Superfamily Cymbaeremoidea
  Cymbaeremeidae (Fig. 21-3H)
  Kodiakellidae
  Micreremidae
Superfamily Ameronothroidea
  Ameronothridae
  Podacaridae

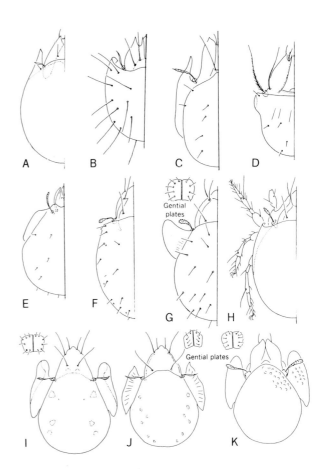

Figure 21-5. Representative oribatids: Dorsal views of (A) Tenuialidae; (B) Oribatellidae; (C) Achipteriidae; (D) Microzetidae; (E) Pelopidae; (F) Oribatulidae; (G) Haplozetidae; (H) Ceratozetidae; (I) Galumnidae; (J) Parakalummidae; (K) Epactozetidae.

Selenoribatidae
Fortuyniidae

Poronoticae (29 families)
Superfamily Passalozetoidea
  Passalozetidae
  Licneremaeidae
  Scutoverticidae
Superfamily Oribatelloidea
  Oribatellidae (Fig. 21-5B)
  Achipteriidae (Fig. 21-5C)
  Tegoribatidae
  Unduloribatidae
Superfamily Microzetoidea
  Microzetidae (Fig. 21-5D)
Superfamily Pelopoidea
  Pelopidae (Fig. 21-5E)
Superfamily Oribatuloidea
  Oribatulidae (Fig. 21-5F)
  Haplozetidae (Fig. 21-5G)
  Oripodidae
  Zetomotrichidae
  Lamellareidae
  Nasobatidae
  Chaunoproctidae
  Neotrichozetidae
  Stelechobatidae
  Symbioribatidae
Superfamily Ceratozetoidea
  Ceratozetidae (Fig. 21-5H)
  Mycobatidae
  Zetomimidae
  Chamobatidae
  Euzetidae
  Mochlozetidae
Superfamily Galumnoidea
  Galumnidae (Fig. 21-5I)
  Ceratokalumnidae
  Parakalumnidae (Fig. 21-5J)
  Galumnellidae
  Epactozetidae (Fig. 21-5K)

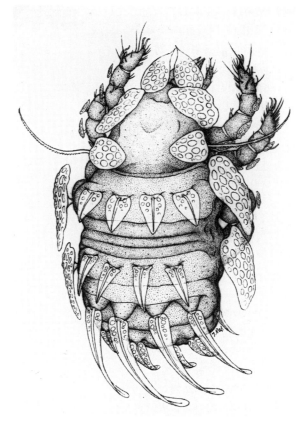

Figure 21-6. Dorsal view of *Atopochthonius artiodactylus* Grandjean (from J. Rafalski).

REFERENCES

Allred, D.M. 1954. Mites as intermediate hosts of tapeworms. *Proc. Utah Acad. Arts Lett.* 31:44–51.

Anderson, J.M. 1971. Observations on the vertical distribution of Oribatei (Acarina) in two woodland soils. *C. R. Colloq. Int. Zool. Sol. 4th, 1970,* pp. 257–272.

Anderson, J.M. 1975. The enigma of soil animal species diversity. *Prog. Soil Zool., Proc. Int. Colloq., 5th, 1973,* pp. 51–58.

Anderson, J.M. 1978. Inter- and intra-habitat relationships between woodland Cryptostigmata species diversity and the diversity of soil and litter microhabitats. *Oecologia* 32:341–348.

Andre, H. 1975. Observations sur les Acariens corticole de Belgique. *Notes Rech. Fond. Univ. Luxemb.* 4:1–31.

Andre, H.M. 1980. Description of *Camisia carrolli* n. sp., with a comparison to two other arboreal *Camisia* (Acari: Oribatida). *Int. J. Acarol.* 6(2):141–146.

Andre, H.M., Ph. Lebrun and S. Leroy. 1984. The systematic status and geographical distribution of *Camisia segnis* (Acari: Oribatida). *Int. J. Acarol.* 10(3):153–158.

Aoki, J.-I. 1967a. Microhabitats of oribatid mites on a forest floor. *Bull. Natl. Sci. Mus. (Tokyo)* 10:133–138.

Aoki, J. 1967b. A preliminary revision of the family Otocepheidae (Acari, Cryptostigmata). II. Subfamily Tetracondylinae. *Bull. Natl. Sci. Mus. (Tokyo)* 10:257–359. (Otocepheoidea)

Aoki, J. 1969. Taxonomic investigations on free-living mites in the subalpine forest on Shiga Heights IBP Area. III. Cryptostigmata. *Bull. Natl. Sci. Mus. (Tokyo)* 12:117–141.

Aoki, J. 1973. Soil mites (Oribatids) climbing trees. *Proc. Int. Congr. Acarol., 3rd, 1971,* pp. 59–65.

Aoki, J. 1983. Some new species of oppiid mites from south Japan (Oribatida: Oppiidae). *Int. J. Acarol.* 9(4):165–172.

Baker, E.W., J.H. Camin, F. Cunliffe, T.A. Woolley, and C.E. Yunker. 1958. *Contrib. No. 3, Inst. Acarol.,* University of Maryland, College Park.

Balogh, J. 1972. *The Oribatid Genera of the World.* Akadémiai Kiadó, Budapest.

Banks, N. 1915. *The Acarina or Mites.* USDA Report No. 108. GPO, Washington, D.C.

Behan-Pelletier, V.M. 1984. *Ceratozetes* (Acari: Ceratozetidae) of Canada and Alaska. *Can. Entomol.* 116:1449–1517.

Bernini, F. 1979. Biogeographic and faunistic data on the oribatids of the Tuscan Archipelago. *In:* Rodriguez, J.G., ed. *Recent Advances in Acarology.* Academic Press, New York, pp. 559–566.

Block, W.C. 1965. The life histories of *Platynothrus peltifer* (Koch 1839) and *Damaeus clavipes* (Hermann 1804) (Acarina: Cryptostigmata) in soils of Pennine moorland. *Acarologia* 7(4):735–743. (Nothroidea, Belboidea)

Bulanova-Zakhvatkina, E.M., B.A. Vainshtein, V.I. Volgin, M.S. Gilyarov, L.D. Golosova, D.A. Krivolutskii, A.B. Lange, V.D. Sevast'yanov, L.G. Sitnikova and E.S. Shaldybina. 1975. Group

Oribatei Dugés. *In:* Gilyarov, M.S. and N.G. Bregetova, eds. *A Key to the Soil-inhabiting Mites (Sarcoptiformes).* Nauka, Leningrad [in Russian].

Clarke, R.T.J. and T. Bauchop, eds. 1977. *Microbial Ecology of the Gut.* Academic Press, New York.

Coleman, D.C. 1970. Food webs of small arthropods of a broomsedge field studied with radio-isotope-labelled fungi. *In:* Phillipson, J., ed. *Methods of Study in Soil Ecology.* UNESCO, Paris, pp. 203–207.

Covarrubias, R. 1968a. Some observations on Antarctic Oribatei (Acarina) *Liochthonius australis* sp. n., and two *Oppia* ssp. n. *Acarologia* 10(2):313–356. (Oppioidea)

Covarrubias, R. 1968b. Observations sur le genre *Pheroliodes.* I-*Pheroliodes roblensis* n. sp. (Acarina, Oribatei). *Acarologia* 10(4):657–695. (Gymnodamaeoidea)

Crossley, D.A.C., Jr. 1977. Oribatid mites and nutrient cycling. *In:* Dindal, D., ed. *Biology of Oribatid Mites.* Syracuse State University, Syracuse, New York, pp. 71–86.

Dindal, D. 1974. The digestive activity of a phthiracarid mite mesenteron. *J. Insect Physiol.* 20:2247–2260.

Engelbrecht, C.M. 1972. Galumnids from South Africa (Galumnidae, Oribatei). *Acarologia* 14(1):109–149. (Galumnoidea)

Engelmann, H.-D. 1968–1981. Bibliographia Oribatologica. Numbers 1–9. *Abh. Ber. Natuurkde Mus. Gorlitz, Leipzig:* 44(1):1–24 (1968); 44(1):1–12 (1969); 45(1):1–15 (1970); 46(1):1–22 (1971); 47(1):1–15 (1972); 48(1):1–19 (1973); 48(1a):1–20 (1973); 49(1):1–18 (1975); 50(1):1–19 (1976); 54(1):1–28 (1980); 55(1):1–24 (1981); 58(1):1–37 (1984); 59(1):1–21 (1985); 60:1–27 (1986).

Evans, G.O., J.G. Sheals and D. McFarlane. 1961. *The Terrestrial Acari of the British Isles. Volume I. Introduction and Biology,* British Museum (Natural History), London, pp. 107–131.

Evison, B.M. 1981. Two new species of the family Brachychthoniidae (Acarina: Oribatei). *Int. J. Acarol.* 7(1–4):225–229.

Foss, E.S. del, H.L. Cromroy and D.H. Habeck. 1975. Determination of the feeding mechanism of the water hyacinth mite. *Hyacinth Control J.* 13:53–55. (Galumnoidea)

Fuhrer, E. 1961. Der Einfluss von Pflanzenwurzeln auf die Verteilung der Kleinarthropoden im Boden, untersucht an *Pseudotritia ardua* (Oribatei). *Pedobiologia* 1:99–112.

Gilyarov, M. S. and N.G. Bregetova, eds. 1975. *A Key to the Soil-inhabiting Mites, Sarcoptiformes* (in Russian). Nauka, Leningrad.

Gjelstrup, P. 1979. Epiphytic cryptostigmatid mites on some beech- and birch-trees in Denmark. *Pedobiologia* 19:1–8.

Grandjean, F. 1928-ff. See Travé and Vachon (1975).

Grandjean, F. 1935. Les poils et les organes sensitifs portes par les pattes et le palp chez les Oribates. *Bull. Soc. Zool. Fr.* 60:6–39.

Grandjean, F. 1957. L'infracapitulum et la manducation chez les Oribates e l'autres Acariens. *Ann. Sci. Nat. Zool.* 11:234–281.

Grandjean, F. 1972–1976. *Oeuvres Acarologique Complètes.* L. Hammen, ed., Vols. 1–7. Junk Publ., The Hague.

Hammen, L. van der. 1952. The Oribatei (Acari) of the Netherlands. *Zool. Verh. Leiden* 17:1–139. (Brachychthonoidea and others)

Hammen, L. van der. 1959. Berlese's primitive oribatid mites. *Zool. Ver. Leiden* 40:1–93. (Mesoplophoroidea, Parhypochthonoidea, Cosmochthonoidea, Brachychthonoidea Phthiracaroidea, Perlohmannoidea, Nothroidea, bibliography)

Hammen, L. van der. 1963. Description of *Fortuynia yunkeri* nov. spec., and notes on the Fortuyniidae nov. fam. (Acarida, Oribatei). *Acarologia* 5(1):152–167. (Ameronothroidea)

Hammen, L. van der. 1968. The gnathosoma of *Hermannia convexa* (C.L. Koch) (Acarida: Oribatina) and comparative remarks on its morphology in other mites. *Zool. Verh. Leiden* 94:1–45.

Hammer, M. 1944. Studies on the oribatids and collemboles of Greenland. *Medd. Groenl.* 141(3):1–210.

Hammer, M. 1952. Investigations on the microfauna of Northern Canada. Part. I. Oribatidae. *Acta Arct.*, Fasc IV: pp. 1–108.

Hammer, M. 1958. Investigations on the oribatid fauna of the Andes Mountains. I. The Argentine & Bolivia. *Biol. Skr.-K. Dan. Vidensk. Selsk.* 10(1):1–129.

Hammer, M. 1962. Investigations on the oribatid fauna of the Andes Mountains. III. Chile. *Biol. Skr.-K. Dan. Vidensk. Selsk.* 13(2):1–65. (Hypochthonoidea, Brachychthonoidea, Cymbaeremaeoidea, Oppioidea, Gymnodamaeoidea, Eremuloidea, Oribatuloidea, Ceratozetoidea, Galumnoidea)

Hammer, M. 1965. Are low temperatures a species-preserving factor? *Acta Univ. Lund.* 2(2):1–10.

Hammer, M. 1966a. Investigations on the oribatid fauna of New Zealand. Part 1. *Biol. Skr.-K. Dan. Vidensk. Selsk.* 15:1–108. (Polypterozetoidea, Ameronothroidea, and others)

Hammer, M. 1966b. Investigations on the oribatid fauna of New Zealand. Part 2. *Biol. Skr.-K. Dan. Vidensk. Selsk.* 15:1–108.

Hammer, M. 1971. On some oribatids from Viti Levu, the Fiji Islands. *Biol. Skr.-K. Dan. Vidensk. Selsk.* 116(6):1–60. (Hypochthonoidea, Phthiracaroidea, Epilohmannoidea, Lohmannioidea, Nothroidea, Nanhermannioidea, Hermannielloidea, Liodoidea, Passalozetoidea, Microzetoidea, Oribatelloidea, and others)

Hammer, M. 1972. Investigations on the oribatid fauna of Tahiti, and on some oribatids found on the atoll Rangiroa. *Biol. Skr.-K. Dan. Vidensk. Selske.* 19(3): 1–65. (Hypochthonoidea, Phthiracaraoidea, Euphthiracaraoidea, Lohmannioidea, Polypterozetoidea, Pelopoidea, and others)

Hammer, M. 1973. Oribatids from Tongatapu and Eua, the Tongan Islands and from Upolu, Western Samoa. *Biol. Skr.-K. Dan. Vidensk. Selsk.* 20(3):1–70. (Otocepheoidea, Carabodoidea, and others)

Hammer, M. 1975. On some oribatids from Central Sahara (Acari: Oribatidae). *Steenstrupia* 3(18):187–196.

Hartenstein, R. 1962a. Soil Oribatei. I. Feeding specificity among forest soil Oribatei. *Ann. Entomol. Soc. Am.* 55(2):202–206.

Hartenstein, R. 1962b. Soil Oribatei. II. *Belba kingi,* new species, and a study of its life history. *Ann. Entomol. Soc. Am.* 55(4):357–361. (Belboidea)

Hartenstein, R. 1962c. Soil Oribatei. III. Studies on the developmental biology and ecology of *Metabelba montana* (Kulcz.) and *Eremobelba nervosa* n.sp. *Ann. Entomol. Soc. Am.* 55(4):361–367. (Belboidea, Eremuloidea)

Hartenstein, R. 1962d. Soil Oribatei. IV. Observations on *Ceratozetes gracilis. Ann. Entomol. Soc. Am.* 55(6):709–713.

Higgins, H.G. 1979. A brief review of the oribatid family Eremaeidae in North America. *In:* Rodriguez, J.G., ed. *Recent Advances in Acarology.* Academic Press, New York, Vol. 2, pp. 541–546.

Hughes, T.E. 1959. *Mites or the Acari.* Athlone Press, London.

Ibarra, E.L., J.A. Wallwork and J.G. Rodriguez. 1965. Ecological Studies of mites found in sheep and cattle pastures. I. Distribution patterns of Oribatid mites. *Ann. Entomol. Soc. Am.* 58(2):153–159.

Jacot, A.P. 1929. *Annotated Bibliography of the Moss Mites.* Catholic Mission Press, Tsingtao.

Jacot, A.P. 1936. Why study the fauna of litter? *J. For.* 34:581–583.

Jacot, A.P. 1939. Reduction of spruce and fir litter by minute animals. *J. For.* 37(11):858–860.

Johnston, D.E. 1967. On the occurrence of two species of *Palaeacarus* in the eastern United States (Acari: Acariformes). *Proc. Entomol. Soc. Wash.* 69(4):301–302. (Palaeacaroidea)

Johnston, D.E. 1982. Oribatida. *In:* Parker, S.P., ed. *Synopsis and Classification of Living Organisms.* McGraw-Hill, New York, pp. 145–146.

Kates, K.C. and C.E. Runkel. 1948. Observations on oribatid mite vectors of *Moniezia expansa* on pastures, with a report of several new vectors from the United States. *Proc. Helminthol. Soc. Wash.* 15(1):18–33.

Knulle, W. 1957a. Die Verteilung der Acari: Oribatei im Boden. *Z. Morphol. Oekol. Tiere* 46:397–432.

Knulle, W. 1957b. Morphologische und Entwicklungsgeschicliche untersuchungen zum phylo genetischen System der Acari: Acariformes Zachv. I. Oribatei: Malaconothridae. *Mitt. Zool. Mus. Berlin* 33(1):97–213.

Kowal, N.E. 1969. Ingestion rate of a pine-mor oribatid mites. *Am. Midl. Nat.* 81:595–598.

Krantz, G.W. 1978. *A Manual of Acarology,* 2nd ed. Oregon State Univ. Book Stores, Corvallis.

Krivolutsky, D.A. 1979. Oribatid mite complexes as bioindicators of radioactive pollution. *In:* Rodriguez, J.G., ed. *Recent Advances in Acarology.* Academic Press, New York, Vol. 1, pp. 615–618.

Krivolutsky, D.A. and Y.A. Druk. 1986. Fossil Oribatids. *Annu. Rev. Entomol.* 31:533–545.

Lebrun, Ph. 1965. Contribution à l'études écologiques des Oribates de la litière dans une foret de Moyenne-Belgique. *Mem. Inst. R. Sci. Nat. Belg.* 153:1–9.

Lebrun, Ph. 1979. Soil mite community diversity. *In:* Rodriguez, J.G., ed. *Recent Advances in Acarology.* Academic Press, New York, Vol. I:603–613.

Lebrun, Ph. and R. Mignolet. 1979. Phenologie des population d'Oribates en relation avec la vitesse d'écomposition des litières. *Proc. Int. Congr. Acarol., 4th, 1974,* pp. 93–100.

Lee, D.C. 1984. A modified classification for oribate mites (Acari: Cryptostigmata). *In:* Griffiths, D.A. and C.E. Bowman, eds., *Acarology VI,* Vol. 1, pp. 241–248. Ellis Horwood, Chichester, England.

Littlewood, C.G. 1969. A surface sterilization technique used in feeding algae to Oribatei. *Proc. Int. Congr. Acarol., 2nd, 1967,* pp. 53–56.

Luxton, M. 1972a. Studies on the oribatid mites of a Danish beechwood soil. I. Nutritional biology. *Pedobiologia* 12:434–463.

Luxton, M. 1972b. Studies on the oribatid mites of a Danish beechwood soil. II. Biomass, calorimetry and respirometry. *Pedobiologia* 15:161–200.

Luxton, M. 1972c. Studies on the oribatid mites of a Danish beechwood soil. V. Vertical distribution. *Pedobiologia* 21:306–334.

Luxton, M. 1979. Food and energy processing by oribatid mites. *Rev. Ecol. Biol. Sol* 16:103–111.

Luxton, M. 1981. Studies on the oribatid mites of a Danish beechwood soil. VII. Energy budgets. *Pedobiologia* 22:77–111.

Luxton, M. 1982. The biology of mites from beech woodland soil. *Pedobiologia* 23:1–8.

McDaniel, B. and E.G. Bolen. 1983. A new species of *Epilohmannia pallida* Wallwork (Oribatida: Epilohmannidae). *Int. J. Acarol.* 9(1):37–41.

McDaniel, B. and Aoki, J-I. 1982. New distribution records of members of the genus *Eohypochthonius* Jacot from Texas and Minnesota. *Int. J. Acarol.* 8(2):115–117.

Michael, A.D. 1884. *British Oribatidae,* Vol. 1, Ray Society, London.

Michael, A.D. 1888. *British Oribatidae,* Vol. 2. Ray Society, London.

Mitchell, M.J. 1977. Population dynamics of oribatid mites (Acari: Cryptostigmata) in an aspen woodland soil. *Pedobiologia* 17:305–319.

Mitchell, M.J. 1979. Energetics of oribatid mites (Acari: Cryptostigmata) in an aspen woodland soil. *Pedobiologia* 19:89–98.

Mitchell, M.J. and D. Parkinson. 1976. Fungal feeding of oribatid mites (Acari: Cryptostigmata) in an aspen woodland soil. *Ecology* 57:302–312.

Murphy, P.W. 1953. The biology of forest soils with special reference to the mesofauna or meiofauna. *J. Soil Sci.* 4:155–193.

Norton, R. 1975. Elliptochthoniidae, a new mite family (Acarina: Oribatei) from mineral soil in California. *J. N. Y. Entomol. Soc.* 83:209–216.

Norton, R. 1977. A review of F. Grandjean's system of leg chaetotaxy in the Oribatei and its application to the Damaeidae. *In:* Dindal, D., ed. *Biology of Oribatid Mites.* Syracuse State University, Syracuse, New York, pp. 33–69.

Norton, R.A. 1978. *Veloppia kananaskis* n. sp., with notes on the familial affinities of *Veloppia* Hammer (Acari: Oribatei). *Int. J. Acarol.* 4(2):71–84.

Norton, R.A. 1979a. Familial concepts in the Damaeoidea as indicated by preliminary phylogenetic studies. *In:* Rodriguez, J.G., ed. *Recent Advances in Acarology.* Academic Press, New York, Vol. 2, pp. 529–534.

Norton, R.A. 1979b. Aspects of the biogeography of Dameidae, sensu latu (Oribatei). with emphasis on North America. *In:* Rodriguez, J.G., ed. *Recent Advances in Acarology.* Academic Press, New York, Vol. 2, pp. 535–540

Norton, R.A. 1979c. Generic concepts in the Damaeoidea (Acari: Oribatei). Part II. *Acarologia* 21(3):496–513.

Norton, R.A. 1980. Observations on phoresy by oribatid mites (Acari: Oribatei). *Int. J. Acarol.* 6(2):121–130.

Perez-Inigo, C. 1968. Acaros oribatidos de suelos de Espana peninsular e Islas Baleares (1.a parte) (Acari, Oribatei). "Graellsi." *Rev. Entomol. Iber.* 24:143–238. (Hypochthonoidea, Brachychthonoidea, Cosmochthonoidea, Phthiracaroidea, Epilohmannioidea, Lohmannioidea, Nothroidea, Hermannioidea)

Perez-Inigo, C. 1970. Acaros oribatidos de suelos de Espana peninsular e Islas Baleares (Acari, Oribatei). Parte II. "Eos." *Rev. Esp. Entomol.* 45:241–317. (Hermannielloidea, Liodoidea, Gymnodamaeoidea, Belboidea, Cepheoidea, Microzetoidea, Zetorchestoidea, Liacaroidea, Eremaeoidea, Eremuloidea)

Perez-Inigo, C. 1971. Acaros oribatidos de suelos de Espana peninsular e Islas Baleares (Acari, Oribatei). Parte III. "Eos." *Rev. Esp. Entomol.* 46:263–350. (Liacaroidea, Carabodoidea, Appioidea, Ameronothroidea, Cymbaeremaeoidea, Passalozetoidea)

Piffl, E. 1972. Zur systematik der Oribatiden (Acari). (Neue Oribatiden aus Nepal, Costa Rica und Brasilen ergeben eine neue Familie der Unduloribatidae und erweitern die Polypterozetidae um die Gattungen *Podopterotegaeus, Nodocepheus, Eremaeozetes* und *Tumerozetes*). *Khumbu Himal* 4(2):269–314. (Oribatelloidea, Polypterozetoidea)

Prusinkiewicz, Z., O. Stefaniak and S. Seniczak. 1975. Wstepne badania nad rola mikroflory przewodu pokarmowego wybranych gatunkow mechowcow (Oribatei, Acarina) w procesash humifikacji i minneralizacji sciolek lesnych. *Mater. Symp. Zool. Gleby, Rogow, PTG, 1st,* pp. 107–116.

Rajski, A. 1959. Mechowce (Acari: Oribatei) jako zywiciele posredni tasiemcow (Cestodes: Anoplocephalata) w swietle literatury. *Zesz. Nauk. UAM Biologia Poznan* 2:163–192.

Ramsay, G.W., J.A. Wallwork and J.G. Rodriguez. 1972. Some observations on the pteromorphs of oribatid mites (Acari: Cryptostigmata). *Acarologia* 13(4):669–674.

Riha, G. 1951. Okologie der Oribatiden im Kalksteinboden. *Zool. Jahrb. (Syst.)* 80:407–450.

Rockett, C.L. 1980. Nematode predation by oribatid mites (Acari; Oribatida). *Int. J. Acarol.* 6:(3)219–224.

Rockett, C.L. and J.P. Woodring. 1966a. Oribatid mites as predators of soil nematodes. *Ann. Entomol. Soc. Am.* 59(4):669–671.

Rockett, C.L. and J.P. Woodring. 1966b. Biological investigations on a new species of *Ceratozetes* and of *Pergalumna* (Acarina: Cryptostigmata). *Acarologia* 8(3):511–520.

Schubart, H. 1975. Morphologische Grundlagen fur die Klarung der Verwandschaftbeziehungen innerhalb der Milbenfamilie Ameronothridae (Acari: Oribatei). *Zoologica (N.Y.)* 123:24–91. (Ameronothroidea)

Schuster, R. 1956. Die Anteil der Oribatiden anden Zerzetzungsvorgangen im Boden. *Z. Morphol. Oekol. Tiere* 45:1–33.

Schweizer, J. 1956. Die Landmilben des Schweizerischen Nationalparkes. 3. Sarcoptiformes Reuter 1909. *Soc. Helvt. Sci. Nat. Parc. Nat.* (N.S.) 5(34):215–377.

Schweizer, J. 1957. Die Landmilben des Schweizerischen Nationalparkes. 4. Ihre Lebensraum, ihre Vergesellschaftung unter sich und ihre Lebensweise. *Soc. Helvt. Sci. Nat. Parc. Nat.* (N.S.) 6(37):11–107.

Sellnick, M. 1928. Formenkreis: Hornmilben, Oribatei. *Tierwelt Mitteleur.* 3(9):1–42. (Nanhermannioidea, Hermannioidea, Belboidea, Oribatuloidea, Ceratozetoidea, and others)

Sellnick, M. and K.-H. Forsslund. 1955. Die Camisiidae Schwedens (Acar. Oribat.). *Ark. Zool.* 8(2):473–530. (Nothroidea)

Sengbusch, H.G. 1977. Review of Oribatid Mite-Anoplocephalan Tapeworm Relationships (Acari:Oribatei: Cestoda; Anoplocephalidae). *In*: Dindal, D., ed. *Biology of Oribatid Mites*. Syracuse State University, Syracuse, New York, pp. 87–104.

Seniczak, S. and O. Stefaniak. 1978. The microflora of the alimentary canal of *Oppia nitens* (Acarina, Oribatei). *Pedobiologia* 18:110–119.

Shereef, G.M. 1976. Biological studies and description of stages of two species: *Papillacarus aciculatus* Kunst and *Lohmannia egypticus* Elbadry and Nasr (Oribatei-Lohmanniidae) in Egypt. *Acarologia* 18(2):351–359. (Lohmannioidea)

Stefaniak, O. and S. Seniczak. 1978. The microflora of the alimentary canal of *Achipteria coleoptrata* (Acarina, Oribatei). *Pedobiologia* 16:185–194.

Stefaniak, O. and S. Seniczak. 1981. The effect of fungal diet on the development of *Oppia nitens* (Acari, Oribatei) and on the microflora of its alimentary tract. *Pedobiologia* 21-202-210.

Stunkard, H.W. 1939. The role of oribatid mites as transmitting agents and intermediate hosts of ovine cestodes. *Int. Kongr. Entomol., 7th, 1938*, Vol. 3; pp. 1671–1674.

Stunkard, H.W. 1941. Studies on the life history of the anoplocephaline cestodes of hares and rabbits. *J. Parasitol.* 27.299 005.

Stunkard, H.W. 1944. Studies on the life history of the oribatid mite, *Galumna* sp., intermediate host of *Moniezia expansa*. *Anat. Rec.* 89(4):550.

Subias, L.-S. 1977. *Taxonomia y Ecologia de los Oribatidos Saxicolas y Arbicolas de la Sierra del Guadarrama (Acarida: Oribatida)*, Catedra de Atropods Trab. No. 24. Univ. Comp. de Madrid.

Tragardh, I. 1932. Palaeacariformes, a new suborder of Acari. *Ark. Zool.* 24B(2):1–6.

Travé, J. 1959. Sur le genre *Niphocepheus* Balogh 1943 les Nephocepheidae, famille nouvelle (Acariens, Oribates). *Acarologia* 8(4):475–498. (Carabodoidea)

Travé, J. 1963. Oribates des Pyrénées-Orientales, 2e série, Zetorchestidae (1re partie): *Saxicolestes pollinivorus* n.sp. *Vie Milieu* 14(2):449–455. (Zetorchestoidea)

Travé, J. 1964. Importance des stases immatures des Oribates en systématique et en écologie. *Acarologia* 6(hors série):47–53.

Travé, J. 1979. Neotrichy in oribatid mites. *In:* Rodriguez, J.G., ed. *Recent Advances in Acarology*. Academic Press, New York, Vol. 2, pp. 515–522.

Travé, J. and M. Vachon. 1975. François Grandjean 1882–1975. (Notice Biographique & Bibliographique). *Acarologia* 17(1):1–19.

Vitzthum, H.G. 1943. Acarina. *Bronn's Klassen* 5:1–1011.

Vossbrink, C.R., D.C. Coleman and T.A. Woolley. 1979. Abiotic and biotic factors in litter decomposition in a semiarid grassland. *Ecology* 69(2):265–271.

Wallwork, J.A. 1958. Notes on the feeding behavior of some forest soil Acarina. *Oikos* 9(2):260–271.

Wallwork, J.A. 1959. The distribution and dynamics of some forest soil mites. *Ecology* 40(4):557–563.

Wallwork, J.A. 1961a. Some Oribatei from Ghana. IV. The genus *Basilobelba* Balogh. *Acarologia* 3(1):344–362. (Eremuloidea)

Wallwork, J.A. 1961b. Some Oribatei from Ghana. VII. Members of the "family" Eremaeidae Willmann (2nd series). The genus *Oppia* Koch. *Acarologia* 3(4):637–658. (Oppioidea)

Wallwork, J.A. 1962a. Some Oribatei from Ghana. IX. The genus *Tetracondyla* Newell 1956 (2nd series). *Acarologia* 4(3):440–456. (Otocepheoidea)

Wallwork, J.A. 1962b. Some Oribatei from Ghana. X. The family Lohmanniidae. *Acarologia* 4(3):457–487. (Lohmannioidea)

Wallwork, J.A. 1963. The oribatei (Acari) of Macquarie Island. *Pac. Insects* 5(4):721–769.

Wallwork, J.A. 1964. Insects of Campbell Island. Appendix. *Campbellobates acanthus* n. gen., n. sp. (Acari: Cryptostigmata). *Pac. Insects Monogr.* 7:601–606. (Scheloribatidae–Ceratozetidae)

Wallwork, J.A. 1965. A leaf-boring galumnoid mite (Acari: Cryptostigmata) from Uruguay. *Acarologia* 7(4):758–764.

Wallwork, J.A. 1966. More oribatid mites (Acari: Cryptostigmata) from Campbell Island. *Pac. Insects* 8(4):849–877. (Palaeacaroidea, Nothroidea)

Wallwork, J.A. 1967a. Some oribatei (Crytostigmata) from Tschad (3rd series). *Rev. Zool. Bot. Afr.* 75(1-2):35–45.

Wallwork, J.A. 1967b. Cryptostigmata (oribatid mites). *Antarct. Res. Ser.* 10:105–122. (Ameronothroidea)

Wallwork, J.A. 1969a. The zoogeography of antarctic Cryptostigmata. *Proc. Int. Congr. Acarol., 2nd, 1967*, pp. 17–20.

Wallwork, J.A. 1969b. Some basic principles underlying the classification and identification of Cryptostigmatid mites. *In*: Sheals, J.G., ed. *The Soil Ecosystem.* Syst. Assoc. Publ. London, No. 8, pp. 155–168.

Wallwork, J.A. 1970. Acarina: Cryptostigmata of South Georgia. *Pac. Insects Monogr.* 23:161–178.

Wallwork, J.A. 1972a. Distribution patterns and population dynamics of the micro-arthropods of a desert soil in Southern California. *J. Anim. Ecol.* 41:291–310.

Wallwork, J.A. 1972b. Mites and other microarthropods from the Joshua Tree National Monument, California. *J. Zool.* 168:9–105.

Wallwork, J.A. 1972c. Distribution patterns of Cryptostigmatid mites (Arachnidae: Acari) in South Georgia. *Pac. Insects* 14(3):615–625.

Wallwork, J.A. 1973. Some aspects of the energetics of soil mites. *Proc. Int. Congr. Acarol., 3rd, 1971,* pp. 129–134.

Wallwork, J.A. 1979. Relict distribution of oribatid mites. *In:* Rodriguez, J.G., ed. *Recent Advances in Acarology.* Academic Press, New York, Vol. 2, pp. 515–522.

Wallwork, J.A. 1983. Oribatids in forest ecosystems. *Annu. Rev. Entomol.* 28:109–130.

Willmann, C. 1931. Moosmilben oder Oribatiden (Oribatei). *Tierwelt Dtsch.* 22:79–200.

Wolf, M.M. and C.L. Rockett. 1984. Habitat changes affecting bacterial composition in the alimentary canal of oribatid mites (Acari: Oribatida). *Int. J. Acarol.* 10(4):209–215.

Woodring, J.P. 1962. Oribatid (Acari) pteromorphs, pterogasterine phylogeny, and evolution of wings. *Ann. Entomol. Soc. Am.* 55:394–403.

Woodring, J.P. 1973. Comparative morphology, functions, and homologies of the coxal glands in oribatid mites (Arachnida: Acari). *J. Morphol.* 139(4):407–429.

Woodring, J.P. and E.F. Cook. 1962. The biology of

Ceratozetes cisalpinus Berlese, *Scheloribates laevigatus* Koch, and *Oppia neerlandica* Oudemans (Oribatei) with a description of all stages. *Acarologia* 4(1):101–137. (Oppioidea, Oribatuloidea, Ceratozetoidea)

Woolley, T.A. 1960. Some interesting aspects of oribatid ecology (Acarina). *Ann. Entomol. Soc. Am.* 53(2):251–253.

Woolley, T.A. 1965. Eutegaeidae, a new family of oribatid mites, with a description of a new species from New Zealand (Acarina: Oribatei). *Acarologia* 7(2):382–388. (Carabodoidea)

Woolley, T.A. 1969. The infracapitulum—a possible index of oribatid relationship. *Proc. Intl Congr. Acarol., 2nd, 1967,* pp. 209–211.

Woolley, T.A. 1972. The systematics of the Liacaroidea (Acari: Cryptostigmata). *Acarologia* 14(2):250–257.

Woolley, T.A. 1979a. The chelicerae of the Gustaviidae. *In:* Rodriguez, J.G., ed. *Recent Advances in Acarology.* Academic Press, New York, Vol. 2, pp. 547–552.

Woolley, T.A. 1979b. The application of SEM in oribatid taxonomy. *Proc. Int. Congr. Acarol., 4th, 1974,* pp. 705–712.

Woolley, T.A. 1982. Mites and other soil microarthropods. *In: Methods of Soil Analysis, Part 2. Chemical and Microbiological Properties,* 2nd, ed., Agron. Monogr. No. 9, Chapter 54, Am. Soc. Agron., Soil Sci. Soc. Am., Madison, Wisconsin, pp. 1113–1141.

Woolley, T.A. and H.G. Higgins. 1968. Megeremaeidae, a new family of oribatid mites (Acari: Cryptostigmata). *Great Basin Nat.* 28(4):172–175. (Eremaeoidea)

Woolley, T.A. and H.G. Higgins. 1979. A new genus and two new species in Damaeidae. *In:* Rodriguez, J.G., ed. *Recent Advances in Acarology.* Academic Press, New York, Vol. 2, pp. 553–558.

PART IV

HISTORICAL AND ECOLOGICAL ASPECTS

22

Of Mites and Men

Die Wissenschaft hat nur ein Vaterland—das ist die Welt—und nur ein Zeil-das is die Wahrheit.

—THIENEMANN

It has been said that the first article written about mites occurred about 300 B.C. Prior to that time Aristotle, the Father of Zoology, had recorded the word *Acari* (ακαρι) in Greek for mites, "little waxies," because of the shiny examples he observed on a honeycomb. This word was taken into Latin as *Acarus,* which means "not cut" and refers to the seemingly undivided body of these tiny creatures. The latter term, however, is erroneous, for the somas of numerous mites, while not having distinctly sutured tagma like insects, in many instances do have superficial divisions that resemble segments.

Howard Evans in his delightful book, *Life on a Little Known Planet,* relates the story that Galileo turned a telescope end-to-end and peered at the compound eye of an insect. Subsequently, a youthful compatriot of his, Marcello Malpighi (1628–1694), used a microscope to describe the external and internal anatomy of several insects. Microscopes are requisite in entomology and the dependence of acarology (and other microscopal disciplines) upon optical and electron lenses for success is implied in the following bit of verse.

> Faith is a fine invention
> When gentlemen can see—
> But Microscopes are prudent
> In an Emergency.
>
> Emily Dickinson[1]

The title of this chapter was paraphrased from lines of the Scottish poet, Robert Burns, because mites do cause plans to "gang aft agley", may leave some "grief" or "pain"—and perhaps give some "promised joy" to those who derive pleasure and esthetic reward from the study of these "wee beasties."

No one can assess the full impact of mites on men. In the face of a burgeoning literature and extensive biographical information, the histori-

---

[1]*Final Harvest, Poems of Emily Dickinson,* 1961. Selected by Thomas H. Johnson, Little Brown, Boston, p. 20.

cal aspects of such a challenge extend beyond the limits of this chapter. It is important, however, to know of some of the workers in the science, both past and present. For example, acarologists specializing in water mites are reviewed by Viets (1955).

Three principal periods in acarological history are recognized: first, the period from Aristotle to Linnaeus; second, a taxonomic-descriptive phase from Linnaeus to the end of World War II; and, third, from World War II to the present (Micherdzinski, 1966). The 1758 edition of Linnaeus' *Systema Naturae* includes 31 species of mites under the genus *Acarus*. The actual collection was lost except for three specimens known today. Each species description was limited to one or two lines with a brief note about the habitat. Eight legs, two lateral eyes, and two "tentacles" distinguished the mites, which he characterized as "apterous insects with eight legs."

The science of acarology was assisted directly by World War II and the Korean conflict as a result of the wartime need to understand the biology of chiggers (Trombiculidae) and to develop control measures. Scrub typhus (Tsutsugamushi disease) and hemorrhagic fever, both chigger-borne diseases, devastated some of the United States forces in the South Pacific, Southeast Asia, and Korea (Micherdzinski, 1966). These deaths resulted in intense investigations, and launched acarology toward its most recent development. Concurrently, an increased emphasis on ticks and tick-borne diseases contributed to this growth. Many entomologists who were involved in these investigations during the war, afterward adopted acarology as a science and aided in its development. Continuation of their efforts since the close of World War II has helped to build the science to its present strength.

The development of the phase-constrast microscope by Zoernicke was a second factor in the growth of the science. Since then TEM and SEM have greatly improved our view and understanding of acarine structures.

Acarology really came of age with the initia-

tion and perpetuation of the International Congresses of Acarology: Fort Collins, 1963; Sutton Bonington, 1967; Prague, 1971; Saalfelden, 1974; East Lansing, 1978; Edinburgh, 1982; and India, 1986. Other meeting of importance were an All-Union Conference, Leningrad, 1966; the first French conference, Banyuls-surmer, 1968; American symposia Cornell, New York, and Waltham, Massachusetts, 1963, 1966, 1968. Publications from these various meetings include the congress proceedings, *Advances in Acarology, Recent Advances in Acarology,* and most recently *Acarology VI.*

The Acarological Society of America was established by charter in 1971 as a subsidiary of the parent Entomological Society of America. This new organization of acarologists established a series of important symposia at the national meetings that has been contributed greatly to the science.

The ninth edition of the *World Directory* (begun by Dr. James Brennan in 1953) names 903 acarologists from 75 countries. The eleventh edition of the *Acarologists of the World* (1979) by Johnston and Wrensch lists 1288 workers from 78 countries: 322 from the United States and 195 from USSR.

The World List (1978) identifies most of the active acarologists of our day. For an appreciation of contributors to acarology, the reader may consult the references at the end of the chapter. Special historical accounts may be found in taxonomic works, for example, Hughes (1959) and Treat (1975). An extensive *History of Acarology* is available (Prasad, 1982) as well as a history of a science in the United States from 1975–1985 (Prasad, 1986). Histories of individuals are also included in the references (e.g., Eyndhoven, 1944; Grandjean, 1966; Hammen, 1972, 1977–78; Lawrence et al., 1959; Motas, 1971; Rack, 1972; Sellnick, 1968; Travé and Vachon, 1975; Vachon, 1966; Woolley, 1979).

The brief summary of acarologists that follows includes many of the principals down to those with birth days prior to 1920. There may have been omissions that some readers feel are unjustified and inclusions for which they would perhaps prefer other substitutions or greater coverage. Because acarology is an international science, the persons named have been selected to represent different nations and specialties. Omissions have not been deliberate slights of people or countries. The historical information is arranged in chronological order based upon year of birth, in 50-year increments down to 1900: 1700–1750; 1750–1800; 1800–1850; 1850–1900. Only a few individuals are listed for the period 1900–1920; that period is left to other writers.

The Age of Microscopy was one of those notable historical peaks of activity that electrified the world of science for a brief moment, only to fade quickly until the technological advances caused its rebirth nearly two centuries later. The early microscopists, Leeuewenhoek, Malpighi, Grew, and Hooke, had little if any communication among themselves, but their letters had great impact and their discoveries had lasting appeal. They observed many things over a wide spectrum of objects. Although Hooke's writings had breadth and depth, we begin this brief history with him because of the specific observations he made on mites.

HOOKE AND MICROSCOPY: ROBERT HOOKE (1632–1703)

It was the classical microscopist Robert Hooke who first figured mites in his famous *Micrographia.* As curator of experiments for the Royal Society he was responsible for a diverse array of displays and objects. The "mite in cheese" (*Acarus geniculatus*) was and is a classical "first" for microscopy and acarology (Frontispiece; Fig. 2-2). In dedicating his book to the king, Hooke stated his purpose, to present "some of the least of all visible things." He wrote of "wandering mites" in cheese, meal, corn, seeds, musty barrels, and leather. As the most distinguished of the classical microscopists, it is said that his illustrations of spiders and insects had no rivals among his contemporaries. The same should be said of his illustrations of mites. Today his drawings exhibit an admirable accuracy. His outstanding drawing of this cheese mite may have been the inspiration for two poems of interest.

> So Naturalists observe a flea
> Has smaller fleas that on him prey.
> And these have smaller still to bite 'em
> And so proceed *ad infinitum.*
> Jonathan Swift (*Poetry, A Rhapsody,* 1733)

The following poem has direct reference to cheese mites, and reflects the social practice of the time. It was fashionable to possess a magnifying lens, usually hung by a chain around the neck. Curiosity spawned observations of various objects that became the basis for many a conversation at court or in social gatherings.

> Dear Madam, did you never gaze
> Thro' optic glass on rotten cheese?
> There, Madam, did you ne'er perceive
> A crowd of dwarfish creatures live?
> The little things, elate with Pride,
> Strut to and fro, from side to side:
> In tiny pomp and portly vein,
> Lords of their pleasing orb they reign;
> And, fill'd with harden'd Curds and Cream,
> Think the whole Dairy made for them.
> So man, conceited Lords of all
> Walks proudly o'er his pendant Ball,
> Fond of their little Spot below,
> Nor greater Beings care to know;
> But think those Worlds which deck the skies,
> Were only form'd to please their eyes.
> Stephen Duck 1736[2]

Even though Swift's poem was written about fleas, it is easy to extend the implications to mites, where a number of similar and peculiar symbiotic relationships have been discovered.

[2]From *Mites or the Acari,* by T. E. Hughes, 1959. University of London, Athlone Press, London.

ACAROLOGISTS AND THEIR MAJOR CONTRIBUTIONS

1700–1750

C. Linnaeus	(1707–1778)	*Acarus*
C. DeGeer	(1720–1778)	Tyroglyphidae, Analgidae, Sarcoptidae, Hydrachnidae

1750–1800

J. C. Fabricius	(1745–1808)	Genera and species
P. A. Latreille	(1762–1833)	Family and generic keys
K. C. L. Koch	(1778–1857)	Water and mud mites, earth, running mites
A. L. D. Dugés	(1797–1838)	Trombidiformes

1800–1850

F. Dujardin	(1801–1862)	Eriophyidae, respiratory systems
F. L. P. Gervais	(1816–1862)	Arachnids, mites
A. A. D. Dugés	(1826–1910)	Water mites, *hypopus*
M. H. Nicolet	(? –1872)	Aerial, terrestrial mites
A. L. Donnadieu	(? – ?)	Tetranychidae
J. P. Megnin	(1828–1905)	Mange, feather mites Dermoglyphidae Proctophyllodidae hetermorphic nymph (=hypopus)
J. L. R. Claparede	(1832–1871)	Claparede's organ (=urstigma), Listrophoridae, development, oribatids
G. Canestrini	(1835–1927)	Listrophoridae, Psoroptidae
A. D. Michael	(1836–1927)	Oribatei, Tyroglyphidae, Hypopus
P. M. Kramer	(1842–1898)	Prostigmata, Hydrachnidae
E. L. Trouessart	(1842–1947)	Halacaridae, Epidermoptidae, Sarcoptidae, Analgidae, Dermoglyphidae, Proctophyllodidae

1850–1900

G. Haller	(1853–1886)	Haller's organ of ticks, Hydracarina
S. Thor	(1856–1937)	Prostigmata, Penthalodidae, Cunaxidae, Hydrachnida, Trombidiidae
A. Nalepa	(1856–1929)	Eriophyidae
A. C. Oudemans	(1858–1943)	General classification, 35 mite families
A. Berlese	(1863–1927)	Berlese funnel, taxonomy of most orders *brevi diagnosi*
H. T. Lohmann	(1863–1934)	Halacaridae, oribatids
N. Banks	(1868–1953)	Halarachnids, tetranychids, oribatids
R. Marshall	(1869–1955)	Water mites
H. G. Tullgren	(1874– ?)	Water mites; Tullgren funnel
H. L. W. Vitzthum	(1876–1947)	Macrochelidae, Pachylaelapidae, Laminiosioptidae

I. O. H. Tragardh	(1878–1951)	Mesostigmata: monogynaspida, trigynaspida; lizard mites, (*Pimeliaphilus*), oribatids
H. C. C. Willmann	(1881–1968)	Moosmilben (oribatids) Poeciliocharidae; rhagidial organs
F. Grandjean	(1882–1975)	*anactinochitinosi, actinochitinosi,* Acariens, Actinedida, Acaridida, Oribatida
K. H. Viets	(1882–1961)	Water mites
H. E. Ewing	(1883–1951)	Parasitic mites; Phytoptipalpidae, Entonyssidae, Ixodorhynchidae, oribatids
E. N. Pavlovsky	(1884–1966)	Nidality, zoonoses; Ixodida
M. Sellnick	(1884–1971)	Hornmilben (oribatids), Zerconidae
I. Sokolow	(1885– ?)	Water mites
L. Szalay	(1887–1970)	Water mites
H. Womersley	(1889–1962)	Trombiculids
A. P. Jacot	(1890–1939)	Oribatids
C. O. Lundbladt	(1890–1970)	Water mites

1900–1920

M. Andre	(1900–1966)	Tetranychidae, trombiculids; *Acarologia*
A. A. Zakhvatkin	(1906–1950)	Tyroglyphidae; *Fauna USSR*
K. O. Viets	(1910–)	Water mites
W. B. Dubinin	(1913–1958)	Parasitic mites
E. W. Baker	(1914–)	Tetranychidae, Tenuipalpidae, Varroidae; *Introduction to Acarology, Revision of Tetranychidae Guide to Families of Mites, Manual of Parasitic Mites, Mites Injurious to Economic Plants*
G. W. Wharton	(1914–)	Trombiculids; Institute of Acarology; *Dermatophaguoides* and house dust allergy, *Manual of Chiggers, Introduction to Acarology*

REFERENCES

Eyndhoven, G.L. van. 1944. In Memoriam Dr. A. C. Oudemans, 12 Nov. 1858–14 Jan. 1943. *Tijdschr. Entomol.* 86(1):1–56.

Eyndhoven, G.L. van. 1965. Some details about the life and work of A.C. Oudemans. *Acarologia* 7(4):589–593.

Grandjean, F. 1966. Marc Andre. *Acarologia* 8(3):397–400.

Hammen, L. van der. 1972. *The Complete Acarological Works of Grandjean,* Vol. 1, pp. 1–37. Junk, The Hague.

Hughes, T.E. 1959. *Mites or the Acari.* Athlone Press, London.

Johnson, T.H. 1961. *Final Harvest, Poems of Emily Dickinson.* Little, Brown, Boston, Massachusetts.

Lawrence, R.F., A. Fain and M. Andre. 1959. W.B. Dubinin. *Acarologia* 1(3):267–284.

Micherdzinski, W. 1966. Historia Akarologii. *Zesz. Probl. Postepow Nauk Roln. Lesn.* 65:7–21.

Motas, C. 1971. Dr. Lazlo Szalay in memoriam. *Acarologia* 13(1):1–3.

Osborn, H. 1937. *Fragments of Entomological History.* Publ. by Author, Columbus, Ohio.

Prasad, V. 1982. *History of Acarology.* Indira Publ., Oak Park, Michigan.

Prasad, V. 1986. *A Decade of Acarology.* Indira Publ., Oak Park, Michigan.

Rack, G. 1972. Dr. Phil. Max Sellnick. *Acarologia* 13(4):545–551.

Sellnick, M. 1957. Dr. Carl Willmann. *Abh. Naturwiss. Ver. Bremen* 34(3):177–180.

Sellnick, M. 1968. Dr. Carl Willmann. *Acarologia* 10(3):395–400.

Shevtshenko, V.G. 1967. To the 110th Anniversary of Doctor Alfred Nalepa. *Acarologia* 9(3):4167–474.

Travé, J. and M. Vachon. 1975. François Grandjean. *Acarologia* 17(1):1–19.

Treat, A. 1975. *Mites of Moths and Butterflies.* Cornell Univ. Press (Comstock), Ithaca, New York.

Vachon, M. 1966 (1967). Marc Andre. *Bull. Mus. Nat. Hist. Nat.* (1) 38(6):763–766.

Viets, K. 1955. *Die Milben des Susswassers und des Meres. Hydrachnellae et Halacaridae (Acari).* Fischer, Jena.

Woolley, T.A. 1979. Of mites and men—The deans of American acarology. *Proc. Int. Congr. Acarol., 4th, 1974,* pp. 17–28.

World List. 1978. Unpublished directory of world acarologists. Institute of Acarology, Ohio State University, Columbus.

23

Ecological Considerations

All the world's a stage and all the [mites] are players.
—SHAKESPEARE

Probably one of the earliest allusions to mite ecology was made by Hooke in 1664: "...there may be no less than a million well grown Mites contain'd in a cubick inch, and five hundred times as many eggs."

In the early part of this century Banks (1915) commented:

> To the ordinary observer of nature mites do not exist. He may walk abroad and see birds and insects about him on every side; occasionally he may notice a tick or a harvest mite; yet a little careful searching would reveal a world of these tiny creatures at his feet. Among the fallen leaves of the forest, in moss or lichens, under stones and loose bark, in fungi, in the loose upper surface of the soil, in the galls of plants, in the streams and ponds, and even in the depths of the sea there are mites innumerable. Hidden is this world of mites to the general naturalist as completely as though it were on another planet.

Next to the taxonomic research on the Acari, ecological studies comprise the larger propor-

tion of acarological investigations. A voluminous amount of literature exists on biological and ecological relationships involving mites as they affect agriculture (food crops, grains, fruit, field crops, ornamentals), stored products (grains and cereals), and human and veterinary medicine. The topic is too extensive for adequate coverage in this volume, except in a somewhat summary fashion, and to direct the reader to cited and general references. In these one can find specific articles and additional bibliographies from which to gain a better understanding of these ecological relationships. Proceedings of international congresses contain many articles on mite ecology. Some references have already been cited in previous chapters where descriptions of biology include ecological aspects. Others will be found in the chapters on classification.

Among the most recent "firsts" in parasitological acarology is the discovery of the allergenic trauma caused by *Dermatophagoides farinae* and *D. pteronyssinus*, house dust mites. Remarkably like their skin-mite cousins, they have become the objects of research in both Europe and the United States as the principal human allergen in house dust. They feed on animal and human dander in over-stuffed furniture, but cannot live in climates where the humidity is less than 60%. No pesticide for them exists, but good housekeeping is a principal factor in their control. A few of the numerous articles about them include Arlian (1976), Arlian et al. (1979), 1984a), Bronswijk (1973, 1979); Haarlov and Alani (1970), Larson et al. (1969), and Wharton (1970, 1976), particularly Wharton's review article (1976).

Venereal transmission of mites from turtle to turtle is another outstanding discovery of a peculiar niche for the family Cloacaridae (Camin et al., 1967). The biology of the moth earmite (Otopheidomenidae) is a remarkable story of ecological relations and behavior (Treat, 1975).

Scabies is an itch-mite infection of humans in crowded environments. It is highly contagious and must be treated medically (Arlian et al., 1984b; Parish et al., 1983). Microscopic follicle mites (Demodicidae) passively infest over 90% of humans (W.B. Nutting, personal communication). Additional papers on medical acarines are found in Rodriguez (1979), Griffiths and Bowman (1984), and Kettle (1984).

Principal agricultural mite pests (Tetranychidae, spider mites) are reviewed (Huffaker et al., 1969) and discussed in two new volumes with extensive bibliographies and an index (Helle and Sabelis, 1985), together with the works about gall mites (Eriophyidae) and mites that affect economic plants (Jeppson et al., 1975), and biological control of mite pests (Hoy and Herzog, 1985; Hoy et al., 1983).

The ecosystem of stored products acarology is summarized in many articles (e.g., Sinha, 1979; OConnor, 1979).

Balashov (1972), Hoogstraal (1970–1982), and Wilkinson (1979) give many resources on tick biology and ecology. The new work of Sauer and Hair (1986) expands our knowledge of these important ectoparasites.

An understanding of the soil acarines is increased in the articles and symposia in the journals *Pedobiologia* (e.g., Tamm et al., 1984; Athias-Binche, 1985) and in *Oikos* (e.g., Schenker, 1984), as well as in reviews by Gilyarov and Bregetova (1975). The biology of oribatid mites as it affects nutrient cycling, soil fertility, and forest ecosystems provides important models of mite ecology as well an understanding of how human relationships influence mite biology (Crossley, 1977; Dindal, 1977; Hammer and Wallwork, 1979; Krivolutksy, 1979; Wallwork, 1983, 1984). A summary of International Biological Programme literature is found in *Oikos* (e.g., Cragg, 1982) along with ecophysiological writings (Somme and Block, 1984).

Tracking patterns in bird and mammal ectoparasites are described by Kethley and Johnston (1975). Mite parasites of vertebrates and relationships with specific host groups are delineated by Price (1975, 1980; Price et al., 1984). Coevolution of parasitic arthropods and mammals is expanded by Kim (1985).

Nutritional ecology of mites and related invertebrates is reviewed in depth by several authors (Slansky and Rodriguez, 1987).

Associations of mites with other arthropods are numerous, as demonstrated in the previous references on the Actinedida and Astigmata (e.g., Fain, 1969).

These relatively few examples of an abundant literature in this important aspect of acarology open the door to innumerable investigations that will fill at least one huge volume, if someone assumes the task of writing about it.

When the public becomes more aware of how mites affect people (e.g., a popular article on the house dust mite (Moser, 1986) is most revealing), the multifaceted conditions that influence mite ecology, and how the morphological and physiological uniqueness of these creatures relate to the circumstances that make them important to human existence, then acarology will become even more firmly established and better understood. Behavioral ecology is just

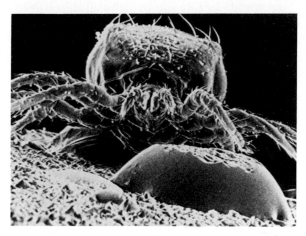

Figure 23-1. *Amblyseius fallacis* SEM (courtesy of Brian Croft).

scratching the surface of the psychological ramifications of human anxiety in the presence of mites (and other animals)—Symbiophobia (Nutting and Beerman, 1983).

We said at the beginning of this work that mites affect man in five ways: health, agriculture, stored products, biological control, and esthetics. It is hoped that these aspects of acarology have been explained. Esthetic or "cultural" acarology, although begun (e.g., Brennan and Nadchatram, 1977), has not developed to the extent that cultural entomology has (Hogue, 1987).

Like the mite in Figure 23-1, seemingly poised for prospective action as if on a lunar landscape, acarology has a bright and inspiring future and "exciting expansions" are expected (Nutting, 1979).

As this volume closes, the omissions, oversights, and possible mistakes lend credence to a statment by Tai L'Ung, the thirteenth century author of *History of Chinese Writing*: "Were I to await perfection my book would never be finished."

REFERENCES

Arlian, L.G. 1976. Mites and house dust allergy. *J. Asthma Res.* 13(4):165–172.

Arlian, L.G., I.L. Bernstein, C.L. Johnson and J.S. Gallagher. 1979. Ecology of house dust mites and dust allergy. *In:* Rodriguez, J.G., ed. *Recent Advances in Acarology.* Academic Press, New York, Vol. 2, pp. 185–195.

Arlian, L.G., D.P. Geis, D.L. Vyszenski-Moher, I.L. Bernstein and J.S. Gallagher. 1984a. Cross antigenic and allergenic properties of the house dust mite *Dermatophagoides farinae* and the storage mite *Tyrophagus putrescentiae. J. Allerg. Clin. Immunology,* 74(2):172–179.

Arlian, L.G., R.A. Runyan and S.A. Estes. 1984b. Cross infectivity of *Sarcoptes scabei. J. Amer. Acad. Dermatology.* 10(6):979–986.

Athias-Binche, F. 1985. Analyses demographiques des populations d'Uropodidae (Arachnides Anactinotrichida) de la hetriae de la Massani France. *Pedobiologia* 28:225–254.

Balashov, Y.S. 1972. *Bloodsucking Ticks (Ixodoidea)—Vectors of Diseases of Man and Animals,* Trans. 500 (T500). Med. Zool. Dept., USNAMRU No. 3, Cairo, Egypt, U.A.R.

Banks, N. 1915. *The Acarina or Mites.* USDA Report No. 108. GPO, Washington, D.C.

Bram, R.A. 1978. Surveillance and collection of arthropods of veterinary importance. *U.S. Dep. Agric. Handb.* 518:1–125.

Brennan, J.M. and M. Nadchatram. 1977. Chigger on a stamp. *Acarologia* 18(3):569.

Bronswijk, J.E.M.H. van. 1973. *Dermatophagoides pteronyssinus* (Trouessart, 1897), in mattress and floor dust in a temperate climate (Acari: Pyroglyphidae). *J. Med. Entomol.* 10(1):63–70.

Bronswijk, J.E.M.H. van. 1979. Houst dust as an ecosystem. *In:* Rodriguez, J.G., ed. *Recent Advances in Acarology.* Academic Press, New York, Vol. 2, pp. 167–172.

Busvine, J.R. 1975. Arthropod vectors of disease. *Inst. Biol. Stud. Biol.* 55.

Camin, J.H., W.W. Moss, J.H. Oliver, Jr., and G. Singer. 1967. Cloacaridae, a new family of cheyletoid mites from the cloaca of aquatic turtles (Acari: Acariformes: Eleutherengona). *J. Med. Entomol.* 4(3):261–272.

Cicolani, R., S. Passariello, and G. Petrelli. 1977. Influenza della Temperatura sull incremento di popolazione in *Macrocheles subbadius*. *Acarologia* 19(4):563-578.

Clark, G.M., C.M. Clifford, L.V. Fadness, and E.K. Jones. 1970. Contributions to the ecology of Colorado tick fever virus. *J. Med. Entomol.* 7(2):189-197.

Collins, N.C. 1975. Tactics of host exploitation by thermophilic water mites. *Misc. Publ. Entomol. Soc. Am.* 9(5):250-254.

Cook, D.R. 1974. Water mite genera and subgenera. *Mem. Amer. Entomol. Inst. No. 21.* Ann Arbor, Michigan, pp. 1-860.

Cragg, J.B. 1982. International Biological Programme. *Oikos* 39(3):286-376.

Crossley, D.A.C., Jr. 1977. Oribatid mites and nutrient cycling. *In:* Dindal, D., ed. *Biology of Oribatid Mites.* Syracuse State University, Syracuse, New York, pp. 71-102.

Dindal, D.L., ed. 1977. *Biology of Oribatid mites.* Syracuse State University, Syracuse, New York.

Dindal, D.L. and R.A. Norton. 1979. Influence of human activities on community structure of soil Prostigmata. *In*: Rodriguez, J.G., ed. *Recent Advances in Acarology.* Academic Press, New York, Vol. 1, pp. 619-628.

Dusbabek, F., V. Cerny, E. Honzakova, M. Daniel and J. Olejnicek. 1979. Differences in the developmental cycle in two closely situated biotypes. *In:* Rodriguez, J.G., ed. *Recent Advances in Acarology.* Academic Press, New York, Vol. 2, pp. 155-158.

Fain, A. 1969. Adaptation to parasitism in mites. *Acarologia* 11(3):429-449.

Fleschner, C.A., M.E. Badley, D.W. Ricker, and J.C. Hall. 1956. Air drift of spider mites. *J. Econ. Entomol.* 49:624-627.

Gilyarov, M.S. and N.G. Bregetova, eds. 1975. *A Key to the Soil-Inhabiting Mites* (in Russian). Nauka, Leningrad.

Griffiths, D.A. and C.E. Bowman, eds. 1984. *Acarology VI,* Vols. 1 and 2. Ellis Horwood, Ltd., Chichester, England.

Haarlov, N. and M. Alani. 1970. House-dust mites (*Dermatophagoides pteronyssinus* (Trt.), *D. farinae* Hughes, *Euroglyphus maynei* (Cooreman) Fain) in Denmark (Acarina). *Ent. Scand 1,* 1970: 301-306.

Hammer, M. and J.A. Wallwork. 1979. A review of the world distribution of Oribatid mites (Acri: Cryptostigmata) in relation to continental drift. *Biol. Skr.-K. Dan. Vidensk. Selsk.* 22:4.

Helle, W. and M.W. Sabelis, eds. 1985. *World Crop Pests. Spider Mites Their Biology, Natural Enemies and Control,* Vols. 1A and 1B. Elsevier, Amsterdam.

Hogue, C.L. 1987. Cultural Entomology. *Ann. Rev. Entomol.* 32:181-199.

Hoogstraal, H. 1970-1982. *Bibliography of Ticks and Tickborne Diseases,* Vols. 1-7, Spec. Publ. USNAMRU No. 3, Cairo, Egypt, U.A.R. (Vol. 1, Oct. 1970; Vol. 2, Dec. 1970; Vol. 3, June 1971; Vol. 4, June 1972; Vol. 5 Pt. I, Aug. 1974; Vol. 5, Pt. II, April 1978; Vol. 6, July 1981; Vol. 7, May 1982).

Hoy, M.A. 1983. Opportunities for genetic improvement of mites as biological control agents. *In:* Hoy, M.A., G.L. Cunningham and L. Knutson, eds., *Biological Control of Pests by Mites.* Agric. Exp. Stn. Univ. of Calif., Berkeley, Spec. Publ. 3304, pp. 141-146.

Hoy, M.A. and D.C. Herzog, eds. 1985. *Biological Control in Agricultural IPM Systems.* Academic Press, Orlando, Florida.

Huffaker, C.B., M. van de Vrie, and J. McMurtry. 1969. Ecology of Tetranychid mites. *Annu. Rev. Entomol.* 14:125-174.

Jeppson, L.R., H.H. Keifer and E.W. Baker. 1975. *Mites Injurious to Economic Plants.* Univ. of California Press, Berkeley.

Kethley, J. and D.E. Johnston. 1975. Resource tracking patterns in bird and mamml ectoparasites. *Misc. Publ. Entomol. Soc. Am.* 9(5):231-236.

Kettle, D.S. 1984. *Medical and Veterinary Entomology.* Wiley, New York, (Acari) pp. 351-448.

Kim, K.C., ed. 1985. *Coevolution of Parasitic Arthropods and Mammals.* Wiley, New York.

Krivolutsky, D.A. 1979. Oribatid mite complexes as bioindicators of radioactive pollution. *In:* Rodriguez, J.G., ed. *Recent Advances in Acarology.* Academic Press, New York, Vol. 1, pp. 615-618.

Lanciani, C.A. 1975. Parasite-induced alterations in host reproduction and survival. *Ecology* 56(3):689-695.

Lanciani, C.A. 1979. The influence of parasitic water mites on the instantaneous death rate of their hosts. *Oecologia* 44(1):60-62.

Larson, D.G., W.F. Mitchell and G.W. Wharton. 1969. Preliminary studies on *Dermatophagoides farinae* Hughes, 1961 (Acari) and house dust allergy. *J. Med. Entomol.* 6(3):295–299.

Lawrence, R.F. 1953. *Biology of Cryptic Fauna of Forests.* A.A. Balkema, Amsterdam.

Lebrun, Ph. 1979. Soil mite community diversity. *In:* Rodriguez, J.G., ed. *Recent Advances in Acarology.* Academic Press, New York, Vol. 1, pp. 603–614.

Liebisch, A. 1979. Ecology and distribution of Q-fever Rickettsia in Europe with special reference to Germany. *In:* Rodriguez, J.G., ed. *Recent Advances in Acarology.* Academic Press, New York, Vol. 2, pp. 225–232.

Lindquist, E.E. 1970. Relationships between mites and insects in forest habitats. *Can. Entomol.* 102:978–983.

Marle, G. 1950. Observations on the dispersal of the fruit tree red spider mite, *Metatetranychus ulmi* (Koch). *Annu. Rep. East Malling Res. Stn., Kent* 1950 (1951): 155–159.

Miller, J.R., G.E. Jones and K.C. Kim. 1973. Populations and distribution of *Steatonyssus occidentalis* (Ewing) (Acarina: Macronyssidae) infesting the big brown bat, *Eptesicus fuscus* (Chiroptera: Vespertilionidae). *J. Med. Entomol.* 10(6):609–613.

Mitchell, R. 1975. Models for parasitic populations. *In:* Price, P.W., ed. *Evolutionary Strategies of Parasitic Insects and Mites.* Plenum, New York, pp. 49–65.

Mitchell, W.F., G.W. Wharton and D.G. Larson. 1969. House Dust, Mites and Insects. *Ann. Allergy* 27(3):93–99.

Moser, P.W. 1986. All the real dirt on dust. *Discover* 7(11):106–115.

Nadchatram, M. 1970. Correlations of habitat environment and color of chiggers and their potential significance in epidemiology of scrub typhus in Indonesia. *J. Med. Entomol.* 7(2):131–144.

Newell, I.M. 1967. Abyssal Halacaridae (Acari) from the Southeast Pacific. *Pac. Insects* 9(4):693–708.

Nilsson, A. and L. Lundquist. 1979. Interspecific relations in small mammal ectoparasites. *In:* Rodriguez, J.G., ed. *Recent Advances in Acarology.* Academic Press, New York, Vol. 1, pp. 451–456.

Nutting, W.B. 1979. Synhospitaly and speciation in the Demodicidae (Trombidiformes). *Proc. Int. Congr. Acarol., 4th, 1974,* pp. 267–272.

Nutting, W.B. and H. Beerman. 1983. Demodicosis and symbiophobia: Status, terminology and treatment. *Int. J. Dermatol.* 22:13.

OConnor, B.M. 1979. Evolutionary origin of astigmatid mites inhabiting stored products. *In:* Rodriguez, J.G., ed. *Recent Advances in Acarology.* Academic Press, New York, Vol. 1, pp. 273–278.

Parish, L.C., W.B. Nutting and R.M. Schwartzman. 1983. *Cutaneous Infestations of Man and Animal.* Praeger, New York, pp. 1–392.

Perring, T.M. and C.A. Farrar. 1986. Historical perspective and current world status of the tomato russet mite (Acari: Eriophyidae). *Misc. Publ. Entomol. Soc. Am.* 63:1–20.

Petrischeva, P.A. 1965. *Vectors of Diseases of Natural Foci* (translated from Russian by B. Hershkovitz, Is. Program Sci. Trans.) Daniel Davey & Co., New York.

Price, P.W., ed. 1975. *Evolutionary Strategies of Parasitic Insects and Mites.* Plenum, New York.

Price, P.W., ed. 1980. *Evolutionary Biology of Parasites.* Princeton Univ. Press, Princeton, New Jersey.

Price, P.W., C.N. Slobodchikoff and W.S. Gaud, eds. 1984. *A New Ecology.* Wiley, New York.

Radovsky, F. 1985. Parasitic Mesostigmata. *In:* Kim, K.C. ed. *Coevolution of Parasitic Arthropods and Mammals.* Wiley, New York, pp. 441–504.

Rodriguez, J.G., ed. 1979. *Recent Advances in Acarology.* Vols. 1 and 2. Academic Press, New York.

Roush, R.T. and J.A. McKenzie. 1987. Ecological genetics of insecticide and acaricide resistance. *Annu. Rev. Entomol.* 32:361–380.

Sauer, J.R. and J.A. Hair, eds. 1986. *Morphology, Physiology and Behavioral Biology of Ticks.* Ellis Horwood, Ltd., Chichester, England.

Schenker, R. 1984. Spatial and seasonal distribution patterns of oribatid mites (Acari: Oribatei) in forest ecosystem. *Oikos* 27:89–97.

Schuster, R. 1979. Soil mites in the marine community. *In:* Rodriguez, J.G., ed. *Recent Advances in Acarology.* Academic Press, New York, Vol. 1, pp. 593–602.

Sinha, R.N. 1979. Role of acarina in the stored grain ecosystem. *In:* Rodriguez, J.G., ed. *Recent Advances in Acarology.* Academic Press, New York, Vol. 1, pp. 263–272.

Slansky, F. and J. G. Rodriguez, eds. 1987. *Nutritional Ecology of Insects, Spiders, and Related Invertebrates.* Wiley, New York.

Somme, L. and W. Block. 1984. Ecophysiology of two intertidal mites at South Georgia. *Oikos* 42(3):276-282.

Sonenshine, D.E. 1979. Zoogeography of the American Dog Tick, *Dermacentor variabilis. In:* Rodriguez, J.G., ed. *Recent Advances in Acarology.* Academic Press, New York, Vol. 2, pp. 123-134.

Sonenshine, D.E. and G.M. Clark. 1968. Field trials on radioisotope tagging of ticks. *J. Med. Entomol.* 5(2):229-235.

Stout, I.J. 1979. Ecology of spotted fever ticks in eastern Washington. *In:* Rodriguez, J.G., ed. *Recent Advances in Acarology.* Academic Press, New York, Vol. 2, pp. 113-122.

Tamm, I.C., H.W. Wittman and S. Woas. 1984. Zur Landmilbenfauna eines jahressperiodische trochenfallenden Stauseebodens. *Pedobiologia* 27:377-393.

Treat, A. 1975. *Mites of Moths and Butterflies.* Cornell Univ. Press, Ithaca, New York.

Vossbrinck, C.R., D.C. Coleman and T.A. Woolley. 1979. Abiotic and biotic factors in litter decomposition in a semiarid grassland. *Ecology* 60(2):265-271.

Wallwork, J.A. 1983. Oribatids in soil ecosystems. *Annu. Rev. Entomol.* 28:109-130.

Wallwork, J.A. 1984. Perspective in Acarine Biogeography. *In:* Griffiths, D.A. and C.E. Bowman, eds., *Acarology VI,* Vol. 1. Ellis Horwood, Chichester, England, pp. 63-70.

Weignian, G. 1973. Zur okologie der Collembolen und Oribatiden in Grenzbereich Land-Meer (Collembol, Insecta—Oribatei, Acari). *Z. Wiss. Zool.* 186(3/4):295-391.

Wharton, G.W. 1970. Mites and commercial extracts of house dust. *Science* 167:1382-1383.

Wharton, G.W. 1976. House dust mites. *J. Med. Entomol.* 12(6):577-621 (review article).

Wilkinson, P.R. 1979. Early achievements, recent advances, and future prospects in the ecology of the Rocky Mountain Wood Tick. *In:* Rodriguez, J.G., ed. *Recent Advances in Acarology.* Academic Press, New York, Vol. 2, pp. 105-112.

Appendix

TECHNIQUES FOR COLLECTION, PRESERVATION, AND PREPARATION

The literature on this topic is voluminous, but it is appropriate to cite references that will lead the reader to specific methods. Some annotations may also prove useful.

Methods of collecting mites vary, but certain principles apply to the repository of collected materials. Collection data should always be with the specimens even though a collection record or data book is kept separately with greater details. Location, date, and collector are imperative to validate collected specimens. The repository collection should be set up in such a way that anyone could step in, study the collection, and deduce the system involved. Collections should require a minimum of maintenance over time. Mites should be maintained in standard containers and in standard fluid. Individual containers for alcoholic specimens should be sealed with good neoprene stoppers. (Stoppers should be nonreactive to ultraviolet light; these can be time-tested on a window sill to determine deterioration.) A year's supply of such expendables is advisable. Alcoholic specimens are best kept in individual vials within larger communal jars with screw caps or with parafilm to prevent evaporation. After desired species are extracted from a major collection, all residues should be retained for future use.

Chelicerates are generally preserved in fluids such as 70–90% ethanol. Mites can be preserved in Oudeman's fluid which has the advantage of extending the appendages. This consists of 87 parts of 70% ethanol, 5 parts of glycerine, and 8 parts of glacial acetic acid.

Preservation of gall mites (Eriophyidae) in alcohol is unsatisfactory. The infested parts of plants should be collected and wrapped in soft tissue paper, then allowed to dry. The dried material can be kept indefinitely and the mites can be recovered by warming them in Kiefer's solution (Keifer, 1975). This solution is made by mixing 50 g of resorcinol, 20 g of diglycolic acid, 25 ml of glycerol, 10 ml of water, and enough iodine to give color to the solution.

Woolley (1982) cites various methods and techniques regarding soil microarthropods together with additional useful references.

REFERENCES

Arlian, L. and T.A. Woolley. 1970. Observations on the biology of *Liacarus cidarus* (Acari: Cryptostigmata, Liacaridae). *J. Kans. Entomol. Soc.* 43(3):297–301.

Atyeo, W.T. and N.L. Braasch. 1966. The feather mite genus *Proctophyllodes*. *Bull. Univ. Nebr. State Mus.* 5:1–353.

Feather mites may be recovered from study skins of birds in museums. Proctophyllodid mites tend to remain intact and in compact clusters along the rachis or on the remiges of the feather. They may be removed with jewelers' forceps, dissecting needles, or other instruments and transferred to 70% ethanol. Before mounting on slides for examination, specimens are rehydrated and cleared in lactophenol. Specimens are heated (200–250°F) in the clearant for 5–10 min to clear more rapidly and to extend the legs for orientation.

Balogh, J. 1959. On the preparation and observation of oribatids. *Acta Zool.* 5(3/4):241–253.

Barr, D. 1973. Methods for collection, preservation and study of water mites (Acari: Parasitengona). *Life Sci. Misc. Publ., R. Ont. Mus.*, pp. 1–28.

Boudreaux, H.B. 1953. A simple method of collecting spider mites. *J. Econ. Entomol.* 46(6):1102–1103.

Boyd, E. 1967. Deutonymphs as endoparasites of the Eastern belted kingfisher and the Eastern green heron in North America. *Proc. Entomol. Soc. Wash.* 69(1):73–81.

Detergent washings of the skin and the carcass and examination of the inner surface of the eyelid. A few were found in the adipose tissue of the axilla and the groin.

Camin, J.H. and P.R. Ehrlich. 1960. A cage for maintaining stock colonies of parasitic mites and their hosts. *J. Parasitol.* 46(109–111.

Cross, H.F. and G.W. Wharton. 1954. Techniques for testing the attachment and feeding rate of mites on living hosts. *J. Econ. Entomol.* 47(6):1153–1154.

Crossley, D.A.C., Jr. and M. Witcamp. 1964. Forest soil mites and mineral cycling. *Proc. Int. Congr. Acarol., 1st, 1963,* pp. 137–146.

Desch, C.E. and W.B. Nutting. 1977. Morphology and functional anatomy of *Demodex folliculorum* (Simon) of man. *Acarologia* 19(3):422–462.

The methods for obtaining demodicid mites from humans, goats, dogs, and marsupial mice are described, as well as methods for fixation, sectioning, and observation with both the light and electron microscopes.

Fain, A. 1980. A method of remounting old preparations of acarines without raising or displacing the cover slip. *Int. J. Acarol.* 6(2):169–170.

Foulkes, G.D. 1983. A new method of mounting acarines and other meiofauna. *Int. J. Acarol.* 9(4):211–213.

Gerson, U. 1967. Rearing of *Hemisarcoptes coccophagus. Acarologia* 9(3):632–638.

Hair, J.A., A.L. Hock, R.W. Barker and J.J. Semtner. 1972. A method of collecting nymphal and adult lone star ticks, *Amblyomma americanum* from woodlots. *J. Med. Entomol.* 9(2):153–154.

Vacuum equipment using a portable generator or power from an automobile cigarette lighter was used for collecting nymphal and adult lone star ticks in Oklahoma. Ticks aggregated in the vicinity of sublimating dry ice and the vacuum apparatus enabled the collection of 3000 adults or 5000 nymphal ticks per man-hour.

Jeppson, L.R., H.H. Keifer and E.W. Baker, eds. 1975. *Mites Injurious to Economic Plants.* Univ. of California Press, Berkeley.

Kaiser, M.N. and H. Hoogstraal. 1968. Simple field and laboratory method for recovering living ticks (Ixodoidea) from hosts. *J. Parasitol.* 54(1):188–189.

Keifer, H.H. 1938–1952. Eriophyid Studies. *Bull. Calif. Dept. Agric* (series of descriptive papers).

Keifer, H.H. 1975. Collection, preservation, slide mounting and illustrating of eriophyid mites. *In:* Jeppson, L.R., H.H. Keifer and E.W. Baker, eds. *Mites Injurious to Economic Plants.* Univ. of California Press, Berkeley, pp. 385–396.

Krantz, G.W. 1978. *A Manual of Acarology.* Oregon State Univ. Book Stores, Corvallis.

Various techniques and methods are reviewed (terrestrial and free-living, aquatic and marine mites, and parasitic mites). Techniques for rearing, clearing, dissecting, and mounting are discussed. Mounting techniques include various media as well as storage methods. Use of scanning electron microscopy is also discussed.

Kuo, J.S. and M.E. McCully. 1969. Preparation of thin sections of mites for high resolution light microscopy. *Can. J. Zool.* 47:737-739.

Lipovsky, L.J. 1951. A washing method of ectoparasite recovery with particular reference to chiggers (Acarina: Trombiculidae). *J. Kans. Entomol. Soc.* 24(4):151-156.

Lipovsky, L.J. 1953a. Improved technique for rearing chigger mites Acarina: Trombiculidae). *Entomol. News* 64(1):4-7.

Lipovsky, L.J. 1953b. Polyvinyl alcohol with lactophenol, a mounting and clearing medium for chigger mites. *Entomol. News* 64(2):41-44.

Livingston, C.H. 1958. A simple low-cost shake culture device. *Turtox News* 36(8):190-191.

MacFayden, A. 1953. Notes on methods for extraction of small soil arthropods. *J. Anim. Ecol.* 22(1):65-77.

MacFayden, A. 1955. A comparison of methods for extracting soil arthropods. *In:* McE. Kevan, D.K., ed. *Soil Zoology.* Butterworth, London, pp. 315-332.

Mahr, D.L. 1979. A method of preparing phytoseiid mites (Mesostigmata) for scanning electron microscopy. *Int. J. Acarol.* 5(1):15-17.

McE. Kevan, D.K., ed. 1955. *Soil Zoology.* Butterworth, London.

Medvedeva, G.I., S.R. Beskina and I.M. Grokhovskaya. 1972. Culture of ixodid tick embryonic cells. *Med. Parasitol. Moscow* 41:39-40 (in Russian; in English: NAMRU 3-T594).

Moser, J.C. and L.E. Brown. 1978. A nondestructive trap for *Dendroctonus frontalis* Zimmerman (Coleoptera: Scolytidae). *J. Chem. Ecol.* 4:1-7.

Murphy, P.W. and C.C. Doncaster. 1957. A culture method for soil meiofauna and its application to the study of nematode predators. *Nematologica* 2:202-214.

Newell, I.M. 1955. An autosegregator for use in collecting soil-inhabiting arthropods. *Trans. Am. Microsc. Soc.* 74(4):389-392.

Nutting, W.B. 1964. Techniques for culturing Demodicids in skins of hamsters and chickens. *Proc. Int. Congr. Acarol., 1st, 1963,* p. 280.

Patrick, C.D. and J.A. Hair. 1970. Laboratory rearing procedures and equipment for multi-host ticks (Acarina: Ixodidae). *J. Med Entomol.* 12(3)389-390.

Patterson, S.G., W.L. Mesner and J.G. Rodriguez. 1982. Monitoring spider mite populations with photography. *Int. J. Acarol.* 8(4):231-232.

Pence, D.B. 1972. The hypopi from subcutaneous tissues of birds in Louisiana. *J. Med. Entomol.* 9(5)435-438.

Mites were collected by skinning the hosts and examining the undersurface of the skin under a dissecting microscope. Skins were also washed with detergent. When mites were found, they were fixed in 70% alcohol and mounted in Hoyer's.

Pounds, J.M. and J.H. Oliver. 1976. Reproductive morphological spermatogenesis in *Dermanyssus gallinae* (DeGeer) (Acari: Dermanyssidae). *J. Morphol.* 150:825-842.

Dissection, fixation, and tissue preparation studies in research on *Dermanyssus gallinae.*

Pudney, M., M.G.R. Varma and C.J. Leake. 1973. Culture of embryonic cells from the tick *Boophilus microplus. J. Med. Entomol.* 10(5):493-496.

A simple and successful method is described for the primary culturing of embryonic cells in the tick *Boophilus microplus.* Oviposited eggs that were sterilized were used to establish one culture from which 14 subcultures were derived. Cells from sterilized preovipositing females were simpler and more reliable than surface sterilization of oviposited eggs. (The reference of Medvedeva et al. (1972) concerning cells cultured from *Hyalomma asiaticum* was also cited by these authors).

Radinovsky, S. and G.W. Krantz. 1961. The biology and ecology of granary mites of the Pacific Northwest. II. Techniques for laboratory observation and rearing. *Ann. Entomol. Soc. Am.* 54(4):512-518.

Rivard, I. 1958. A technique for rearing tyroglyphid mites on mould cultures. *Can. Entomol.* 90(3):146-147.

Rohde, C.J., Jr. 1956. A modification of the plaster-charcoal technique for the rearing of mites and other small arthropods. *Ecology* 37(4):843-844.

Rohde, C.J., Jr. 1959. Studies on the biologies of two mite species, predator and prey, including some effects of gamma radiation on selected developmental stages. *Ecology* 40(4):575-579.

Sengbusch, H.G. 1970. Culture methods for cryptostigmatid mites (Acari: Oribatei). *Proc. Int. Congr. Acarol., 4th, 1974,* pp. 83–87.

Singer, G. 1967. A comparison between mounting techniques commonly employed in acarology. Mounting techniques and solutions described. *Acarologia* 9(3):475–484.

Singer, G. and G.W. Krantz. 1967. Nematodes and oligochaetes for rearing predatory mites. *Acarologia* 9(3):485–487.

Solomon, M.E. and A.M. Cunningham. 1964. Rearing acaroid mites. *Proc. Int. Congr. Acarol., 1st, 1963,* pp. 399–403.

Sonenshine, D. and R.R. Slocum. 1979. The development of laboratory reared radio labeled tick populations for use in studies of tick ecology and disease problems. *Proc. Int. Congr. Acarol., 4th, 1974,* p. 595.

Tadkowski, T.M. and K.E. Hyland. 1979. The developmental stages of *Aplodontopus sciuricola* (Astigmata) from *Tamias striatus* L. (Sciuridae) in North America. *Proc. Int. Congr. Acarol., 4th, 1974,* pp. 321–326.

Treat, A.E. 1975. *Mites of Moths and Butterflies.* Cornell Univ. Press (Comstock), Ithaca, New York.

Equipment and techniques for examining and studying mites associated with Lepidoptera. Chapter 2, pp. 33–45.

Vercammen-Grandjean, P. 1973. Of techniques and ortho-iconography. *Proc. Int. Congr. Acarol., 3rd, 1971,* pp. 321–328.

Watson, G.E. and A.B. Amerson, Jr. 1967. *Instructions for Collecting Bird Parasites,* Inf. Leaf. 477. Mus. Natl. Hist, Smithsonian Inst., Washington, D.C.

Wharton, G.W. and W. Knulle. 1966. A device for controlling temperature and relative humidity in small chambers. *Ann. Entomol. Soc. Am.* 59(4):627–630.

Woolley, T.A. 1982. Mites and other soil microarthropods. *In: Methods of Soil Analysis, Part 2. Chemical and Microbiological Properties,* 2nd ed., Agron. Monogr. No. 9, Chapter 54, pp. 1131–1142. Am. Soc. Agron., Soil. Sci. Soc. Am., Madison, Wisconsin.

Wu, K. W. 1986. Review of the polyethylene bottle applicator technique for sealing microslides preparations of mite. *Int. J. Acarol.* 12(2):87–89.

Yao, D.S.C. and D.A. Chant. 1982. A method of confining mites for population studies in the laboratory. *Int. J. Acarol.* 8(4):205–209.

Young, J.H. 1968. The morphology of *Haemogamasus ambulans. J. Kans. Entomol. Soc.* 41(1):101–107.

Includes microtechniques for dissection. Integument of *H. ambulans* is both tough and flexible. Dissection needles consisted of Minuten nadeln and cactus spines fastened to glass rods with sealing wax. Drawn glass tubes were used as syringes to withdraw body fluids and to inject various dyes and India ink.

In dissection, mites placed in 60–62 MP paraffin oriented with venter embedded, legs serving as anchors, covered with saline solution (85%) NaCl). If organs were to be removed and processed later, they were preserved *in situ* with formol–alcohol (30% formalin–70% ethanol).

For transparent structures, vital staining with Evans' blue (2% aqueous solution) brought greatest contrast. Internal structures were not removed from fluid without fixation, for they invariably collapsed into an amorphous mass.

CONTROL

Various control methods have been used to change the physical or biological environment to make it less habitable for mite species. Chemicals affect acarines by inhibiting metabolic functions, neurotoxicity, and possibly cytotoxicity, determined by the nature of the substance. Ionizing radiation affects the physiology of reproduction. Natural predation controls pest species and such biological control is in the forefront of the methods used. Genetic control is among the newest of control methods in insects, but little has been determined for this type of control in mites.

Effective control involves several techniques, a modern method termed *integrated control.* When the problem is complex and has multiple effects on the health of people or animals, crops, stored products, and so on, different types of control are applied in an integrated way. A shotgun approach is not used, but knowledge of the pest's life cycle, habitat, reproduction, biology, and behavior is used to determine what resources and methods are applicable.

Literature on the control of mites is increasing rapidly. The references and brief annotations below will enable the reader to find materials pertinent to the topic.

REFERENCES

AliNiazee, M.T. 1979. Mite populations on apple foliage in western Oregon. *In:* Rodriguez, J.G., ed. *Recent Advances in Acarology.* Academic Press, New York, Vol. 1, pp. 71–80.

Allen, J.R. 1979. The immune response as a factor in management of acari of veterinary importance. *In:* Rodriguez, J.G., ed. *Recent Advances in Acarology.* Academic Press, New York, Vol. 2, pp. 15–24.

Baker, J.A.F. and G.D. Stanford. 1979. Slow-release devices as aids in the control of ticks infesting the ears. *In:* Rodriguez, J.G., ed. *Recent Advances in Acarology.* Academic Press, New York, Vol. 2, pp. 71–78.

Bosch, R. van den, P.S. Messenger and A.P. Gutierrez. 1982. *An Introduction to Biological Control.* Plenum, New York.

Bowley, C.R. and C.H. Bell. 1981. The toxicity of twelve fumigants to three species of mites infesting grain. *J. Stored Prod. Res.* 17:83–87.

Bruce, W.A. and G.L. LeCato. 1979. *Pyemotes tritici:* Potential biological control agent of stored-product insects. *In:* Rodriguez, J.G., ed. *Recent Advances in Acarology.* Academic Press, New York, Vol. 1, pp. 213–220.

Bruce, W.A. and G. L. LeCato. 1980. *Pyemotes tritici:* A potential new agent for biological control of the red imported fire ant, *Solenopsis invicta* (Acari: Pyemotidae). *Int. J. Acarol.* 6(4):271–274.

Biological control of the red imported fire ant is effected by populations of the straw itch mite (*Pyemotes tritici*) (Bruce and LeCato, 1980). Nests of the fire ant are opened and cultures of the mite, raised on cigarette beetles, are sprinkled on the top of the mound. Ants carry the mites into the nest where the mites attack the ants, inactivate the nest, and destroy the colony in over 50% of the cases.

Chen, C.N., C.C. Cheng and K.C. Hsaio. 1979. Bionomics of *Steneotarsonemus spinki* attacking rice plants in Taiwan. *In:* Rodriguez, J.G., ed. *Recent Advances in Acarology.* Academic Press, New York, Vol. 1, pp. 111–118.

Childers, C.C. and G.C. Rock. 1981. Observations on the occurrence and feeding habits of *Balaustium putmani* (Acari: Erythraeidae) in North Carolina apple orchards. *Int. J. Acarol.* 7(1–4):63–68

Cone, W.W. 1975. Crown-applied systemic acaricides for control of the twospotted spider mite and hop aphid on hops. *Int. J. Acarol.* 68(5):684–686.

Cone, W.W. and J.C. Maitlen. 1976. Systemic activity of aldicarb against twospotted spider mites on hops and aldicarb residues in hop cones. *J. Econ. Ent.* 69(4):533–534.

Croft, B.A. 1975a. Integrated control of apple mites. *Coop. Ext. Serv., Mich. State Univ., Ext. Bull.* E-825:1–11.

Croft, B.A. 1975b. Integrated control of orchard pests in the U.S.A. *C.R. Symp. Lutte Integree Vergers, 5th* OILB/SROP:109–124.

Croft, B.A. 1975c. Tree fruit pest management. *In:* Luckmann, W.H. and R.H. Metcalf, eds. *Introduction to Insect Pest Management.* Wiley, New York, pp. 471–506.

Croft, B.A. 1976. Establishing insecticide-resistant phytoseiid mite predators in deciduous tree fruit orchards. *Entomophaga* 21(4):383–399.

Croft, B.A. and E.J. Blyth. 1979. Aspects of the functional, ovipositional, and starvation response of *Amblyseius fallacis* to prey density. *In:* Rodriguez, J.G., ed. *Recent Advances in Acarology.* Academic Press, New York, Vol. 1, pp. 41–48.

Croft, B.A. and D.L. McGroarty. 1977. The role of *Amblyseius fallacis* (Acarina: Phytoseiidae) in Michigan apple orchards. *Mich. State Univ. Agric. Exp. Stn. Res. Rep.* 333, 22.

Croft, B.A. and R.H. Meyer. 1973. Carbamate and organophosphorus resistance patterns in populations of *Amblyseius fallacis. Environ. Entomol.* 2:691–695.

Croft, B.A. and E.E. Nelson. 1972. Toxicity of apple orchard pesticides to Michigan populations of *Amblyseius fallacis. Environ. Entomol.* 1:576–579.

Darrow, D.I., W.J. Gladney and C.C. Dawkins. 1976. Lone star tick mating behavior of gamma-irradiated males. *Ann. Entomol. Soc. Am.* 69(1):106–108.

Drummond, R.O. and O.H. Graham. 1964. Insecticide tests against the tropical horse tick, *Dermacentor nitens*, on horses. *J. Econ. Entomol.* 57(4):549–553.

Drummond, R.O. and J.M. Ossorio. 1966. Additional tests with insecticides for the control of the tropical horse tick on horses in Florida. *J. Econ. Entomol.* 59(1):107–110.

Drummond, R.O., T.M. Whetsone and S.E. Ernst. 1967a. Insecticidal control of the ear tick in the ears of cattle. *J. Econ. Entomol.* 60(6):1735–1738.

Drummond, R.O., T.M. Whetstone and S.E. Ernst. 1967b. *Control of the Lone Star Tick on Cattle*, pp. 1735–1738. Entomol. Res. Div., Agric. Res. Serv., USDA, Kerrville, Texas.

Drummond, R.O., W.J. Graham, S.E. Ernst and J.L. Trevino. 1969. Evaluation of insecticides for the control of *Boophilus annulatus* (Say) and *B. microplus* (Canestrini) (Acarina: Ixodidae) on cattle. *Proc. Int. Congr. Acarol., 2nd, 1967*, pp. 493–498.

Drummond, R.O., W.J. Gladney and O.H. Graham. 1974. Recent advances in the use of ixodicides to control ticks affecting livestock. *Bull. Off. int. Epizoot.* 81(1–2):47–63.

Drummond, R.O., S.E. Ernst, J.L. Trevino, W.J. Gladney and O.H. Graham. 1976. Tests of acaricides for control of *Boophilus annulatus* and *B. microplus*. *J. Econ. Entomol.* 69(1):37–40.

Elbadry, E.A. 1979. Management of mite pests of cotton in Egypt. *In:* Rodriguez, J.G., ed. *Recent Advances in Acarology. Academic Press, New York, Vol. 1, pp. 49–58.*

Friese, D.D. and F.E. Gilstrap. 1982. Influence of prey availability on reproduction and prey consumption of *Phytoseiulus persimilis*, *Amblyseius californicus*, and *Metaseiulus occidentalis* (Acarina: Phytoseiidae). *Int. J. Acarol.* 8(2):85–89.

Gladney, W.J., S.E. Ernst, C.C. Dawkins, R.O. Drummond and O.H. Graham. 1972. Feeding systemic insecticides to cattle for control of the tropical horse tick. *J. Med. Entomol.* 9(5):439–442.

   Six systemic insecticides for *Anocenter nitens* exert a high degree of control in smaller numbers of engorged females. They cause reduced weight of females and the egg masses produced, as well as a lower percentage of eggs that hatch. Some systemics were lethal and others gave poor control.

Gladney, W.J., S.E. Ernst and R.O Drummond. 1974. Chlordimeform: a detachment-stimulating chemical for three-host ticks. *J. Med. Entomol.* 2(5):569–572.

Gladney, W.J., M.A. Price and O.H. Graham. 1977. Field tests of insecticides for control of the gulf coast tick on cattle. *J. Med. Entomol.* 13(4):579–586.

Glass, E.H. and S.E. Lienk. 1971. Apple insects and mite populations developing after discontinuance of insecticides: ten year record. *J. Econ. Entomol.* 64:23–26.

Good, E.A.M, L.M. Stables and D.R. Wilkin. 1977. The control of mites in stored oilseed rape. *Proc. Br. Crop Prot. Conf.—Pests Dis.* 1977(1):161–168.

Gould, H., J. Hussey and W.J. Parr. 1969. Large scale commercial control of *Tetranychus urticae* Koch on cucumbers by the predator *Phytoseiulus persimilis* A.H. *In:* Evans, G.O., ed. *Proc. 2nd Int. Congr. Acarology* 1967 (Akad Kaido) Budapest, pp. 383–388.

Graham, O.H. and R.O. Drummond. 1964. Laboratory screening of insecticides for the prevention of reproduction of *Boophilus* ticks. *J. Econ. Entomol.* 57(3):335–339.

Graham, O.H., R.O. Drummond and G. Diamant. 1964. The reproductive capacity of female *Boophilus annulatus* collected from cattle dipped in arsenic or coumaphos. *J. Econ. Entomol.* 57(3):409–410.

Griffiths, D.A. and C.E. Bowman, eds. 1984. *Acarology VI*, Vol. 2. Ellis Horwood, Ltd., Chichester, England.

   Biological control, chemical control, control of mites of medical and veterinary importance and control of stored products mites.

Grothaus, R.H., G.A. Mount, J.M. Hirst and W.T. Keenan. 1975. ULV and HV lone star tick control. *Pest Control Magazine* 43(11):18–19, 24.

Haarlov, N. 1979. Mites from plots supplied with different quantities of manures and fertilizers. *In:* Rodriguez, J.G., ed. *Recent Advances in Acarology*. Academic Press, New York, Vol. 1, pp. 125–128.

Hartenstein, R.C. 1960. The effects of DDT and Malathion upon forest soil microarthropods. *J. Econ. Entomol.* 53(3):357–362.

Helle, W. and M.W. Sabelis, eds. 1985. *World Crop Pests. Spider Mites: Their Biology, Natural Enemies and Control*, Vols. 1A and 1B. Elsevier, Amsterdam.

Heller-Haupt, A., M.G.R. Varma, S. Crook and A. Radalowicz. 1979. The effect of synthetic pyrethroids on some African Ixodidae. *In:* Rodriguez, J.G., ed. *Recent Advances in Acarology.* Academic Press, New York, Vol. 2, pp. 79–84.

Hern, D.H.C., J.E. Cranham and M.A. Easterbrook. 1979. New acaricides to control resistant mites. *In:* Rodriguez, J.G., ed. *Recent Advances in Acarology.* Academic Press, New York, Vol. 1, pp. 95–104.

Hoy, M.A. and N.F. Knop. 1979. Studies on pesticide resistance in the phytoseiid *Metaseiulus occidentalis* in California. *In:* Rodriguez, J.G., ed. *Recent Advances in Acarology.* Academic Press, New York, Vol. 1, pp. 89–94.

Hoy, M.A. 1983. Opportunities for genetic improvement of mites as biological control agents. *In:* Hoy, M.A., G.L. Cunningham and L. Knutson eds., *Biological Control of Pests by Mites.* Agric. Exp. Stn. Univ. of Calif., Berkeley, Spec. Publ. 3304, pp. 141–146.

Hoy, M.A. and D.C. Herzog, eds. 1985. *Biological Control in Agricultural IPM Systems.* Academic Press, Orlando, Florida.

Hoyt, S.C. 1969. Integrated chemical control of insects and biological control of mites on apple in Washington. *J. Econ. Entomol.* 62(1):74–86.

By selecting chemicals that affected insects but conserved the predaceous *Typhlodromus occidentalis* it was possible to effect control of the spider mite *Tetranychus mcdanieli.* If the apple rust eriophyid *Aculus schechtendali* was available as food *T. occidentalis* fed upon it, but did not control the species. The predator also prevented a resurgence of populations of the European red mite (*Panonychus ulmi*) after chemical controls had affected the population. Understanding of population dynamics, relationships of spray practices, integrated chemical and biological control, and experience help to provide effective integrated control programs.

Hoyt, S.C. and E.C. Burts. 1974. Integrated control of fruit pests. *Annu. Rev. Entomol.* 19:231–252. (Phytoseiidae)

Huffaker, C.B., M. van de Vrie and J.A. McMurtry. 1969. The ecology of tetranychid mites and their natural control. *Annu. Rev. Entomol.* 14:125–174.

Hussey, N.W., W.H. Read, and J.J. Hesling, eds. 1969. *The Pests of Protected Cultivation.* Edward Arnold, London.

Krantz, G.W. and E.E. Lindquist. 1979. Evolution of phytophagous mites (Acari). *Annu. Rev. Entomol.* 24:121–158.

Investigations are being made to determine if the oribatid mite *Orthogalumna terebrantis* Wallwork, which feeds as larvae and nymphs in the parenchymous tissue of water hyacinth leaves, could be used to control this aquatic weed.

Kunz, S.E. 1978. Highlights of veterinary entomology 1952–1957. *Bull. Entomol. Soc. Am.* 24(4):401–406.

Marshall, V.G. 1979. Effects of the insecticide Diflubenzuron on soil mites of a dry forest zone in British Columbia. *In:* Rodriguez, J.G., ed. *Recent Advances in Acarology.* Academic Press, New York, Vol. 1, pp. 129–134.

McCosker, P.J. 1979. Global aspects of the management and control of ticks of veterinary importance. *In:* Rodriguez, J.G., ed. *Recent Advances in Acarology.* Academic Press, New York, Vol. 2, pp. 45–54.

McMurtry, J.A. 1984. A consideration of the role of predators in the control of acarine pests. *In:* Griffiths, D.A. and C.E. Bowman, eds. *Acarology VI,* Vol. 1. Ellis Horwood, Chichester, England, pp. 109–121.

McMurtry, J.A., C.B. Huffaker and M. van de Vrie. 1970. Ecology of tetranychid mites and their natural enemies: Their biological characters and the impact of spray practices. *Hilgardia* 40:331–390.

Meleney, W.P. and I.H. Roberts. 1979. Trials with eight acaricides against *Psoroptes ovis,* the sheep scabies mite. *In:* Rodriguez, J.G., ed. *Recent Advances in Acarology.* Academic Press, New York, Vol. 2, pp. 95–104.

Motoyama, N., G.C. Rock and W.C. Dautermann. 1970. Organophosphorus resistance in an apple orchard population of *Typhlodromus (Amblyseius) fallacis. J. Econ. Entomol.* 63:1439–1442.

Navvab-Gojrati, H.A. and N. Zare. 1979. Resistance of different varieties of sunflower and safflower to *Tetranychus turkestani* in southern Iran. *In:* Rodriguez, J.G., ed. *Recent Advances in Acarology.* Academic Press, New York, Vol. 1, pp. 77–80.

Nolan, J. 1979. New acaricides to control resistant ticks. *In:* Rodriguez, J.G., ed. *Recent Advances in Acarology.* Academic Press, New York, Vol. 2, pp. 55–64.

Nolan, J. and W.J. Roulston. 1979. Acaricide resistance as a factor in the management of acari of medical and veterinary importance. *In:* Rodriguez, J.G., ed. *Recent Advances in Acarology.* Academic Press, New York, Vol. 2, pp. 3–14.

Norizumi, S. 1979. Geographical distribution of the citrus red mite *Panonychus citri* and European red mite *P. ulmi* in Japan. *In:* Rodriguez, J.G., ed. *Recent Advances in Acarology.* Academic Press, New York, Vol. 1, pp. 81–88.

Obenchain, F.D. 1979. Non-acaricidal chemicals for the management of acari of medical and veterinary importance. *In:* Rodriguez, J.G., ed. *Recent Advances in Acarology.* Academic Press, New York, Vol. 2, pp. 35–44.

Parent, B. 1967. Populations studies of phytophagous mites and predators on apple in southwestern Quebec. *Can. Entomol.* 99:771–778.

Penman, D.R., C.H. Wearing, E. Collyer and W.P. Thomas. 1979. The role of insecticide-resistant phytoseiids in integrated mite control in New Zealand. *In:* Rodriguez, J.G., ed. *Recent Advances in Acarology.* Academic Press, New York, Vol. 1, pp. 59–70.

Petrishcheva, P.A. 1965. *Vectors of Diseases of Natural Foci* (translated from Russian by B. Hershkovitz, Isr. Program Sci. Transl.). Daniel Davey & Co., New York.

Poe, S.L., W.E. Noble and R.E. Stall. 1979. Acquisition and retention of *Pseudomonas marginata* by *Anoetus feroniarum* and *Rhizoglyphus robini. In:* Rodriguez, J.G., ed. *Recent Advances in Acarology.* Academic Press, New York, Vol. 1, pp. 119–124.

Pruszynski, S. and W.W. Cone. 1972. Relationships between *Phytoseiulus persimilis* and other enemies of the two-spotted spider mite on hops. *Environ. Entomol.* 1(4):431–433.

Rockett, C.L. 1980. Nematode predation by oribatid mites (Acari: Oribatida). *Int. J. Acarol.* 6(3):219–224.

Rodriguez, J.G., P. Singh and B. Taylor. 1970. Manure mites and their role in fly control. *J. Med. Entomol.* 7(3):335–341.

Rodriguez, J.G., ed. 1979. *Recent Advances in Acarology,* Vols. 1 and 2. Academic Press, New York. Pest management of agricultural mites and stored products mites; control of mites of medical and veterinary importance; disease transmission.

Roush, R. and J. McKenzie. 1987. Ecological genetics of insecticide and acaricide resistance. *Annu. Rev. Entomol.* 32:361–380.

Senff, W.A. and J.R. Gorham. 1979. The food and drug administration and regulatory acarology. *In:* Rodriguez, J.G., ed. *Recent Advances in Acarology.* Academic Press, New York, Vol. 1, pp. 317–322.

Solomon, K.R., H. Heyne and J. van Kleef. 1979. The use of a population approach in the survey of resistance to pesticides in ticks in southern Africa. *In:* Rodriguez, J.G., ed. *Recent Advances in Acarology.* Academic Press, New York, Vol. 2, pp. 65–70.

Sonnenshine, D.E. 1984. Pheromones of Acari and their potential use in control strategies. *In:* Griffiths, D.A. and C.E. Bowman, eds., *Acarology VI,* Vol. 1. Ellis Horwood, Chichester, England, pp. 100–108.

Welch, S.M. 1979. The application of simulation models to mite pest management. *In:* Rodriguez, J.G., ed. *Recent Advances in Acarology.* Academic Press, New York, Vol. 1, pp. 31–40.

Westigard, P.H. 1971. Integrated control of spider mites on pears. *J. Econ. Entomol.* 64(2):496–501. (Phytoseiidae)

White, N.D.G. and R.N. Sinha. 1979. Natural regulation of *Tarsonemus granarius* numbers in stored wheat ecosystems—a multivariate assessment. *In:* Rodriguez, J.G., ed. *Recent Advances in Acarology.* Academic Press, New York, Vol. 1, pp. 291–298.

Wicht, M.C., Jr. and J.G. Rodriguez. 1970. Integrated control of muscid flies in poultry houses using predator mites, selected pesticides and microbial agents. *J. Med. Entomol.* 7(6):687–692.

Wilkin, D.R. 1975. The control of stored product mites by contact acaricides. *Proc. Br. Insectic. Fungic. Conf., 8th.* 1:335–364.

Wilkin, D.R. 1979. The control of mites in cheese stores. *In:* Rodriguez, J.G., ed. *Recent Advances in Acarology,* Vol. 1. Academic Press, New York, pp. 221–229.

Wilkinson, P.R. 1979. Ecological aspects of pest management of ixodid ticks. *In:* Rodriguez, J.G., ed. *Recent Advances in Acarology.* Academic Press, New York, Vol. 2, pp. 25–34.

Index